67.50
100K

HEAVY ION
INERTIAL
FUSION

AIP CONFERENCE PROCEEDINGS 152

RITA G. LERNER
SERIES EDITOR

HEAVY ION INERTIAL FUSION
WASHINGTON, DC 1986

EDITORS:
MARTIN REISER
UNIVERSITY OF MARYLAND

TERRY GODLOVE
U.S. DEPARTMENT OF ENERGY

ROGER BANGERTER
LAWRENCE LIVERMORE NATIONAL LABORATORY

AMERICAN INSTITUTE OF PHYSICS   NEW YORK 1986

Copy fees: The code at the bottom of the first page of each article in this volume gives the fee for each copy of the article made beyond the free copying permitted under the 1978 US Copyright Law. (See also the statement following "Copyright" below.) This fee can be paid to the American Institute of Physics through the Copyright Clearance Center, Inc., 21 Congress Street, Salem, MA 01970.

Copyright © 1986 American Institute of Physics

Individual readers of this volume and non-profit libraries, acting for them, are permitted to make fair use of the material in it, such as copying an article for use in teaching or research. Permission is granted to quote from this volume in scientific work with the customary acknowledgment of the source. To reprint a figure, table or other excerpt requires the consent of one of the original authors and notification to AIP. Republication or systematic or multiple reproduction of any material in this volume is permitted only under license from AIP. Address inquiries to Series Editor, AIP Conference Proceedings, AIP, 335 E. 45th St., New York, NY 10017.

L.C. Catalog Card No. 86-73185
ISBN 0-88318-352-8
DOE CONF-860510

Printed in the United States of America

Contents

Preface	ix
The U.S. Magnetic Fusion Program J. F. Decker	1
U.S. Inertial Fusion Program Overview R. L. Schriever	13
Overview of Heavy Ion Fusion Accelerator Research Program at the U.S. Department of Energy R. Gajewski	20
The German Heavy Ion Fusion Program R. Bock	23
Overview of HIF Program in Japan Y. Hirao	39
RF Linac for Heavy Ion Fusion Driver I. M. Kapchinskiy, V. V. Kuschin, N. V. Lazarev, V. G. Schevchenko, V. S. Artemov, V. A. Batalin, E. N. Damiltzev, A. Ju. Djadin, D. D. Iosseliani, A. M. Kozodaev, A. R. Kurs, I. M. Lipkin, I. O. Parshin, S. V. Plotnikov, V. S. Skachkov, S. B. Ugarov, and A. B. Zarubin	49
Experiments and Prospects for Induction Linac Drivers D. Keefe	63
Development of RF-Linac and Storage Ring System for High-Current Experiments at GSI I. Hofmann	74
Review of Target Studies for Heavy Ion Fusion J. D. Lindl, R. O. Bangerter, J. W-K. Mark, and Y.-L. Pan	89
Beam Plasma Interactions in HIF T. D. Beynon	100
Heavy Ion Fusion Systems Assessment Study D. J. Dudziak and W. B. Herrmannsfeldt	111
Heavy Ion Acceleration and Cooling Ring TARN-2 T. Katayama	126
Experiments on the ISIS Synchrotron G. H. Rees	138
Preliminary Results from MBE-4: A Four Beam Induction Linac for Heavy Ion Fusion Research T. J. Fessenden, D. L. Judd, D. Keefe, C. Kim, L. J. Laslett, L. Smith, and A. I. Warwick	145
Status of the Heavy-Ion RFQ Linac "MAXILAC" R. W. Müller, U. Kopf, R. Keller, P. Spädtke, and J. Bolle	156
Emittance Growth from Charge Density Changes in High-Current Beams T. P. Wangler, K. R. Crandall, and R. S. Mills	166

The Maryland Transport Experiment—General Perspectives and Recent Results with Off-Centered Beams 186
    M. Reiser, J. McAdoo, D. Kehne, K. Low, J. D. Lawson, and C.R. Prior

An Intense Metal Ion Beam Source for HIF 207
    I. G. Brown

HIF Transport Issues for $P > 10^{-3}$ Torr and $Z > 1$ 215
    C. L. Olson

Studies on Longitudinal Beam Compression in Induction Accelerator Drivers 227
    J. W-K. Mark, D. D-M. Ho, S. T. Brandon, C-L. Chang, A. T. Drobot, A. Faltens, E. P. Lee, and G. A. Krafft

Threshold for Emittance Growth of A-G Focused Intense Beams 236
    I. Haber

Longitudinal Compression of Heavy-Ion Beams with Minimum Requirements on Final Focus 243
    D. D.-M. Ho, R. O. Bangerter, J. W.-K. Mark, S. T. Brandon, and E. P. Lee

Some Mechanisms and Time Scales for Emittance Growth 253
    O. A. Anderson

Head-to-Tail Velocity Tilt in an Ion Induction Linac 264
    C. H. Kim, D. L. Judd, L. J. Laslett, L. Smith, and A. I. Warwick

Beam Test of the RFQ Linac 'TALL' 271
    N. Ueda, A. Mizobuchi, S. Yamada, S. Arai, M. Olivier, T. Nakanishi, T. Fukushima, S. Tatsumi, and Y. Hirao

Simulation of Transverse Combining of Space-Charge Dominated Beams 278
    C. M. Celata

Theory and Simulation of Emittance, Space Charge, and Electron Pressure Effects on Focusing of Neutralized Ion Beams 287
    D. S. Lemons and M. E. Jones

Transport of Intense Proton Beams in an Induction Linac by Solenoid Lenses ... 295
    W. Namkung, J. Y. Choe, and H. S. Uhm

Particle Tracking in Solenoid Channels 302
    C. R. Prior

Ponderomotive Confinement of Ion Beams in a Cylindrical Waveguide 309
    C. Grebogi and H. S. Uhm

Grid-Controlled Metal Ion Sources for Heavy Ion Fusion Accelerators 314
    L. K. Len, S. Humphries, Jr., and C. Burkhart

Progress on the Los Alamos Heavy-Ion Injector 323
    D. C. Wilson, K. B. Riepe, E. O. Ballard, E. A Meyer, R. P. Shurter, F. W. Van Haaften, and S. Humphries, Jr.

Charge and Current Neutralization Physics of a Heavy Ion Beam During Final Transport 330
    G. R. Magelssen and D. W. Forslund

Beam Transport Experiments Using an RFQ Channel 338
    N. Zoubek, K. Langbein, P. Junior, A. Schempp, H. Klein, and G. Riehl

RFQ Development and Sparking Experiments in Frankfurt 346
    A. Schempp, H. Klein, M. Ferch, A. Gerhard, and M. Kurz

Charge State Evolution and Energy Losses in a Beam-Plasma Interaction
Experiment 360
    R. Dei-Cas, J. M. Guihaumé, M. Beau, M. A. Beuve, J. F. Glicenstein,
    J. P. Laget, C. Moreau, J. P. Mosnier, M. Renaud, R. Barchewitz,
    and M. Cukier

Energy Deposition of Heavy Ions in Plasma 371
    J. Meyer-ter-Vehn, Th. Peter, and R. Arnold

Heavy Ion Beam Transport and Interaction with ICF Targets 381
    G. Velarde, J. M. Aragonés, J. A. Gago, L. Gámez, M. C. González,
    J. J. Honrubia, J. M. Martínez-Val, E. Mínguez, J. L. Ocaña, R. Otero,
    J. M. Perlado, J. M. Santolaya, J. F. Serrano, and P. M. Velarde

Beam Plasma Interaction Experiments at GSI 391
    D. H. H. Hoffmann, J. Jacoby, H. Wahl, K. Weyrich, R. Noll, C. R. Haas,
    and B. Weikl

Beam Losses in the Storage Ring Due to Charge Changing Collisions within a
Beam of Xe^+ Ions 400
    F. Melchert, K. Rinn, K. Rink, and E. Salzborn

SPQR II: A Beam-Plasma Interaction Experiment 408
    R. Bimbot, S. Della-Negra, D. Gardés, M. F. Rivet, C. Fleurier, B. Dumax,
    D. H. H. Hoffman, K. Weyrick, C. Deutsch, and G. Maynard

Radiative Transfer in Statistically Heterogeneous Mixtures 416
    C. Deutsch and D. Vanderhaegen

Gain Scaling Laws for HIF 428
    G. R. Magelssen

Substantial Reductions of Input Energy and Peak Power Requirements in
Targets for Heavy Ion Fusion 435
    J. W-K. Mark and Y.-L. Pan

Symmetry Issues in a Class of Ion Beam Targets Using Sufficiently Short
Direct Drive Pulses 441
    J. W-K. Mark and J. D. Lindl

On Radiation Energy Transport in Ion Driven Targets 448
    J. Meyer-ter-Vehn, R. Schmalz, and R. Ramis

Hydrodynamic Simulation Code for Target Implosion Including Shock Waves
and Calculated Target Gain Irradiated by Ion Beams 454
    K. Niu

Accelerator and Final Focus Model for an Induction Linac Based HIF System
Study 461
    E. P. Lee

The Cost of Induction Linac Drivers for Inertial Fusion for Various Target Yields 474
    J. Hovingh, V. O. Brady, A. Faltens, D. Keefe, and E. P. Lee

Survey of System Options for Heavy Ion Fusion 482
    L. M. Waganer, D. E. Driemeyer, D. S. Zuckerman, and K. W. Billman

Performance and Cost Modeling of a Linac-Driven HIF Power Plant 489
    D. S. Zuckerman, D. E. Driemeyer, L. M. Waganer, J. Hovingh, E. P. Lee, and K. W. Billman

Influence of System Optimization Considerations on LIA Drivers 498
    D. E. Driemeyer, L. M. Waganer, and D. S. Zuckerman

Repetition Rates in Heavy Ion Driven Fusion Reactors 507
    R. R. Peterson

Economic Studies for Heavy-Ion-Fusion Electric Power Plants 515
    W. R. Meier, W. J. Hogan, and R. O. Bangerter

Energy Analysis of HIF Reactor "HIBLIC-I" 523
    T. Nagai, H. Obayashi, and T. Yamaki

Heavy-Ion Fusion Reactor Concepts, Requirements, and Attractive Features   531
    J. H. Pendergrass

Inertial Confinement—Concept and Early History 539
    J. G. Linhart

Targets for Laser and Ion Beam Drivers 547
    R. O. Bangerter

Present Status and Future Prospects of Gas Lasers as Drivers for ICF 553
    Y. Kato

Present Status and Future Prospects for Direct Drive Laser Fusion 561
    S. E. Bodner

Present Status and Future Prospects of Light Ions as Drivers for Inertial Fusion 569
    J. P. VanDevender

Present Status and Future Prospects of Heavy Ion Beams as Drivers for ICF   579
    T. F. Godlove

Reactor Design Aspects for Inertial Confinement Fusion 591
    G. Kessler, U. von Möllendorff, G. A. Moses, and R. Peterson

Ten Years of HIF Research 599
    J. D. Lawson

Author Index 605

List of Participants 609

Preface

The 1986 International Symposium on Heavy Ion Inertial Fusion (HIF) was held May 27–29 at the L'Enfant Plaza Hotel in Washington, D.C., with 105 registered participants from 32 countries. This meeting was the third in a series of biennial symposia begun in 1982, that continue a prior series of four annual workshops in the United States started in 1976. The Symposium was sponsored by the U.S. Department of Energy (DOE), and was organized by the University of Maryland with assistance from the Naval Research Laboratory and the National Bureau of Standards. Previous symposia were held at the Institute for Nuclear Studies (INS) of the University of Tokyo (1984) and at the Gesellschaft f. Schwerionenforschung (GSI), Darmstadt, West Germany (1982).

The purpose of the Symposium was to present current research on the application of heavy ion accelerators to inertial fusion. Invited and contributed papers were divided into four topical areas:

- Summaries of national HIF programs.
- Accelerators, beam dynamics, beam transport and focusing.
- Atomic physics, beam-plasma interactions, energy deposition in high-temperature target materials, and target studies.
- System studies.

In addition, several special invited papers were included to provide an overview of related fusion programs. Among these were reviews by J. F. Decker on the U.S. magnetic fusion program and by R. L. Schriever on the inertial fusion program administered by DOE's Defense Programs. J. G. Linhart reminisced about the early history of inertial fusion. A special invited session on Thursday, chaired by J. D. Lawson, provided a lively forum for discussing the present status and future prospects of the major inertial fusion driver concepts. A retrospective summary entitled "Ten Years of HIF Research," written by Lawson after the Symposium, is included as the final paper of the proceedings.

The use of high-power, high-energy heavy ion beams for igniting inertial fusion targets was first suggested by A. W. Maschke in 1974–75. A persuasive argument for the proposal was the fact that high-energy accelerators are based on decades of technology development and offer most of the required characteristics of an inertial fusion driver. These include excellent electrical efficiency, reliability, repetition rate and the ability to focus beams to small spot size. That same year—1975—J. Shearer of Livermore and M. Clauser of Sandia published the first serious target design studies based on ion beams.

Following these proposals a small, broadly-based HIF program was established in several DOE laboratories. During the late seventies four workshops in the U.S. supported the scientific soundness of the heavy ion fusion concept. Meanwhile HIF research programs were also initiated in the Federal Republic of Germany and in Japan. These two countries adopted a RF linear accelerator/storage ring scheme for the driver while the U.S. HIF program opted for the development of a single-pass induction linac (proposed by D. Keefe) to generate the heavy ion beam. A comprehensive study of the German concept called HIBALL was presented at the first International HIF Symposium in Darmstadt, W. Germany, in 1982. The Japanese followed with a similar study called HIBLIC at the 1984 International Symposium in Tokyo.

A major highlight of the 1986 International Symposium in Washington reported here was the Heavy Ion Fusion Systems Assessment Study (HIFSA) of the induction-linac based U.S. HIF scenario. The results of the HIFSA study were very encouraging since they showed that a broad range of system parameters could be used to produce competitive electricity in HIF power plants. Especially intriguing was the proposal for a higher ion charge state to reduce costs dramatically.

Overall, it would appear that HIF is slowly becoming recognized as a major fusion contender. The national programs are still clearly underfunded compared with other fusion programs, but HIF came on the international scene at an awkward time—oil supplies have been plentiful, science budgets have been mercilessly squeezed, and HIF is perceived as needing a lot of work to catch up with existing projects. Thus it is not clear what one might expect for the new upstart.

While admitting that many years of R&D remain, HIF proponents claim to have come a long way toward proving their case for being a major fusion contender. At this Symposium some remaining pieces fell into place. In addition to the HIFSA study, progress was reported in several other areas. For example, a rather fundamental understanding of the limiting current in beam transport systems has developed during the last few years. And, unlike the frequently recalcitrant behavior of nature, the final result in this case is quite favorable. The HIFSA study combined with higher allowed beam currents led the Berkeley group to propose the use of higher charge states as noted above. Further, the important question of emittance growth at high currents is better understood with the advent of recent theories of "field energy" redistribution.

Progress in other areas has been steady and reassuring. Many details are being filled in for the design based on RF linacs and storage rings. An interesting paper from the Soviet Union is included in this connection, although the Soviet representatives could not make it to the meeting. The Radio-Frequency-Quadrupole (RFQ) accelerator, invented by Kapchinskiy in the Soviet Union, plays an important role as a high-current, low-energy injector for RF linacs. Heavy ion RFQ's are being pursued with vigor in Japan and West Germany as well as in the Soviet Union. G. Rees reported the first results of experiments to study instability limits in storage rings using the Rutherford synchrotron.

It was good to hear that the synchrotron system (SIS) for the GSI heavy ion nuclear physics program has been approved for funding. The commonality and overlap between the GSI nuclear physics needs and HIF requirements are clearly being exploited to the fullest. Likewise the Japanese have made substantial progress in the construction of the versatile facility TARN-2, a synchrotron/storage ring designed for beam accumulation, beam cooling, nuclear and atomic physics as well as HIF experiments.

The highlight of the U.S. progress, aside from the HIFSA study, was the report of initial results from Berkeley on the so-called Multiple Beam Experiment (MBE-4). The precise measurement of four separate beams, threaded through common accelerating gaps, with a modest amount of axial compression, all at the space-charge limit, was impressive. We look forward to the full MBE-4 system results.

New at this Symposium were the extensive reports of the French and German projects to study the deposition of heavy ions in hot, dense plasma. Some experiments have begun, others are being planned. Many of us were unaware of the innovative methods, extensive diagnostics, and depth of planning that have gone into these difficult but important experiments. There was also a report of interesting related theoretical work by the group at Madrid, Spain.

The conditions and activities outside of the conference room also contributed to the success of the Symposium. The weather cooperated beautifully throughout the meeting, and the L'Enfant Plaza provided excellent accommodations. Many guests took advantage of the convenient downtown location to visit the museums and other attractions of the Nations's Capital. (In case you did not hear the story, Alessandro Pascolini actually found a painting in one of the art galleries that was mis-labelled. After some discussion and study, the curator agreed with Pascolini!)

The traditional banquet on Wednesday evening was well attended and enjoyed by all. Music was provided by a quartet from the Montgomery Chamber Orchestra, and after-dinner talks were given by Ron Davidson (MIT), who described the results of the 1985 review of DOE's inertial fusion program by the National Academy of Sciences, and by Ben Cooper, who gave a "tell-it-like-it-is" discussion of the political process in Congress as it relates to science policy and funding priorities. Dr. Cooper is a staff member of the Energy and Natural Resources Committee of the Senate.

The Symposium could not have been a success without the tireless efforts of many people. The editors would particularly like to thank Mrs. Rita Ricketts for the patient and cooperative spirit with which she has carried out the thousand details of the Symposium and subsequent editing. Special thanks are also due to Dorothy Godlove, Inge Reiser and Ellen Haber for organizing sightseeing arrangements for accompanying guests, and to Irving Haber, Mark Wilson, John McAdoo and Bill Jenne for assisting with the local arrangements. Last but not least, financial support for the Symposium and these proceedings was shared by three DOE Offices: High Energy and Nuclear Physics; Basic Energy Sciences; and Inertial Fusion.

While these proceedings were being edited, we were saddened to learn of the tragic death of Kenneth Riepe in a traffic accident in Los Alamos, New Mexico. Ken had worked on heavy ion fusion for several years and was the lead designer of the multi-beam injector under development at Los Alamos National Laboratory.

Martin Reiser,
Terry Godlove,
and Roger Bangerter
Editors

THE U.S. MAGNETIC FUSION PROGRAM

J. F. Decker
U.S. Department of Energy, Washington, DC 20545

ABSTRACT

This paper is an overview of the U.S. Magnetic Fusion Program. A new program plan, issued in February 1985, defines four key issues for fusion: developing an improved confinement system, understanding burning plasmas, materials, and nuclear technology. The implementation of the program will rely strongly on foreign collaboration. Some recent technical highlights are also presented.

INTRODUCTION

The Office of Fusion Energy, supports work at over eighty institutions that include national laboratories, universities and industrial contractors. Figure 1 shows several of the participants. Most of the annual budget of $365M in FY 1986 supports major experiments and research at six sites: the Princeton Plasma Physics Laboratory, GA Technologies in San Diego, Massachusetts Institute of Technology and the Oak Ridge, Lawrence Livermore and Los Alamos National Laboratories.

Through all these efforts the technical agenda for the fusion program set in the mid-1970's has been or is now being completed, and a new technical agenda is emerging. Our program can cite great achievements over the last decade but must also reckon with current circumstances which include significant changes in recent years in the public and governmental perceptions of energy needs and great uncertainties and pressures associated with the national and world economies.

In this overview of the magnetic fusion program I will speak mainly about where the program will be going in the next fifteen years. Let me begin by briefly reviewing some of our recent accomplishments and describing important aspects of the strategy with which our future plans are being formulated.

TECHNICAL PROGRESS

The new generation of tokamaks, the most scientifically advanced of the toroidal confinement systems, have produced temperatures of one hundred million degrees and average energy confinement times of nine tenths of a second. The experimental campaign to achieve energy breakeven conditions has just begun in the Tokamak Fusion Test Reactor (TFTR) at Princeton University (Figure 2) and success is anticipated during 1986. Figure 3 shows the progress in confinement parameters for hydrogen plasmas and

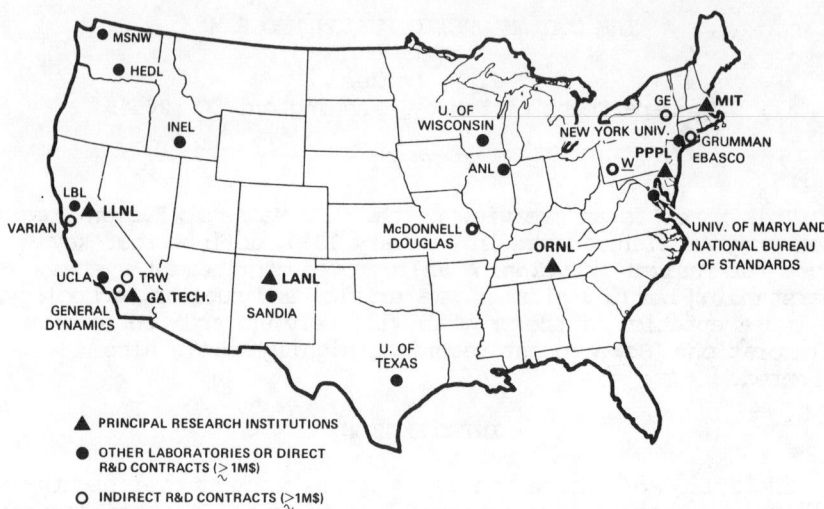

Figure 1. Locations of many of the participants in the US Magnetic Fusion Program.

Figure 2. The Tokamak Fusion Test Reactor at the Princeton Plasma Physics Laboratory in 1985 with (all) five neutral beams.

deuterium plasmas toward the conditions required for a self sustaining D-T plasma.

The general scientific progress of the last decade has also advanced the performance and the future prospects of a number of other magnetic confinement systems. The principle of plugging the ends of linear magnetic systems using the tandem mirror concept has been demonstrated. Several toroidal systems -- both near and distant relatives of the conventional tokamak -- have also achieved a state of scientific maturity, such that a logical development program can be planned. Several of these systems, as well as improved versions of the conventional tokamak, appear capable of exceeding the minimum power densities considered necessary for competitive economic power systems.

Two recently started (non-tokamak) projects are the ATF and the RFP. The Advanced Toroidal Facility (ATF) is a torsatron, a confinement experiment with helical coils around the toroid, being built at Oak Ridge National Laboratory and the first plasma is expected in 1990. The Reversed Field Pinch (RFP) concept is a toroidal, axisymmetric plasma-magnetic field confinement configuration that employs external toroidal magnetic fields and large plasma currents to produce a near-minimum-energy equilibrium state.

On the technology side, the International Large Coil Project at Oak Ridge National Laboratory (ORNL), and the Mirror Fusion Test Facility (MFTF-B) at Lawrence Livermore National Laboratory (LLNL) have operated and demonstrated the operation of reactor scale superconducting magnets. The 42 superconducting coils for MFTF-B (see Figure 4) can store 1200 MJ, about an order of magnitude larger than the largest MHD magnet. The largest coil weighs 175 tons. The Tritium Systems Test Assembly at Los Alamos National Laboratory is an operating, reactor level tritium fuel clean up and processing system. Radio frequency heating modules at reactor power levels, and pellet fueling systems have been developed and applied to our largest experiments. A 140 GHz gyrotron that can deliver 100kW of continuous power was recently demonstrated by Varian Associates. The feasibility of materials, which are long lived in a fusion environment, has been established for two classes of alloys. Initial work on the development of second generation low activation materials shows good promise.

Our progress has been good. Although some of the problems have proven to be more difficult than anticipated, excellent and sustained progress has been made in solving them. It is generally agreed that the remaining problems can be solved and that practical magnetic fusion systems can be developed.

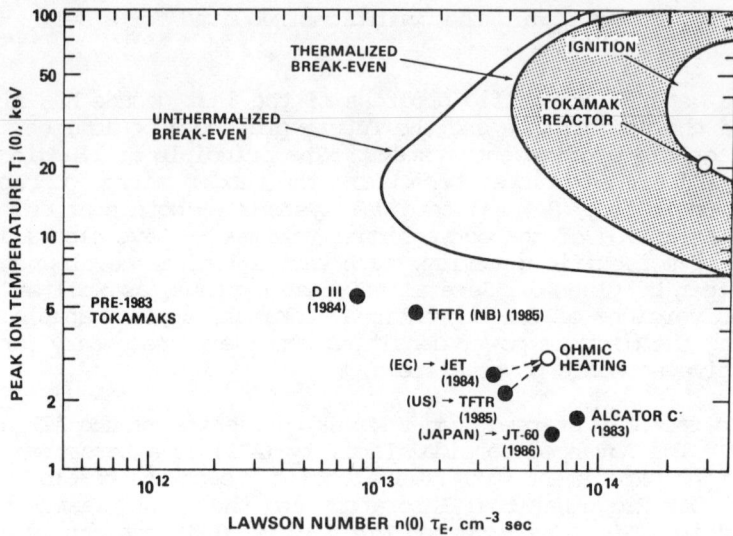

Figure 3. Progress in confinement represented on the familiar plot of ion temperature versus density times confinement time.

Figure 4. End view of superconducting magnet set at one end of the mirror facility MFTF-B, taken during installation.

THE MAGNETIC FUSION PROGRAM PLAN

During the early period (1973-1976) of national concern over the then looming energy crisis, the Magnetic Fusion Energy Program laid out a plan based on a goal of having a working fusion energy reactor system by the year 2000. This goal and the supporting plan were used to guide the Program's activities and to determine the support levels until the past few years.

The need for an energy supply system based on fusion has not diminished; nor has its desirability. This need is even acknowledged now by most policy makers. Nonetheless, it's hard to convince most people of the urgency to develop such a system by the year 2000.

More recently, we felt the need of a revised program plan that takes into account the present economic conditions, the present energy situation and its possible changes in the future, the present and future role of technical progress in general, as well as in fusion, the role of the program in training scientists and engineers, the contributions to basic research in the fields of plasma physics and atomic physics, and technology transfer.

The revised Magnetic Fusion Program Plan was issued by DOE in February 1985. The plan aims at an intermediate stage in the ultimate realization of commercial fusion power -- the point where the government, and the private sector, will have sufficient information about fusion to enable decisions on whether fusion deserves further development as part of the national response to the energy realities of the next century. Our goal is to establish the scientific and technological base required for fusion energy within the next fifteen years. Included in this goal is the successful transfer of technology to industry.

The essence of the program strategy is to maintain a broad domestic research and development program with emphasis on establishing the basic elements (components or subsystems) of the science and engineering technology required for fusion. No matter what the priorities of society, these technical elements will not change, and a program based on this technical reality should be able to maintain coherence and direction. The schedule for completion of magnetic fusion development should be directly related to the technical, economic, and political uncertainties of energy supply that are likely to persist for several decades.

The knowledge of how to employ fusion power, of its economic costs, and of its environmental effects should be available when major decisions must be made on the deployment of new energy systems in the United States. Based on the time scales for resolving uncertainties with respect to other energy sources, the scientific and technological base for fusion should be substantially available by the turn of the century.

Let us now turn to two fundamental questions regarding this challenging goal. What key technical issues must be resolved to accomplish this goal? How can the work be done?

TECHNICAL ISSUES IN MAGNETIC FUSION

Although the research and development in the magnetic fusion program is addressing a wide spectrum of detailed scientific and technological problems, the remaining work that must be accomplished to reach the program goal can be summarized by defining four key issues. These four key issues lie at the heart of the Magnetic Fusion Program Plan.

The first of these issues concerns magnetic confinement systems. The particular type of magnetic confinement used in a fusion reactor has a profound influence on its economics. Although significant technical progress continues to be made, it is not possible to design a fusion reactor at this time, and development of a variety of magnetic confinement systems is essential to achieving the program's goal.

It is quite possible that further development of the tokamak concept will lead to an economic fusion system. However, it is likely that the most desirable fusion reactor systems, in the next century, will require features from other types of magnetic confinement systems. Furthermore, the history of fusion development has shown that progress in all systems will be more rapid because of the cross fertilization possible between related concepts. Therefore, it is essential to explore an appropriate range of possibilities for magnetic confinement.

Over the past few years, we have been shifting resources from the base scientific issues of present tokamak operation to the more advanced ideas that offer hope of substantial improvement in the future. We have begun investigation of "bean-shaped" plasmas on a rebuilt PBX experiment at Princeton. We have built the AFT (torsatron), mentioned previously, which will improve our knowledge of toroidal confinement, as well as allow investigation of the second stability regime. The rebuilt Doublet-III-D, a joint U.S./Japanese project at GA Technologies in San Diego, is equipped with advanced long pulse neutral beams and 60 GHz gyrotrons to address confinement, high beta, and stabilization in shaped plasmas. All of these efforts, as well as initiatives on spheromaks, field reversed mirrors, and reversed field pinches, are aimed at achieving our scientific objective -- obtaining a predictive capability of plasma behavior in fusion relevant magnetic configurations.

The second key technical issue is understanding any significant effects from self heating when energetic alphas are produced in a burning plasma. In present experiments we drive heat into the plasma with beams and rf heating and the effects of this external heating on temperature, density and confinement may differ from alpha heating. Achievement of the program goal requires that at least one burning plasma be produced and studied to complete the scientific base. Furthermore, since a burning plasma will require much of the technology required for a fusion reactor, such an experiment could offer the additional benefit of advancing the technological portion of the goal.

The Alcator series of experiments have provided an experimental basis to investigate the use of high magnetic field tokamaks to address the essential scientific issues associated with ignition at a reasonable cost. A national design effort, headed by the Princeton Plasma Physics Laboratory, has identified what they believe to be a workable device that could be built in five years for less than three hundred million dollars. We call the group of design concepts being investigated the Compact Ignition Tokamak or CIT.

The third key technical issues concerns materials for fusion systems. Since materials play a central role in determining the environmental characteristics of a fusion reactor, they are the key to realizing the benefits of fusion. Materials in the first wall, blanket, shield and magnets of a fusion reactor will have to maintain their integrity for significant lifetimes under severe neutron radiation. The components close to the plasma used for controlling impurities will also be bombarded by energetic particles escaping the plasma.

Achievement of the program goal requires the development of new materials with properties that enhance the economic and environmental potential of fusion. Of great benefit would be the development of structural materials that minimize long lived radioactivity and permit manageable interim storage, for example via shallow land burial, of retired components. Our materials scientists have identified several promising materials that can reach this goal. We have shifted resources into developing these new materials and are planning for the facilities that will be needed to test them.

The fourth key issue concerns nuclear technology of fusion systems. This last issue requires advances in basic fusion materials research. Fusion blankets are a good example; they have a generic function in all D-T fusion systems and also impose some of the more challenging technological requirements in fusion. We are exploring the possibilities of blankets with liquid metals, with ceramics and with molten salts as tritium breeding materials. Each approach has its problems. For example, many liquid metal concepts face difficulty in providing adequate heat transfer from the first wall into the liquid metal. One impediment in resolving this problem is the lack of knowledge about liquid metal MHD, and we have started the appropriate liquid metal MHD tests at the level of very basic science.

To achieve our program goal, a certain breadth of research in fusion technology must be maintained in order to be able to optimize the economic, safety, and environmental performance of fusion reactors. This research program will be completed only when integrated blankets are tested in a nuclear environment. In other words, a major test facility will be needed.

The Magnetic Fusion Program Plan issued in 1985 outlined the key issues and broad strategy for magnetic fusion. Within various technical areas we have working groups of experts that have developed more detailed technical objectives. A comprehensive activity of this type, the Technical Planning Activity (TPA), under the direction of Argonne National Laboratory, was started in mid-1985 to identify the detailed scientific and technical tasks that have to be performed to arrive at our program goal. They have already accomplished a great deal in this difficult assignment and will be issuing a report at the end of FY 1986. An important aspect of the TPA's mission is to document the logical interrelationships among the technical tasks. One of the conclusions from the preliminary work in the TPA is that a large volume fusion neutron source is needed to complete the development of fusion nuclear technology.

The U.S. is creating a comprehensive program to address all of the issues of magnetic fusion. From our planning activities, it appears that clear technical paths can be defined to resolve the issues. From the technical point of view, there seems to be no reason why the world cannot have the basis for the development of a new energy option by the turn of the century. However, this is only part of the answer to the question of whether we will in fact have that basis? The rest of the answer involves the politics of energy development in the 1980's. Will the priorities of energy lead to the needed resources? Given the long term need for new energy sources, fusion R&D will be pursued. Given the lack of attention on energy following from today's energy glut, it is unlikely that fusion can be pursued rapidly, unless we pool our national resources with those of other nations interested in advancing magnetic fusion.

INTERNATIONAL COLLABORATION

The United States has maintained an active program of international cooperation for many years. However, the importance of international collaboration to the success of the program has increased dramatically in recent years. The world is now nearing the time when fusion will proceed into the stage where significant fusion nuclear technology must be included in the major devices that will carry research toward a useful commercial goal. These fusion systems will cost more than previous experiments. Like the U.S., the world community in fusion also sees finite resources. We all recognize the desirability, in terms of priorities and control, of each nation or community of nations having its own independent program, but we also all recognize the benefits of combining resources by a judicious program of international collaboration that can permit each of the participants to preserve the character of its program and draw upon the strengths of other participants.

The magnetic fusion programs of the world currently have a vigorous program of international cooperation and collaboration that includes scientific and technological exchanges, joint experiments, and joint planning activities. Through our common efforts, with the encouragement of the Economic Summit, these activities are being strengthened to emphasize international involvement, at an early stage, in planning for major new activities and facilities. This should enable us to identify mutually advantageous opportunities for collaboration and, on a case by case basis, to negotiate agreements that will minimize duplication and accelerate the achievement of our common goals.

Lest this 'world approach' seem overly optimistic to you, let me offer some evidence that the time is right. Agreement on a world view of fusion and even a world program does seem to be coalescing. Magnetic fusion was identified as an area of increased international cooperation for the western allies when the Economic Summit Process was launched in Versailles in 1982, and the visibility of magnetic fusion was enhanced by its inclusion as a topic in the Reagan-Gorbachev Summit in the fall of 1985.

Progress toward establishing a common set of objectives among the United States, the Soviet Union, the European Community, and Japanese fusion programs has been remarkable. There is substantial international agreement on the nature of the four key technical issues. Planning aimed at eliminating duplication, enhancing complementary tasks, and establishing more extensive task sharing is well advanced under the auspices of the Economic Summit Process. Agreements have been reached on particular tasks in each of these categories.

Let us look now at some of the ongoing processes. For the past three years, a group of policy level officials has been working under the auspices of the Economic Summit to identify ways of sharing the costs and benefits of fusion development,. This group, called the Fusion Working Group (FWG), includes representatives from the U.S., Canada, Japan, Germany, the UK, Italy and France, and another, a more technically oriented group called the Technical Working Party (TWP) have provided a systematic framework for the discussion of international collaboration and consideration of activities that could follow the Geneva Summit initiative in fusion.

In Munich this winter, the FWG reached a remarkable consensus. The group acknowledge that a common, medium term goal for all fusion programs is an Engineering Test Reactor (ETR) and that the joint design, construction, and operation of this ETR is a reasonable objective for international collaboration. They also agreed it is necessary to maintain the breadth of the overall fusion program through strong scientific and technical programs within each of the participants. Within these individual programs are sizable projected efforts supportive of an ETR that could also be the objective of international cooperation, including the previously mentioned U.S.-sited CIT (ignition experiment).

Among the remarkable products of this winter meeting of the FWG was a schematic diagram (Figure 5) prepared as an illustration of the kind of world program envisioned by the group. The items shown in this diagram are the activities related directly to the ETR and FWG clearly recognized that the activities indicated would address only two of the four key issues required to establish the feasibility of fusion. The issues of confinement systems development and long term development of materials require additional work. The common efforts on the ETR path were viewed as a means to free resources to complete these other necessary activities.

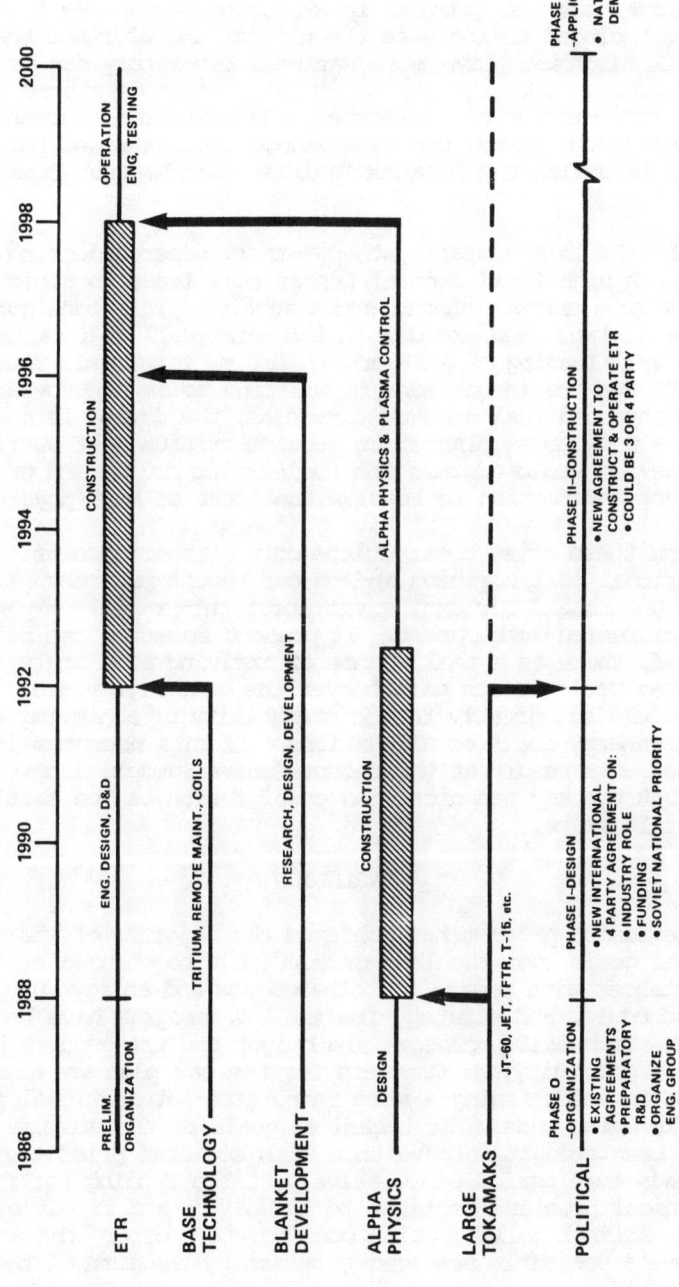

Figure 5. Schematic illustration of the world fusion program.

The ETR clearly has a central role in the world program. As part of our internal preparation for these international discussions, the U.S. program is exploring innovative ideas that might be included to increase the utility and decrease the cost of an ETR. Lawrence Livermore National Laboratory has adapted an earlier fusion reactor study to include recent developments in magnet configuration and performance limits and in current drive concepts. Their design for an advanced tokamak based test facility is called the Tokamak Ignition Burn Reactor Experiment (TIBRE-II).

Tibre-II is a compact, steady state, superconducting tokamak. A high field central pusher coil leads to a high beta plasma with a cresent shaped cross section. This configuration puts the maximum wall loading on the outer wall at a maximum neutron wall loading of 4 MW/m^2. Lifetime first wall fluences of 30 to 100 dpa are indicated. In addition to ample provision for testing advanced fusion blanket modules, the device is expected to have a work horse blanket to provide tritium self sufficiency. Such a device would address the nuclear technology and provide sufficient information to resolve the issue of burn physics.

From these brief observations on the importance of international collaboration and on our recent progress, I hope you can understand our excitement about this rapidly expanding aspect of fusion development. If present momentum can be sustained, there is a real chance of arriving at a truly integrated world fusion effort over the next five years. Such an effort would add greatly to the probability of achieving the domestic energy goals of all nations. If this momentum is increased as a result of the recent Geneva Summit, I have little doubt that the key technical issues of fusion can be settled in the next 15 years.

CONCLUSION

The fusion program has achieved the majority of its technical goals over the last decade. Due to changed external circumstances with regard to both budgets and energy supply, the goal, schedule, and strategy for the U.S. program have been revised. Technical prospects are bright and the support of the technical community and Congress for the new plan are gratifying. Progress in establishing a more integrated international fusion effort is being made. The technical goals of the program appear to have been brought into balance with societal priorities. If this leads to a period of relatively stable funding for fusion, if technical problems continue to be solved, and if our efforts at international collaboration continue to prosper, the world may well have a desirable new energy option by the turn of the century.

U. S. INERTIAL FUSION PROGRAM OVERVIEW

R. L. Schriever
Director, Office of Inertial Fusion, U.S. Department of Energy

ABSTRACT

The status of laser and light ion beam research and development is discussed in the context of the recently completed review of the U. S. program by a committee of the National Academy of Sciences.

INTRODUCTION

The U.S. program has recently been the subject of a review by a committee of the National Academy of Sciences (NAS). Its report, just published, provides an overview of program activities.[1] The committee found that the program had made substantial progress in a number of priority areas recommended by the 1979 Foster review panel:

o Shorter wavelength lasers.

o Understanding of laser coupling issues.

o Design of efficient targets, and attention to target fabrication.

o Ion drivers.

The requirements for igniting targets can be predicted with increasing confidence. However, the technical performance and cost implications of scaling the various driver technologies up to multi-megajoule size still need thorough investigation. Even for the best understood experimental driver, frequency-converted neodymium glass lasers, there is much uncertainty about performance and cost at one to two orders of magnitude higher beam energies.

SHORT WAVELENGTH LASERS

The effort during the 1970's to identify and develop "new lasers" -- gas lasers operating at less than a micron rather than the 10 microns of carbon dioxide -- has produced one serious candidate, the krypton fluoride laser. At Los Alamos a very large aperture KrF laser module has been developed (Fig. 1). This project essentially scaled up the e-beam pumping technique and demonstrated control of the laser kinetics. As a next step, the AURORA project involves multiplexing a beam of KrF laser light to form a 96-beam array, and then demonstrating the alignment system required to maintain beam separation and control through amplification and recombination of the beams onto a target. This demonstration is scheduled for the fall of 1987.

Fig. 1 KrF Large Aperture Module

The NAS panel noted that there are important technical uncertainties concerning the use of KrF lasers for fusion. Furthermore, they did not recommend a large follow-on laser project of any kind until results were available from HALITE/CENTURION and the NOVA experiments, and until the performance of pulsed power technology could be assessed. Even with these results, there remain very large questions of system cost, which for any driver technology except possibly pulsed power is at present too high by a factor of perhaps twenty to be affordable.

During the next two years or so the AURORA project should progress to the target shooting stage and provide the basis for resolving some of the technical uncertainties. The manufacturing and cost factors are being addressed in a parallel activity, the Power Amplifier Module or PAM, which involves scaling the technology demonstrated by the large amplifier module to 100-200 kilojoules.

Frequency converted Nd:glass lasers may be capable of being scaled to the requirements of a single-pulse high-gain target driver. Progress has been made in reducing costs by increasing amplifier efficiency, lowering the unit cost of glass, and raising the damage threshold of optical elements. Novel architectures are also being studied as a means of reducing the cost per joule.

Converting glass lasers from one-micron operation to the half, third, or quarter-micron has greatly improved the efficiency of absorption of beam energy by the target but, particularly for direct drive experiments, we require also symmetrical deposition of beam energy and a high degree of beam uniformity. The Naval Research Laboratory has pioneered a technique, induced spatial incoherence (ISI), by which each beam is divided into a large number of beamlets. Each beamlet is given a random phase, and all then illuminate the target surface. Any spatial nonuniformities are averaged out (Fig. 2). This technique, and a related approach developed at Osaka, have both demonstrated improved uniformity.

Focal Distributions with and without I.S.I.

(green laser)

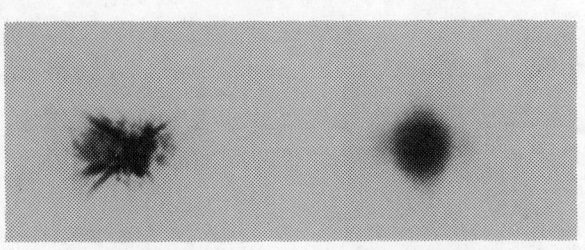

Fig. 2

no I.S.I.   with I.S.I.
(overlapped beamlets)

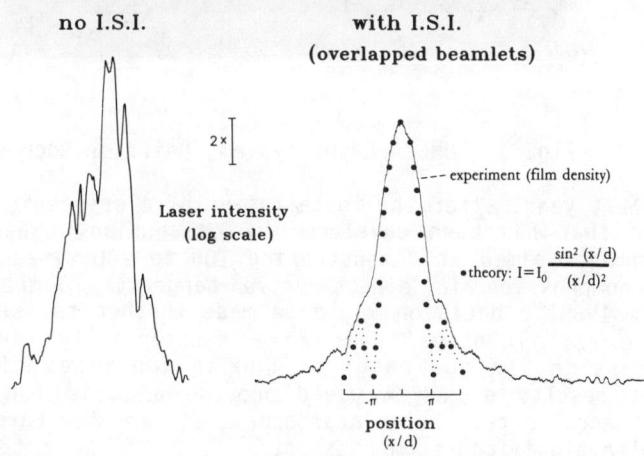

Laser intensity (log scale)

--- experiment (film density)

• theory: $I = I_0 \dfrac{\sin^2(x/d)}{(x/d)^2}$

position (x/d)

The development of the ISI technique illustrates one of the overlaps between present experimental and driver development efforts, since ISI can and should be applied to a KrF driver as well, either to the Los Alamos system or to a separate one. This is now under discussion.

Another candidate for the ISI technique is the OMEGA laser system at Rochester. The 24-beam OMEGA glass laser system operating at up to 2.5 kilojoules can support spherical implosion experiments.

Until recently OMEGA experiments had registered the highest thermonuclear neutron yield from an inertial fusion experiment. Efforts are now under way to improve the drive symmetry and demonstrate fuel compression to 100 times liquid density or greater. The PHAROS system at NRL is limited by its lower energy to flat target experiments. However, useful data on energy absorption phenomena, fluid instabilities, and target acceleration are gained in these experiments. These two efforts constitute the main U. S. effort in direct drive target experiments.

Fig. 3   OMEGA Laser System, Univ. of Rochester

Next year, after the installation of a cryogenic target support system that has been developed by KMS Fusion, OMEGA will support experiments aimed at demonstrating 100 to 200 times liquid density fuel compression in direct drive targets. Thereafter, perhaps during 1988, a decision would be made whether to use ISI on OMEGA. If the ISI technique leads to a substantially higher degree of compression, in the range of 200 to 400 times liquid density, target results on OMEGA would provide a basis for comparing the direct and indirect drive approaches, at very low target energy gain but with significant compression.

An experimental campaign just completed at KMS Fusion featured polymer shell targets containing a highly uniform layer of cryogenic fuel, illuminated by about 500 joules of green light. Drive symmetry was provided by KMS's unique double-bounce illumination system. A measurement of the implosion symmetry was obtained by x-ray backlighting of the imploding target. An initial look at the data indicates that very good compression was achieved for the available energy. In addition, a wealth of data is obtained on the laser performance, the target and cryogenic fuel integrity, temperatures, and some of the dynamics of the implosion.

The NOVA laser at Livermore is routinely operating at the second and third harmonics with 20 to 50 kilojoules on target. Various physics issues are being addressed in experiments using only two of the ten beams, which are switched into a small target chamber in the former SHIVA bay. Both the two and ten-beam systems will be used to continue the x-ray laser experiments begun on NOVETTE in 1984. For inertial fusion, target characterization at the second and third harmonics and the first intermediate density experiments are scheduled this year. The results of the initial

Fig. 4 NOVA Laser, Lawrence Livermore National Laboratory

experiments were recently announced: a new record neutron yield of 10^{13} neutrons from a glass shell target driven directly by ten beams in the third harmonic, with about 20 kilojoules on target. Over the next several years NOVA will be used to demonstrate key aspects of the performance that will be required of high gain targets: high densities of compressed D-T, high implosion velocities, the required stability of the imploded shell, and the appropriate drive conditions for short wavelength laser light and indirect drive targets. Although NOVA will not achieve ignition, it will demonstrate that the conditions for high gain can be brought about in a target of tested design and a laser driver of appropriate energy, wavelength and flexibility.

LIGHT ION FUSION

The light ion accelerator effort has recently met a major milestone in the completion and operation of the Particle Beam Fusion Accelerator II. Pulsed power tests over the next year are expected to bring the machine to its full operating capacity: 100 terawatts and 4.2 megajoules from the pulse forming section, and module simultaneity with a standard deviation of less than three nanoseconds. Power flow experiments will demonstrate the operation

Fig. 5 Particle Beam Fusion Accelerator II, Sandia National Laboratories, Albuquerque

of the plasma switch. Diode experiments will begin to explore the major features of high voltage, high current ion diode behavior. Four ion sources are being carried through the proof-of-principle stage, before making a decision on the optimum design. Finally, deposition experiments will be conducted with various materials in preparation for target experiments beginning in 1988.

NATIONAL ACADEMY OF SCIENCES REVIEW

A Congressionally mandated review of the Department of Energy's inertial fusion program was accomplished during 1984 and 1985 by a committee of the National Academy of Sciences. The Committee, under the chairmanship of Prof. William Happer of Princeton University, was favorably impressed by the quality of research facilities, the work being carried out, and the caliber and motivation of the research teams, and by the program's technological achievements in many fields. The Committee found that the "main uncertainties are the nature and practicality of the driver needed to ignite high-gain pellets; the minimum mass of DT fuel that can be imploded and efficiently burned, and how much energy is required to accomplish this; and the degree to which laser-plasma interactions and hydrodynamic instabilities such as Rayleigh-Taylor can be controlled." Finally, there was the question whether the cost of a high-gain system would be commensurate with the objectives and the potential benefits of inertial fusion.

The Committee judged that the program could, if adequately funded so that it held to its schedule, provide in about five years a clear indication as to whether it should be stopped or whether it should be pushed forward with renewed interest and support. The majority of the Committee felt the program identity should be maintained and that a continuous oversight committee should be established to help evaluate and guide the program.

REFERENCES

1. Review of the Department of Energy's Inertial Confinement Fusion Program, Commission on Physical Sciences, Mathematics, and Resources, National Academy of Sciences (Wash. D.C., 1986).

OVERVIEW OF THE HEAVY ION FUSION ACCELERATOR RESEARCH PROGRAM AT THE U. S. DEPARTMENT OF ENERGY

Ryszard Gajewski, Director
Division of Advanced Energy Projects
Office of Basic Energy Sciences
U. S. Department of Energy
Washington, D. C. 20545

ABSTRACT

A brief description is given of the Heavy Ion Fusion Accelerator Research program at the U.S. Department of Energy from a management perspective. Included are the rationale for support, budgetary considerations, and an overview of the current and proposed programs.

INTRODUCTION

It gives me a very special pleasure to present an overview of the Heavy Ion Fusion Accelerator Research (HIFAR) program in the Department of Energy (DOE). As you will be able to find out for yourself in these Proceedings, it is a program blessed with the talents, skills and unmatched dedication of a group of truly outstanding scientists; we, at the DOE, are tremendously proud of them.

And, while I am on the subject of tributes, let me mention one person who has guided the U.S. Heavy Ion Fusion program with wisdom and skill through thick and thin (mostly through thin...) since the program's inception. Terry Godlove, the HIFAR Program Manager, has just retired from full-time civil service, but he has graciously agreed to continue, on a part-time basis, to help us manage the program. I think I can speak for everyone in the U.S. Heavy Ion Fusion program if I say that we appreciate what Terry has done for the program and are looking forward to his continued involvement in the years to come!

WHY IS HEAVY ION FUSION SUPPORTED?

Before proceeding to tell you how we are going about supporting Heavy Ion Fusion research, let me tell you why we support it in the first place.

It is our agency's mission to care about the nation's energy future. That overall concern is reflected in a multitude of activities, with various time scales attached to them. Thus, for example, we are concerned about the clean burning of coal--a problem of immediate urgency--but we are also seeking energy sources capable of filling our nation's long-term energy needs. For reasons we are all familiar with, the process of nuclear fusion appears to offer an attractive long-range option.

Now, nuclear fusion is a process which can be effected in a number of ways. Magnetically confined plasmas offer one approach. Inertially confined plasmas offer another. Yet another, and very intriguing approach is that of catalyzed ("cold") fusion.

As we all know, fusion utilizing magnetically confined plasmas, or "magnetic fusion", was the first approach proposed. Over the last 35-40 years it attracted, world-wide, a very high degree of attention and support. Magnetic fusion constitutes a major component of the DOE research program, and we are very proud of its progress.

Given that, why look for alternatives? Because it is just conceivable that inertial fusion may offer advantages over magnetic fusion which are too important to be overlooked. (One day, it would certainly be interesting to see a study comparing the relative promise of the two approaches!)

Within the inertial fusion approach, different particle beams have been proposed as "drivers": photons ("laser fusion"), electrons (an approach pursued for a while but now abandoned), light ions, and heavy ions. In the United States substantial efforts have been invested in laser fusion and light ion fusion. However, again, heavy ion fusion appears possibly to offer advantages over the other approaches which are too important to ignore.

Just to remind you of the specific characteristics which, in combination, could make the heavy ion method an attractive fusion option, they are:
- high driver (i.e. accelerator) efficiency, repetition rate and reliability;
- focus at reactor distance;
- efficient deposition of particle energy; and
- cost projections in an acceptable range.

HOW IS HEAVY ION FUSION RESEARCH SUPPORTED?

Having presented to you the general rationale for supporting heavy ion fusion research, let me now tell you how it is done at the DOE.

DOE's "mainline" effort related to fusion is in magnetic fusion. That work is conducted under the auspices of the Office of Energy Research, by the Office of Fusion Energy. The total budget of that activity in fiscal year 1986 is $365 million. The vast majority of work related to inertial fusion is conducted under the auspices of the Assistant Secretary for Defense Programs, specifically by the Office of Inertial Fusion, with a current budget of $155 million. That program comprises work on both laser and light-ion drivers, and also work on target physics and development, to a large extent common to all three approaches to inertial fusion (laser, light-ion, and heavy-ion). The other-than-target portion of the heavy ion fusion program concentrates on accelerators, hence the program's name, HIFAR, for Heavy Ion Fusion Accelerator Research. The HIFAR program is managed within the Office of Basic Energy Sciences, part of the Office of Energy Research.

WHAT ARE WE SUPPORTING?

The primary emphasis in the HIFAR program is given to demonstrating the orders-of-magnitude increase in beam intensity that is required to make a compelling case for heavy ion fusion.

The program activities derive largely from a 1983 Program Plan which called, first, for research and development based on what was called a Multiple Beam Experiment, to be followed by a scale-up to a significant fraction of key accelerator parameters. So far, only the first phase is being implemented.

The major activities conducted under HIFAR are: single and multiple beam experiments, along with work on accelerator theory and design-- conducted mostly at the Lawrence Berkeley Laboratory (LBL); the development of a multi-beam source injector--at Los Alamos National Laboratory (LANL); and a systems assessment study coordinated by LANL and supported mostly out of the Office of Program Analysis and the Office of Inertial Fusion, with additional funding from the Electric Power Research Institute.

The allocation of fiscal year 1986 HIFAR funds is as follows: LBL - $3.8 million, LANL - $1.1 million, other - $0.3 million.

WHERE DO WE GO FROM HERE?

At present, the HIFAR program is at cross roads. There seems to be every indication that the heavy ion fusion approach may be as viable an option in the pursuit of fusion power as any of the other options presently explored. No "show stopper" has been identified. What, then, should the next step in the development of the heavy ion fusion option be?

The consensus of workers in the field is that the next logical step ought to be the demonstration of accelerator capability at some fraction of the ultimate requirements. What fraction? Clearly, the larger the fraction, i.e. the closer the proposed demonstration approximates "the real thing", the more credibility will a successful demonstration lend to the heavy ion fusion option. At the same time, the larger the fraction, the higher the cost. Back in 1983 it was estimated that a 1/3 scale demonstration (1/3 in key accelerator physics parameters) would cost about $70 million. More recently, a suggestion has been made of a 1/10 scale demonstration, using the concept of an Induction Linac System Experiment (ILSE). A program plan based on that concept has been developed. The price tag for ILSE is estimated to be about $26 million, with an implementation time of four years. Add to this the base R & D activity at the present level of about $6 million a year and you arrive at a $50 million HIFAR package, a roughly 100% increase over the current levels. In today's budget climate, with budgets for fusion generally declining, with the overwhelming pressure of other, ongoing DOE research programs competing for essentially the same research dollar, I see reaching such increased funding levels as a major problem - not insurmountable, perhaps, but still major.

While I cannot offer you any grounds for optimism in the short run, I am still a firm believer in the basic soundness of our system of supporting science. I am confident that the day will come when heavy ion fusion will get its "day in court", will receive a rational, dispassionate and fair evaluation, and will be funded in proportion to the promise it offers to the nation and, in fact, to mankind. My colleagues and I are committed to work towards accomplishing this goal.

THE GERMAN HEAVY ION FUSION PROGRAM

R. Bock
GSI, P.O. Box 110541, D-6100 Darmstadt 11, FRG

ABSTRACT

The German Heavy Ion Fusion Program is reviewed. Results obtained in the past funding period have increased our confidence that a heavy ion driver based on the linac/storage ring concept can be built which fulfills the specifications and the conditions of reliability and economics. Many critical issues have been identified which deserve detailed *experimental* studies, to be carried out in the second funding period which started this year. The GSI synchrotron/storage ring facility funded last year will play a crucial role in the future program. In addition, an RFQ high current injector will raise the intensity considerably. The first 5 out of 12 sections are already in operation. With this facility both the interaction of heavy ion beams with dense matter plasmas and beam dynamics issues of the HIBALL driver concept can be investigated.

INTRODUCTION

It was already mentioned by Professor Reiser, that this conference coincides with the 10. anniversary of the first workshop on Heavy Ion Fusion, which was held in Berkeley, July 19-30, 1976, in order to discuss the feasibility of heavy ion beams for inertial confinement fusion and the prospects for a fusion power plant based on a heavy ion driver. This historical event, from which I see many of the old pioneers in the audience, was the igniter for many activities on this subject in some countries, and subsequently some governmental funding programs were established.

Our program in W. Germany started at the end of 1979, and the first six years' funding period expired some months ago, end of 1985. The budget during this time was about 2 M$ per year. Its title was *Studies on the Feasibility of Heavy Ion Beams for Inertial Confinement Fusion* and its goal was two-fold:

1. The identification of key issues of the heavy ion fusion concept, and
2. the experimental and theoretical investigation of some of the issues in the field of accelerator and target physics as far as they could be investigated with existing facilities and with the modest funds available.

This period can be characterized as a first *exploratory approach* to the scientific and technical problems of the heavy ion fusion concept in order to define and pin down the essential issues and difficulties and to find out which direction would be most promising.

Concerning the driver, its realization looked rather straightforward on first sight, and at that time it was considered a more technical rather than scientific challenge. At the end of this first phase things look different now. On the one hand, many important results have been obtained which have increased our confidence that heavy ions are the most promising direction to ICF power generation. On the other hand, crucial physics issues have been revealed, both in the field of accelerator physics as well as in target physics, which necessarily have to be investigated experimentally, before the next major step, the design and construction of a dedicated facility, can be made.

Phase 2 of our program, which already started during the last year, will be dominated, therefore, by experimental activities. Experiments on accelerator issues and on problems of beam-target interaction as well as on matter at high energy density and heavy ion generated dense plasmas, will be the focus of the program. The new synchrotron/storage ring facility of GSI (SIS/ESR) which is funded with 120 M$ to be built in the next four years, will open access to such experiments in the near future (1989).

But even before that date, the recently accomplished first stage of the RFQ injector MAXILAC at GSI can already be used in a limited way for such experiments. In addition, an experiment on beam-plasma interaction is being prepared. All these experiments will be supported by theoretical activities.

The new title of this program is *High Energy Density in Matter Produced by Heavy Ion Beams* and the level of funding is about the same as in previous years. It is our present understanding that only when we will have obtained appropriate experimental information about these problems can we consider to build a dedicated facility in a subsequent phase.

My talk is organized as follows: In the first part I will give a brief account of the major achievements of the first funding period. The second part is devoted to the description of the new GSI facility. Finally, results of the ongoing work will be discussed and the philosophy of the future program and the plans for experiments with the new facilities will be sketched.

1. RESUME OF ACHIEVEMENTS 1979-85

The main emphasis of the research during this period was laid on four different fields:

1. The systems study HIBALL
2. Accelerator and ion source development
3. Theory on beam dynamics and on target issues
4. Experiments on atomic physics issues

All of them were covered in a previous paper[1] and in a brief overview on "Inertial Confinement Fusion with Heavy Ion Beams"[2]. The results and the present status[3] can be summarized as follows:

1. The Systems Study HIBALL. After a concentrated effort during the first two years the HIBALL conceptual design study was finished in 1981.[4] It was based on an rf-linac driver and a liquid first wall of Pb-Li and it was the first study of this kind dealing with all the large components and all the essential aspects of an ICF power station. Its goal was to demonstrate the compatibility of physics and engineering design in the areas of driver, target and reactor chamber through a self-consistent design, and to identify the key problems of this concept.

Based on new results, most of them presented at a workshop in Darmstadt in 1982, an upgrading was started resulting in a new set of the HIBALL parameters. A new report (HIBALL II)[5] including all our present knowledge, was published recently with the following major changes of the driver (Fig. 1):

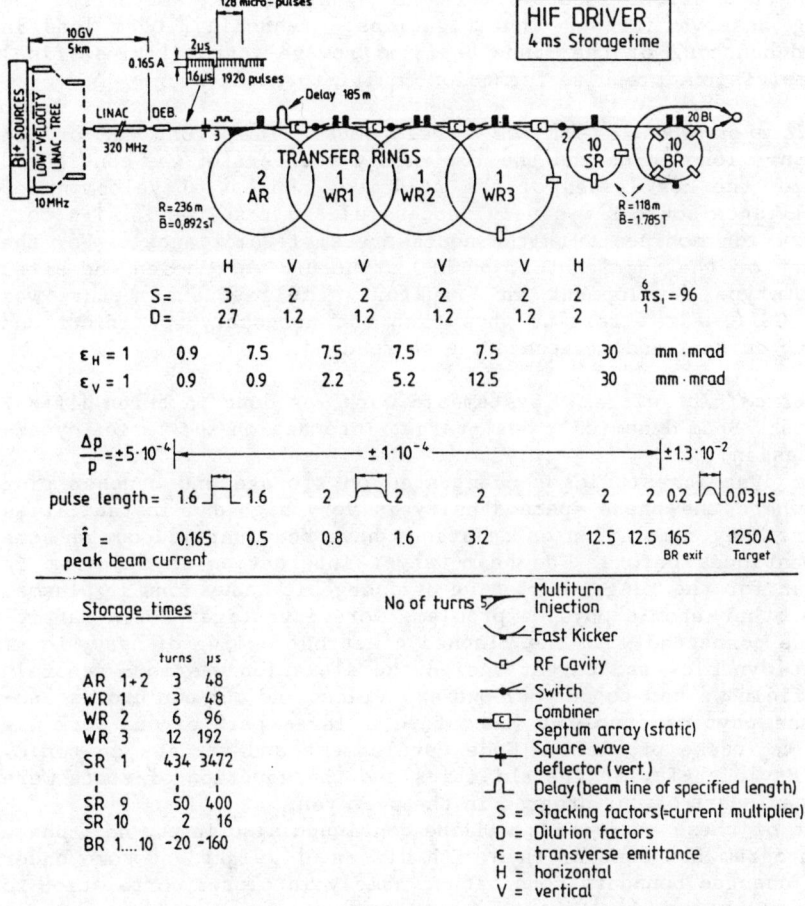

Fig. 1 Concept of the heavy ion fusion driver (HIBALL-II)

a. Because of space charge limitations the double-charged Bi^{2+} was replaced by Bi^{1+}, resulting in a linac of two times the previous length.
b. The storage time in the storage rings had to be reduced by a factor of 10 down to ≤ 4 ms due to micro-wave instabilities.
c. For the final bunching the induction linac bunchers were replaced by buncher rings because of cost considerations.
d. The large "transfer"-ring was changed into several smaller rings because better longitudinal and transverse emittance can be achieved and the momentum blow up during debunching can be omitted.

In our opinion, the HIBALL study has now reached a status in which our present knowledge is taken into account. It will be frozen for some time at the present level until new data will be available allowing a significant improvement.

Among the driver issues investigated during the last period the following deserve further investigations: funneling, beam load in linacs, debunching of the linac beam, microwave instabilities, final bunch compression, prepulse formation, multi-turn stacking.

2. Accelerator and Ion Source Development. The front end of the accelerator, ion source and the low-energy accelerator was considered as one of the key issues of the rf-driver concept. Development of high-brilliance sources was very successful and resulted in high-current heavy ion sources with the necessary small emittance[6]. For the first part of the accelerator, an RFQ structure was chosen and after some prototype development in Frankfurt a full size structure was built at GSI[7]. First results show that the necessary specifications concerning current and emittance can be reached.

3. Theoretical Activities. Systematic work was done in three different fields: Beam dynamics[8], beam-target interaction and target dynamics and design[9].

Beam dynamics studies concentrated on storage and buncher ring issues, where the phase space density is very high and instabilities may occur. Many simulation calculations have been carried out on some topics mentioned before. The beam-target interaction is relevant in particular for the range shortening of energetic heavy ions in plasma, but also other atomic physics problems were investigated, in particular those connected with the planned electron cooling of heavy ions. In target dynamics and target design the situation was more general. In this field we had to get our own experience and our own understanding of the physics involved, therefore a large part of our work was devoted to these problems. Code development and studies on topics such as Rayleigh-Taylor instabilities and the equation-of-state were an important part of our program in the past year.

Most of these activities will be continued also in the next phase of the program but some of them with different weight and some under somewhat changed boundary conditions, namely in closer correlation to the needs of the experiments.

4. Experimental Activities. Some experiments have been carried out, in particular on energy-range correlations for heavy ions in cold matter, an information which was available before only for some selected cases. Systematic and very detailed data have been measured at the UNILAC between 2 and 10 MeV/u for ions up to uranium. In the future this kind of measurements will be continued for hot matter on a much broader scale. Another experiment on ion-ion cross sections at low energy was made with great effort in order to calculate the loss rates for the heavy ion beam in storage rings due to beam-beam interaction. It turned out that the loss rates are not too serious. Also this experiment, because of its importance for the cross section measurements, will be continued with heavier ions. More details will be given in a contribution to this conference and in section 3.

2. THE GSI FACILITIES FOR HEAVY ION FUSION EXPERIMENTS

As reported at the last meeting in Tokyo[1], the German heavy ion fusion community had been discussing for quite some time the question of how to use the heavy ion synchrotron SIS planned at GSI for heavy-ion fusion experiments. The solution found in 1983 was based on the idea of increasing the phase space density of the heavy ion beam by electron cooling in order to approach the phase space limit and to obtain high energy density and high power density by depositing such a beam in matter[10]. An additional storage and cooler ring was designed (SITAR)[11] for this purpose. Achieving high phase space density and the possibility of beam manipulations by this two-ring facility made this project very attractive for heavy ion fusion accelerator and target experiments. The nuclear physics community soon realized the advantages of such a facility for their own research and it was last year that the whole project, now named SIS/ESR[12], was funded and will be finished in 1989/90 at GSI.

The whole new facility will consist of 4 different accelerators: (1) The existing heavy ion linear accelerator UNILAC, (2) the high-current injector MAXILAC, an RFQ structure consisting of about 12 modules, 5 being already in operation, (3) the heavy ion synchrotron SIS of 18 Tm using the UNILAC as an injector and (4) the cooler and storage ring ESR ('ESR' for 'Experimental Storage Ring') of 10 Tm. The UNILAC accelerates already now ions up to uranium to energies up to 20 MeV/u with intensities of about 10^{10} to 10^{12} ions/s limited by the multi-charge ion source (e.g. 8+ for uranium). With an RFQ injector the intensity can be raised by factor of 100 to 1000. The SIS will accelerate heavy ions such as uranium up to about 1 GeV/nucleon. The beam quality can be greatly improved by the ESR, thus providing fine-focussed beams with high intensity and small emittance.

What are the features of this new project with respect to heavy ion fusion and what kind of experiments can be achieved ?

The RFQ High-Current Heavy-Ion Injector
An RFQ structure is particularly appropriate for high-intensity low-velocity acceleration because it has a high focussing power for

very slow ions. This feature makes the RFQ structure very attractive for the low-energy end of a heavy ion fusion driver. We decided therefore, some years ago, to develop such a structure in the framework of the German heavy ion fusion program. The design and construction of this RFQ structure has been made at GSI. The special feature of this design (to be used as an injector into UNILAC) is its low rf frequency and the large mass to charge ratio of $A/q \leq 130$. The frequency was chosen half of the Wideröe frequency (13.55 MHz) in order to enable, at a later stage, the operation of two RFQ accelerators in parallel and to combine the two beams by funneling. Though a little more restricted in parameter space than an electrostatically focussed rf linac, this new scheme promised better technical solutions because of the absence of insulators in the high-voltage area.

Most RFQ devices designed for protons use the four-vane structure operating at frequencies of 100 to 400 MHz. For 13.5 MHz this mode is no more adequate. The tanks either get too big or the shunt impedance is too low. A new type of cavity was developed by R.W. Müller (contribution to this conference), which avoids these shortcomings. It is the so-called split coaxial cavity, a TM mode where the magnetic flux inducing the rf voltage toroidally includes the accelerating system which is contained within an inner conductor.

The Darmstadt high-current RFQ injector for UNILAC will consist of about 12 structures of this kind. Five of them are finished and in operation (Fig. 2). The rest will be finished in 1989. With its

Fig. 2
The first stage of the RFQ injector MAXILAC, consisting of 5 modules (45 keV)

mass-to-charge ratio of 130, singly charged ions up to I^{1+} can be accelerated. The expected space charge limit, occuring when the transverse oscillation frequency is lowered by space-charge forces by 30% is 0.2 x (A/q) mA, e.g. 25 mA for I . If the rf conductivity losses are as designed, 60% of the rf power is transmitted into the beam. The injection energy of the present 5-module accelerator is 2.3 keV/u, the final energy 45 keV/u, the total length 9.5 m. The whole set of five sections came into operation only some months ago, therefore the maximum intensity has not yet been reached. Intensities for various beams are up to about 10 mA. Intensities obtained so far and a set of parameters will be given in another contribution to this conference by R.W. Müller. A schematic view of the whole injector MAXILAC which will be finished in 3 years from now is shown in Fig. 3.

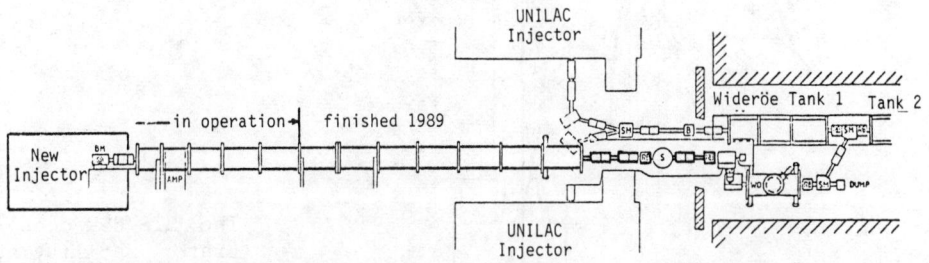

Fig. 3 Scheme of the MAXILAC injector. On the right: Front end of the UNILAC.

The High-Current Heavy Ion Synchrotron/Storage Ring Facility

A synchrotron with a bending power of 18 Tm (SIS 18) is under construction, with MAXILAC and UNILAC as an injector (Fig. 4). The maximum energy of fully stripped uranium ions will be 1.3 GeV/nucleon. The intensities of various heavy ions are given in Fig. 5. The storage and cooler ring (ESR) will be operated in combination with SIS 18. This two-ring accelerator complex shown in Fig. 4 can be used as a test facility for the investigation of accelerator and target issues of heavy ion fusion. In particular many dynamics problems at high space charge density can be investigated.

Due to calculations and simulations, it should be possible to obtain a very small focus spot with the cooled beams and, consequently, very high energy density in a small target volume. The beam diameter should be <0.3 mm and it is expected to reach a power deposition in the target of about 5 TW/g with a pulse length of about 70 ns. If shorter pulses can be achieved (a factor of 3 would be desirable), higher specific power density can be obtained. Some data for heavy ion beam parameters are given in Tab. 1.

After completion this facility will enable the investigation of a number of important target issues relevant to heavy ion fusion, depending on the temperature which can be reached. Fig. 6 exhibits the temperature of the plasma as a function of the specific deposition power. Temperatures up to about 20 eV should be reached and

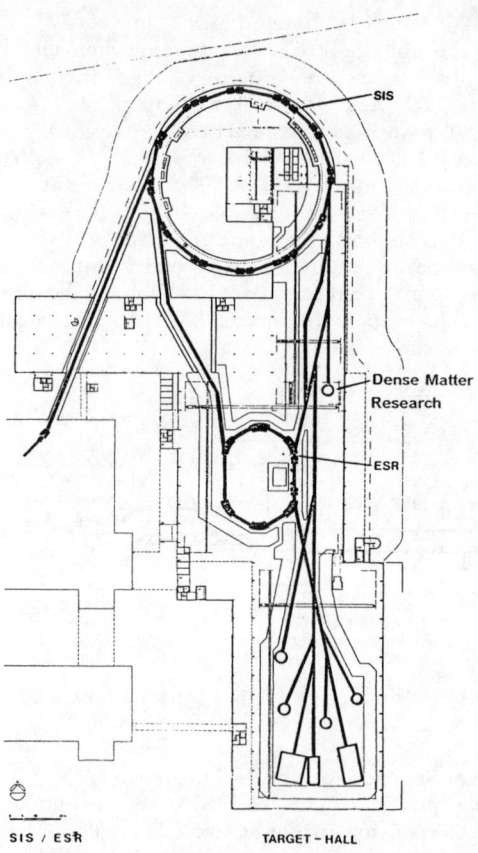

Fig. 4
The SIS/ESR accelerator facility at GSI. The diameter of the synchrotron (SIS) is 65 m. A special area is provided for high-current experiments ('Dense Matter Research')

Tab. 1 SIS/ESR Parameters	
Max. SIS Energy for Uranium	1 GeV/u
Ring Diameter (SIS)	65 m
Bending Power	18 Tm
Intensities	
Uranium (at 1 GeV/u)	$4 \cdot 10^{10}$ p/s
(at <500 MeV/u)	10^{11} p/s
Neon (at 1 GeV/u)	$3 \cdot 10^{11}$ p/s
Stored Ions in the ESR	$10^8 - 10^{10}$ ions
Cooled beams: $\Delta p/p \sim$	$10^{-4} \ldots 10^{-5}$
$\varepsilon_r \simeq$	$10^{-6} \ldots 10^{-7}$ $\pi \cdot$m$\cdot$rad

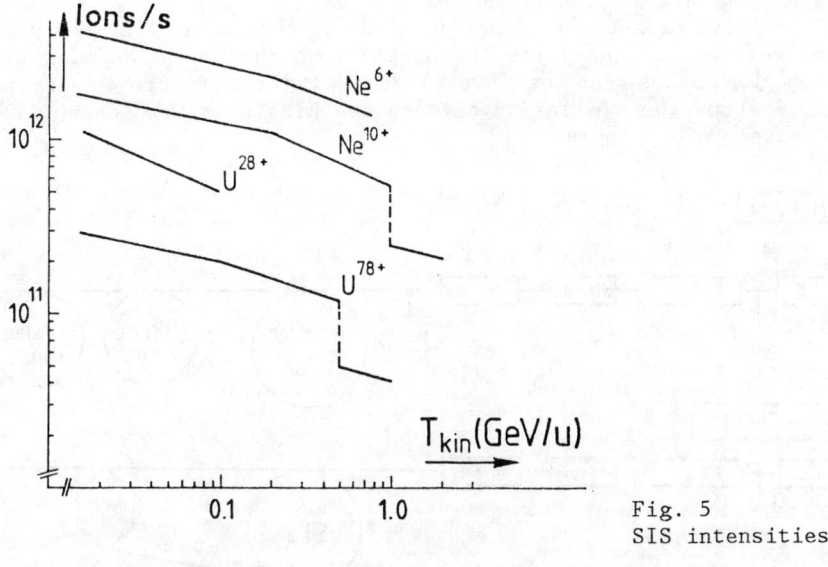

Fig. 5 SIS intensities

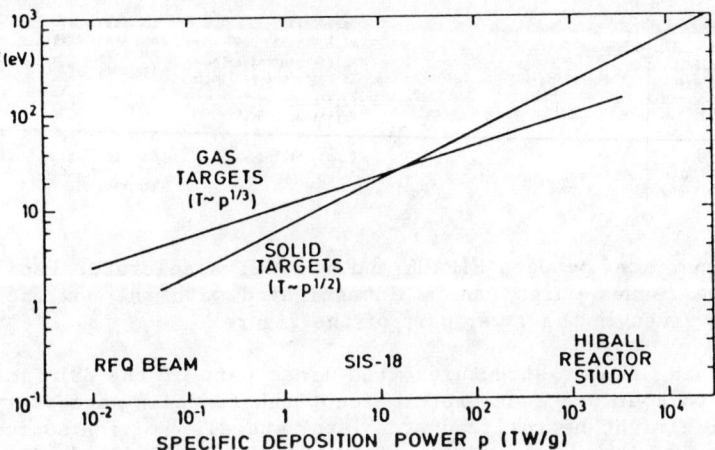

Fig. 6 Temperature reached in a heavy ion beam generated plasma as a function of the specific deposition power

investigations about ion deposition in hot matter, heavy ion generated plasmas, emitted radiation, short wave length laser and the equation-of-state can be carried out.

Comparison between HIBALL and the GSI Test Facility

In the previous section it was pointed out that a large number of problems relevant to heavy ion fusion and to high-current acceleration of heavy ion beams can be investigated with the future GSI facility. Fig. 7 shows the similarity between the HIBALL driver concept and the GSI accelerator complex.

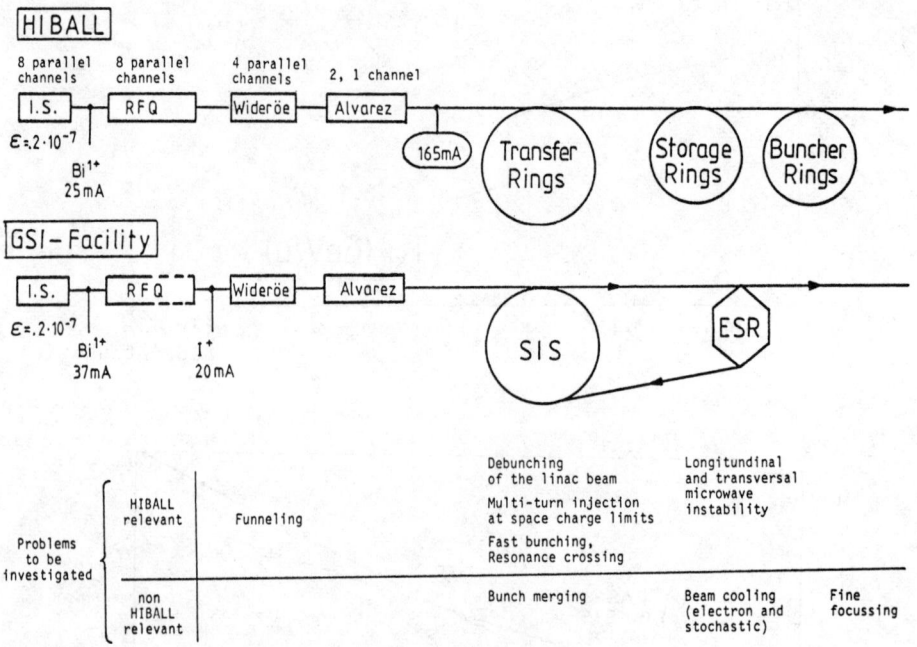

Fig. 7   Comparison between HIBALL and the GSI Accelerator Facility. Some issues which can be investigated with the new facility are given in the lower part of the figure.

Concerning the rf structures, the linac part of the GSI facility is similar to a driver accelerator except the funneling necessary for HIBALL (which might be realized at a later stage). For ion source and RFQ, the part which is already realized at GSI, the HIBALL design values for current and emittance can be reached soon. As mentioned before, the two-ring complex SIS/ESR will allow many HIBALL-relevant problems of beam dynamics to be investigated experimentally, in particular the debunching of the linac beam, multi-turn injection at the space charge limits, fast bunching and resonance crossing, the longitudinal and transversal micro-wave instabilities and fine focussing of heavy ion beams. In addition, a number of non-HIBALL-relevant issues will be investigated, such as bunch merging, electron and stochastic cooling. It is evident that this facility for the first time will provide high-intensity heavy-ion beams for fusion-relevant studies and will comply with the needs for beam handling under such extreme conditions.

3. EXPERIMENTAL ACTIVITIES: PRESENT AND FUTURE

Four different types of experimental activities are going on and will be continued. Some more are in preparation to be carried out at the UNILAC, at the RFQ and, in particular, at SIS/ESR in some years from now.

1. Ion-Ion Charge Exchange

The charge exchange in intra-beam ion-ion collisions due to betatron oscillations is still a key issue for the rf driver/storage ring concept. Most critical are the storage rings because of the long storage time (max. 4 ms). The longitudinal oscillations are negligible, the transversal relative motion is up to 10^5 m/s corresponding to an energy of 130 keV.

A setup for measuring charge exchange cross sections in a crossed beam experiment has been built at Gießen University[13]. Several ion-ion systems have been measured, the heaviest being $Xe^+ + Xe^+$. This is the first measurement of a medium heavy system, in which ionization and capture cross sections have been determined separately in the relevant energy range (Fig. 8).

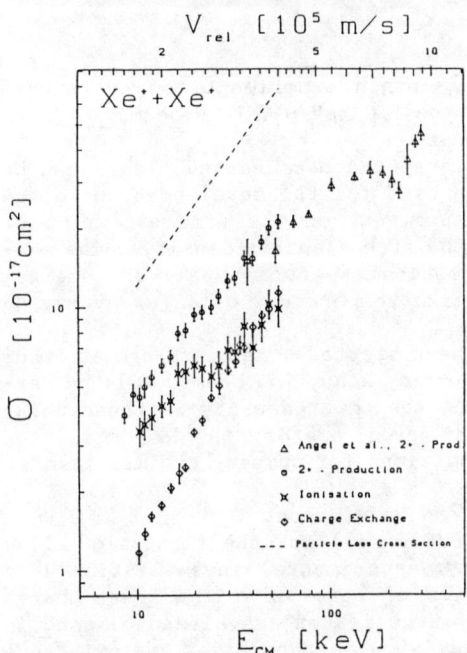

A preliminary calculation of beam losses based on the measured cross sections and on the most unfavourable assumptions concerning the velocity spectrum of the relative motion results in a loss rate of less than 5% for the longest storage time of 4 ms. It is expected that the real loss rates are much smaller. But even a loss of 1% would be a problem. On the other hand the situation for very heavy ions (on which HIBALL is based) might be much different. Consequently an experiment with Bi^+ on Bi^+ is in preparation. In these measurements the metastable ions, which can give a large contribution to the cross section, have been separated.

Fig. 8 Cross section for electron capture, ionization and the total Xe^{2+} production in $Xe^+ + Xe^+$ collisions (Salzborn *et al.*, contribution to this conference).

2. Beam-Plasma Interaction

Range shortening in hot matter, predicted by theory[14][15], is a key issue for pellet design. Therefore an experiment for stopping power measurements on heavy ions in plasma is being prepared using the UNILAC beam and a Z-pinch as plasma target. The setup of the Z-Pinch and the diagnostics and measuring devices are shown in Fig. 9. The plasma target is under construction at Aachen University. The line density to be achieved during maximum compression is in the order of $10^{19} cm^{-2}$ and the dimensions of the fully ionized plasma will be 200 mm in length and 14 mm in diameter. The current rise is designed for 5×10^{12} A/s and the maximum current 8×10^5 A. Before entering the target the beam is deflected by $15°$ in order to use a laser collinear to the ion beam for diagnostics. A second magnet is provided for the analysis of charge states. The experiment will be assembled at the 1.4 MeV/nucleon beam of the UNILAC and should be in operation in 1987.

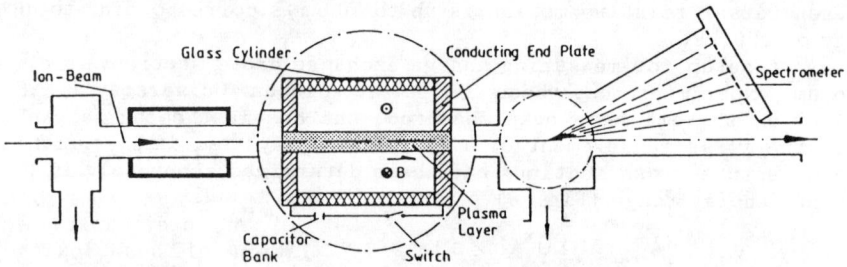

Fig. 9 Conceptual set-up of the Z-pinch experiment to investigate beam-plasma interaction at the 1.4 MeV/u UNILAC beam

In addition, at the RFQ accelerator a beam heated plasma can be produced already now which will be used for the development of diagnostic tools and techniques. Measurements on the time evolution of such a plasma are in preparation, the free electron density being measured by Thomson scattering, the optical spectra by streak cameras. In a gas target of 1 mm in diameter a temperature of a few eV may be obtained.

Another experiment which has been started some years ago at other accelerators[16], the pumping of lasers by heavy ion beams, will be carried out at the new facilities. As the power density is the crucial parameter for getting to shorter wave lengths, the RFQ and later SIS/ESR will provide favourable conditions for pumping excimer lasers.

3. Accelerator Experiments, Theory and Development

The high-current performance of SIS/ESR will for the first time allow a broad spectrum of fusion relevant experimental investigations, in particular those on the manipulation of beams near the space charge limit. Some of the problems to be investigated are already listed in Fig. 7. Most important are the studies on instabilities and emittance growth, for which this two-ring facility provides excellent conditions. During the next three years techniques and concepts for this kind of investigations will be developed.

In this context the ongoing theoretical studies are particularly valuable because they provide the guide lines for these experiments. Present work is focussing on current and phase space density limits in the SIS and ESR. The limitations by intra beam scattering and the resistive microwave instability have been examined for the ESR lattice. Preliminary results indicate that the microwave instability can be overcome by the "stabilizing tail" concept. Work on resonance crossing in the presence of space charge has been extended to a non-linear resonance. In emittance theory a new set of equations has been derived which allows to predict emittance growth in 3 dimensional bunched beams in linacs from excess non-linear field energy. With this theory it has become possible to derive from basic principles the emittance growth in high-current 3d-beams. This work will go on aiming at an experimental verification.

A number of studies is concerned with the electron cooling in the ESR and other issues, not relevant to the fusion driver but very important for the production of high-intensity heavy-ion beams. In this connection an experiment is being prepared to study dielectronic recombination of partially stripped ions as a principal mechanism that limits the lifetime of the ions during cooling.

With a beam transport line, the emittance growth and beam matching has been investigated experimentally and has confirmed the theoretical value of the minimum output emittance for vanishing input emittance[17]. The role of space charge compensation in beam transport has been studied.

Transmission studies with the Frankfurt split coaxial RFQ structure resulted in 85% transmission.

Whereas at GSI the construction of the MAXILAC has been pushed forward and experiments aiming at the improvement of its performance are carried out, the present and future Frankfurt activities cover a broad scope of topics. Some of them are listed in Table 2, and will be covered in more detail in another contribution to this conference.

4. Target Experiments and Target Theory

The important parameter for high-power target experiments is the specific power deposition in matter. What power density can be achieved with this two-ring accelerator and what are the temperatures obtained with this power density ?

Tab. 3 shows a set of parameters for a Iodine beam which is one of several options considered by I. Hofmann[18]. With 10^{11} ions per pulse, an emittance of $\varepsilon = 1.5 \cdot 10^{-6}$ m·rad and $\Delta p/p = \pm 4 \cdot 10^{-3}$ a power density of 40 TW/cm^2 should be obtained, resulting in a specific power density of 5 TW/g in solid gold. This number is based on a relatively long pulse length of 70 ns. For target dynamics reasons the optimum pulse length is about 20 ns or less and there are plans for more bunching power, which would raise the power density accordingly.

Tab. 2 Accelerator Development at Frankfurt University
Matching of High Current Beams to the RFQ
Beam Transport Experiments with High Space Charge
Funneling of Particle Beams
Multi Channel Accelerators (RFQ and MEQALAC)
Experimental Investigation of RF-Breakdown Phenomena in Vacuum (Sparking)
Development of Accelerator and Buncher Cavities for the High Current Injector at GSI (MAX-ILAC)
Beam Neutralisation
Dielectronic Recombination

Tab. 3 Expected Beam Parameters for High-current Experiments with SIS/ESR	
Ion (as an example)	$^{127}I^{20+}$
Number of Stored Particles	10^{11}
Particle Energy	330 MeV/u
Stored Energy	600 J
Pulse Length	<70 ns
Focal Spot (Diameter)	0.2 mm
Beam Power	12 GW
Power Density	40 TW/cm²
Range in Au	7 g/cm²
Specific Power Density	5 TW/g
Specific Energy	0.25 MJ/g
Plasma Temperature	15 eV

According to simulations with hydrodynamic codes carried out by the group at Garching a temperature of 15 eV should be obtained with 5 TW/g in solid gold. It is expected that under certain conditions temperatures up to 50 eV may be obtained in solid material. Extensive calculations have been made by the Garching group with 1-d and 2-d hydrodynamic codes to get a clear picture of the thermodynamic and hydrodynamic regimes which can be reached with SIS/ESR. Also different target scenarios have been investigated. Fig. 6 shows the predicted temperature dependence as a function of the specific power deposition. It will be discussed in a later talk at this conference by J. Meyer-ter-Vehn. Results on the dynamic behaviour of such a target are shown in Fig. 10.

Several experiments to be carried out with the SIS/ESR beam are discussed at present. One direction will be the investigation of the hydrodynamic and thermodynamic evolution of beam heated targets of various design. Another field, the pumping of short wave length laser will be further pursued. Some new results on temperature scaling of gas targets and on the attempts to reach x-ray lasing conditions look very promising. High-energy heavy ion beams create an elongated needle shaped plasma volume, required for x-ray laser, in a natural way. The crucial question for most of these investigations is whether high enough temperatures can be reached.

Considerations about the experimental set-up have been started, but it is too early to report here on the details of planning. Fig. 11 shows the planned target area at SIS/ESR for these experiments. Its location is shown in Fig. 4.

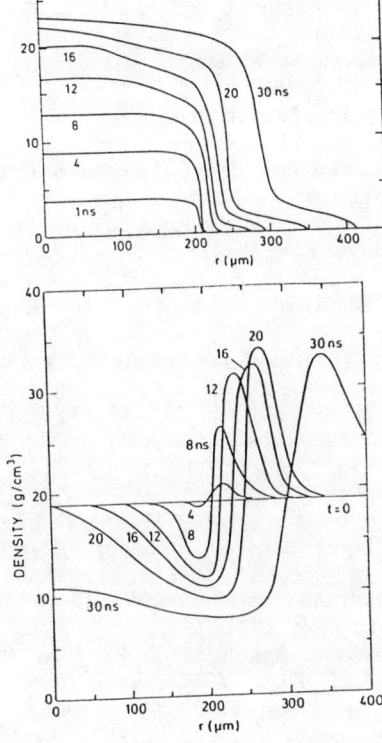

Fig. 10
Temperature and density evolution of a heavy ion beam heated cylindrical target (beam radius 0.2 mm)

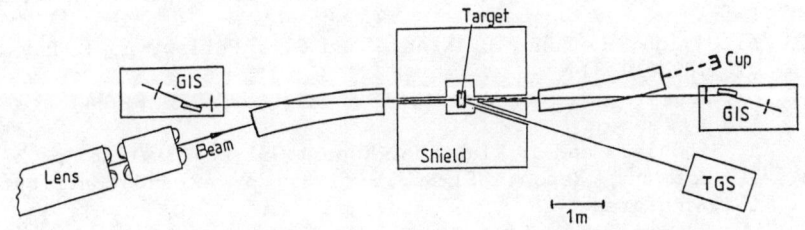

High-Intensity Target Station
equipped for Observation of coherent Radiation

GIS = Grazing incidence spectrometer
TGS = Transmission grating

Fig. 11 Planned target area at SIS/ESR for research on high energy density in matter

In conclusion, I would like to emphasize that with the new facilities at GSI a new experimental era for research on fundamental problems of heavy ion physics may evolve and at the same time, based on the results of this phase the further direction of heavy ion fusion can be defined more clearly.

REFERENCES

1. R. Bock, The German Heavy Ion Fusion Research Program
   Laser and Particle Beams 2, 395 (1984)
2. R. Bock, I. Hofmann and R. Arnold, Inertial Fusion with Heavy Ion Beams, Nucl. Sci. Applications 2, 97 (1984)
3. Annual Reports of the German Heavy Ion Fusion Program:
   GSI-83-2, GSI-84-5, GSI-85-21, GSI-86-13
4. B. Badger et al., HIBALL - a Conceptual Heavy Ion Beam Driven Fusion Reactor Study, KfK-3202/UWFDM-450 (1981)
5. B. Badger et al., HIBALL II - an Improved Conceptual Heavy Ion Beam Driven Fusion Reactor Study, KfK-3840/FPA-84-4/UWFDM-625 (1985)
6. R. Keller, P. Spädtke and F. Nöhmayer, Proc. Int. Ion Engin. Congress ISIAT '83, Kyoto 1983
7. R.W. Müller, Report GSI-86-13 (1986) and contribution to this conference
8. I. Hofmann et al., Annual Reports GSI-85-21 and GSI-86-13;
   Proc. Int. Symp. on Heavy Ion Accelerators (Eds. Y. Hirao and T. Katayama) Tokyo 1984; Report GSI-85-19 and 86-11
9. J. Meyer-ter-Vehn, Report MPQ-103 (1985)
10. K. Beckert, R. Bock, D. Böhne, I. Hofmann, R.W. Müller, I. Brezina, H. Wollnik, R. Arnold and J. Meyer-ter-Vehn, Report GSI-83-10 (1983)
11. I. Hofmann, Report GSI-SIT-INT/83 (1983)
12. Die Ausbaupläne der GSI, GSI-Report März 1984;
    P. Kienle, The Heavy Ion Cooler and Synchrotron Ring at GSI. Proceedings of the Visby Conf., Nucl. Phys. A 447, 419c (1986)
13. K. Rinn, F. Melchert, E. Salzborn, J. Phys. B 18, 3783 (1985)
14. E. Nardi and Z. Zinamon, Phys. Rev. Lett. 49, 1251 (1982)
15. T. Peter, Report MPQ-105 (1985) and to be publ. in Phys. Rev. Lett.
16. A. Ulrich, H. Bohn, P. Kienle and G.J. Perlow, Appl. Phys. Lett. 42, 782 (1983)
17. M. Reiser et al., Particle Accel. 14, 227 (1984) and 15, 47 (1984)
    A. Schönlein and J. Klabunde, Report GSI-86-13 (1986) p. 37-39
18. I. Hofmann, Report GSI-86-13 (1986) p. 42 and contribution to this conference

OVERVIEW OF HIF PROGRAM IN JAPAN

Y. Hirao

Institute for Nuclear Study, University of Tokyo,
Tanashi, Tokyo 188, Japan

ABSTRACT

In the last HIF Symposium at Tokyo, the HIBLIC-I was proposed by Japanese group. Further discussions on the design problems have been continued to improve or modify the conceptual design in the future. At the same time, near-term experiments are being planned for the related physics and engineering which the proposal is based on. In this paper, the outline of these approaches is given.

INTRODUCTION

As a feasibility study of the inertial confinement fusion driven by heavy ion beam, a conceptual design of HIF power plant "HIBLIC-I" has been performed since 1982.[1,2] This work has been carried out as a collaboration of several institutes and universities with a financial support of the Special Grant-in-Aid for Nuclear Fusion Research by the Ministry of Education, Science and Culture. It is aimed at defining the critical issues to be appeared in a process of making up a consistent picture of HIF power plant. In this paper, the present status of HIF studies after proposing the HIBLIC-I and then some plans in the future program are described briefly.

PRESENT STATUS OF HIF STUDIES

CONCEPTUAL DESIGN STUDY

The layout of HIBLIC-I power plant and the concept of reactor cavity are shown in Fig.1 and 2, respectively.

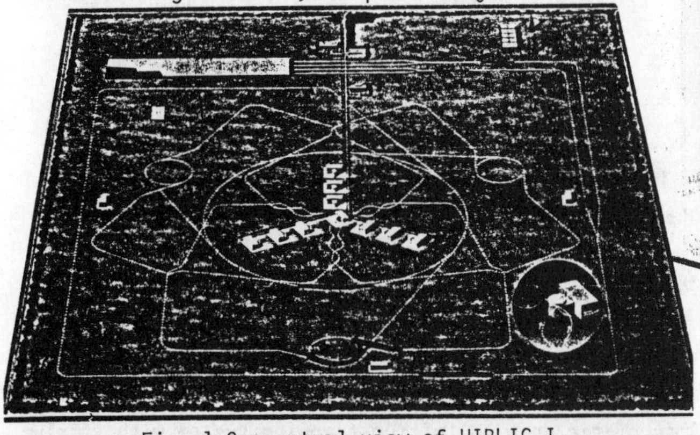

Fig. 1 Conceptual view of HIBLIC-I

Fig. 2 Reactor cavity in HIBLIC-I

Main parameters of HIBLIC-I are summarized in Table I. Some further efforts in design study have been made, as follows.

Table I. Major Parameters of HIBLIC-I

DT thermal power	4000 MW	Target	Pb-Aℓ-DT
Gross thermal power	4700 MW	DT mass/target	7.37 mg
Gross electrical output	1800 MWe	Target yield	400 MJ
Net electrical output	1500 MWe	Target gain	100
Plant availability factor	75 %		
Driver type	RF-Linac	Target shot rate/cavity	1 Hz
Driver efficiency	25 %	No. of cavity	10
Ion Species	^{208}Pb$^+$	Coolant and breeder	Li
Ion energy	15 GeV	Maximum coolant temp.	470°C
		Tritium breeding ratio	1.55
Beam energy	4 MJ	Tritium inventory	5 kg
Beam power	160 TW	Tritium recovery	Y-getter
Driver repetition rate	10 Hz	Structural material	HT-9
No. of beams/cavity	6	First wall (Sacrificial wall)	
Current intensity/beam	1.78 kA	Protection scheme	Li-Curtain
		Max. dpa rate in 1st wall	36.6/FPY
Cavity pressure at RT	10^{-4} Torr	Max. dpa rate in 2nd wall	0.93/FPY
		Life time of 1st wall	2 FPY
		Life time of 2nd wall	30 FPY
		Wall loading (1st wall)	5 MW/m^2

Consideration for Tritium Recovery

As a method of tritium recovery from liquid lithium, yttrium getter is used in HIBLIC-I.[3] In the previous proposal, yttrium of about 1,000 kg/cavity is needed to remove ^{3}H, ^{2}D, ^{1}H isotopes from liquid lithium loop. It seems to be tightly difficult in a view point of engineering. It is, then, found that the idea of sponge-like getter shown in Fig.3, will reduce remarkably the amount of yttrium down to 7.7 kg/cavity.

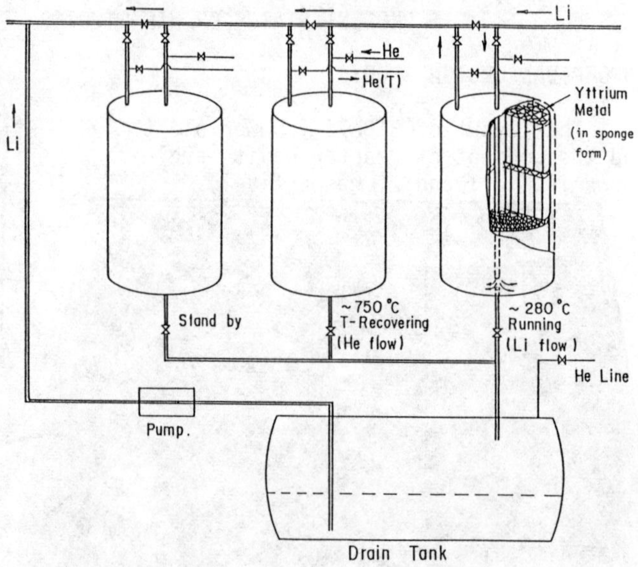

Fig.3 A schematic drawing tritium trapping system with yttrium

Energy Analysis of HIBLIC-I

To check the feasibility of HIBLIC-I in economical aspect, an energy analysis of the plant has been crudely made.[4] The required input energies for driver system and reactor cavity were calculated on the basis of energy intensity estimated for component materials, while those for other subsystems such as turbine equipments, which are mostly similar to or common with the conventional fission power plants, were taken up from the energy analyses for LWR power plants. The result of the analysis is shown and compared with MCF power plant "STARFIRE" in Table II.

Table II. Energy balance of HIBLIC-I

	HIBLIC-I	STARFIRE
Net electric power (MWe)	1500	1200
Availability factor (%)	75	75
	Energy required, unit: 10^9 kcal(%)	
Initial requirements		
Reactor	810 (7)	2377 (35)
Energy driver*	4696 (41)	—
Others	4026 (35)	3292 (48)
Subtotal	9532 (83)	5669 (83)
Maintenance	1906 (17)	1134 (17)
Total	11438 (100)	6803 (100)
Energy ratio	22.2	29.9

* including buildings and tunnels for the energy driver.

Energy ratio is defined as electrical energy output for 30 years/ total input energy for 30 years. This crude analysis has shown that the HIF power plant is feasible in terms of energy economy.

Considerations on Target Design

A new calculation scheme which treats shock waves in the fuel layer as discontinuous surface was proposed instead of introducing the artificial viscosity.[5] Calculated result shows that the optimum energy of ion beam of $A \sim 200$ is about 3 GeV for simple single shell target.

This result implies more sophisticated target design is necessary in an aspect of matching with accelerator enegineering.

ACCELERATOR RESEARCHES

Inductive Postacceleration of a Charge- and Current-Neutralized Intense Pulsed Ion Beam

In order to reduce difficulties of the light ion beam (LIB) fusion (beam propagation and focusing) and heavy ion beam (HIB) fusion (beam intensity), the study of inductive postacceleration of an intense medium mass ion beam (MIB) produced by ion diode has been started at Nagaoka Technical University.

As a first step of the multistage acceleration of space charge- and current-neutralized ion beam, a single stage induction accelerator system, MALIA-I (medium-mass Atom Linear Induction Accelerator) was constructed as shown in Fig.4 and the first observation of highly neutralized LIB and MIB was made.[6] An annular ion beam of 90 keV, 5.7 kA, 750 nsec extracted from an applied-Br magnetically insulated diode is injected into the induction accelerator, where a short pulse of 210 kV, 50 nsec was applied.

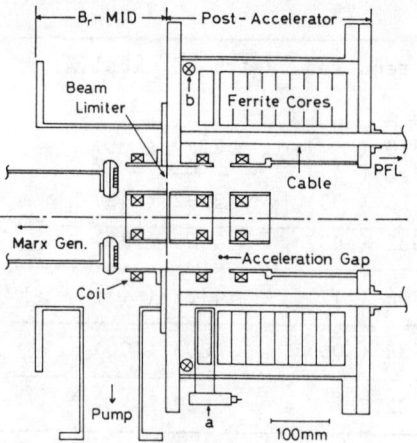
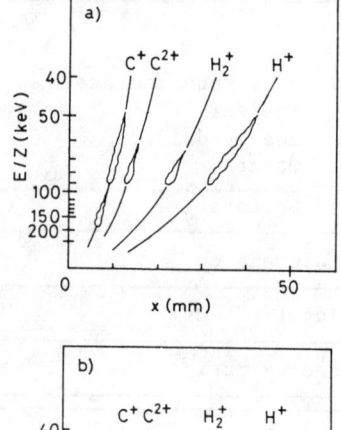
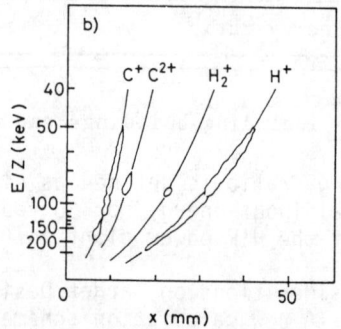

Fig.4 Schematic of induction-accelerator system, MALIA-I, where a and b denote the voltage divider (CuSO$_4$) and the Rogowski coil, respectively.

Measurements of beam energy by a Thomsom-parabola spectometer before and after the postacceleration have confirmed the increase in beam energy of H$^+$ up to ~ 240 keV as shown in Fig.5. Further progress of LIB and MIB postacceleration experiment is expected.

Fig.5 Thomson-parabola spectrometer data (a) before and (b) after postacceleration.

TARN II Project

TARN (Test Accumulation Ring for Numatron)[7] is a storage ring, built for developing accelerator technologies of beam accumulation and stochastic cooling with a view of future application to the proposed high-energy heavy-ion accelerator complex, so called Numatron.

Beam stacking of about 20 times each in the horizontal and the longitudinal phase spaces was achieved, and overall stacking number and momentum spread reached to more than 300 revolutions and 1 %, respectively. Life time of several thousand seconds was achieved for 7 MeV proton beam because of an ultrahigh vacuum reaching to the region of 10^{-12} Torr. Molecular hydrogen ions, alpha particles and carbon ions were also accumulated. The stochastic cooling experiment was performed on the rf stacked beams of proton and alpha particles to decrease their momentum spreads. Obtained momentum resolutions were the order of 10^{-4} for protons and alpha particles of about 10^8 as shown in Fig.6.

On the basis of these successful achievements on accelerator studies, construction of the improved TARN, TARN II,[8] was started in May 1985, which will have functions of accumulator accelerator, cooler and stretcher. As shown in Table III, this ring has the magnetic rigidity of 7 Tm and can accelerate up to 450 MeV per nucleon for fully stripped ions of medium mass nuclei.

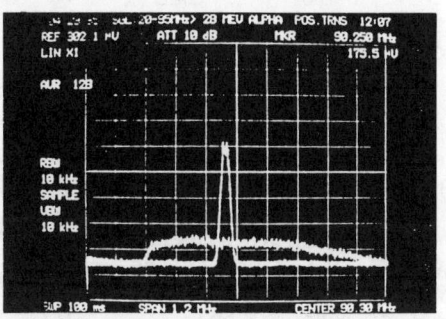

Fig.6 Schottky signals at the 80th harmonic before and after cooling of 420 s, for 28 MeV α particles.

Table III. Specification of TARN II

Maximum Beam Energy	proton	1300 MeV
	ions with $\varepsilon = 1/2$	450 MeV/u
Circumference		77.76 m
Average Radius		12.38 m
Radius of Curvature		3.82 m
Focusing Structure		FBDBFO
Superperiodicity (Synchro. Mode/Cooler Mode)		6/3
Betatron Tune Value ν_h (Synchro./Cooler)		1.75/1.75
" ν_v (")	1.25 or	1.75/1.25
Transition γ (")		1.86/2.97
Repetition Rate for Synchrotron Mode		1/2 Hz
Maximum Field of Dipole Magnets		18 kG
Deflection Angle of Dipole Magnets		15
Maximum Gradient of Quadrupole Magnets		70 kG/m
Acceleration Frequency for Adiabatic Capture		0.71-7.02 MHz
" Synchronous Capture		0.86-8.00 MHz
Harmonic Number		2
Maximum RF Voltage		6 kV
Vacuum Pressure		better than 10^{-10} Torr

Figure 7 shows the installed dipole and quadrupole magnets. The injector is the existing SF cyclotron (K = 68) at the commissioning stage and additional heavy ion linac will be constructed in the next stage. The first part of the linac is the RFQ-TALL and has been completed[9] in the last summer as shown in Fig.8.

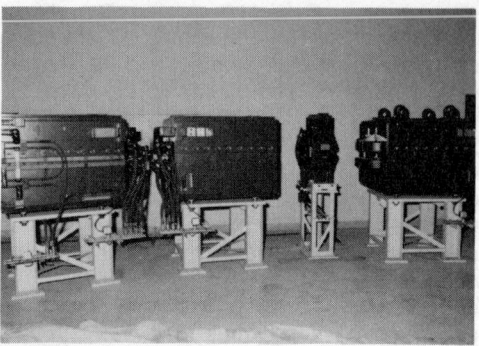

Fig. 7 Installed dipole and quadrupole magnets in TARN II

In the improved ring, the electron beam cooling device will be installed in addition to stochastic cooling one, both of which are complemental to each other. By the use of the combination of both methods expected beam emittance and momentum resolution are ~ 1 mm·mrad and $\sim 10^{-5}$, respectively, in a wide energy range. Main parameters of the electron beam cooling system are shown in Table IV.

Fig. 8 Completed RFQ-TALL, shorter one in the right is RFQ-LITL.

Table IV. Electron cooling parameters

Maximum working energy	ions	200 MeV/A
	electrons	120 keV
Cooled ions		$H^+ - {}^{20}Ne^{10+}$
Gun optics	Pierce type + resonance focussing electrodes	
Length of interaction region		1.8 m
Maximum electron current density		0.5 A/cm^2
Cathode diameter		50 mm
Maximum current		10 A
Maximum solenoid field		1000 G

Most of the device has been completed, as shown in Fig.9.
    Using the TARN II, internal and external beam experiments in various kinds of research fields are being planned in 1987.

Besides the TARN II, a heavy ion medical synchrotron (HIMETRON) project is going to start at National Institute of Radiological Science, Chiba which can accelerate up to argon to 800 MeV per nucleon. The first beam is expected in 1992. For the purpose of future nuclear physics research, a big hadron accelerator project is at a discussion stage which includes high intensity protons of 3 GeV and very heavy ions such as lead ions of 1 GeV per nucleon, while the perspective of this project is still quite opaque.

Fig. 9 Field measurement of the electron cooling device

CONSIDERATIONS OF FUTURE PROGRAM

PROGRAM FOR HEAVY-ION-BEAM EXPERIMENTS

In the development of HIF approach, the most fundamental and critical issue is to examine heavy-ion-beam incident implosion physics experimentally.

Table V Experimental Program with Heavy Ion Beams

Phase	I		II
Ion Species	$^{40}Ar^{18+}$	$^{84}Kr^{18+}$	
Ion Energy (MeV/u)	75	75	
Number of Ions	10^{11}	10^{12}	×10
Number of Beams	1	1	2
Pulse Width (ns)	10	10	
Beam Diameter (mm)	0.3	0.3	
Range in Gold (g/cm^2)	1.5	0.9	
Specific Power Deposition (TW/mg)	0.005	0.18	>1
Plasma Temperature (eV)	10-20	60-80	>100
Remarks	Injection from Heavy Ion Linac instead of SF Cyclotron		Addition of Accumulator Cooler and Buncher Ring

Although no full-sized facility for this purpose is available up to date, there are some plans of energetic heavy ion accelerators even in our country as mentioned above, which are possibly applicable to the basic studies in HIF. The TARN II has been considered as one of such programs. From the HIF point of view, its upgrading in two phases[8] might be of great use to obtain the final beam conditions as listed in Table V and as shown in Fig.10. Such the beam of ^{84}Kr is expected to realize high-temperature target plasma experiments of ~ 100 eV to examine elementary processes in ion beam induced implosion physics. At the same time, acceleration and manipulation experiments on high-intensity heavy-ion beams to give the basis of HIF driver system will also be performed.

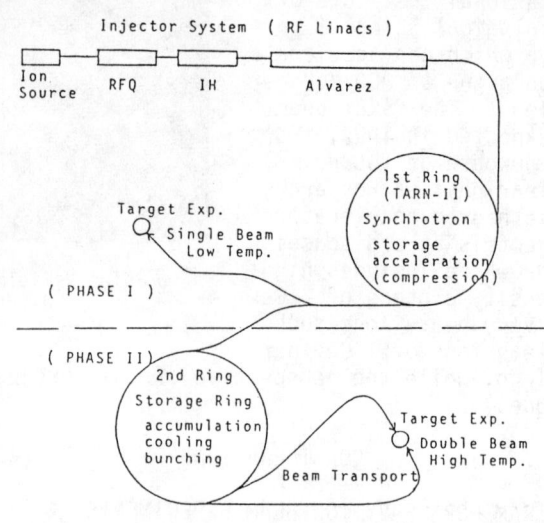

Fig.10 Heavy Ion Fusion Exp. on TARN-II (Phase I & II)

PROPOSAL OF AN EXPERIMENT FOR LITHIUM CURTAIN DYNAMICS

In the design of reactor cavity chamber of HIBLIC-I, concept of liquid-lithium water-fall type ICF reactor cavity has been proposed to protect the first structural wall. However, behavior of the liquid-lithium stream at microexplosion is not yet established both experimentally and theoretically. Repetitive formation and fragmentation processes of the lithium curtain, by which the residual vapor pressure inside the chamber is controlled within a required period of successive target implosions, should be investigated experimentally with some suitable simulator. A provisional design of such an experimental device, called ALLICE (Apparatus for Liquid Lithium Curtain Experiment)[10], has been proposed schematically as shown in Fig.11. This device is composed of an electron beam generator, a liquid lithium circulation loop of 400 kg lithium inventory, and a test chamber in which a small and flat curtain is formed as a blow-down stream from a nozzle. In order to give an energy diposition to simulate the implosion heat load in the

HIBLIC-I, a pulsed electron beam from a Marx generator of 5 MV, 60 kJ 50 nsec is applied to the area of 10 ~ 20 cm φ on the curtain. The expected amount of energy is 100 ~ 1000 J/g and well to be compared with 200 J/g in the HIBLIC-I case.

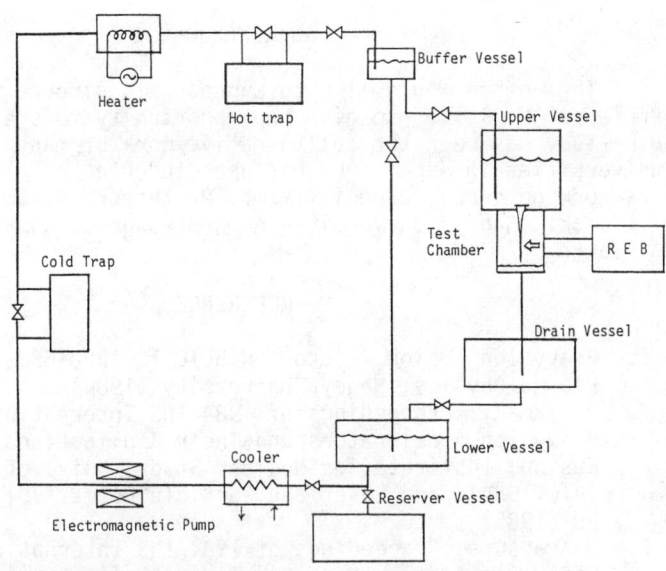

Fig. 11 Apparatus for Liquid Lithium Curtain Experiments

The physical and technical problems to be verified with ALLICE are:
1) Formation of lithium curtain,
   Efficient way of curtain formation, Hydrodynamical stability of curtain,
2) Fragmentation of curtain at pulsed energy deposition,
   Obsevation of fragmentation, Pressure propagation within curtain, Change in surface temperature, Size distribution of resultant liquid droplets,
3) Evaporation and condensation processes,
   Transient change in residual vapor prssure, Development in size of liquid droplets.

These are of fundamental importance for the detailed cavity design of HIBLIC-I and other similar systems.

SUMMARY

Since the last HIF Symposium, experimental studies in target physics and reactor engineering have been emphasized as the most important subjects to develop the HIF study. However, activities of the cooperative group of HIBLIC study has not well increased mainly because of the budgetary limitation. Possible stimulation will be brought about by commissioning of new experimental facilities as mentioned in this paper. At the same time, brushing up of the concept of HIBLIC is continued in every section including numerical analyses on target physics by computer simulations.

ACKNOWLEDGMENT

The author would like to express his sincere thanks for Professor M. Reiser to give the opportunity to present this paper. HIF study has been the collaborative work of many institutes and universities in Japan. He has been indebted to all the members of the working group, especially, to Professors H. Obayashi and T. Yamaki, Nagoya University and Professor T. Katayama, University of Tokyo.

REFERENCES

1. Heavy Ion Fusion Reactor "HIBLIC-I" IPPJ-663, Institute of Plasma Physics, Nagoya University (1984)
2. T. Yamaki, Proceedings of 1984 INS International Symp. on Heavy Ion Accelerators and Their Applications to Inertial Fusion, Institute for Nuclear Study, Univ. of Tokyo, 141 (1984), and Laser and Particle Beams, Vol.3, part 1, 29 (1985)
3. H. Katsuta, Proceedings of 1984 INS International Symp. on Heavy Ion Accelerators and Their Applications to Inertial Fusion, Institute for Nuclear Study, Univ. of Tokyo, 834 (1984)
4. T. Nagai, H. Obayashi and T. Yamaki, contributed paper to this Symp.
5. K. Niu, contributed paper to this Symp.
6. T. Tanabe et al., Phys. Rev. Letters, Vol.56, No.8, 831 (1986)
7. A. Noda et al., Proceedings of 1984 INS International Symp. on Heavy Ion Accelerators and Their Applications to Inertial Fusion, Institute for Nuclear Study, Univ. of Tokyo, 265 (1984)
8. T. Katayama, invited paper to this Symp.
9. N. Ueda et al., contributed paper to this Symp.
10. Y. Fuji-ie et al., Report of Working Group on Fusion Reactor, Research Information Center, IPP, Nagoya Univ. Nov. (1985) in Japanese

RF LINAC FOR HEAVY ION FUSION DRIVER

I. M. Kapchinskiy, V. V. Kuschin, N. V. Lazarev,
V. G. Schevchenko, V. S. Artemov, V. A. Batalin,
E. N. Damiltzev, A. Ju. Djadin, D. D. Iosseliani,
A. M. Kozodaev, A. R. Kurs, I. M. Lipkin, I. O. Parshin,
S. V. Plotnikov, V. S. Skachkov,
S. B. Ugarov, and A. B. Zarubin

Institute for Theoretical and Experimental Physics,
117259, Moscow, USSR

ABSTRACT

A possible structural design and approximate parameters of a rf linac for HIF driver are given. Bi^{2+} ions will be accelerated up to an energy of 20 GeV/nucleus with a beam current of 500 mA and beam duration of 2 ms. Some types of ion sources and their performance are considered.

A prototype of the first section of the RFQ linac for acceleration of Bi^{2+} ions to 2.15 MeV with a working frequency of 6.2 MHz was constructed at ITEP. The calculated parameters of this section with lumped inductors and the results of its examination with Xe^{2+} ions accelerated from 130 keV to 1.35 MeV at a pulsed beam current of 5 mA are discussed. The structure allows to raise the electric field to levels high enough for acceleration of Bi^{2+} ions.

INTRODUCTION

For a number of years ITEP has been engaged in feasibility studies for an ICF power facility with a heavy ion driver. The proposed driver scheme will include an rf linac followed by systems for beam storage, compression and transport.[1]

Estimates made in ITEP[2] show that the rf linac has to provide a possibility to hit the fusion target with 20 GeV Bi beams and to deliver to it a total energy of 9 MJ (400 TW total power). To attain this goal the rf linac should have the following parameters:

Bi^{2+} Beam Energy	20 GeV
Pulse Beam Current	500 mA
Pulse Duration	2 ms
Momentum Spread	$\pm 3 \cdot 10^{-5}$
Normalized Emittance	0.2 cm-mrad

For a driver being coupled with four fusion reactors the linac will consume 810 MW, assuming the linac efficiency to be about 50%. The block diagram for the proposed rf linac and the linac performance characteristics are discussed below. Also, some experimental results obtained during start-up of the first section prototype with a 5 mA Xe^{2+} accelerated from 130 keV to 1.35 MeV are presented.

RF LINAC BLOCK DIAGRAM

The main advantages of linacs in obtaining an ion beam with predetermined kinetic energy, as compared to cyclic accelerators, are:

- Possibility of accelerating beams of any desired pulse duration;

- Relative simplicity of distributed rf power feeding and heat removal;

and, finally,

- Simple and reliable collimation and total beam extraction at all energy levels.

Difficulties in accelerating high current, low charge state ion beams are primarily caused by the low intensity of ion sources, low ion velocity during injection and low acceptance of the accelerating section. These difficulties were overcome after invention in the USSR of the RFQ-principle and the introduction in our country and abroad of multibeam arrays for the initial part, in which the number of beam lines decreases as particle energy increases.

It is expected that in each of 16 injection beam lines 50 mA ion beam of Bi^{2+} will be obtained. Initial acceleration up to an energy of 10 MeV will take place in the RFQ at a frequency of 6 MHz. Design and performance of the prototype of such a section (6 m long) tested at nominal rf field levels are discussed below. It seems that the results of beam parameter studies and investigation of accelerating sections for the initial part will permit to judge the validity of the accelerator scheme as a whole because the design of the subsequent sections has been tested at operating facilities for ions with higher charge states.

A list of the main parameters for the rf linac (see Fig. 1) is given in Table 1.[3]

All beam lines of the first three sections can be located in one vacuum tank. According to estimates, 30mA of beam current would be captured for acceleration in each beam line of the first

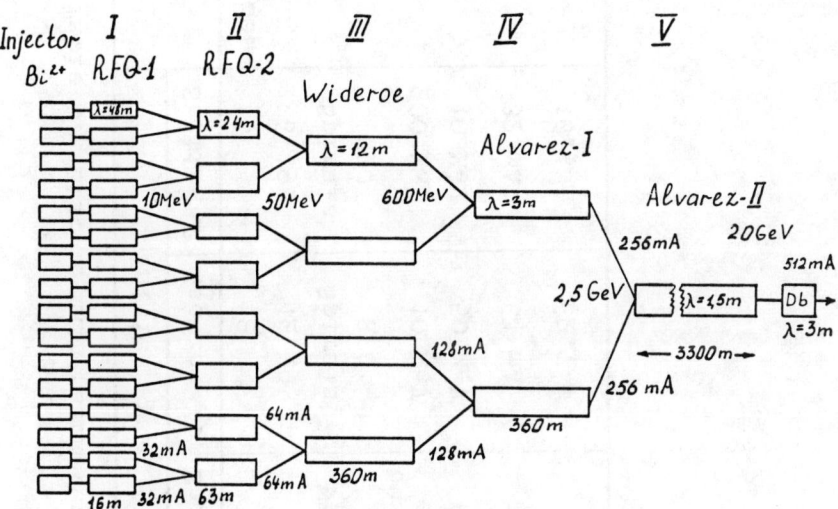

Fig. 1. Block scheme of linac proposed for HIF driver in ITEP.

Table 1. List of main parameters for rf linac

Parameter	Initial part				Main part
	Section I	Section II	Section III	Section IV	
Number of beam-lines	16	8	4	2	1
Frequency f (MHz)	6.19	12.39	24.78	99.10	198.2
RF structure	RFQ	RFQ	Wideröe	Alvarez	Alvarez
Focusing system	Spirals	Spirals	REC quadrupoles	REC quadrupoles	REC quadrupoles
Input ion energy	200 keV	10 MeV	50 MeV	600 MeV	2.5 GeV
Output ion energy	10 MeV	50 MeV	600 MeV	2.5 GeV	20 GeV
Mean current per beam line	32 mA	64 mA	128 mA	256 mA	512 mA
Section length	16.4 m	62.7 m	360 m	360 m	3296 m

section. In the main part not less than a 500 mA beam pulse will be accelerated, each second separatrix being filled. It should be noted that the relatively high output ion energy requires to increase the acceleration rate; this leads to momentum spread increase. As in the main part only each second separatrix is filled, and debunching at the output can be improved. In principle, doubling of the number of beam lines in each section of the initial part could allow to increase the mean beam current at the output of the linac, but the momentum spread will increase accordingly.

The frequency of the accelerating field in resonators is optimized according to ion velocities. Exact frequency values in various sections are chosen in such a way as to use in the main part of the linac the frequency established for the first part of the meson factory being constructed by the Academy of Science of the USSR.

ION SOURCE

For the linac injector a duopigatron ion source is being developed. The main feature of its design is vacuum separation of the duoplasmatron and duopigatron chambers by a pulsed electromagnetic valve (shutter), which opens only for a period necessary for penetration of ions and electrons from one chamber to another. A similiar valve is used for the connection of the pigatron chamber with the vacuum chamber of the injector. Because of the valves, injection of vapors and gases into each chamber can be made independently. So ballast gas can be injected into the duoplasmatron chamber while Bi vapors will occupy the pigatron chamber. Among the elements with atomic mass $A \gtrsim 200$ Bi is the most suitable as a working medium in the source. The pressure in each chamber can be set independently. The valve opening is 3.5 mm in diameter, the delay time is 3 ms, the holding time in opened position is 1-3 ms. In addition to the axial magnetic field, there is the possibility to apply a peripherial magnetic field in the pigatron chamber. Estimates show that at the source output we can expect to have a Bi^{2+} beam with a current up to 50 mA and a normalized emittance E_N = 0.03 cm-rad. At present the source is being prepared for tests.

At the first stage it was decided to test the RFQ structure with a Xe^{2+} beam. These ions were produced in a duoplasmatron designed to use noble gases as a working medium. The main features of the source are a cold cathode in the form of a hollow cylinder[4] and a pulse shutter.[5] The main advantage of the cold cathode is long life time (many thousands of hours) that permits, inter alia, to maintain stable injection conditions for a long time. The output valve allows to receive the desired high density of the working gas in the discharge chamber and at the same time to minimize leakage into the high vacuum chamber of the linac. For the purpose of time-of-flight analyses the ion optics systems was equipped with a control grid. High voltage short front pulses provided ion pulses of duration from 0.5-10 μs. The charge state of the ions was determined by measuring the transit time between beam transformer and Faraday cup. Xe^+, Xe^{2+}, and Xe^{3+} ions at 65 kV injector output (corresponding to 100 kV for Bi) were reliably determined, inspite of the fact that each peak on the oscillogram corresponded to a $Xe^{129-136}$ mixture. Beam transportation through the matching channel and test run of the RFQ structure were performed with a 16 mA beam and, later on with a 28 mA beam measured at the injector output.

RFQ STRUCTURES

As is known, the increase of average ion current leads to substantial sparking limit decrease in the accelerating tube (column) of electrostatic injectors. Because of these voltage limitations the velocity of low charge state heavy ions at the linac input is so low that the use of magnetic focusing is practically impossible. The RFQ is the most effective in this case. The beam accelerated in the RFQ structure can be bunched sufficiently as required for the most effective capture in Wideroe and Alvarez structures.

Our estimates show that in order to obtain 30-40 mA accelerated Bi beam with phase density j = 0.3 A/cm-mrad the working frequency should be chosen to be about 6 MHz. As it can be seen from Table 1, the frequency of the accelerating field in the first RFQ section with lumped resonant elements[6] was chosen to be quite low (6.19 MHz). This permitted to optimize the focusing regime and to inject Bi^{2+} ions at a voltage as low as 100 kV. The main RFQ parameters are given in Table 2.

The first section of the initial part consists of three parts: shaper, buncher and accelerator. In the second RFQ section the beam is accelerated at constant synchronous phase. The rigidity of the quadrupole channel is proportional to the wavelength of the accelerating field.[7]

$$K = \frac{\lambda}{2a} \left[\kappa \frac{ZeU_L}{AE_o} \right]^{1/2} .$$

For the accelerating part of the first RFQ section we have K^2 = 2.19 and a defocusing factor γ_s = 0.092. This permits to attain optimum values for transverse oscillation frequencies.

As to the sparking limit of the gaps, it was found[8] that breakdown levels determined during tests made on vacuum gaps in the frequency range of 25-150 MHz, are similar to that measured at 6 MHz. Breakdown with copper, aluminum and duraluminum electrodes began at exactly the same levels of electric field strength E_s = 200 kV/cm that is 2.5 times higher than those

Table 2. Main RFQ parameters

Parameter	Symbol	The first RFQ section			The 2nd RFQ section
		Shaper	Buncher	Accelerator	
Input energy	$W_s(inp)$	196 keV	204 keV	1.6 MeV	10 MeV
Output energy	$W_s(out)$	204 keV	1.6 MeV	10 MeV	50 MeV
Voltage between poles	U_L	190 kV	190 kV	190 kV	320 kV
Surface field (max)	E_s	122 kV/cm	122 kV/cm	122 kV/cm	250 kV/cm
Average radius	R_0	8.5–2.2 cm	2.2 cm	2.2 cm	1.8 cm
Aperture	a	8.5–2.1 cm	2.1–1.47 cm	1.47 cm	1.44 cm
Modulation depth	m	1–1.095	1.095–2.0	2.0	1.5
Equilibrium phase	$\|\varphi_s\|$	90°–85°	85°–35°	35°	37°
Rigidity of the focusing channel	K	0.396–1.536	1.536–1.490	1.490–1.460	1.197–1.193
Transit time factor	T	0–0.0317	0.0317–0.432	0.432–0.465	0.288–0.299
Focusing efficiency	$æ$	1–0.915	0.915–0.420	0.420–0.403	0.620–0.616
Phase advance per period	$\sqrt{M}$		1.02–0.938	0.938–0.981	0.622–0.645
Acceptance	V_k cm–mrad		0.53–0.25	0.25–0.26	0.39–0.40
Relative frequency of the phase oscillations	$\underline{\Omega}{\omega}$		0.097	0.097–0.040	0.046–0.021

corresponding to the Kilpatrick criterion for 6.2 MHz. For stainless steel electrodes E_s was found to be 1.5 times higher (for titanium 3-4 times) than for copper ones. RF training of copper electrodes permitted to increase E_s up to 500 kV/cm and even higher. As for titanium electrodes, training does not lead to any increase, and to reach the maximum level $E_{s,max}$ after the first breakdown is impossible.

A number of problems has arisen in the development of the accelerating structure. It was impossible to use a split coaxial resonator[9] (for λ = 50 m) because of its insufficient stiffness. Indeed, the quarter-wave vibrators, fixed at one end, should be 5-6 m long, and the half-wave vibrators, fixed at both ends, should be 10-12 m long. It was also impossible to use coaxial stubs as a support structure (a method widely used in Wideroe-type resonators) because of their length. That is why the possibility to use lumped inductors (flat or cylindrical spirals) was investigated.

The resonant structures of both sections are specially modulated four-wire lines (80 pF/m specific capacity) which are fixed on spirals made of copper tubes (see Fig. 2). These inductive support elements have the form of a symmetric triangular star that ensures mutual compensation of pondermotive forces arising in spirals. Damping of swinging forces is essential because the frequency of the mechanical resonance of the structure and the proposed linac repetition rate are quite close. A certain part (6 m long, 1.2 m in diameter) of the first section was manufactured and tested (see Fig. 3). At a resonance frequency of 6.19 MHz the structure parameters are:

Quality Factor	800
Shunt Impedance	20 kΩ

and the mechanical parameters are:

Stiffness	1500 N/mm
Frequency	9 Hz
Quality Factor	400

After the section was evacuated to a pressure of 5×10^{-6} Torr, rf power was supplied and the voltage between the electrodes reached 210 kV without breakdowns while for acceleration of Bi^{2+} ions, the calculated value was 190 kV.

At the startup, the ion optics was tuned so that the crossover of the Xe^{2+} beam was located at the inlet of the 2 m matching channel equipped with two electrostatic lenses and two steering devices. During the first experiments we have received 7 mA of

Fig. 2. Resonant structure of RFQ consisting of four-wire lines connected to spiral-shaped inductive support elements in triangular configuration.

Fig. 3. Part of the first RFQ section that was built and tested at resonant frequency of 6.19 MHz.

beam current at the inlet of the RFQ section. After the threshold level of the rf field was exceeded, the main fraction of the Xe^{24+} peak (> 95%) measured at the output of the RFQ section shifted for 5 μs; this corresponds to an acceleration of these ions to a calculated energy of 1.35 MeV. The output current was about 5 mA, i.e. the capture factor was close to 70%. Further research to be conducted with the first part of the 6 m long RFQ structure will allow to define more exactly the parameters of the second part of the 6.2 MHz RFQ and of further parts of the driver prototype.

SECTIONS WITH QUADRUPOLE MAGNET FOCUSING

Consideration of low charge state heavy ion focusing features show that transverse stability should be provided by a rather large number of one-sign lenses in the focusing period. That is why the acceptance of the channel would be rather low. Nevertheless as the Coulomb repulsion in $Z/A \ll 1$ ions beams is low it is possible to reach a rather high value of the beam current limit. As is shown in Table 1, we are planning to use rare-earth quadrupoles in the Wideroe and Alvarez structures.

The design of the resonant structure in the Wideroe section is similar to the design developed for the appropriate section of the UNILAC facility. After the 600 MeV level is reached it is possible to use Alvarez resonators which allow to increase the acceleration rate.

The rf linac should contain about 9450 drift tubes. Each electromagnetic quadrupole will dissipate approximately 2 kW. It means that in total 20 MW would be needed not taking into account efficiency of the rectifiers and the power consumed by the water cooling system. The rather high duty factor and the strict requirements for gradient stability of the magnetic lenses make a pulse power supply ineffective. Rare-earth quadrupoles would permit to exclude the stabilized power supply as a whole from the scheme of focusing channels, to avoid related operational difficulties and to save the above-mentioned 20 MW. In order to reduce the costs of the focusing channel it is reasonable to use samarium-cobalt alloy only for some lenses and for most of them to use alnico alloy which is cheaper.

RF POWER SUPLY

The active rf losses in resonators were estimated in ITEP by the following methods:

a) For RFQ sections, experimental data obtained for real resonators were used;

b) For Wideroe sections, appropriately scaled data for active losses in UNILAC sections were used;

c) For Alvarez sections, codes developed at ITEP were used in which the specific conductivity of copper was taken equal to $5.71 \times 10^7 \, \Omega^{-1} \mathrm{m}^{-1}$.

Data on the active rf pulse losses, the rf pulse power delivered for beam acceleration, and the number of generators are given in Table 3.

Total rf pulse (peak) power delivered for beam acceleration will be 5120 MW and total pulse (peak) losses will be 1400 MW. Thus, the electronic efficiency of the linac will be 79%. If we assume the generator efficiency to be equal to 65%, the rf system efficiency as a whole will be 50%. For a driver coupled with one reactor, the average rf power consumed by the linac at the repetition rate of 10 Hz and pulse duration of 2 ms will be 130.4 MW, and the total average power consumed by the power supply system (taking into account generator efficiency) will be 200 MW. The power consumption by other technical systems is rather negligible in comparison with the rf system so that the total linac-driver efficiency could be approximately 50%.

One linac could accomplish the irradiation of targets in four or five reactors in turn if the repetition rate and the power supply would be increased (this, of course, should be provided in the rf system design).

CONCLUSION

Wide-ranging research in several accelerator centers made it possible to determine the basic scheme of the rf linac designed for a driver. The main tasks involved in the development of the scheme proposed in ITEP are:

a) Further optimisation of accelerating and focusing structures at the frequencies of 6-25 MHz;

b) Development of a 50 mA Bi^{2+} injector with beam emittance of about 0.02 cm-mrad;

c) Development of an injector-linac matching scheme;

d) Development of a funneling scheme without beam losses and emittance growth;

e) Further development of effective rare-earth quadrupoles for drift tubes;

f) Development of highly reliable, high-power rf generators and RF power supply equipment for resonators.

Table 3. The data on pulse losses, RF power and number of generators for linac

Parameter		RFQ I	RFQ II	Wideröe	Alvarez I	Alvarez II
Accelerating field frequency	MHz	6.19	12.39	24.78	99.1	198.2
Active losses per beam line	MW	2.0	15.0	8.0	68	1080
Number of beam lines		16	8	4	2	1
Total active losses	MW	32	120	32	136	1080
RF power for beam	MW	2.56	10.24	140.8	486.4	4480
Total pulse RF power	MW	34.6	130	173	622	5560
Generator pulse power	MW	10	10	5	5	5
Number of generators		4	13	35	125	1112

REFERENCES

1. P. R. Zenkevitch, I. M. Kapchinskiy, D. G. Koschkarev, "Accelerating complex for HIF," Preprint ITEP-143, Moscow, 1980.
2. P. R. Zenkevitch, et al., "ITEP researches on application of heavy ion beams for ICF," Proc. of the 8th All-Union Acc. Conf., Dubna 1983, Part I, p.92.
3. I. M. Kapchinskiy, et al., "Low-charged heavy ion linacs for ICF," Voprosi atomnoi nauki i techniki, Ser. Technika physicheskogo eksperimenta, Vol. 2(23), Moscow, 1985, p.10.
4. V. A. Batalin, et al., "Duoplasmatron ion source with cold cathode," IEEE Trans. Nucl. Sci. NS-23 (2), 1097 (1976).
5. V. A. Batalin, et al., "Operation performances of exhaust valve for proton source of Linac I-2," Preprint ITEP-97, Moscow, 1979.
6. E. N. Daniltzev, et al., "RFQ accelerating structure for heavy ions on frequency of 6 MHz," Proc. of the 9th All-Union Acc. Conf., Dubna, Part I, p.218.
7. I. M. Kapchinskiy, "The theory of resonance linacs," Moscow 1982, p.140.
8. A. B. Zarubin, et al., "The investigation of spherical electrodes sparking limit on frequency of 6 MHz," Preprint ITEP-188, Moscow, 1984.
9. R. W. Müller et al., "Experimental results with a very-heavy-ion RFQ accelerating structure at GSI," GSI-84-11, 1984, p.77.

EXPERIMENTS AND PROSPECTS FOR INDUCTION LINAC DRIVERS*

Denis Keefe
Lawrence Berkeley Laboratory
University of California, Berkeley, CA 94720

ABSTRACT

In the last three years, the U.S. program in Heavy Ion Fusion has concentrated on understanding the induction linac approach to a power-plant driver. In this method it is important that the beam current be maximized throughout the accelerator. Consequently, it is crucial to understand the space-charge limit in the AG transport system in the linac and, also, to achieve current amplification during acceleration to keep pace with the kinematical increase of this limit with energy. Experimental results on both these matters and also on the use of multiple beams (inside the same accelerating structure) will be described.

A new examination of the most attractive properties of the induction linac for a fusion driver has clearly pointed to the advantage of using heavy ions with a charge-state greater than unity -- perhaps $q = 3$ may be an optimum. This development places even greater importance on understanding space-charge limits and mechanisms for emittance growth; also, it will require a new emphasis on the development of a suitable ion source.

INDUCTION LINAC DRIVER

The general concept of a heavy-ion induction linac using current amplification has been reported on often in this Symposium series (preceded by the "Workshop" series).[1] The basic idea is to inject a long beam bunch (many meters in length, several microseconds in duration), and to arrange for the inductive accelerating fields to supply a velocity shear so that, as the bunch passes any point along the accelerator, the bunch tail is moving faster than the head. As a consequence, the bunch duration can be made to decrease and the current can be amplified from amperes at injection to kiloamperes at the end of the accelerator (~ 10 GeV). The current is further amplified by a factor of about 10, and the pulse length further shortened to about 10 nanoseconds, in the drift section between the accelerator exit and the final focussing lenses. Transverse space charge forces are large enough that some sixteen parallel beams are needed to handle the beam in the drift-compression and focus sections. In the drift section one is relying on the longitudinal space-charge self-force in the beam to remove the velocity shear so that chromatic aberration does not spoil the final focussing conditions.[2]

With the passage of time, improvements in the design have taken place, and confidence has grown as a result of experimental and theoretical

---

*This work was supported by the Office of Energy Research, Office of Basic Sciences and Office of Program Analysis, U.S. Department of Energy, under Contract No. DE-AC03-76SF00098.

studies. A significant design improvement, made in 1981, was the incorporation of several independently-focussed beams inside the same accelerating structure.[3,4] Apart from the physics advantage that each beamlet could have a smaller emittance than that of an equivalent single large beam -- and so allow focussing to a smaller spot -- a design with multiple beams was also shown to be cost effective if the number was in the range of 8 to 16 beams. At any point in the accelerator the larger the number of beams the larger the current that can be transported and the shorter the acceleration pulse length. The economic optimum balances the decrease in the acceleration unit costs against the increase in cost of the transport lenses.

Experiments relevant to the Induction Linac method can be thought of in two broad categories (which are not completely independent):

(1) Proof-of-Principle Experiments, where the design is believed to be based on sound principles but where one wishes to be sure that there are no unmanageable surprises;

(2) Discovery Experiments, where there is no satisfactory theoretical understanding and answers have to be arrived at empirically.

Our present experiment (MBE-4) at LBL belongs in the first category, and others to be discussed below (SBTE, MEVVA) in the second. A proposed new experiment (ILSE) would have elements of both.

MULTIPLE BEAM EXPERIMENT (MBE-4)

About 50% of the planned apparatus[5] has now been assembled and results of measurements to date are given at this meeting in two reports presented by Warwick[6] and by Kim.[7] The experiment is to prove the principle of current amplification while keeping the longitudinal and transverse beam dynamics under control and, in addition, to face the additional complication of handling multiple beams (four in MBE-4).

The transverse dynamics is strongly space-charge dominated in that the betatron phase-advance per focussing-lattice period is strongly depressed -- from $\sigma_0 = 60°$ down to about $\sigma \sim 12°$. (See Fig. 1) For a mono-energetic beam without acceleration the SBTE (see below) has shown stable beam behavior to lower values of σ (7° - 8°), but new issues in transverse dynamics arise in MBE-4 because of (a) the difference in velocity along the bunch as it passes through a given lens which results in values for σ_0 and σ that vary along the bunch length, and (b) the discrete accelerating kicks which can cause envelope-mismatch oscillations.

For the longitudinal dynamics two separate features arise in MBE-4. Space charge effects throughout the body of each long bunch (about 100 cm long and 1 cm radius) are strong enough that the dynamical response to velocity kicks or acceleration errors is described in terms of space-charge (Langmuir) waves rather than in single-particle terms. Secondly, the tapered charge density that occurs at the ends of the bunch will cause collective forces that are accelerating at the head and decelerating at the tail and, if not counteracted, will cause bunch spreading both in length and in momentum. A major part of the experimental effort is centered on designing and successfully employing the electrical pulsers to handle the correcting fields at the bunch ends.

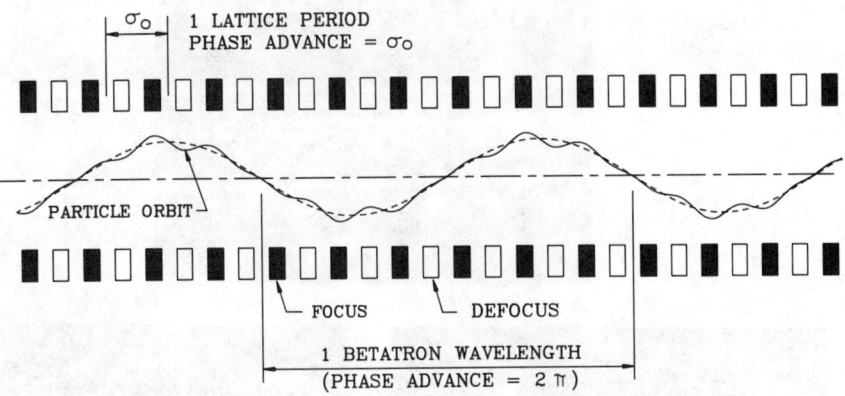

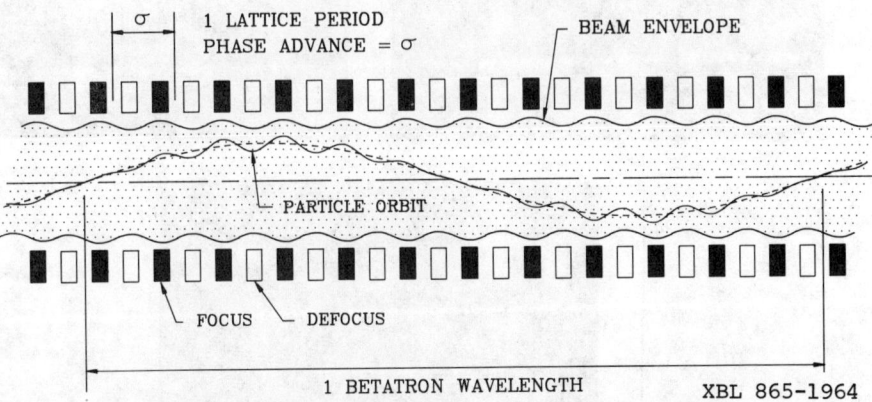

Fig. 1: Transverse motion of a particle in an alternating gradient focussing lattice. A lattice period corresponds to a focussing lens, a drift, a defocussing lens and another drift (FODO). The definition of phase advance per period of the quasi-sinusoidal motion is shown for cases in which space-charge effects are negligible (top, σ_0), and strong (bottom, σ).

Figure 2 shows an example of current amplification results obtained to date, where it can be seen that the pulse duration has been shortened by nearly a factor of two and the current correspondingly increased. Because MBE-4 operates at relatively low energy (accelerating from 200 keV to 1 MeV), we can try rather aggressive schedules for current amplification, which correspond to setting up a large velocity shear, $\Delta\beta/\beta$. We do not have a firm argument for exactly how high a velocity-shear can be and still be considered tolerable. An experiment with $\Delta\beta/\beta = 0.4$ is described in the poster paper by Kim;[7] this is more than will be needed in a driver.

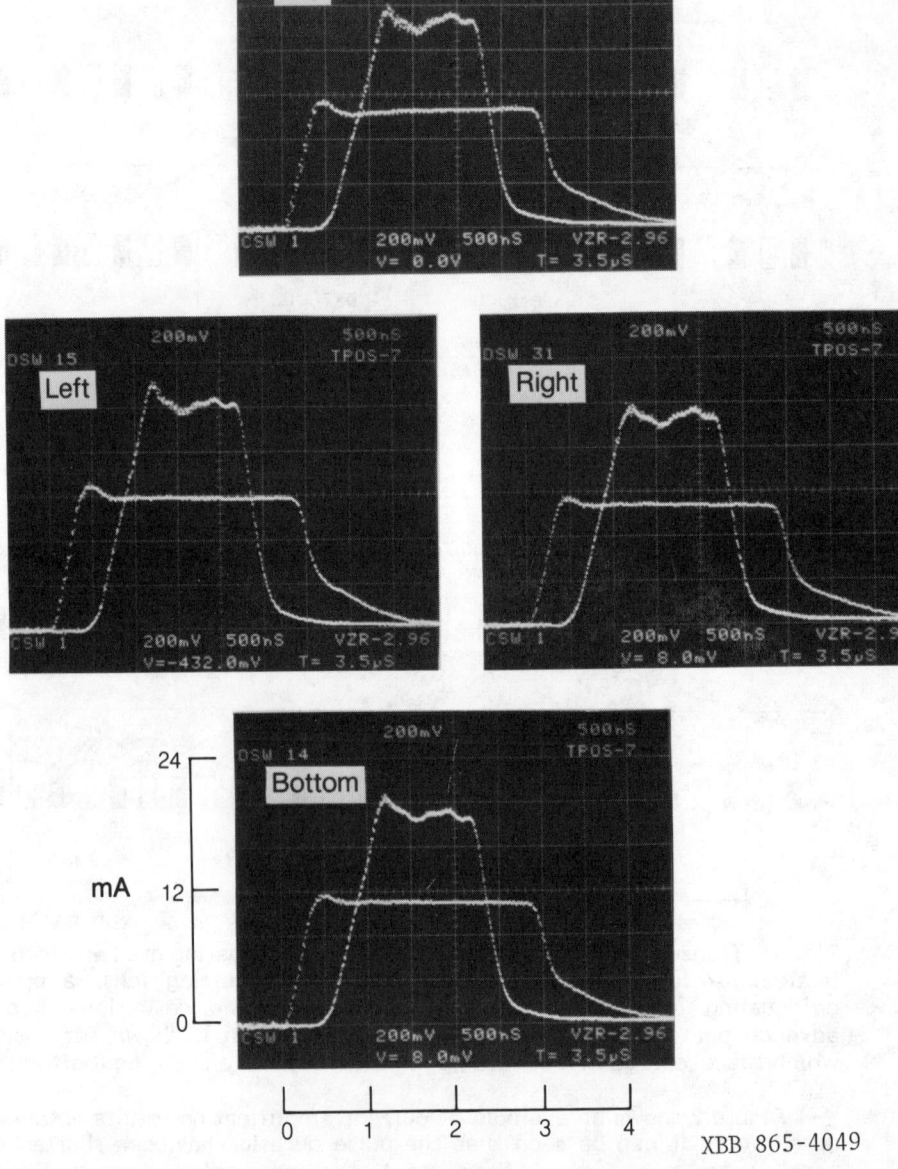

Fig. 2: Current profiles for the four beams in MBE-4 measured after the eighth accelerating gap, with the pulsers off (lower amplitude trace), and on (higher amplitude trace). The current amplification accompanying acceleration is clearly visible.

SINGLE BEAM TRANSPORT EXPERIMENT (SBTE)

Since the IEEE Particle Accelerator Conference in Vancouver in May 1985 the results on high-current beam transport limits in the 87-quadrupole SBTE have been refined and more careful calibrations made. The results, shown in Figure 3, are substantially unaltered; at the highest currents and lowest emittance values obtainable from the 120-200 kV cesium injector, no detectable growth in emittance was observed in the 41-period transport section provided σ_0 did not exceed 85°. A threshold value of current above which emittance growth occurred could, however, be measured for values of σ_0 in excess of 85°. Since the transportable current is greatest for $\sigma_0 < 85°$, the design of drivers will be restricted to σ_0 values in this range. (Tiefenback has found that beyond $\sigma_0 = 85°$ the threshold corresponds rather well to the empirical condition that the beam plasma period equals the beam transit time through three lattice periods).[8]

Earlier theoretical work on beam current limits in AG focussing systems utilizing an idealized distribution (K-V) indicated that it could be dangerous to use σ_0 greater than 60° and that σ could probably be depressed from that value down to 24°, but not below.[9] The experimental limits from SBTE shown in Table I can be seen to be much more encouraging:

Table I - Experimental Limits on σ_0, σ

σ_0	60°	78°	83°
σ	< 7°	< 11°	< 15°

EMITTANCE GROWTH IN HIGH-CURRENT BEAMS

In his original consideration of high current limits in magnetic AG systems Maschke showed that the limiting particle current could be written (non-relativistically) as:

$$I_p = K (\eta B)^{2/3} (\epsilon)_N^{2/3} V^{5/6} / q^{1/2} A^{1/2} \quad (1)$$

with B the limiting pole-tip field, η the fraction of length occupied by magnetic lenses, qV the ion kinetic energy, and A, q, the ion mass and charge state, respectively. (Two other equations, involving lattice-period and radius, must be simultaneously obeyed for Eq. 1 to hold). The coefficient, K, first suggested by Maschke was given for an implicit assumption that σ/σ_0 was equal to 0.7. In the smooth approximation discussed by Reiser,[10,11] the following expression holds:

$$K \propto \left[1 - (\sigma/\sigma_0)^2\right] \sigma_0^{2/3} / (\sigma/\sigma_0)^{2/3} \quad (2a)$$

In light of the improved knowledge from experiment and simulation that σ/σ_0 can be small, it is useful to write the explicit dependence of K on σ, σ_0, to a good approximation as:

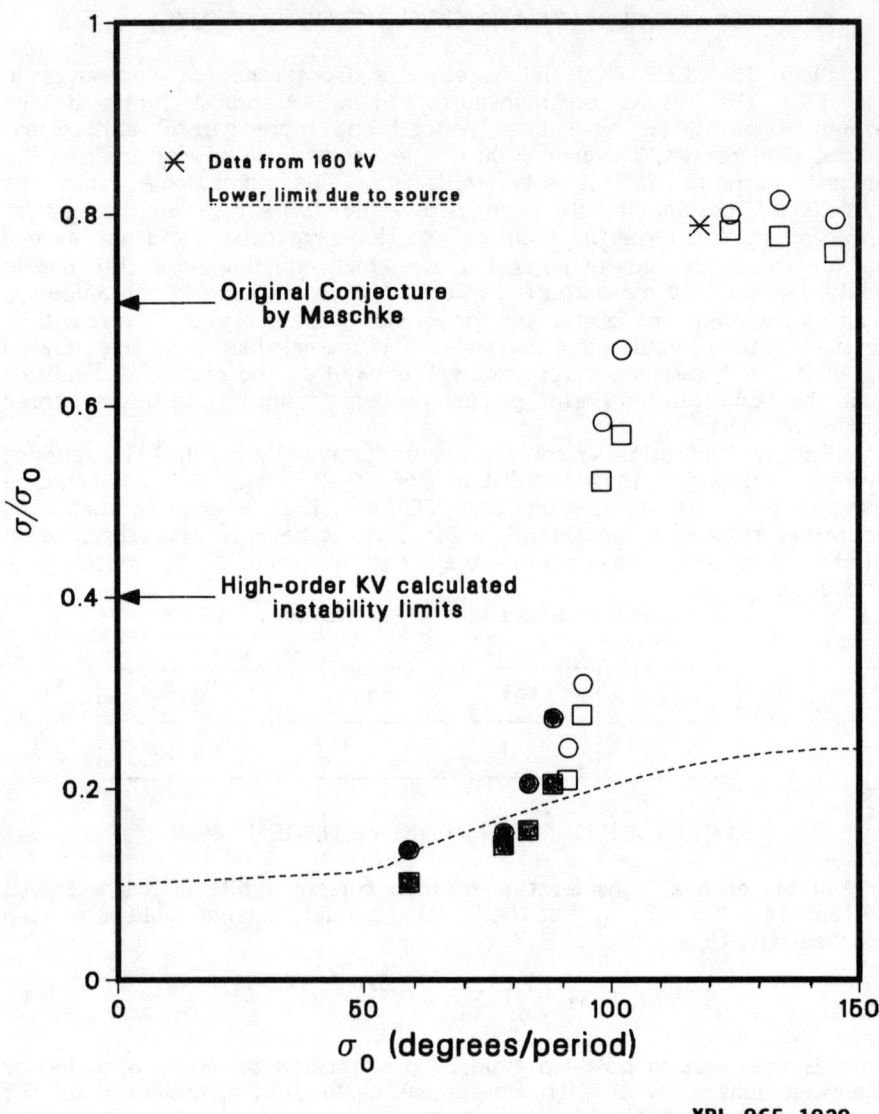

Fig. 3: Values of σ/σ_0 for various σ_0 reported by Tiefenback.[8] Calculated values of σ are based on the emittance for 95% (squares) and 100% (circles) of the beam. The line indicates the lower observable bound set by the source brightness. Emittance growth was observed only for $\sigma_0 \geq 88°$.

$$K \propto \sigma_0^{2/3} \Big/ (\sigma/\sigma_0)^{2/3} . \qquad (2b)$$

If, in fact, there were no lower bound on σ/σ_0 the transportable current could grow very large (the required aperture, however, would do likewise).

Just as the SBTE measurements were beginning, Hofmann and Haber, each using simulation codes for well-centered beams without images, reported that for $\sigma_0 = 60°$, σ could be allowed to go lower than 24° without emittance growth occurring. During the course of SBTE measurements further simulations showed that values of σ down to 1° or 2° might be alright.

The situation changed, however, with Celata's simulation studies of an off-axis beam, which corresponds to the real-world situation. For a beam with $\sigma/\sigma_0 = 6°/60°$, no growth was detected. If either a dodecapole component in the field or the effect of images in the electrodes was introduced, r.m.s. emittance oscillations and steady growth showed up clearly. When both images and the right amount of dodecapole component were included, however, the surprising result emerged that the emittance did not grow.[12]

A pessimist might thus argue that it is dangerous to consider designing a system with σ/σ_0 as low as 6°/60° because we have identified two mechanisms that can cause trouble. The optimist, on the other hand, could argue that self-cancelling effects such as described, even if they are not understood, can be used to carry us below this value. For the moment it is probably only prudent to consider that designs with σ lower than experimentally established may be on shaky ground, at least until we have a more thorough understanding of all the physics.

The growth in emittance due to the beam distribution in configuration space alone has been a topic of much discussion in the past few years. An intense beam with a non-uniform spatial distribution will usually readjust itself in a fraction of a plasma period to an almost exactly uniform distribution. The change in electrostatic field energy is always such that energy is fed into the thermal motion of the beam particles thus causing emittance growth. For given initial and final distributions Struckmeier et al. have given a prescription for determining the amount of growth if the final distribution is assumed to be uniform.[13] (This result is implicit in earlier work by Lee and Yu).[15] This work has been extended by Wangler,[14] and more recently a report at this meeting by Anderson has made a significant advance in the theory by describing how the growth evolves without assuming what the final distribution will be.[16]

This mechanism for emittance growth clearly can occur just after an ion source which is emitting a non-uniform beam. But it is also of importance in combining (or splitting) beams that are round or elliptical by means of a septum. Simulation results on emittance growth in the case where four beams are stacked side by side by septa to form one are given by Celata.[17]

NEW CONSIDERATIONS FOR DRIVER DESIGN

Much of the early design work for induction linac drivers was restricted to considering that ions with charge state $q = 1$ were most suitable and, also, that $\sigma/\sigma_0 = 24°/60° = 0.4$ was an optimum value.[4] The driver design program, LIACEP[18], did, however, indicate that cost savings could accrue if either condition could be relaxed, but at the cost of additional complications, namely:

(i) Reduced current at any point (V) in the driver (see Eq. 1).

(ii) Generating ions with q > 1, which was visualized to be done by stripping from a beam with q = 1 at some intermediate energy.

(iii) An increased number of beam lines in the drift-compression section.

(iv) Neutralization after the final lenses to prevent focal-spot enlargement by space-charge.

The results from SBTE have altered thinking and encouraged us to re-open the matter of using ions with charge state q > 1. As an illustration, consider the reference case given in 1981 for V = 10 GV, q = 1. We could build only the first 5 GV part and use charge state q = 2 to give the same final kinetic energy, 10 GeV. We could still maintain the same particle current at each voltage point provided the product $q^{1/2} (\sigma/\sigma_0)^{2/3}$ is kept constant, i.e. $\sigma/\sigma_0 \propto q^{-3/4}$. (This can be seen from Eqs. 1 and 2). Since we know that very low values are permitted for σ/σ_0, we can in principle continue this argument to higher charge-states, dropping σ/σ_0 in value and shortening the accelerator at each step. A limitation occurs, however, beyond q = 3 (for A = 200) because the increased perveance (i.e. space-charge) in the final drift lines rises as q^2 and the cost of the very large number of final beam lines that will be needed overrides the cost reduction in the accelerator. This argument is given in more detail in the invited paper by Lee.[19]

It now appears that the direct generation of adequately high-currents of ions with q > 1 from a source is possible as a result of work by Brown with the MEVVA source.[20] Using a similar source, Humphries has shown how to avoid plasma pre-fill of the extraction region and thus has solved the problem of rapid turn-on of the source (< 1 μsec) needed for an induction linac driver.[21]

Since the SBTE has shown that σ_0 can exceed 60° safely (but not 85°) present driver designs have benefitted by using σ_0 = 80°, resulting in a somewhat greater beam current limit (see Eq. 2).

With ions of q = 1 the low velocity end of the linac (< 250 MeV) represented only 10% of the cost.[4] With ions of q = 3 the bulk of the accelerator has been shortened from 10 GV down to 3.3 GV and the cost of the front-end represents a much more significant fraction of the overall cost and, hence, is now receiving much more design attention. If electrostatic lenses are used in the low velocity end, the mapping argument given earlier (for magnetic transport) from equal voltage points in a q = 1 to a q > 1 case no longer holds unless the number of beams is increased. With higher charge-state, therefore, we visualize a driver starting with as many as 64 beamlets from the injector, which are then combined, perhaps at 250 MeV, to create the 16 beamlets that undergo the bulk of the acceleration (See Fig. 4). Before this strategy can be established as a

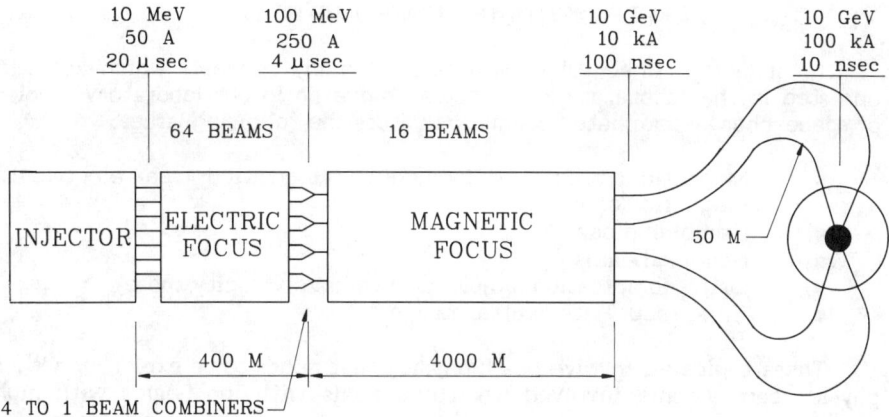

Fig. 4: Schematic of current concept for a 3.3 MJ driver that uses ions with A = 200, q = 3.

THE HEAVY ION FUSION SYSTEMS STUDY (HIFSA)

The first systems assessment for a power plant based on an induction linac driver has been in progress for a year and a half under the auspices of EPRI and the DOE Office of Program Analysis and Office of Basic Energy Sciences. The major participants include McDonnell-Douglas (MDAC), LANL, LBL, and LLNL. The main emphasis as expressed in the term "Assessment" is not on developing a point design such as HIBALL[18] but on exploring a broad range of parameters to establish general conclusions (A wide variety of point designs can, of course, be generated from the results). The preliminary results to date are given in the paper by Dudziak and Herrmannsfeldt,[19] and further details appear in other reports at this meeting from the groups at MDAC, LANL, LLNL and LBL.[20]

Four different reactor types and five different target designs are included in the examination. The driver parameters range from 5 GeV to 20 GeV and from 1 MJ to 10 MJ. Results to date show that a cost of electricity of 5.5 cents/kW-hr seems quite reasonable to expect for a 1000 MWe plant that uses ions with A = 200, q = 3. The familiar "economy-of-scale" effect is also apparent, with the cost of electricity being less (4.5 cents/kW-hr) if a 2000 MWe plant is considered, or more (9.5 cents/kW-hr) for a 500 MWe plant. One of the more interesting results is that such values of electric energy cost can be realized for a very broad range of driver parameters and for several choices of both reactor and target designs.

EXPERIMENTS THAT NEED TO BE DONE

An induction linac driver will rely on many concepts which are still untested in the laboratory. We need to move on in our laboratory studies of space-charge-dominated beams to address the following issues:

- Magnetic transport (including the transition from electric to magnetic focussing)
- Combining beams
- Bending beams
- Drift-compression (adjusted to remove velocity shear)
- Final focus with neutralization

These topics all involve beam physics that needs to be explored. Other physics can become involved when one deals with ion beams with high energy and high power, e.g. surface emission of gas, ions, or electrons, due to beam loss from a beam "halo", or unanticipated surprises analogous to the "brickwall" effect seen at the CERN ISR. Thus it is desirable to move up from the present laboratory level of ~ 1 joule, 1 MW, to maintain credibility that extrapolation to driver parameters is believable. We add therefore to the list:

- Handling ion beams with high energy and high power.

Most of the issues to be addressed require or benefit from use of magnetic elements and hence an ion with velocity, $\beta > 0.04$ (where $\underline{v} \times \underline{B} \gtrsim \underline{E}$). At low energies the cost of induction acceleration is relatively high, so that to explore features such as the above at affordable beam voltage (say 10 MV) and cost, will require the use of a light ion in the range from A = 12 to 27.

At LBL we are in the process of designing and proposing to the U.S. DOE an experiment which would move forward to address the topics listed earlier. A central component of the 10 MV medium-weight ion experiment will be the unique 2 MV mutiple-beam injector being developed at LANL, which will be reported on at this meeting.[21]

SUMMARY

Experimental progress to date has strengthened our belief in the soundness and attractiveness of the heavy ion method for fusion. What surprises that have shown up in the laboratory (e.g. in SBTE) have all been of the pleasant kind so far.

The systems assessment has supported the view that the heavy ion approach can lead to quite economically attractive electric power and that a wide variety of options exists in all parameters. The systems work has also been of great help in pointing the way for the research and development activities.

Several more experiments related to driver physics need urgently to be done, and they can be addressed on a laboratory scale.

REFERENCES

1. D. Keefe, Proc. ERDA Summer Study of Heavy Ions for Inertial Fusion, 1976, Lawrence Berkeley Laboratory Report LBL-5543 (1976).
2. D. Ho et al. in this Proc.
3. A.W. Maschke, Brookhaven National Laboratory Report BNL-51029 (1979).
4. A. Faltens, D. Keefe, and E. Hoyer, Proc. 4th Int. Top. Conf. on High-Power Electron and Ion Beam Research and Technology (ed. Doucet and Buzzi, Paris, Ecole Polytechnique), 751 (1981).
5. R.T. Avery, C.S. Chavis, T.J. Fessenden, D.E. Gough, T.F. Henderson, D. Keefe, J.R. Meneghetti, C.D. Pike, D.L. Vanecek, and A.I. Warwick, IEEE Trans. Nuc. Sci. 32, 3187 (1985).
6. T.J. Fessenden, D.L. Judd, D. Keefe, C. Kim, L.J. Laslett, L. Smith, and A.I. Warwick, elsewhere in this Proc.
7. D.L. Judd, C.H. Kim, L.J. Laslett, L. Smith, and A.I. Warwick, elsewhere in this Proc.
8. M.G. Tiefenback, "Space-Charge Limits on the Transport of Ion Beams in a Long A.G. System" (Ph.D Thesis), Lawrence Berkeley Lab. Rep. No. LBL-21611 (1986).
9. I. Hofmann, L.J. Laslett, L. Smith and I. Haber, Part. Accel., 13, 145 (1983).
10. M. Reiser, Part. Accel., 8, 167 (1978).
11. M. Reiser, IEEE Trans. Nuc. Sci., 26, 3026 (1979).
12. C.M. Celata, I. Haber, L.J. Laslett, L. Smith, M.G. Tiefenback, IEEE Trans. Nuc. Sci., 32, 2480 (1985).
13. J. Struckmeier, J. Klabunde and M. Reiser, Part. Accel., 15, 47 (1984).
14. T.P. Wangler, Proc. Workshop on High-Brightness, High-Current, High Duty-Factor Ion Injectors (San Diego, 1985), AIP Conf. Proc. No. AIP-139, 133 (1986); see also, F.W. Guy, R.H. Stokes, and T.P. Wangler elsewhere in this Proc.
15. E.P. Lee and S.S. Yu, Lawrence Livermore Laboratory Report No. UCID-18330 (1979).
16. O. Anderson, elsewhere in this Proc.
17. C.M. Celata, elsewhere in this Proc.
18. A. Faltens, E. Hoyer, D. Keefe, and L.J. Laslett, Proc. HIF Workshop, 1978, Argonne Natl. Lab. Rep. No., ANL79-41 (1979).
19. E.P. Lee, elsewhere in this Proc.
20. I. Brown, elsewhere in this Proc.
21. S. Humphries, Jr., Particle Accelerators, Vol. 20, 1986 (in press).
22. B. Badger, et al., "HIBALL-II, An Improved Heavy Ion Beam Driven Furion Reactor Study", Univ. of Wisconsin Rep. No. UWFDM-625 (1984).
23. D. Dudziak and W.B. Herrmannsfeldt, elsewhere in this Proc.
24. J. Hovingh et al.; L. Waganer et al.; D. Zuckerman et al.; D. Driemeyer et al.; R. Peterson; R. Bangerter et al., elsewhere in this Proc.
25. D.C. Wilson, K.B. Riepe, E.O. Ballard, E.A. Meyer, R.P. Shurter, F.W. Van Haaften, and S. Humphries, Jr., elsewhere in this Proc.

Development of RF-Linac and Storage Ring System for High-Current Experiments at GSI

I. Hofmann

GSI Darmstadt, P.O.B. 110541, 6100 Darmstadt, West-Germany

ABSTRACT

The heavy ion facility under construction at GSI is suitable for the generation of high-current beams. It consists of a new high-current injector into the Unilac and a synchrotron and storage ring. We discuss the capability of this system to produce short pulses of heavy ions with a specific energy of the order of 0.1 MJ/g. Under these conditions the system allows to perform a first generation of heavy ion driven target experiments and to test most of the critical issues of a large scale heavy ion fusion driver facility.

INTRODUCTION

At the 1984 Tokyo symposium the feasibility of heavy ion fusion experiments at GSI using a synchrotron (SIS) and a storage ring with the intensity upgraded UNILAC as injector has been discussed [1,2]. The early plans of a relatively small test storage ring "SITAR"[3] employing electron cooling led to the ESR ("Experimental Storage Ring") as a facility, which promised the maximum benefit for the nuclear physics community. With its 10 Tm magnetic rigidity the ESR also yields a sufficiently good high-current performance to warrant a first generation of target experiments ("high-temperature" or "high-energy-density-in-matter"), where intense short pulses of heavy ions heat solid state density matter up to plasma temperatures (> 10 eV). The new SIS/ESR complex received its funding in May 1985 and completion is expected for 1988/89.

In this status report we first review briefly the design of the new GSI facility[4] and then present the high-current performance expected on the basis of theoretical estimates. Machine experiments relevant to high currents and heavy ion fusion conditions have been described at the previous symposium[1].

STATUS OF MAXILAC/SIS/ESR

We consider two main options for the generation of high-current beams:

(A) Single-turn injection into SIS with maximum current from the Maxilac/Unilac injector. No electron cooling and use of the ESR, hence the actual bottleneck for high current and phase space density is space charge at the low energy end of the injector and transfer into SIS.

(B) Multi-turn injection into SIS with subsequent phase space cooling by electrons in the ESR. Phase space limitations during cooling are a critical issue here. There is still a need for the Maxilac injector, yet with relaxed demands on brightness (by a factor 5-10).

In Fig. 1 we show a layout of the facility including the existing Unilac. Five modules of the RFQ High Current Injector "Maxilac" are already operational, the remaining seven modules will be completed until 1989. The design goal of Maxilac is to reach a space charge limited current of

$$I \leq 0.2 \frac{A}{q} [e\ mA] \qquad (1)$$

which is equivalent to 26 mA for the maximum A/q = 130. For detailed results with the existing section we refer to the contribution by R.W. Müller et al., these proceedings.

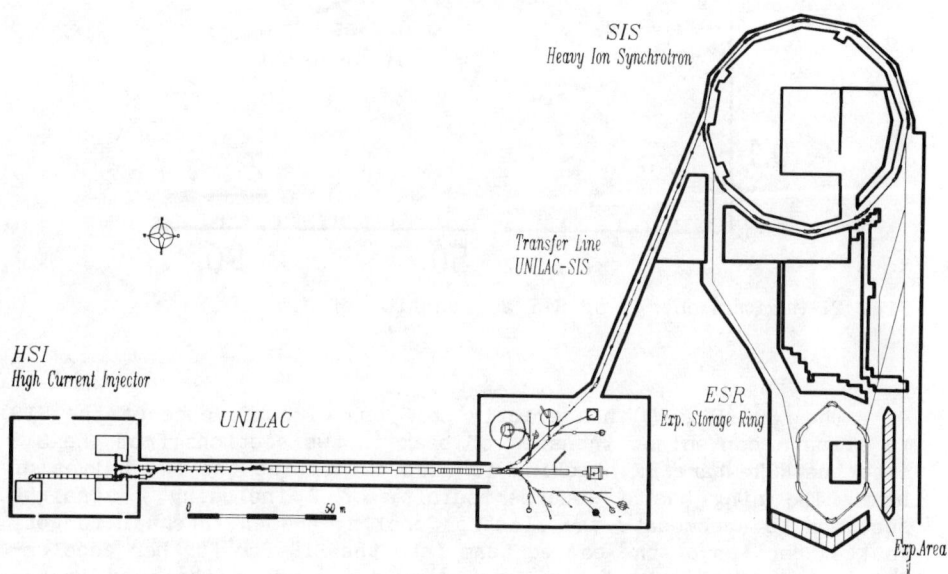

Fig. 1: Layout of GSI heavy ion facility

The SIS 18 is a rapid cycling synchrotron with an energy of 1.3 GeV/u for the heaviest projectiles (up to Uranium). The circumference is 216 m and the repetition rate 1...3 Hz. The acceptance is 200 mm mrad horizontally and 50 mm mrad vertically with a maximum momentum acceptance of $\Delta p/p = \pm 3$ % (at low emittance). In Fig. 2 we show the dependence of maximum energy on Z. Two candidates of ions suitable for heavy ion fusion experiments, $^{127}\text{I}^{20+}$ and $^{132}\text{Xe}^{44+}$, are indicated with circles.

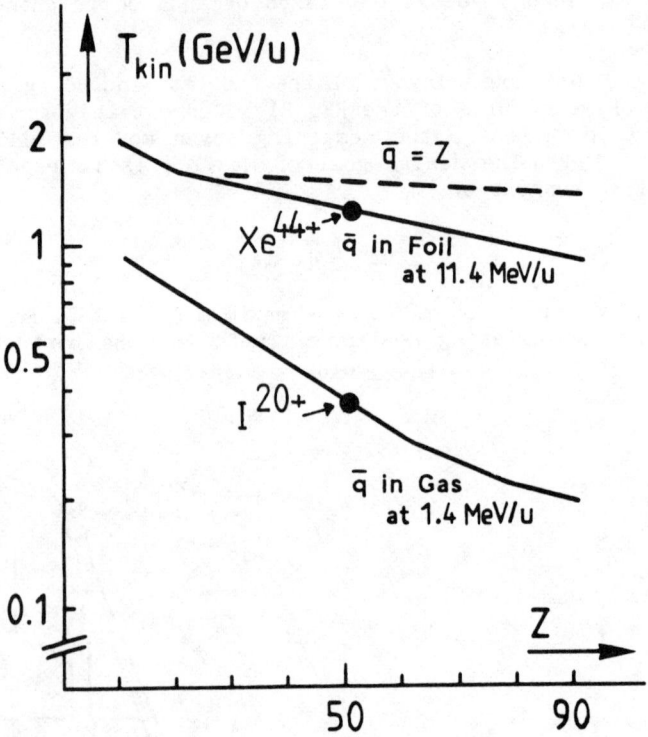

Fig. 2: Maximum energy of SIS as function of Z.

The ESR (Fig. 3) has exactly half the circumference of the SIS to allow a convenient transfer of beam in two sections from the SIS into the ESR where RF stacking is used to combine the two halves of beam. The ring has a superperiodicity of 2 including 2 straight sections to accomodate the electron cooling and an internal target. Fast extraction of the cooled beam into the SIS for further acceleration is possible. In Table I we summarize some of the most important machine parameters.

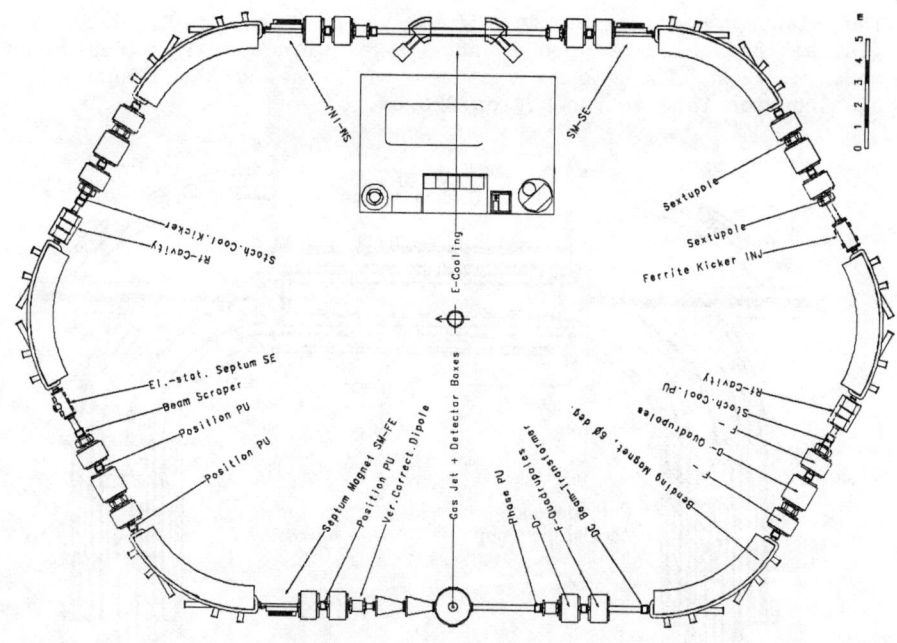

Fig. 3: Layout of ESR

Table I. Parameter list of ESR

Ion	Ne^{10+}	Kr^{36+}	Xe^{54+}	U^{92+}
Energy in MeV/u	3 - 834	3 - 656	3 - 609	3 - 556
Stripping efficiency	-	100 %	100 %	60 %

Modes of Operation

D = const	$\Delta p/p = \pm 2\%$	$\varepsilon_\perp$ (max) = 140 π mm mrad
D = 0	$\Delta Q/Q = \pm 1.7\%$	(U^{89+} to U^{92+})
D = large	crossing of two co-circulating beams	

Electron Cooling: Range: 30 - 560 MeV/u ($J_e \approx$ 1 A/cm, U_e = 16.5 - 310 kV)

Ring + Magnets:

Circumference (= 1/2 SIS)   108.4 m
Dipoles: 6 × 60°, B = 1.6 T, ϱ = 6.25 m
Quadrupoles: 20 × L = 0.75 - 1.17 m, B' = 6.2 T/m

RF:

Revol. Frequency	0.21 - 2.4 MHz
Harmonic Number	2 or 4
Accel. Frequency	0.85 - 5 MHz
Cavities × Voltage	2 × 5 kV

Injection: Fast
Extraction: Fast, to SIS
Slow, to Targ. Hall

Beam Accumulation:

Precooling	stochastic
Stacking	by RF
Vacuum Pressure:	10^{-11} mbar

The electron cooling device is shown in Fig. 4. Electron energies can be varied between 2 and 320 keV with a maximum design current of 10 A. The e-beam diameter is 5 cm and the length of the electron-ion interaction region 250 cm.

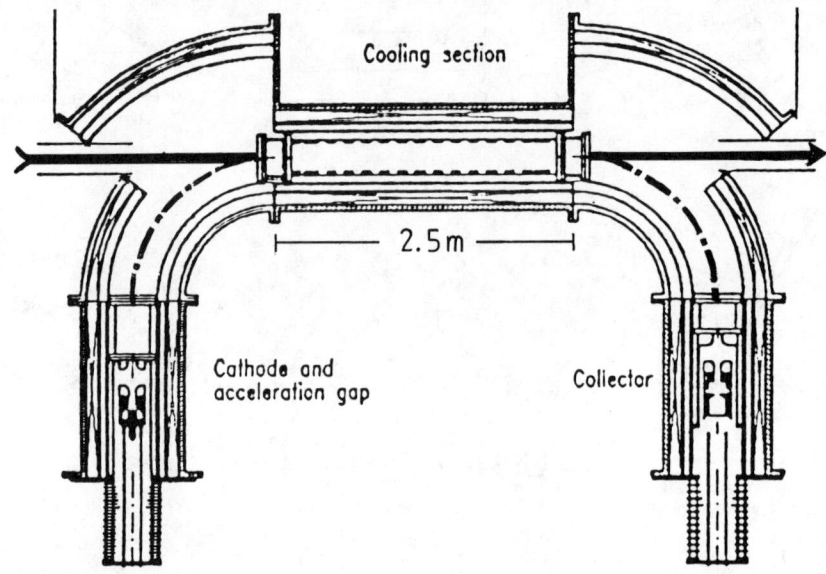

Fig. 4: Electron cooling device

HIGH CURRENT PERFORMANCE

Beam Manipulations and Parameters

A principal goal of the high-current machine experiments is the simulation of accelerator issues of a heavy ion fusion driver (see the contribution by R. Bock, these proceedings). Similar issues have to be well understood to achieve the desired performance of a significant target experiment[5]. The main difference with demands from nuclear and atomic physics experiments is that we need a large number of stored ions ($\sim 10^{11}$) simultaneously with small emittance and momentum spread.

The bunch manipulations leading to a single high-current bunch are shown schematically in Fig. 5 with the following phases[6]:

1. Injection of a single turn (option A) or several turns (option B) of beam into the SIS. The single turn injection has minimum emittance dilution and this leads to maximum phase space density (no use is made of electron cooling).

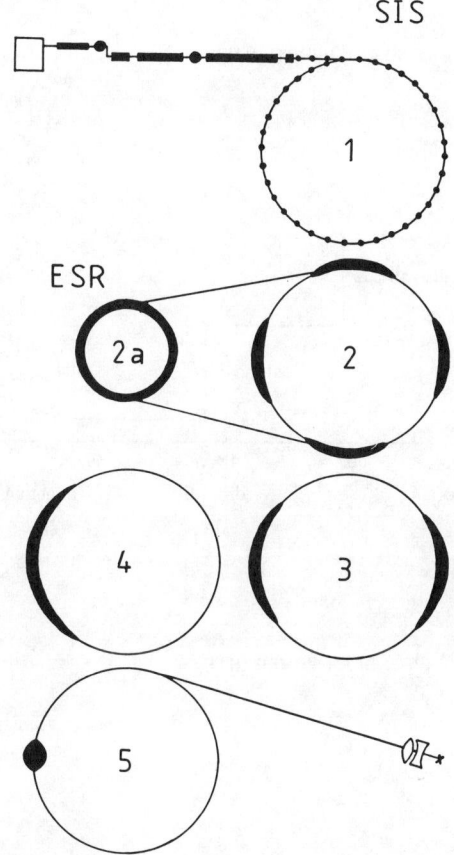

Fig. 5: Bunch manipulations for high-current experiments

2. After debunching and rebunching on the 4th harmonic the beam is accelerated to either the full energy (option A); or the energy of the ESR (option B), where it is debunched, cooled, re-injected into SIS, rebunched and accelerated to full energy.

3., 4. Merging of four bunches into two and finally one single bunch by switching of RF harmonics[7].

5. Fast bunch compression and extraction to the target beam line.

Fig. 6a shows the change of particle current from Maxilac to target, where the main current amplification is due to final bunching. The corresponding rise of beam power is shown in Fig. 6b. The values for current shown here have been found on the basis of several conditions, which are the most limiting to our present understanding.

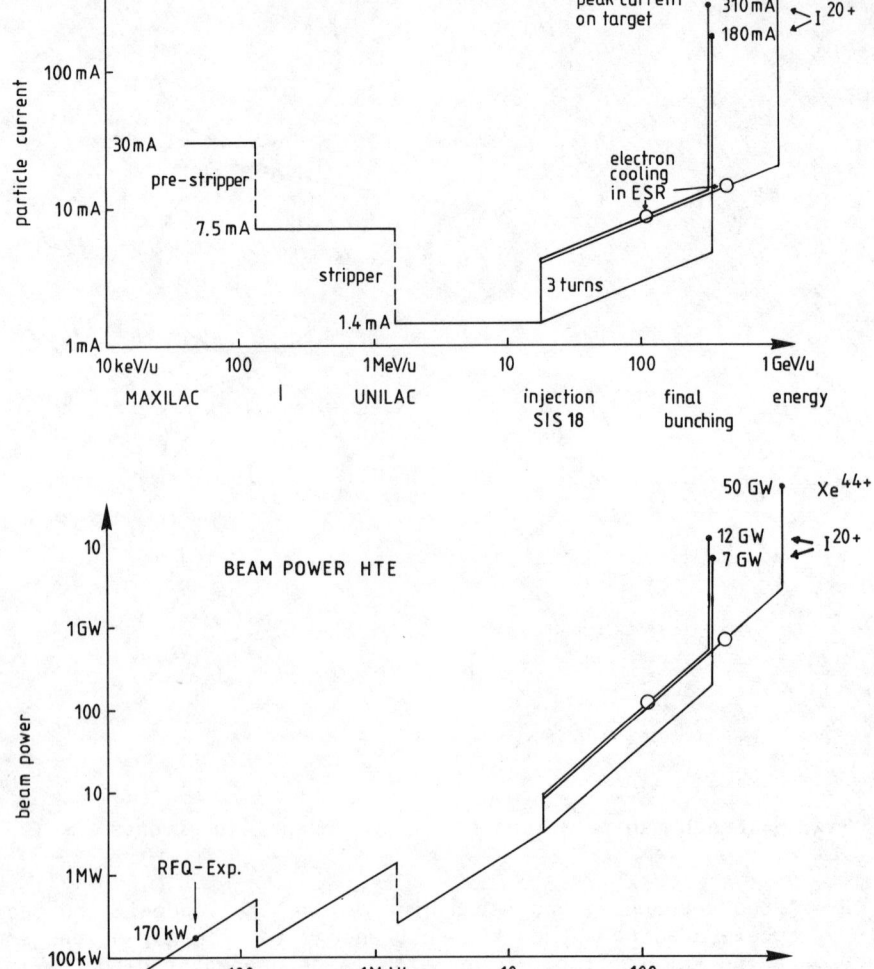

Fig. 6: a) Build-up of beam current
b) Build-up of beam power

In Table II we summarize the parameters of the three distinct cases shown in Fig. 6, before we discuss in the next section in more detail the principles that have led to this particular choice. Here we note that all three cases have in common the same final focusing system, which provides a spot of 200 µm diameter calculated to third order[8].

Table II. Summary of parameters on target for option A (first column) and option B (second and third column).

Final Parameters on Target: 200 μm diameter,
   ε = 1.5 mm mrad $\Delta p/p$ = ± 0.4 %

Ion	I^{20+}	I^{20+} (e-cooled)	Xe^{44+}
N	$3 \cdot 10^{10}$	$9 \cdot 10^{10}$	$1.2 \cdot 10^{11}$
MeV/u	330	330	1100
Joules	200	600	2500
Δt (nsec)	40	70	70
Gigawatt (peak)	7	12	50
TW/cm²	20	40	160
R (g/cm²)	4.5	4.5	25
TW/g	4.5	8	6
T (eV) (gold)	15	20	17

The actual figure of merit for heavy ion fusion oriented target experiments is the specific power deposition

$$P = \frac{E \cdot I}{R \cdot F} \quad \left[\frac{Watt}{g}\right] \qquad (2)$$

with E the kinetic energy [eV], I the beam particle current [A], R the range [g/cm²] and F the focal spot area [cm²]. The range-energy relationship and the dependence of target temperature on the specific power deposition have been taken from Ref. 9. In the present regime of parameters we can make use of the approximate scaling relationship for the target temperature achieved by direct beam heating

$$T \sim P^{1/2} \qquad (3)$$

It is obvious that T depends most strongly on the final pulse length Δt (via $I \sim N/\Delta t$) and on F. Unfortunately these two parameters are not independent due to final lens aberrations[5] (smaller Δt requires a larger $\Delta p/p$ and thus F); it is thus not obvious how to gain considerable improvement without a much larger effort in final focusing design.

Discussion of Constraints

(a) Choice of ion

In principle the highest P can be achieved for the heaviest ions and the lowest charge state compatible with the injector. With $A/q \leq 130$ for the RFQ, this leads to the choice of $^{127}I^{20+}$ or any other ion close to this. For option B there is some concern that an ion with partially filled M-shell might have a large cross section for dielectronic recombination with the cooling electrons. This might limit beam life time below what is needed for effective cooling. We have therefore also considered Xe^{44+} which has an empty M-shell. In this case capture of a free electron into the M-shell requires simultaneous excitation of a bound electron from the K or L shell; the energy difference in this resonant process must come from the kinetic energy of the free electron, which turns out much larger in this case than what one actually has from the relative motion between ions and cooling electrons (typically ≤ 10 eV).

(b) Final lens constraint

A final focusing system consisting of 7 quadrupole lenses with maximum field of 1.9 T and an overall length of 12.4 m has been considered for a uniform beam of ε = 1.5 mm mrad and $\Delta p/p$ = ± 0.4 % (see Fig. 7). The system has been calculated for a first order spot of 160 µm diameter[8]. The trapping efficiency on the nominal 200 µm diameter spot is found as 61 %, which is dominated by chromatic rather than geometric aberrations. A $\Delta p/p$ = ± 0.2 % would raise the efficiency to 88 %. Doubling of the emittance reduces the efficiency to 28 % for $\Delta p/p$ = ± 0.4 %.

(c) Space charge limit of SIS

A given final emittance leads to a total number of particles via the Laslett tune shift constraint

$$N = \frac{2\pi A \beta^2 \gamma^3 \varepsilon B_f \Delta Q}{q^2 r_p} \tag{4}$$

where we assume ΔQ = 0.25 and a bunching factor of B_f = 1/3 (possibly requiring a second harmonic of the RF).

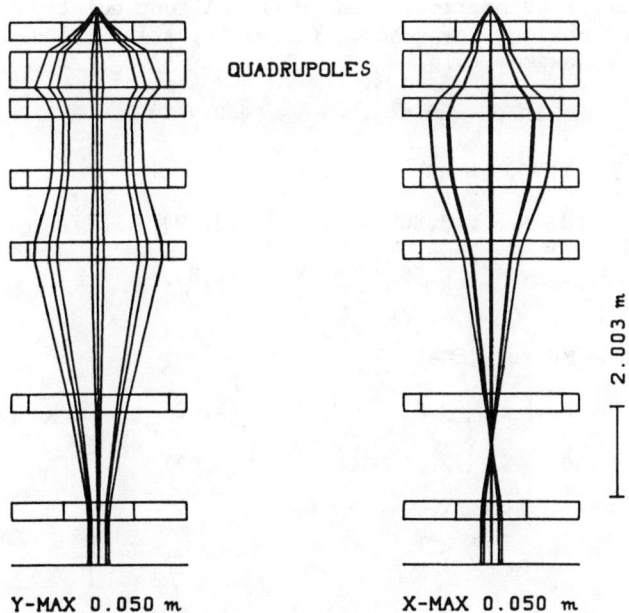

Fig. 7: Final lens system design

Eq.(4) is evaluated at the most critical point, which is the injection energy into SIS (option A) or the re-injection energy from ESR after cooling (option B). The respective emittances are larger than the final emittance by the factor $(\beta\gamma)_{final}/(\beta\gamma)$, which yields

$$\begin{array}{l}\epsilon/_{inj.} = 6.8 \text{ mm mrad} \quad (I^{20+}, \text{ option A}) \\ \epsilon/_{re-inj.} = 2.7 \text{ mm mrad} \quad (I^{20+}, Xe^{44+}, \text{ option B})\end{array} \quad (5)$$

We thus obtain the values of N and the total stored energy contained in Table II.

During the final bunching B_f becomes quite small and thus Eq.(4) predicts a large increase of ΔQ. A calculation of ΔQ for the SIS lattice and the final bunching parameters yields the results of Table III.[10] We observe that most of the tunes are crossing an integer, although there is a noticeable reduction of ΔQ_h caused by the dilution of space charge with finite $\Delta p/p$. On the other hand, computer simulation has not indicated a problem with dipole resonances, whenever Q is driven across an integer by means of space charge[11]. We thus ignore the effect of Q crossing an integer, during final bunching. Beam envelopes are only slightly changed by the effect of space charge even at maximum compression (Fig. 8).

Table III. Shift of horizontal and vertical tune due to space charge at maximum compression (in parenthesis including effect of maximum $\Delta p/p$).

	$\Delta p/p = 0$	(± 0.4 %)
I^{20+} 330 MeV/u; 6 eA		
Q_h = 4.20 →	3.40	(3.99)
Q_v = 3.40 →	2.56	(2.85)
Xe^{44+} 1100 MeV/u; 18 eA		
Q_h = 4.20 →	3.65	(4.06)
Q_v = 3.40 →	2.81	(3.02)

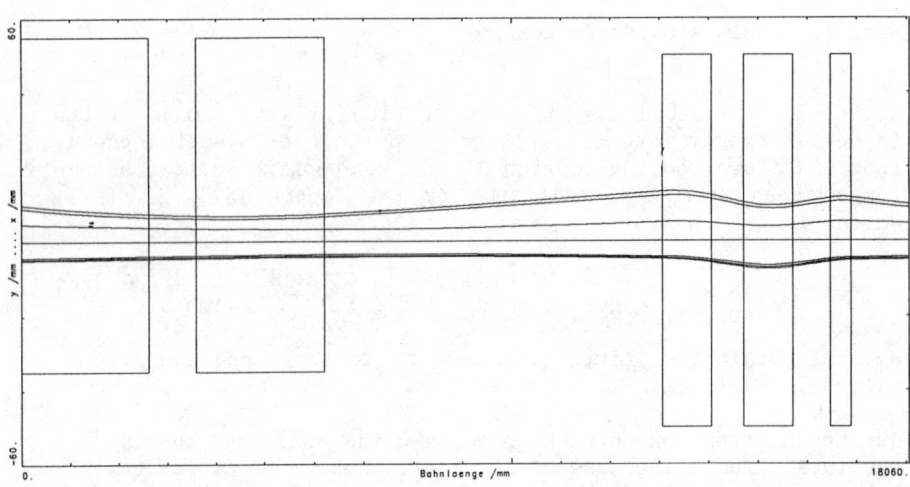

Fig. 8: Beam envelopes (x: horizontal; y: vertical) in unit cell of SIS for I^{20+} at maximum bunch compression with (6 eA) and without current and assuming $\Delta p/p = \pm 0.4$ % (outer pairs of envelopes); also shown are the $\Delta p/p = 0$ envelopes with current (innermost curves).

(d) Minimum Momentum Spread

Here we have to consider options A and B separately.

Linac De-bunching (option A)

$\Delta p/p$ in the SIS is determined in the absence of cooling by what the Unilac delivers and by what space charge does during debunching in the transfer line and the first few revolutions in SIS. A major problem is the increase of $\Delta p/p$ by space charge forces during this transfer. In the limit of the de-bunching dominated by space charge, which is applicable here, the final $\Delta p/p$ only depends on the current and the bunching factor B_f (more precisely, the local maximum current given by I/B_f) according to [2]:

$$\left(\frac{\Delta p}{p}\right)^2_{deb.} \approx 2 \frac{qI}{A} B_f^{-1} \frac{Z_0 g/(2\beta\gamma^2)}{\pi \, mc^2/e \, \eta\gamma\beta^2} \qquad (6)$$

For 30 mA of I^{20+} at 18 MeV/u we obtain

$$\left(\frac{\Delta p}{p}\right)_{deb.} \approx 7 \cdot 10^{-4} \, B_f^{-1/2} \qquad (7)$$

For a bunch length of $\Delta t = \pm 1/2$ nsec and a frequency of 13.5 MHz we have $\Delta p/p = 7 \cdot 10^{-3}$, which is far above what can be tolerated. A 27 MHz rebuncher cavity would improve the bunching factor to 1/12 and the de-bunched moment spread to $2.5 \cdot 10^{-3}$.

A coasting beam momentum spread of this size is a factor of 6 larger than what is needed for option A ($4 \cdot 10^{-4}$). A possible cure would be a saw- tooth rebuncher cavity or a bunch-into-bucket injection scheme; both possibilities are technically demanding due to the high voltages required and need a detailed study.

IBS and LMI in the ESR (option B)

The electron cooling force is balanced by heating through intrabeam scattering (IBS) and the longitudinal microwave instability (LMI).[12] To first approximation the cooling limit can be defind by setting the e-folding time for cooling in the longitudinal direction equal to that for IBS heating. For constant emittance and the lattice of the ESR we obtain the curves shown in Fig. 9a, b.

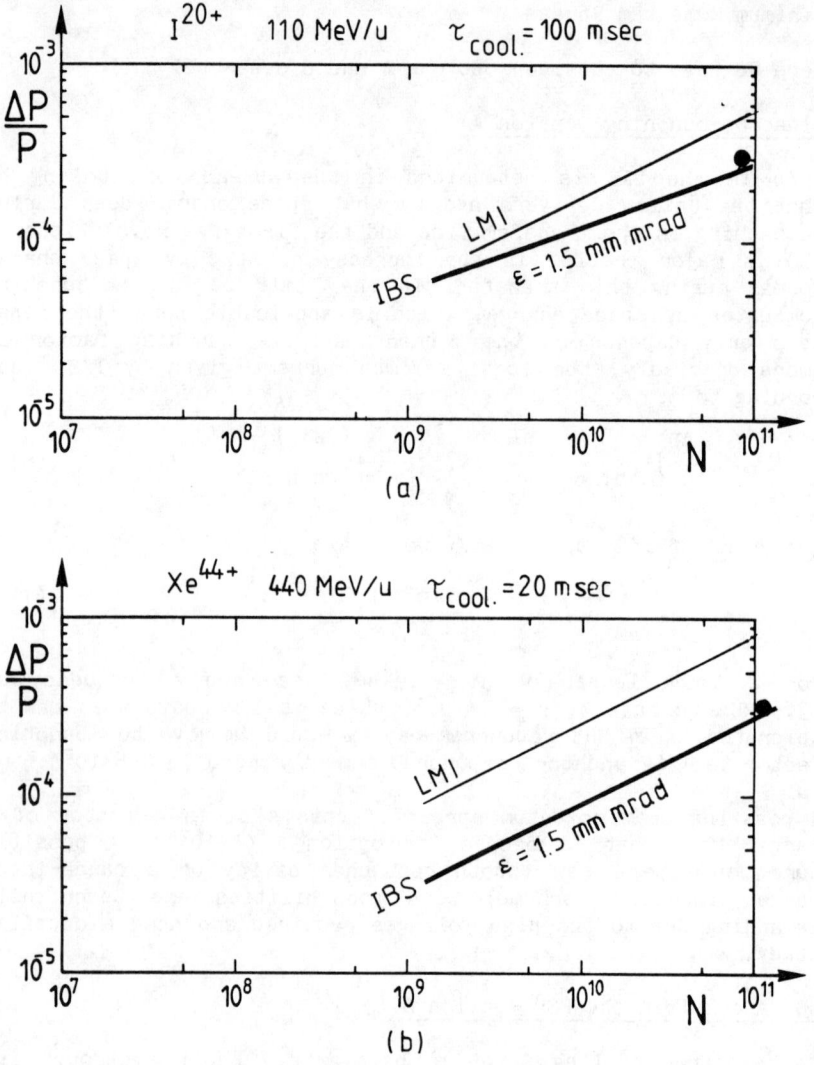

Fig. 9 a, b: Cooling limits in ESR due to intrabeam scattering and longitudinal microwave instability.

Cooling e-folding times are consistent with an electron current of 4 A in this case[13] (see Fig. 10). The emittance assumed in this calculation is a factor 1.8 smaller than what is actually needed at the ESR energy (keeping in mind that ε_n is constant during post-cooling acceleration in the SIS), hence we have this as a safety factor for possible uncertainties in IBS calculations.

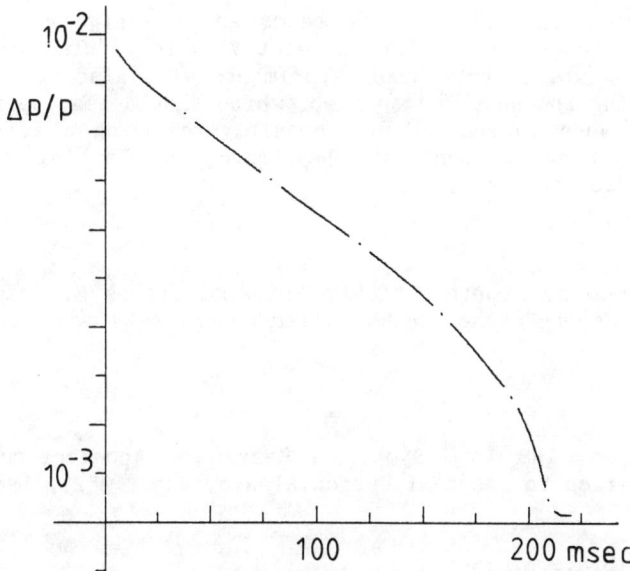

Fig. 10: Momentum cooling versus time for I^{20+} at 110 MeV/u and 4 A electron beam current.

It will be noted from Fig. 9 that the cooling equilibrium investigated has an N exceeding the Keil-Schnell limit of the LMI by typically a factor of 5. Here we take advantage of the theoretical result that beyond the Keil-Schnell limit stabilization of the LMI occurs through formation of a thin tail in the momentum distribution involving only a few percent of the total intensity[14]. Investigation of this mechanism and of stability during rebunching in the SIS (bunched beam microwave instability) will be an essential part of the experimental program.

CONCLUSION

It is expected that after mid of 1989 the new GSI facility allows first extensive machine studies. The full injector intensity will become available about a year later. Due to the nonrelativistic energies and the high phase space density we anticipate that this machine complex will allow to study under proper conditions almost all of the accelerator issues identified for HIF drivers. For establishing such a machine program as part of the R & D it is important to realize that high phase space density is of great benefit to many nuclear physics experiments as well.

As a back up to some of the uncertainties involved we have discussed three possibilities - with and without cooling - of generating the beam power necessary for a first generation of target experiments on "high energy density in matter" with plasma temperatures of 15 to

20 eV. Particular features of these beams are the high repetition rate (~ 1 Hz) and on the other hand a relatively long pulse duration (40-70 nsec). We expect that these experiments will also serve as a basis for defining the next larger step, which should lead to plasma temperatures between 50 and 100 eV, possibly by using a superconducting storage ring as part of the long-term planning of GSI towards a heavy ion collider.

Acknowledgement:

The author appreciates contributions by his colleagues B. Franczak, J. Klabunde, R. Meyer-Prüßner, R.W. Müller and F. Nolden.

References:

[1] D. Böhne, Proc. INS Int. Symp. on Heavy Ion Accelerators and their Application to Inertial Fusion, Tokyo, Jan. 13-27, 1984, p. 173

[2] I. Hofmann, ibid., p. 184

[3] I. Hofmann, Internal Report GSI-SIT-INT/83 and B. Franzke, GSI-SIS-INT/83-5 (1983)

[4] K. Blasche. D. Böhne, B. Franzke and H. Prange, IEEE Trans. Nucl. Sci. NS-32, 2657 (1985)

[5] K. Beckert et al., GSI Report 83-10 (1983)

[6] I. Hofmann, HIF Annual Report 1985, GSI 86-13, p. 42

[7] I. Bozsik, I. Hofmann, A. Jahnke and R.W. Müller, Proc. Conf. on Computing in Accelerator Design and Operation, Berlin, Sept. 20-23, 1983, p. 128

[8] I. Brezina, private communication (1985)

[9] R.C. Arnold and J. Meyer-ter-Vehn, private communication (1986)

[10] B. Franczak, private communication (1986)

[11] I. Hofmann and K. Beckert, IEEE Trans. Nucl. Sci. NS-32, p. 2264 (1985)

[12] I. Hofmann and R. Meyer-Prüßner, GSI Report 85-19 (1985)

[13] F. Nolden, private communication (1986)

[14] I. Hofmann, Laser and Particle Beams 3, 1 (1984)

REVIEW OF TARGET STUDIES FOR HEAVY ION FUSION*

J. D. Lindl, R. O. Bangerter, J. W-K. Mark, Yu-Li Pan
Lawrence Livermore National Laboratory, Livermore, CA 94550

ABSTRACT

We present an updated set of gain curves for radiation driven ion beam targets. The improved target performance calculated with nuclear spin polarized fuel will also be discussed. We discuss the conditions required for efficient conversion to x-rays of ion beam energy. These requirements are compared with those obtained for lasers. Recent results on symmetry requirements for direct drive ion beam targets are presented.

INTRODUCTION

Classification restrictions prevent a comprehensive discussion of progress and issues for the radiation drive approach to ICF. However, the national ICF program under congressional mandate underwent an extensive review by the National Academy of Sciences in 1985. The Academy has published both a classified and unclassified summary of their findings. This review provides a favorable, independent evaluation of the progress of ICF.

Quoting from the report: "The classified Centurion-Halite program involves weapons and ICF groups in a theoretical and experimental effort to investigate design characteristics of efficient ICF targets. This program has thus far been very successful. The Centurion-Halite program involves very difficult experiments, and the program should be given sufficient time and funding to provide definitive results. We believe that it will be possible to complete the Centurion-Halite program within five years, with sufficient resources. A longer period of time may be required if there are funding limitations or serious technical problems."

The NAS committee also concluded that the planned Nova experiments were essential to assessing the potential of radiation driven ICF. Since the committee wrote its report, Nova has indeed begun to fulfill its potential. In its first high density implosion experiment, the Nova laser imploded a capsule to conditions which represent a significant advance in the fuel conditions attained in laboratory ICF experiments. These

---

*Work performed under the auspices of the U. S. Department of Energy by the Lawrence Livermore National Laboratory under contract number W-7405-ENG-48.

experiments, which had a neutron yield of $Y = 5 \times 10^{10}$ and a measured ion temperature of $\theta = 1.7 \pm 0.3$ keV had an estimated value of $n\tau\theta \sim 2-4 \times 10^{14}$ s·keV/cm^3 where τ is the confinement time and n the particle density. This value is as good as has ever been achieved on any laboratory fusion device and is particularly significant since the experimental results were very nearly equal to the 1-D implosion calculations. This is in contrast to earlier high density experiments in which experimental neutron yields were often lower than the calculated values by an order of magnitude or more.

Although the results of the Nova experiments were achieved with a laser, the implosion results for a radiation driven capsule apply to any driver which can provide the required radiation environment. This is particularly important to the HIF efforts since there is no target experiments program for these drivers.

X-RAY CONVERSION EFFICIENCY AND TARGET GAIN

To implode efficiently a capsule with x-rays, it is first necessary to convert efficiently the driver energy to x-rays. Both lasers and ion beams can achieve high x-ray conversion efficiency but they have opposite intensity scaling as shown in Fig. 1.

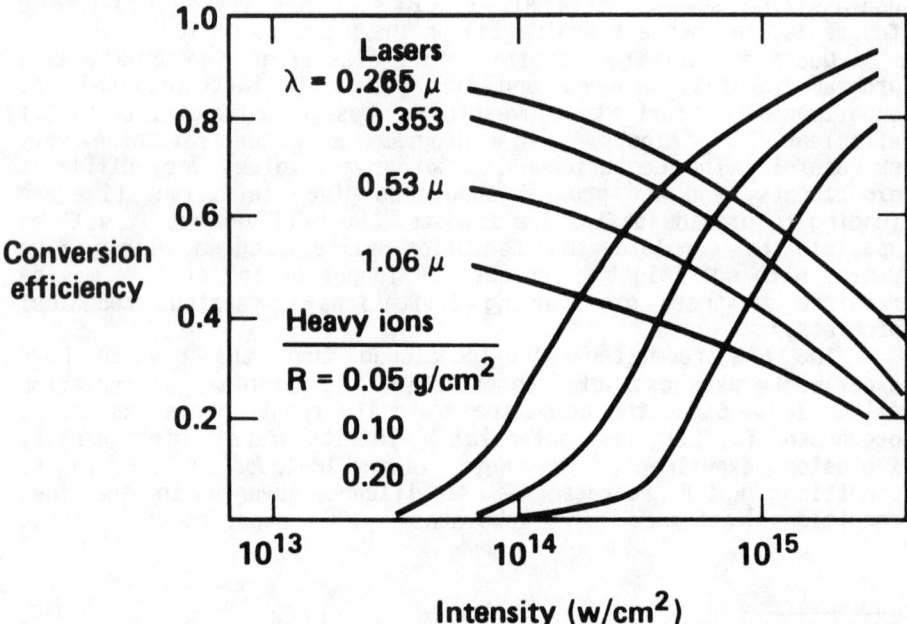

Fig. 1. Both lasers and ion beams can achieve high x-ray conversion efficiency but have opposite intensity scaling.

Plotted are the experimental conversion efficiencies on gold for wavelengths from 1/4μm to 1μm for lasers and theoretical results for three different ion ranges, in low-Z material. The ion conversion efficiency curves are from a simple analytical model which is a good match to optimized LASNEX calculations. The ion beam analog to the gold disk experiment, from which the laser data are obtained, is shown in Fig. 2a. Here, a low-Z radiator, an ion range thick, is surrounded by a lossless tamper whose function is to constrain the hydrodynamic motion to one-dimension.

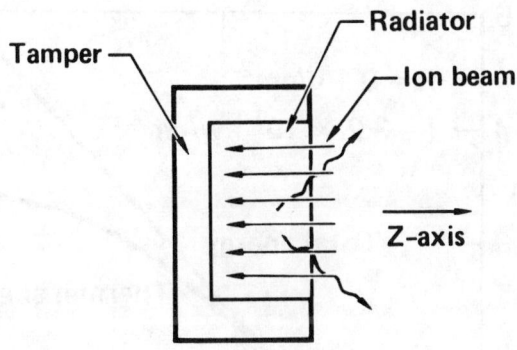

Fig. 2a. A simple model provides an estimate of ion to x-ray conversion efficiency.

The incident ion beam intensity I (W/cm^2) is then partitioned into three components given by

$$I = R \, d\varepsilon/dt + dK/dt + \sigma T^4 \tag{1}$$

where R is the ion range, ε is the specific energy of the radiator material, K is the kinetic energy and T is the temperature. The quantity K is obtained from an approximate solution of

$$\frac{Dv}{Dt} = -\frac{1}{\rho}\frac{dP}{dZ} \tag{2}$$

and can be minimized by choosing a low density radiator material such as a foam. Conversion efficiency η is defined by

$$\eta_{conv} = \int \sigma T^4 dt / \int I dt \tag{3}$$

If the intensity is too low, the heat capacity losses dominate because the temperature for reasonable pulse lengths never gets high enough for efficient radiation. An example with $I = 3 \times 10^{14}$ W/cm^2 and $R = 0.1$ g/cm^2 with a 10 ns pulse is shown in Fig 2b. At sufficiently high intensity, the radiator is quickly heated to a temperature at which σT^4 dominates and reradiates energy at the incident power. For example, in Fig. 2c at $I = 3.2 \times 10^{15}$ W/cm^2 with $R = 0.1$ g/cm^2, an overall conversion efficiency of nearly 90% is achieved.

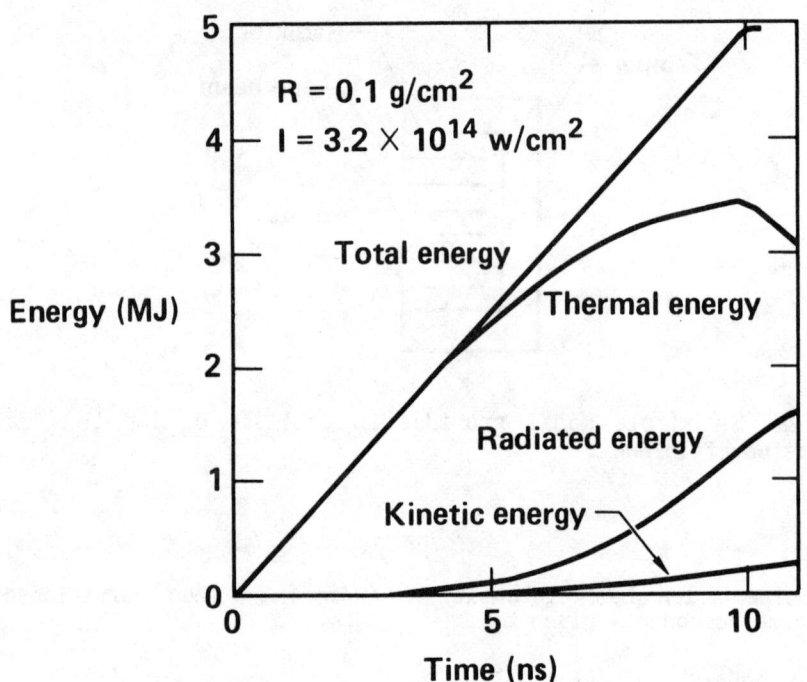

Fig. 2b. Low intensity gives low radiation temperature and high thermal losses.

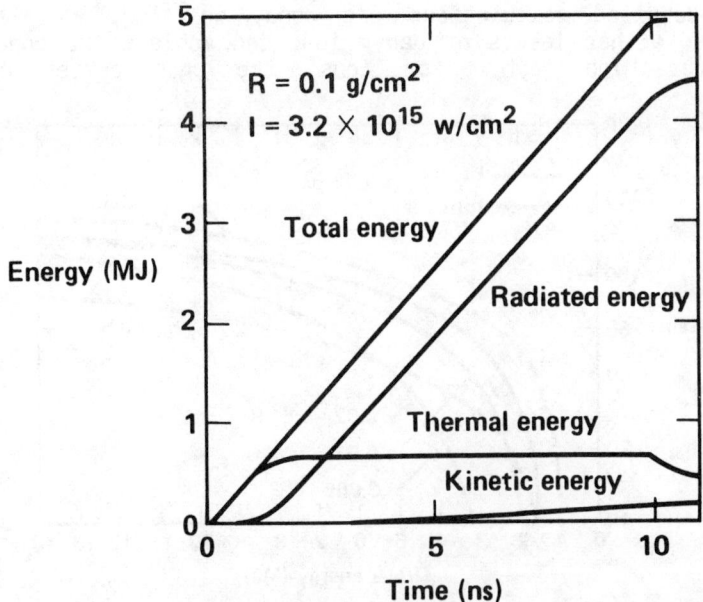

Fig. 2c. High conversion efficiency requires high intensity which results in low kinetic and thermal energy losses.

Because of the differences in conversion efficiency versus intensity for lasers and ions, there are some important differences for various applications. Lasers get conversion efficiencies η > 70% for I < 10^{14} W/cm^2 for λ < 1/3μ. This is advantageous from a plasma physics viewpoint since all important instabilities are below threshold at this intensity. But the laser system would require many beams covering a large fraction of solid angle at the target to achieve this low intensity. Such architectures are cumbersome and poorly matched to the most attractive reactor designs. Laser experiments, using higher-Z and lower density radiator materials, are underway to increase the intensity at which η > 70% is achieved. For ion beams on the other hand, conversion efficiency increases with intensity and reaches values in excess of 80-90% if intensities of several times 10^{15} can be reached with typical beam parameters. This intensity need not be reached by each beamlet if there are many beams, but only in a small number of spots on target where many beams could overlap. Beams can cover a very small solid angle at the target which is ideally matched to one or two sided power production scenarios. On the other hand, as is clear from Fig. 1, it is extremely important that the ion beams come very close to reaching the design spot size. For typical design points, if the spot size were a factor of three larger than planned, so the intensity were an order of magnitude lower, x-ray conversion efficiency would drop to a very low value and the accelerator would no longer be useful for x-ray driven capsules.

When detailed calculations are done, we find, as shown in Fig. 3, that either lasers or heavy ions can achieve the conditions required for high gain. For ions, the gain curves can be

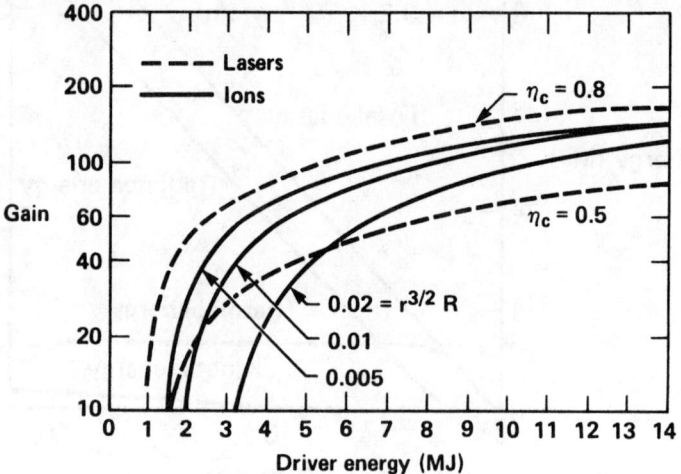

Fig. 3. Either lasers or heavy ion beams can achieve the conditions required for high gain.

approximately represented as a function of $r^{3/2}R$ and driver energy where r is the spot radius (cm) and R is the ion range (g/cm^2). A more accurate representation requires that r and R be treated as independent variables. Compared to the optimized laser designs, current heavy ion targets, because of lower x-ray transport efficiency, have a lower gain for a given x-ray conversion efficiency. Hence the maximum gain for ions is less than for the 80% conversion efficiency laser curve.

POLARIZED FUEL

A variety of advanced target approaches exist to increase target gain beyond those indicated in Fig. 3. One of these is the use of nuclear spin polarized fuel.[1,2] The fusion cross section σ for DT reactions is given by

$$\sigma_{pol} = (1 + 1/2\ P_D P_T)\ \sigma_{unpol} \tag{4}$$

where $P_{D,T}$ are the polarization fraction for D or T. With complete polarization, σ increases by a factor of 1.5. But because the increase is quadratic in the polarization fraction, 50% polarization only gives a 12.5% increase in cross section. If one ignores depletion, the burn efficiency is given by $\phi \propto \rho r \langle \sigma v \rangle$ where ρr is the product of the fuel density ρ and its radius r and $\langle \sigma v \rangle$ is the Maxwell averaged cross section. Gain is proportional to $m \rho r \langle \sigma v \rangle / E_{in}$ which is proportional to $\rho r \langle \sigma v \rangle$ because $E_{in} \propto m$ where E_{in} is the incident driver energy and m is the fuel mass. For a fixed target then, with 100% polarization, the gain increases by a factor of about 1.5. For fixed gain, ρr can be reduced by a factor of 1.5. Assuming constant density, $\rho r \propto r \propto m^{1/3} \propto E^{1/3}$. Therefore, the driver energy might be reduced by a factor of $(1.5)^3 \sim 3$. The results of detailed calculations of ion beam targets with and without polarized fuel are shown in Fig. 4. We do find the 1.5 increase in target gain

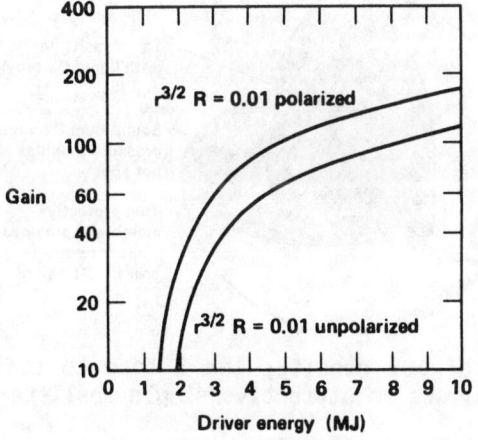

Fig. 4. Spin polarized fuel gives higher gain.

for a fixed target but the energy reduction for fixed gain is less than a factor of 3. Because of a variety of non-hydrodynamic effects, the performance of smaller targets drops off more rapidly than $E^{1/3}$. Highly polarized DT has not been produced but studies are underway at Princeton, MIT, Syracuse and LLNL. Many depolarization mechanisms during an implosion have been studied[2] and no problems have been found.

DIRECT DRIVE

Assessing the prospects for direct-drive heavy-ion targets is more difficult than for radiation drive because much less of the information from current experiments is applicable. However, potential gains are substantially higher than those for radiation driven targets if the energy loss involved in obtaining symmetry is small.

An exciting development in the past year has been the emergence of low density, low-Z foams wetted with DT.[3] These targets provide a nearly ideal direct drive target, similar to those originally proposed by Nuckolls for ICF. As shown in Fig. 5, this foam, wetted with liquid DT provides both the ablator and fuel. The vapor pressure of DT just above the triple point provides a nearly optimum gas mass for a hot spot. The main penalty for using this target, compared to a pure DT shell, is an increase in the ignition temperature as shown in Fig. 6. This temperature is defined by the

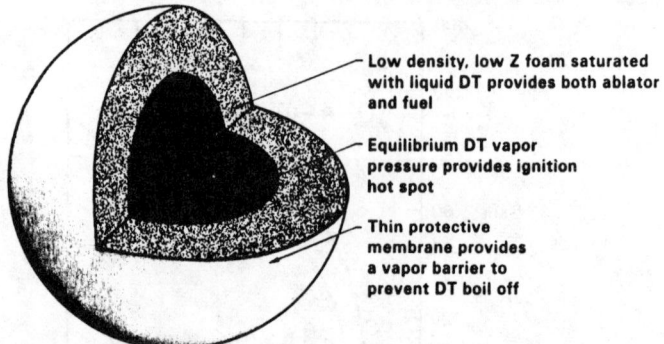

Fig. 5. Use of low density, low Z foam to define and stabilize the DT shell allows an attractive single shell target design.

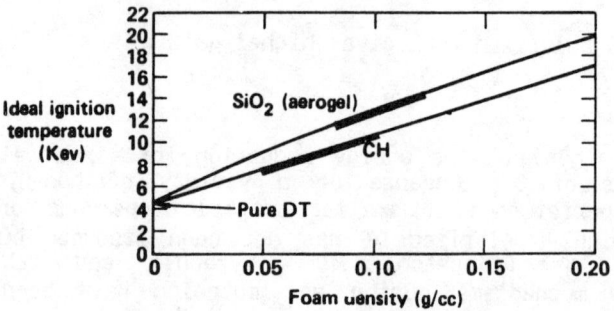

Fig. 6. The ideal ignition temperature for wetted foams depends primarily on the foam density and composition. Highlighted bands are the ranges of densities where the materials shown are expected to meet the target and fabrication requirements.

point at which the alpha particle production rate for thermonuclear burn equals the bremsstrahlung rate. For $\rho = 0.1$ g/cc aerogel, a currently available SiO_2 foam, the ignition temperature increases from 4.5 keV for DT to about 13 keV. However for materials under development such as $\rho = 0.05$ CH, the increase is quite small and results in little change in target gain.

Typical direct drive targets have a convergence ratio, the ratio of the initial target radius to the final compressed radius, large enough that peak to minimum intensity variation must be held to about 1-2%. A variety of focusing schemes exists to achieve this kind of uniformity in principle, with adequate beam quality. However, these schemes typically produce optimal uniformity at a single target radius. If the radius of deposition varies strongly during the implosion, as is the case with short wavelength lasers, it becomes difficult to achieve uniformity without a very large number of beams. For ions, we have been able to design a target which largely avoids this problem as shown in Fig. 7. The target has a radial excursion of the peak in the energy deposition of only ±8%. This is accomplished by making use of the relatively low energy loss of ions in high-Z material to provide a tamper on the explosion of the wetted foam target. By adjusting the mass of this tamper, we control the motion of the location of the energy deposition.

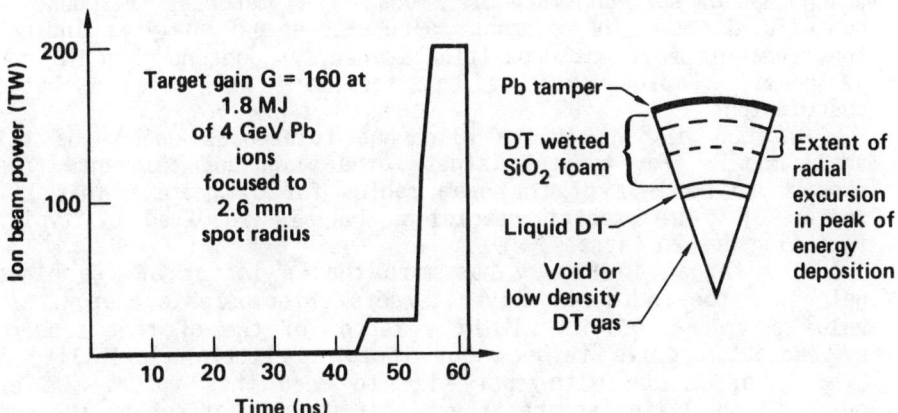

Fig. 7. Tamped direct drive ion beam targets can have low radial motion of peak deposition.

We have studied two methods, the minimum potential and Gaussian quadrature approaches to obtain symmetry for direct drive ion targets. The minimum potential approach has been widely used for direct drive laser targets. In this approach, beams are imagined to be charges on the surface of a sphere. The charges are allowed to move around on the surface until their potential energy is minimized. In the Gaussian quadrature approach[4], beams are arranged on the rims of cones whose relative powers are given by Gaussian quadrature weights and whose angles are the zero's of the Legendre polynomials. This approach geometrically eliminates the lowest $2n-1$ perturbation modes in θ where n is the number of rings of beams and $m-1$ perturbations in φ where m is the number of beams on a ring. This method is amenable to analytic analysis but unequal beam powers are required to minimize the number of beams. Both methods give comparable symmetry in static, thin-shell models. LASNEX calculations indicate that results of the analytical thin shell model are reasonable for the Gaussian quadrature approach. The LASNEX calculations are done by taking a snapshot of the 1-D implosion at several times. At each of the selected times, the calculation is expanded to 2-D to obtain deposition as a function of angle and depth for a single beam. This deposition is then added analytically for all beams to produce the total deposition versus angle and depth. These calculations are not entirely self consistent since the flows are assumed to be spherically symmetric at each point. We are in the process of performing a 2-D approximation to the hydrodynamics which would self-consistently model the material response to nonuniformities. The current calculations and analyses indicate that the uniformity of deposition is nearly good enough with about 32 beams. Improvements to this estimate will require better calculations.

We can also calculate alignment tolerances using the thin shell model. For typical targets, the alignment tolerance (rms) appears to be 1-3% of the beam radius for 32 beams. This is a factor of 5-10 greater precision than is required by typical radiation driven targets.

The target in Fig. 7 has more than a factor of two higher gain than the radiation driven targets of comparable energy. Its value of $r^{3/2}R\sim.005$ is within a factor of two of that required by radiation drive targets but there is little flexibility in trading off range with spot size to meet this value. We are currently analyzing whether the higher gain is offset by the more complex beam geometry, more restrictive beam parameters and more stringent focusing requirements.

CONCLUSIONS

In conclusion, progress in radiation driven targets in the U.S. ICF program has been very encouraging over the past couple of years. Since heavy ion ICF targets can make use of much of this information, there is a strong motivation for a continued or accelerated machine development program. The critical issue, for heavy-ion accelerators, for radiation driven targets, is their ability to achieve focused intensities of order 10^{15} W/cm^2 for typical machine parameters. Direct drive heavy ion targets in principle have much higher gain than radiation driven targets but have more restrictive pointing requirements, less flexibility in accelerator parameters, and a more complex geometry. Without a heavy ion target experiment program, it will be nearly impossible to obtain sufficiently accurate answers to these questions for direct drive targets.

REFERENCES

1. R. M. Kulsrud, H. P. Furth, E. J. Valeo, and M. Goldhaber, Princeton Plasma Physics Laboratory, Princeton, NJ, Report 1912 (1982).

2. R. M. More, Nuclear Spin-Polarized Fuel in Inertial Fusion, Phys. Rev. Letters 51, 396 (1983).

3. R. A. Sacks and D. H. Darling, Direct Drive Cryogenic ICF Capsules Employing DT Wetted Foam, Lawrence Livermore National Laboratory, Livermore, CA, UCRL-94381 (1986).

4. J. W-K. Mark, Near Spherical Illumination of Ion-Beam and Laser Targets, Physics Letters, 114A, 458 (1986).

BEAM PLASMA INTERACTIONS IN HIF

T.D. Beynon,

University of Birmingham, Birmingham B15 2TT, U.K.

ABSTRACT

A review is presented of some recent studies on the effects of beam characteristics (energy, beam spot size, radial particle distribution, charge state) on the predicted performance of directly driven HIF targets. Phenomena of particular importance in such beam-plasma interaction are (i) the effects of plasma density and temperature on ion range, and (ii) the effects of charge exchange and accretion due to the beam-target interaction. Range effects are shown to be important for energy-coupling, Rayleigh-Taylor instability and fission fragmentation effects. Effects associated with the charge state of the incident ion beam are shown to lead to a reduction in target energy input, beam coupling efficiency and fusion yield, thus demanding some degree of charge neutralization.

INTRODUCTION

A wide variety of interesting and important physics phenomena become apparent when one considers, in a self-consistent fashion, the effects on target performance of various ion beam characteristics such as beam energy, beam spot size, radial particle distribution and ion charge state. In this paper results are presented of some recent studies of the effects of ion-range shortening and beam widths on the energy coupling of a beam into a target and on the Rayleigh-Taylor instability of the imploding plasma. Also discussed are the effects of beam-induced nuclear fission on target performance and how this is also dependent on ion range effects and beam widths.

Of particular importance to an overall HIF system design is how the charge state of the ions in the incident focused beams can affect target performance. Results of a detailed study of the electrodynamic coupling of the beam and target are given which indicate the upper limits of the charge state which can be tolerated.

All the numerical simulations in the paper are based on one dimensional hydrodynamic simulations of directly driven targets using the 1D code MEDUSA[1] which has been modified for HIF studies to include high and low-Z opacity data, a full transport theory radiation transport package, a corrected equation of state, a full slowing down description of beam ions (which includes density-temperature effects and a velocity-dependent effective charge), and which has the ability to deal with arbitrary beam configuration in a fashion discussed below. The burn-physics allows for fusion-product charged particle and neutron transport.

The 1D constraint of MEDUSA means that a representation of a multiple beam irradiation must always result in a uniform surface irradiation on the target. With this constraint, the diagram in Fig. 1 illustrates how the effective radial energy deposition $\epsilon(r)$ is obtained for a large number (formally, an infinite number) of beams on the target. The solid lines indicate the ion trajectories in a single beam, giving a contribution $\Delta\epsilon(r,\theta)$ in dV at (r,θ). An adjacent beam, (dashed lines) displaced clockwise by an angle ψ, will give a contribution $\Delta\epsilon'(r,\theta)$ which is the same as the contribution in the original beam computed at a position $(r, \theta-\psi)$ i.e. by a counter clockwise rotation. Hence the total energy deposition at a radius r is obtained by fixing a single beam in the coordinate system of the target and then, for a fixed r value, sweeping through the beam to obtain the total deposition $\epsilon(r)$. This algorithm allows an arbitrary radial density distribution of each beam to be defined as a function of the beam radial coordinate (x in Fig. 1). All our present studies have been made for Gaussian distributions, i.e. $\exp(-x^2/2\sigma^2)$ where σ is the beam spot side. A narrow beam implies $\sigma = 0$ and the expression 'wide beam' means $\sigma = r$, the initial target radius.

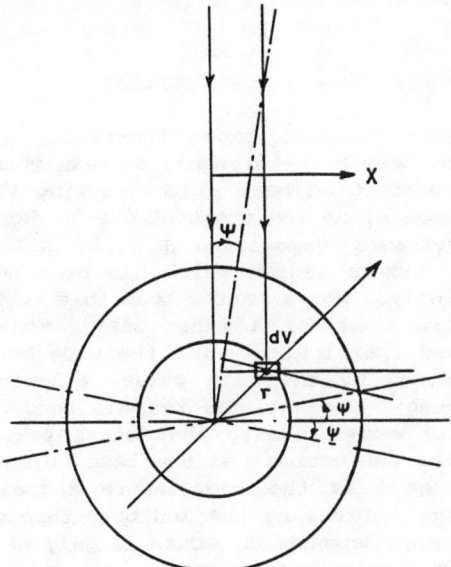

Fig. 1
The Multibeam irradiation of a target

Except where otherwise indicated the target used throughout is that of Fig. 2. Driven by U-ions just able to penetrate the carbon layer (corresponding to an ion energy of about 15 GeV) the target has a constant-level prepulse representing 0.5% of the power level of the main pulse typically about 1×10^{15}W, which has a pulse length which switches off when the shock wave has reached the inner edge of the D-T or when 20 ns has elapsed.

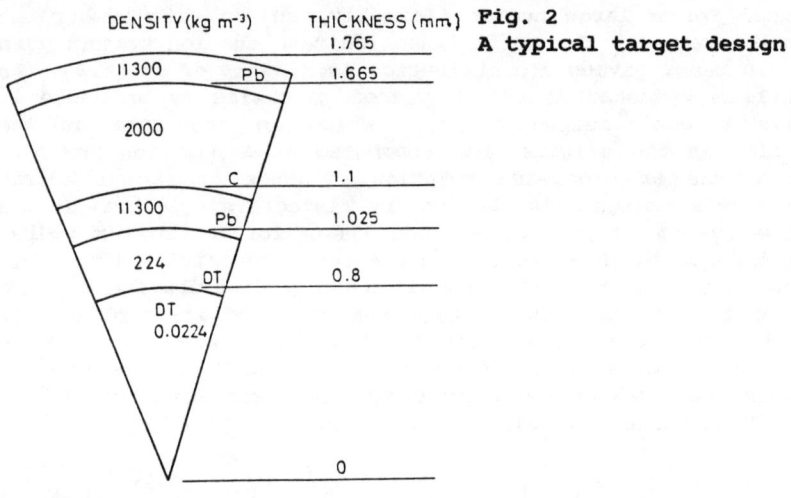

Fig. 2
A typical target design

BEAM BURN-THROUGH AND BEAM COUPLING

An important consequence of ion range shortening in dense plasma is the phenomenon of beam burn-through[2], or beam tunnelling, the extent of which is dependent on beam width. During the early part of a prepulse, the range of an ion shortens due to density and temperature. The time-averaged deposition profile due to the prepulse will, therefore, show a region which has been heated by the ions for a short time only. For a narrow beam this region will be heated considerably more than for a wider beam. This cooler region which is left behind, particularly for the wide beam case, is now appreciably compressed by the main pulse, allowing range lengthening ie. burn-through. Because the thermal conduction is high enough this region is warmed during the latter part of the main pulse, due to both the conduction and the beam burn-through, resulting in high pressures. As the temperature increases the range continues to lengthen, increasing the ion burn-through. For narrow beams, where the energy deposition occurs largely at the end of the ion range, this final burn-through can be very significant, as is illustrated in Fig. 3 where the energy deposition has occurred across the carbon-lead interface of the target of Fig. 2. In the cold material the ions would have stopped entirely in the carbon. The effect of the narrow beam is essentially one of radial focusing and can have important consequences for the interface stability, as is demonstrated later.

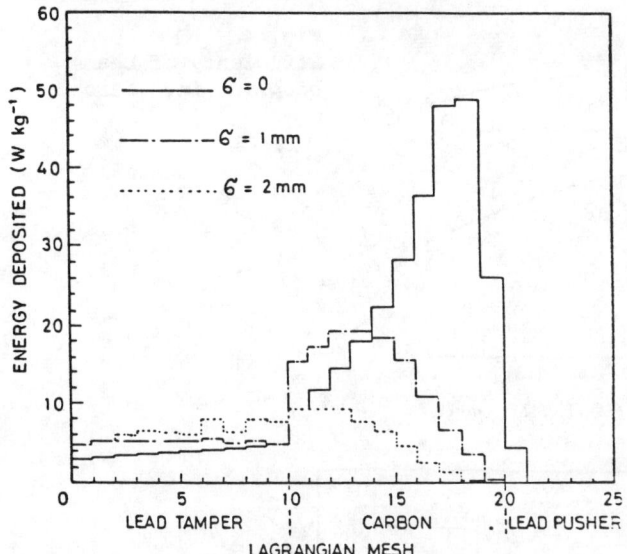

Fig. 3
Time-averaged energy deposition profiles

An example of the importance of the effects of burn-through and beam widths is shown in Fig. 4. A coupling efficiency parameter for the target plasma, defined as the maximum attained inward value of the kinetic energy (which is achieved when the lead shell starts to slow down due to the main reflected shock reaching its inner surface), is shown as a function of beam spot size for two different power levels. At $\sigma=0$, burn-through across the carbon-lead interface degrades the target performance because of the radial focusing effect described above. In the case of the wide beam (σ = 2mm) the uniform radial distribution of energy deposition (cf. Fig. 3), together with the effect of part of the wide beam missing the target, also results in a degraded performance.

RAYLEIGH-TAYLOR INSTABILITY AND BEAM WIDTHS

A detailed study of the effects of ion beam widths on Rayleigh-Taylor (RT) instabilities in targets has been published elsewhere[3] and just one example is discussed here to put some of the points made above into perspective. All calculations were made in an inconsistent fashion using the MEDUSA HIF code to compute the instability growth across the ablator-pusher interface of the target of Fig. 2 driven at a power level of 1.0×10^{15}W. For an initial perturbation of 5×10^{-8}m the growth of the instability,

$\eta(t)$, is shown in Fig. 5 for beam spot sizes $\sigma = 0$, 1 and 2 mm, for a spherical harmonic order $\ell = 20$ corresponding to a 20-beam irradiation.

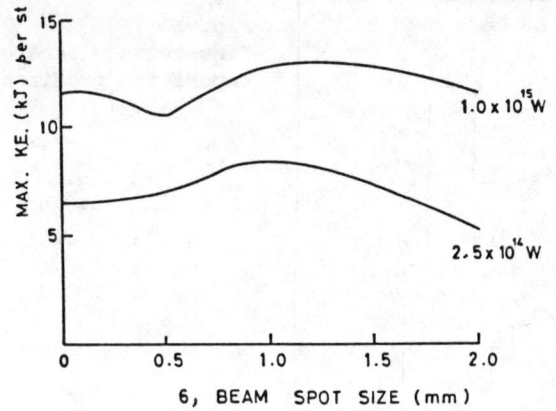

Fig. 4
Efficiency of beam coupling into target

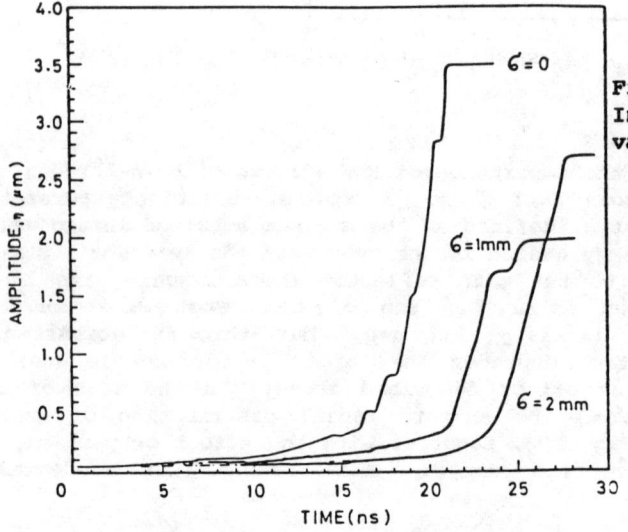

Fig. 5
Instability growth at various beam widths.

It is seen that the use of a narrow beam results in the greatest instability growth. In this case, greater beam coupling to the target and a higher energy deposition density result in a rapid imposion with a higher acceleration of the interface. The reduction in the density of the carbon layer just adjacent to the pusher is due to the high energy-deposition in this region which results in a higher Atwood number for the narrow beam. Switching to full power, the beam undergoes substantial burn-through across the carbon-lead interface which has the effect of reducing the density gradient and hence the Atwood number during the period of maximum growth.

For the medium-width beam ($\sigma = 1.0$ mm) the flatter energy distribution profile results in a less rapid acceleration of the interface and smaller burn-through. The Atwood number is still considerably reduced during this period of maximum growth and these two factors result in the smallest instability amplitude.

In the case of the wide beam ($\sigma = 2.0$ mm) the interface acceleration and burn-through are reduced even more. For this case, however, the total energy deposition is in the carbon layer and there is an increasingly larger fraction of the beam missing the target as compression occurs and there is therefore only a small reduction in the Atwood number as the beam switches to full power. The resultant instability growth is consequently increased as the materials interface density gradient is left substantially intact.

FISSION FRAGMENTATION AND BEAM WIDTHS

It has recently been shown[4] that fission fragmentation in certain combinations of beam-ion and target material can result in significantly degraded target performance. The primary beam of energetic ions undergoes fission as it slows down in the target plasma, producing two fission fragments of equal mass and charge but different energies. In the range of incident ion energy of interest to HIF applications, the fragments will be moving in the same direction as the beam in the laboratory system but with different (E^2/Z^2A) signatures and hence different ranges.

The combined effects of beam widths and fission fragmentation are shown in Fig. 6 for a target driven with 10 GeV U^{+1} ions. The target configuration is a spherical single shell of outer radius 2.23 mm with a Li D ablator region of thickness 1.13 kg m^{-2} covering a Pb region of thickness 11.3 kg m^{-2} and a D-T zone of thickness 4.48×10^{-2} kg m^{-2}. Fig. 6 shows the maximum value of the density-radius product, ρr, attained by the D-T as a function of power level and beam spot sizes. The combined effects of burn through and beam width are clearly illustrated. The preheating of the inner regions of the target by fission fragmentation at each of the two power levels (1.0×10^{15}W and 2.5×10^{14}W) is seen to have a marked effect on target performance, reducing ρr-values by about a factor 2. The effect at a given beam size is larger at the higher power level when the preheat is large in comparison with the Fermi energy of the cold material. At a given power level the effect is lessened with increasing beam size as a larger fraction of the primary beam trajectories does not point towards the central region of the target.

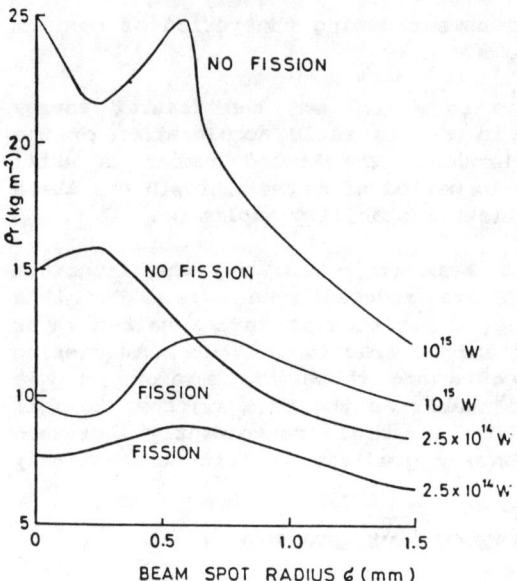

Fig. 6
Effects of fission fragmentation at various beam widths and power levels

ELECTRODYNAMIC COUPLING OF BEAM TO TARGET

Three basic problems have been addressed in this topic, namely can the accumulation of charge on the target for a given beam width and ion charge state

i) significantly bend the beam off the target?

ii) produce electric fields inside the target which significantly effect the hydrodynamic performance of the target?

iii) produce a significant potential barrier on the target to effect energy deposition and hence target performance?

All three questions have received some attention during the past decade but no systematic electrodynamic analysis, as opposed to an electrostatic analysis, is evident in the literature. In the final part of this review, some results are presented of recent studies which have performed fully coupled electrodynamic-hydrodynamic calculations in an attempt to elucidate the above questions. Full details of these calculations will be given elsewhere[5].

The charge accumulated on a target, $Q(t)$, due to multiple beam irradiation of power level $P_o(t)$ of ions in charge state Z_i with kinetic energy E_i is such that

$$\frac{dQ}{dt} = Z_i \frac{P_o(t)}{E_i} \quad (1)$$

so that
$$Q(t) = \frac{Z_i}{E_i} \int_o^t P_o(t') \, dt' \quad (2)$$

The amount of kinetic energy that an ion transfers to potential energy PE(t) of the target is simply, for a target of radius R(t)

$$PE(t) = V\{R(t),t\} \, Z_i = Q(t)Z_i/4\pi\epsilon_o R(t) \quad (3)$$

where V {R(t),t} is the corresponding potential of the target at time t.

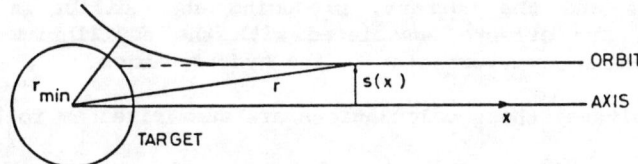

Fig. 7 Geometry for orbit equation

To compute the fraction of ions which miss the target due to the electric field associated with V(t), the orbit equation,

$$m\ddot{r} = (\ell^3/mr^3) - Z_i e \left(\frac{\partial V}{\partial r}\right) \quad (4)$$

must be solved. Here m denotes the ion mass, $\ell^2 = s^2 \, 2mE_i$ where ℓ is the orbital angular momentum and s is the impact parameter. The orbit trajectories are sketched in Fig. 7. If s(x) is smaller than the beam outer radius then the corresponding part of the beam misses the target. Eqn. (4) is solved using (3) self-consistently with the MEDUSA code to obtain R(t). In the results given in table

I, 15 GeV ions at $P_o(t) = 1.0 \times 10^{15}$ W were used in the target of Fig. 2 for $\sigma = 2.0$ mm, ie. a wide beam.

Charge State	+1	+3
% beam loss	~0	~1

Table I Percentage beam losses for U-ion beam in different charge states.

The effect is seen to be negligible, a result which is largely due to the expansion of the target during the pulse. It is worth noting that calculations based on static target radii can overestimate the effect by an order of magnitude.

The approach to the second question, whether or not the charge deposited by the beam can effect the hydrodynamic performance of the target, needs a considerably more detailed analysis and here we shall only outline the steps and the physics arguments. An ion injected into the target loses most of its electrons near the surface giving a negative surface charge distribution. The ion then picks up electrons as it slows down, producing a positive charge distribution in the pellet. An E-field is set up, driven by this charge distribution, which it tries to cancel. A dynamic equilibrium is reached between the charge source due to the injected ions and the current, producing an equilibrium charge distribution. The effects associated with the equilibrium charge may be treated as a perturbation to the hydrodynamics.

The results of these calculations are summarized as follows

a) Hydrostatic forces are in all cases found to orders of magnitude greater than the forces associated with the induced E-field.

b) Ohmic heating associated with the equilibrium current density is of order $10^7 - 10^9$ W/m^3 compared with the beam energy deposition of $10^{13} - 10^{15}$ W/m^3, and is therefore negligible.

c) Perturbations of the electron density are negligible and cannot effect dE/dx for the incident ion beam.

d) Calculations to estimate whether the induced electric field would effect electron exchange cross sections gave values of $E \sim 10^6$ V/m compared with the Coulomb values of 10^{10} V/m in the outer orbits of a U^{+1} ion.

In conclusion, no significant perturbation could be found which affected the hydrodynamic behaviour of the targets.

More positive results have been obtained in seeking the answer to the third question, ie. does the charging of the target by the beam affect the beam and hence the target performance? The build up of positive charge transported into the target by the beam produces a system with a high positive potential barrier and associated potential energy. This energy comes from the kinetic energy of the ions according to eqn. (3). and is therefore unavailable for target implosion work. The analysis performed allows a) for the charge state of the ion during its slowing down and b) for the beam width, as discussed earlier, of the incident ions. The targets are assumed to be in a perfect vacuum. As the charge on the target builds up with time, the effective beam power available for the target is P(t) where

$$P(t) = [E_i - PE(t)] P_o(t)/E_i$$

The process has two important consequences: a) part of the energy input is not used for implosion and b) the reduced kinetic energy of the ions results in a range shortening and hence a change in the energy deposition profile.

Table II summarizes the results for our typical target. Charge state zero implies that the hydrodynamic simulations have been performed without making any allowance for this effect.

Table II   Effects of beam charging of target on target performance

---

Narrow beam
($\sigma = 0$)

charge state	0	1	2	0	1	2
input energy (MJ)	9.13	9.13	9.13	45.64	45.64	45.64
utilized energy(MJ)	9.13	8.55	6.71	45.64	35.19	8.2
gain	12	12	$\sim 10^{-3}$	4	$\sim 10^{-2}$	$<10^{-2}$

Wide beam
($\sigma = R(0)$)

charge state	0	1	2
input energy(MJ)	12.16	12.16	12.16
utilized energy(MJ)	12.16	11.32	8.64
gain	7	5×10^{-3}	1×10^{-3}

No target design was found which had a gain greater than unity for charge states greater than unity. For charge states > 2, a very large fraction of beam energy is effectively lost, as it to be expected from a comparison of Eqns. (2) and (3) where $P(E) \propto Z_i^2$. We must therefore conclude that charge neutralization is essential for charge states greater than or equal to 2, and indeed is desirable even for charge state unity.

REFERENCES

1. J.P. Christiansen, D.E.T.F. Ashby and K.V. Roberts. Comput. Phys. Common. $\underline{7}$, 271 (1974).

2. T.D. Beynon and E.H. Smith. Proc. 1984 INS Intern. Symp. on Heavy Ion accelerators and their applications to inertial fusion (Inst. for Nuclear Study, University of Tokyo, 1984).

3. T.D. Beynon and D.P. Edwards, J. Phys. D. Appl. Phys. $\underline{19}$, 427 (1986).

4. T.D. Beynon and E.H. Smith, Phys. Lett. $\underline{109A}$, 163 (1985).

5. T.D. Beynon and A.M. Zodiates. To be published.

HEAVY ION FUSION SYSTEMS ASSESSMENT STUDY*

Donald J. Dudziak
Los Alamos National Laboratory
University of California
Los Alamos, New Mexico 87545

and

W. B. Herrmannsfeldt
Stanford Linear Accelerator Center
Stanford University, Stanford, CA 94305

Abstract

The Heavy Ion Fusion Systems Assessment (HIFSA) study was conducted with the specific objective of evaluating the prospects of using induction linac drivers to generate economical electrical power from inertial confinement fusion. The study used algorithmic models of representative components of a fusion system to identify favored areas in the multidimensional parameter space. The results show that cost-of-electricity (COE) projections are comparable to those from other (magnetic) fusion scenarios, at a plant size of 1000 MWe. These results hold over a large area of parameter space, but depend especially on effecting savings in the cost of the accelerator by using ions with a charge-to-mass ratio about three times higher than has been usually assumed. The feasibility of actually realizing such savings has been shown: (1) by experiments showing better-than-previously-assumed transport stability for space charge dominated beams, and (2) by theoretical predictions that the final transport and compression of the pulse to the target pellet, in the expected environment of a reactor chamber, may be sufficiently resistant to instabilities, in particular to streaming instabilities, to enable neutralized beams to successfully propagate to the target. Neutralization is assumed to be required for the higher current pulses that result from the use of the higher charge-to-mass ratio beams. The study was conducted jointly by the Lawrence Berkeley Laboratory, the Lawrence Livermore National Laboratory, and the Los Alamos National Laboratory, and also by the McDonnell Douglas Astronautics Company with funding from the Electric Power Research Institute.

---

* Work supported by the Department of Energy, contract DE-AC03-76SF00515.

Objectives

The Heavy Ion Fusion Systems Assessment (HIFSA) study was organized to deal with a specific premise, and had as its charge, a specific set of objectives. The premise, most directly stated, is that fusion in general, and the Heavy Ion Fusion (HIF) approach to Inertial Confinement Fusion (ICF) in particular, appears to be so costly and requires scaling to such large power plants that it has not been possible to design a program that would be attractive to the electric utility industry.

The most concise statement of the objectives of the study, which is intended to find a solution to this programmatic dilemma, was drafted by the three DOE offices that are funding the HIFSA study:

"Briefly stated, the objective of the study is to perform an assessment of heavy ion inertial fusion systems based on induction accelerators, including representative reactor systems, beam focusing and final transport, target design, and system integration. Emphasis will be given to systems for electric power production and to design innovations and parameter ranges which offer credible promise of reducing system size and cost. No attempt will be made to review heavy ion fusion as a whole, nor current programs, except by inference and in summaries of previous studies. Rather, effort will be concentrated on system and subsystem conceptual design and analysis, including cost/performance models for studying and exhibiting major system parameter variations. Identification of needed R&D will be included. It is expected that the study will be used to guide the direction of future heavy ion fusion programs in the U.S., as well as fill a major gap in current fusion program studies."

Note especially the last requirement, "(to) fill a major gap in current fusion program studies." At each of the two previous symposia in this series results were presented, by the laboratories of the host nation, from comprehensive design studies, HIBALL[1] and HIBLIC[2], respectively. In sharp contrast to the rf accelerator technology featured in both of these studies, the US program[3] has for several years concentrated on the single-pass induction linac approach. It seems incumbent upon us to present a study that will fairly examine the systems aspect of the induction linac as a driver for HIF.

Background

In recent years, various critics[4] have expressed their opinion that, "...even if fusion is found to be technically feasible, at the costs and with the complexities indicated by current estimates, no one would want it." The standard arguments in favor of HIF have always included the economic advantages of high efficiency drivers, the technical simplifications resulting from the separation of driver and reactor, the advantages of the extensive experience with charged particle accelerators, etc. However, the cost of the accelerator system, added to the cost of the reactor, balance of plant (BOP), etc., means a total cost that requires a large power production capacity in order to achieve adequate economy of scale. For example, the HIBALL plant was designed to include four reactor chambers and had a total capacity of nearly 4000 MWe. Even at this size, the cost per kWh of produced electricity was about the same as projected by studies for magnetic fusion[5].

It is likely that the large system studied for HIBALL was, in fact, a result of assumptions in the point design and not just a derived conclusion of the study. In a conceptual study for a point design, the initial design criteria can predetermine the results. An objective of the present study is to find parameters for smaller sized power plants by examining a broad range of parameters to determine the cost implications of new technical innovations that might permit extending the valid parameter space. The logic here is that, unless one can demonstrate the possible advantage of such an extension, it is hard to get anyone interested in studying the problems that it causes.

A second example of the effects of choosing the initial design criteria can be found in a somewhat earlier study by Westinghouse Electric Corporation[6]. Here the potential advantage of a high repetition rate was shown by the results which tended toward lower power costs at the 10 pps limit that the project used for the upper bound for pulse repetition rate for the particular technology that was selected. Because it was clear that power costs more for a lower repetition rate system, one would like to see the result for a higher repetition rate. However, both the accelerator system (an rf accelerator with storage ring current multiplication) and the reactor system (a 10- to 20-m-radius dry wall chamber) were designed for the 10 pps limit.

In contrast to the various point designs, there was one very important systems assessment led by K. A. Brueckner for the Electric Power Research Institute (EPRI)[7]. In this report, Brueckner et al. examined the anticipated cost of electricity for a range of parameters for different drivers. The conclusions, based on the limited technical information available in 1979, was that ion beam drivers are promising candidates for commercial fusion power plants. A much more detailed assessment should be possible now using the new data available from target, reactor, and accelerator studies.

In light of the present economic situation of the utility industry, with nuclear plants being cancelled and virtually all previous projections for future power needs being grossly too high, there is understandably no enthusiasm for large-scale fusion scenarios. Even though any long-range energy forecast will conclude that eventually the world must stop burning vast amounts of fossil fuel and turn to an inexhaustible energy resource, the place of fusion as the preferred power source of the future is certainly not enhanced by these expensive scenarios.

Thus it is incumbent upon proponents of HIF to document the purported advantages of their technology. To make a significant impact, it is necessary to depart from conventional approaches. To reduce the capital cost of a projected plant, which is the largest single stumbling block, the total power rating has to be smaller, and the cost of the accelerator must be reduced.

Parameter Space

There is a very large parameter space available to a systems designer. The usual way of considering a commercial ICF system is to divide it into three parts: the driver, the targets, and the reactor, plus BOP. There are at least two major sub parts, which are, in effect, the interfaces with the first three parts: the beam transport system and the target factory.

The BOP, of course, provides the interface between the reactor and the utility customer. The principal plant performance parameter of interest from the BOP is the thermal-to-electricity conversion efficiency. An important secondary role of the heat exchangers in the BOP is to provide a barrier to prevent diffusion of tritium into the environment. With the exception of the magnetically protected dry-wall concept, no attempt has been made in this study to employ direct, or MHD-type, conversion techniques. The thermal conversion efficiency is principally affected by the temperature of the neutron absorbing material in the reactor wall and by the type of heat exchangers needed.

There have been several ICF reactor concepts studied and reported in varying detail over the last several years. The approach in this study was to choose representative reactors from those available; in particular, those with which the participants in the study were most familiar (usually, the concept they had invented). The risk of significantly biasing the study in this way was offset by the presence of reactor designers from two centers, the Lawrence Livermore National Laboratory (LLNL) and the Los Alamos National Laboratory (LANL), and by employing a systems integration team from the McDonnell Douglas Astronautics Company. As anyone familiar with these laboratories knows, they are known for "keeping each other honest." The reactors that were included in the study were a "CASCADE" type in which a spinning drum holds lithium-based ceramic granules against the outer wall by centrifugal force, a "wetted wall" type in which a thin jet of liquid lithium is kept against the wall of a spherical chamber by the centrifugal force of the jet, a magnetic-field-protected dry wall chamber, which was introduced primarily as a generic high repetition rate example (it being recognized that the magnetic field scheme might interfere with beam transport), and the HYLIFE liquid lithium waterfall reactor. The HYLIFE reactor was introduced as an example of a low repetition rate concept that would require a minimum of 0.5 seconds for clearing the chamber between shots. The dry-wall concept could operate up to about 20 pps while the other two concepts could conceivably operate in the range of 5-10 pulses per second.

The issue of reactor repetition rate is important because of its significance as a systems parameter that directly determines many other parameters. For example, a 1000 MWe power plant might reasonably need ~4000 MW of fusion power, equivalent to 2000 J/shot at 2 pps or 800 J/shot at 5 pps. Obviously, these would be very different plants in almost all respects. Repetition rate can be used to illustrate some of the complexities of a systems study. Among the advantages usually cited for heavy ion induction linacs is the intrinsic ability to operate at any reasonable pulse repetition rate. The repetition rate for any reasonable system is thought to be limited by reactor and target injection requirements, usually by reactor clearing time[8].

A simple (and incorrect) illustration is as follows: suppose one builds a 1 MWe HIF power plant designed for 5 pps. If after tests it turns out that all components will operate as well at 6 pps, then it would appear that the plant could produce 20% more power for the same capital cost, and the cost of electricity will be (almost) 20% less. Superficially this may appear to be correct, but it is wrong because, from a systems standpoint, it is no longer the 1 MWe plant that was designed. Because the specification for BOP equipment was for 1 MWe, if the system is to operate at 6 pps, the per shot yield must be reduced. This implies lower driver energy and lower target gain (because the gain curve is assumed to be a monotonically rising function of driver energy). The result is that the product ηG is reduced (where η is driver efficiency and G is overall fusion gain). However, the lower energy driver costs less assuming, as we have, that the driver is a heavy ion accelerator easily capable of the higher repetition rate. Without knowing the specific dependence of both the ηG function and the driver cost, it is not possible to say whether the increased repetition rate will increase or decrease the COE. It is possible to say that there can be an optimum repetition rate, above or below which the COE is higher. The happy result from this study, which we will examine in the next section, is that the nearly optimum repetition rates lie in a broad range for COE, and are around 5 to 10 pps, where feasible reactor concepts exist. An interesting sidelight is the issue of cost of the driver for higher repetition rate. Some people have expressed concern about higher cost for a higher repetition rate heavy ion accelerator. As we have seen, the higher repetition rate accelerator will cost less for fixed electric power, because it is, in fact, a less powerful machine.

It was recognized quite some time ago that the key to reduced cost for HIF was to reduce the cost of the accelerator. A program known as LIACEP was written at the Lawrence Berkeley Laboratory (LBL) to find optimum design parameters for the induction linacs being studied. In an earlier paper given at the Palaiseau Conference[9], a number of options for reducing the cost of the linac were examined. Several of these, such as increasing space charge limited current by decreasing the allowed minimum betatron tune, were based on the hope that future experiments would confirm the feasibility of the idea. The lower minimum tune is, in fact, one of the important experimental advances of the HIF program.

Study Results

Two important computational tools were developed for the study:

1. The linac optimization program LIACEP was extensively rewritten, and

2. The program ICCOMO was written to permit examination of large areas of commercial plant parameter space to find local optima.

Probably the most important technical results of the study came from re-examining the cost-saving ideas that were in the Palaiseau paper. The report by Lee[10] at this symposium shows how some of these ideas, modified by newer experimental results, make it possible to envision very significant cost reductions by (especially) using higher charge-to-mass ratios. Most of the study was done for $q = +3$, $A = 130$. In fact, the results would be scarcely affected if $A = 200$ were used.

The methods and results from the systems study are extensively reviewed in other papers at this symposium[11-14]. Readers are referred to these papers for the assumptions and methods that were employed. We would like to single out some of the most significant results.

In Fig. 1, we display the results from the study for the "wetted wall" reactor concept. The data plotted are COE vs. repetition rate for five different types of targets. A number of fondly cherished ideas are quickly demolished by this plot:

1. High repetition rate is always better. This turns out to be a misconception because the ηG product suffers at high repetition rate, as was discussed earlier. The result is that the cost of providing for recirculating power, and also the cost of targets, begin to dominate the COE. On the other hand, 3

pps is a lot better than 1 pps, which is the design repetition rate for the lithium waterfall reactor concepts.

2. Symmetric targets, which may use the beam energy more efficiently, can result in lower COE. This concept suffers because of the cost of the required transport system.

3. Higher gain targets are important. Higher gain is always an advantage but in this case it is not a very significant advantage (note the depressed scales). What advantage may exist can disappear quickly if the higher gain targets cost more to produce or need better beam quality. The benefit is small (~5%) because the old standard single shell target still should have an adequate ηG product. Since "Advanced Concept" is a euphemism for "using some untested concepts to improve target performance", it is important to note that such hopes, while potentially useful, are not necessary for competitive COE from HIF.

In Figs. 2 and 3, we display two sets of bar charts showing the "near optimum parameter ranges" for different target concepts and reactor designs. Each bar covers the lowest 5% of COE for that combination. Note that this is for $q = 3$, so the accelerator voltage is reduced by a factor of three compared to an accelerator for $q = 1$. The accelerators are thus much shorter than had been assumed previously, and hence the driver cost is reduced by a factor of about two. The lowest COE results for the granular wall and the wetted wall. Both of these are near optimum in the broad range 3-9 pps and 6-12 GeV.

Finally, in Fig. 4 we display the comparison of cavity types for the single shell target with 16 beams in a two sided illumination scheme. Note that the wetted wall and granular wall types are very close in minimum COE (too close for anyone except their designers to get very excited about). Perhaps here the real message is absolute COE. This study was performed by the McDonnell-Douglas company, under EPRI funding, using methods they had applied previously to magnetic fusion studies. In spite of the requirement for (only) 1000 MWe, the COE is quite competitive with other fusion studies, and with other technologies.

It was recognized long ago that the one HIF accelerator can sequentially drive pellets in several reactors. In HIBALL, four reactors were used. Thus, these results from the HIFSA study penalize HIF by limiting the requirement to 1000 MWe. The study looked at the COE for a 500 MWe plant, and found it ~40% higher. For a 2000 MWe plant the COE is reduced by ~25%. One would not expect anything else (the rules of economy of scale cannot be repealed) but it is encouraging to find that even at 500 MWe, the COE is not out of sight. This result suggests a staged construction plan in which expansion options are designed in from the beginning.

Conclusions and Recommendations

Among the accomplishments of the HIFSA study are:

1. The development of the codes LIACEP and ICCOMO for future optimization of HIF systems is a significant and tangible product from this study.

2. The discovery that broad minima in the COE can be found for most key parameters, including especially the repetition rate in the range 3-9 pps, kinetic energy in the range 6-12 GeV, and ion masses in the range 100-230 AMU.

3. The understanding that potentially major reductions in the cost of induction linacs result from using higher charge-to-mass ratios and multiple beams. The potential savings result directly from the experimental progress made in stable beam transport for intense ion beams in the SBTE and MBE-4 experiments at LBL.

It is worthwhile to note that, although the study was done mostly for $q = 3$, $A = 130$ or $A = 200$, very similar results can be obtained for $q = 1$, $A = 67$. The reasoning is that, while there is good progress with MEVVA sources for multiple-charged heavy ions, it may be that some price must be paid (for example, in higher emittance). With the same electrical current, a beam of $q = 3$, $A = 200$ ions would have the same beam properties as a beam of $q = 1$, $A = 67$ ions, except that the latter would have slightly longer range. The range difference becomes less noticeable at lower kinetic energies, corresponding to the shorter range favored for better target performance. Thus, the accelerator R&D should continue now without necessarily concentrating on how to make a good charge-state-plus-three ion source.

One important conclusion of the study, not discussed here yet is that with the higher currents it is certainly necessary to invoke neutralization during final transport. Work by Stroud[15] gives confidence that streaming instabilities will not destroy the emittance during transport through the target chamber.

One of the principal objectives of this study was to help define future directions for the HIFAR program. We noted that the significant cost savings identified by the study are based on experimental results in the SBTE and MBE-4 experiments at LBL. Both of these are very small scale experiments. It is most important to move into significant beam power and particle velocity, if for no other reason than just to gain more relevant experience. History has taught us to expect new phenomena when key parameters, such as beam power, are extended by orders of magnitude. The LBL group has proposed a machine called

ILSE (Induction Linac System Experiment) which has scaled power and higher particle velocity as its chief goals.

Historically, HIF was considered to be the ion beam approach that could use vacuum transport to hit the pellet, and avoid all the messy complexities of plasma physics. The current understanding of reactor chamber physics, and the use of higher currents (higher charge state, lower kinetic energy) makes this old hope wishful thinking. Beams will neutralize and neutralization must be invoked just to hit the target. The neutralization phenomena must be studied and any possible relevant experiments must be planned. Also, the handling of intense beams in bending and focusing systems must be demonstrated. The high intensities needed at the pellet require longitudinal compression of the pulse as it nears the target. The expectation is that longitudinal space charge forces will control the longitudinal momentum spread, and permit adequate control of chromatic aberrations. This needs verification both by simulation and by experiments. Fortunately, it should be possible to perform relevant experiments at low kinetic energies.

The other areas in which R&D is especially needed have all been known for some time. The cost advantages of multiple beams in the accelerator, for example, are well known, and MBE-4 has demonstrated that at least four beams can be accelerated together. Techniques for instrumenting a multiple beam accelerator are needed for orbit diagnostics and corrections.

The largest number of beams is needed in the low velocity part of the linac. After the injector area, merging of beams can make the magnetic transport system much more economical. Experiments with merging are planned for the ILSE program.

Significant cost savings can be made with good engineering, especially of induction cores and pulsers. Except for the areas noted above (merging and final transport) most of the physics issues for HIF are in hand. What is most needed now is practical experience with the engineering and operation of high intensity systems.

Any list of HIF R&D contains ion source development. Although a good start has been made, much work remains on the 16-beam, 2-megavolt injector that is being built at Los Alamos.

References

1. "HIBALL-II, An Improved Heavy Ion Beam Driven Fusion Reactor Study," Kernforschungszentrum, Karlsruhe, Report KfK-3480; Fusion Power Associates, FPA-84-4; University of Wisconsin, UWFDM-625, July 1985.

2. "HIBLIC-I, Conceptual Design of a Heavy Ion Fusion Reactor," Research Information Center, Institute for Plasma Physics, Nagoya University, Report IPPJ-663, January 1985.

3. R. O. Bangerter (compiler) with Heavy Ion Fusion Staffs at Los Alamos National Laboratory and the Lawrence Berkeley Laboratory, "Heavy Ion Fusion Accelerator Research Program Plan for FY84-FY89," Los Alamos National Laboratory, LA-UR-83-1717 (1983).

4. For example, L. Lidsky, "The Trouble With Fusion," MIT Technology Review, October 1983.

5. C. C. Baker et al., "STARFIRE – A Commercial Fusion Tokamak Power Plant," Argonne National Laboratory, ANL/FDP/80-1, September 1980.

6. E. W. Sucov et al., "Inertial Confinement Fusion Central Station Electric Power Plant," Westinghouse Electric Corporation, Pittsburgh, PA 15236, WFPS-TME-81-001 (1981).

7. K. A. Brueckner et al., "Assessment of Drivers and Reactors for Inertial Confinement Fusion," Electric Power Research Institute, Palo Alto, CA 94304, EPRI AP-1371 (1980).

8. R. Peterson, "Repetition Rates in Heavy Ion Driver Fusion Reactors," in these proceedings.

9. A. Faltens, E. Hoyer and D. Keefe, "A 3-megajoule Heavy Ion Fusion Driver," Proc. IV International Topical Conference on High Power Electron and Ion Beam Research and Technology, Palaiseau, France (1981).

10. E. Lee, "Induction Accelerator Model for HIF Systems," in these proceedings.

11. J. Hovingh et al., "The Cost of Induction Linac Drivers for Inertial Fusion for Various Target Yields," in these proceedings.

12. D. Zuckerman et al., "Performance and Cost Modelling of a Linac Driver HIF Power Plant," in these proceedings.

13. D. Driemeyer et al., "Influence of Systems Optimization Considerations on LIA Drivers," in these proceedings.

14. L. Waganer et al., "Survey of Systems Options for Heavy Ion Fusion," in these proceedings.

15. P. Stroud, "Heavy Ion Beam Transport in ICF Reaction Chambers," in these proceedings.

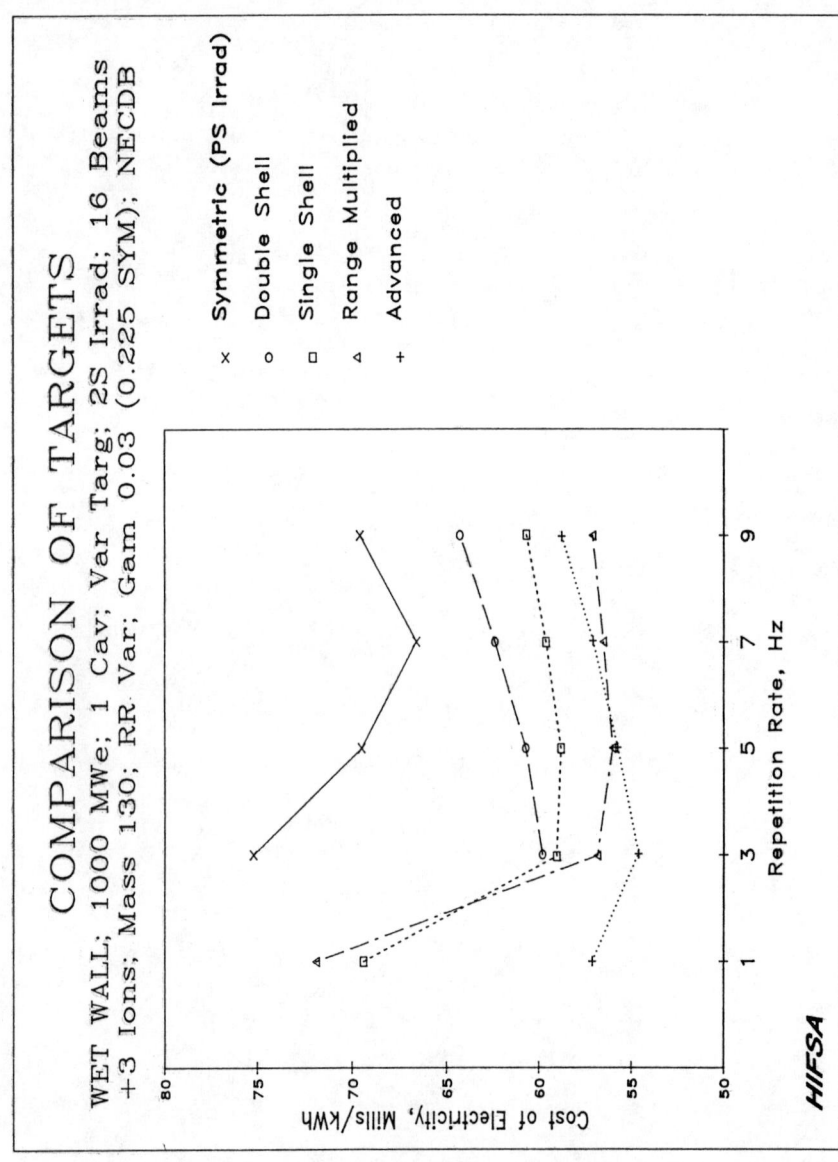

Fig. 1. COE as a function of repetition rate for different target types. The wetted wall chamber is assumed and 16 beams are used in a two sided illumination scheme.

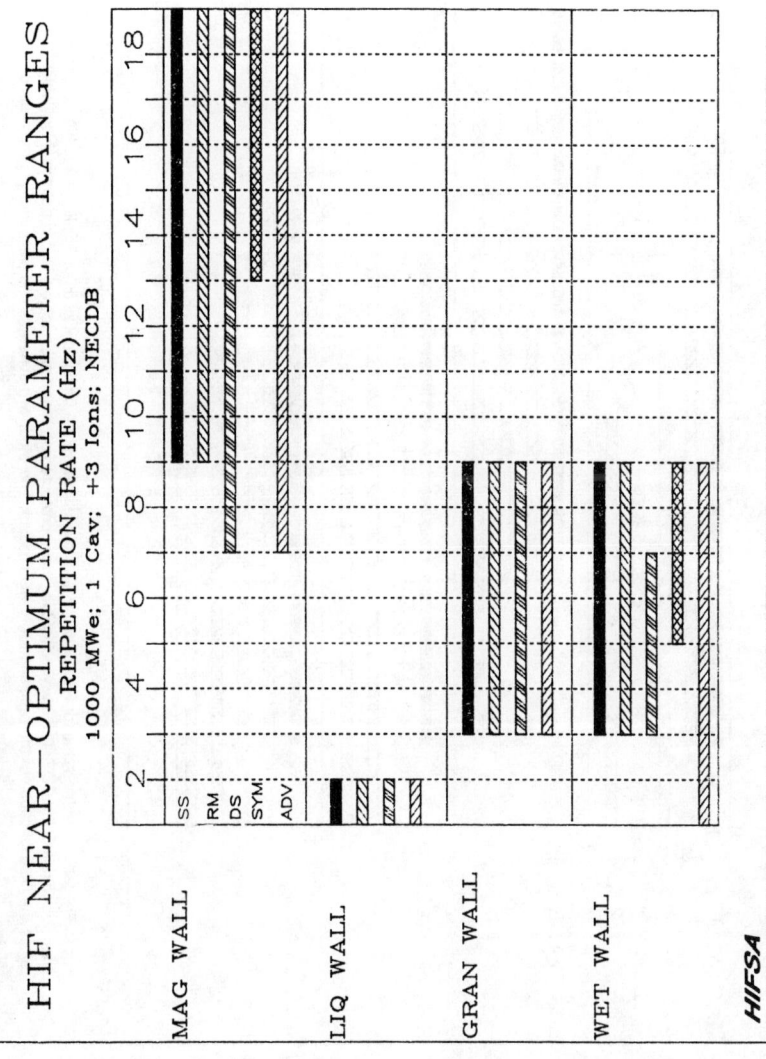

Fig. 2. The "near optimum ranges" are shown for COE within 5% of the minimum for each combination of parameters. The lowest power costs are for the wetted wall and granular wall. The bars refer to target types: SS for single shell, RM for range multiplied, DS for double shell, SYM for symmetric illumination (which cannot be used in the granular wall concept) and ADV for advanced design.

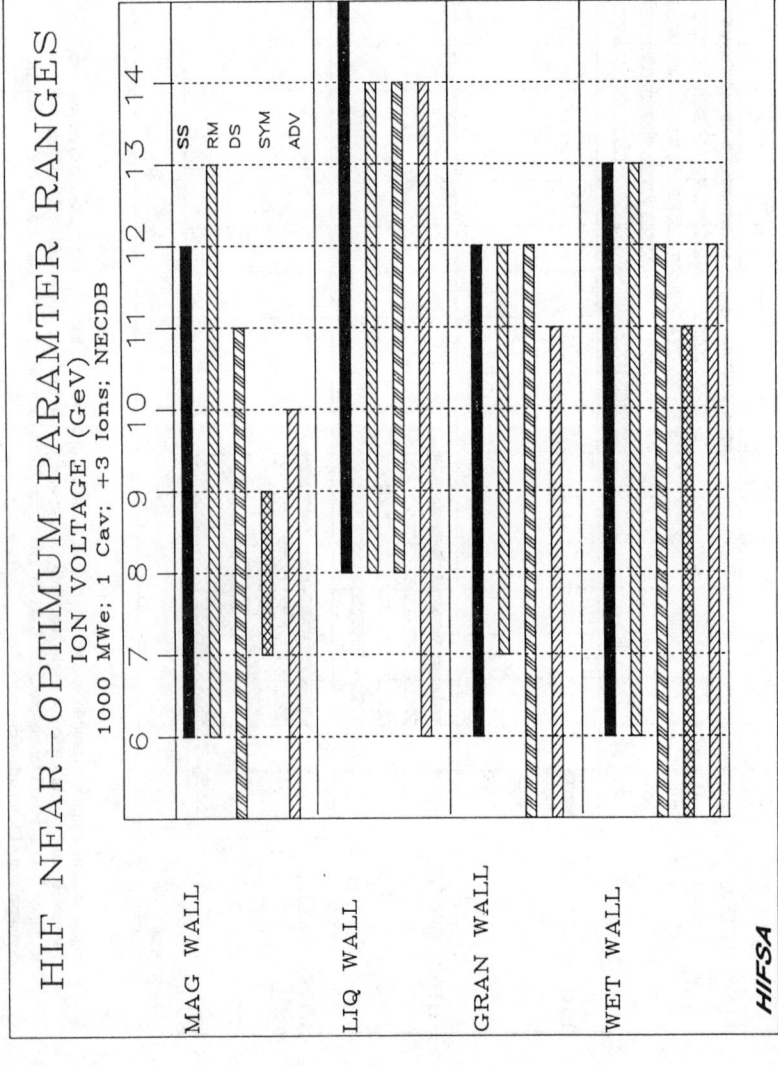

Fig. 3. The labels are the same for Figs. 2 and 3. The low repetition rate for liquid wall (HYLIFE) limits it to high beam energy, and raises the COE. The low yield limit for the unprotected dry wall limits it to very high repetition rates, which raises the COE because of the resulting lower ηG product. The other two schemes find the middle of the optimum range.

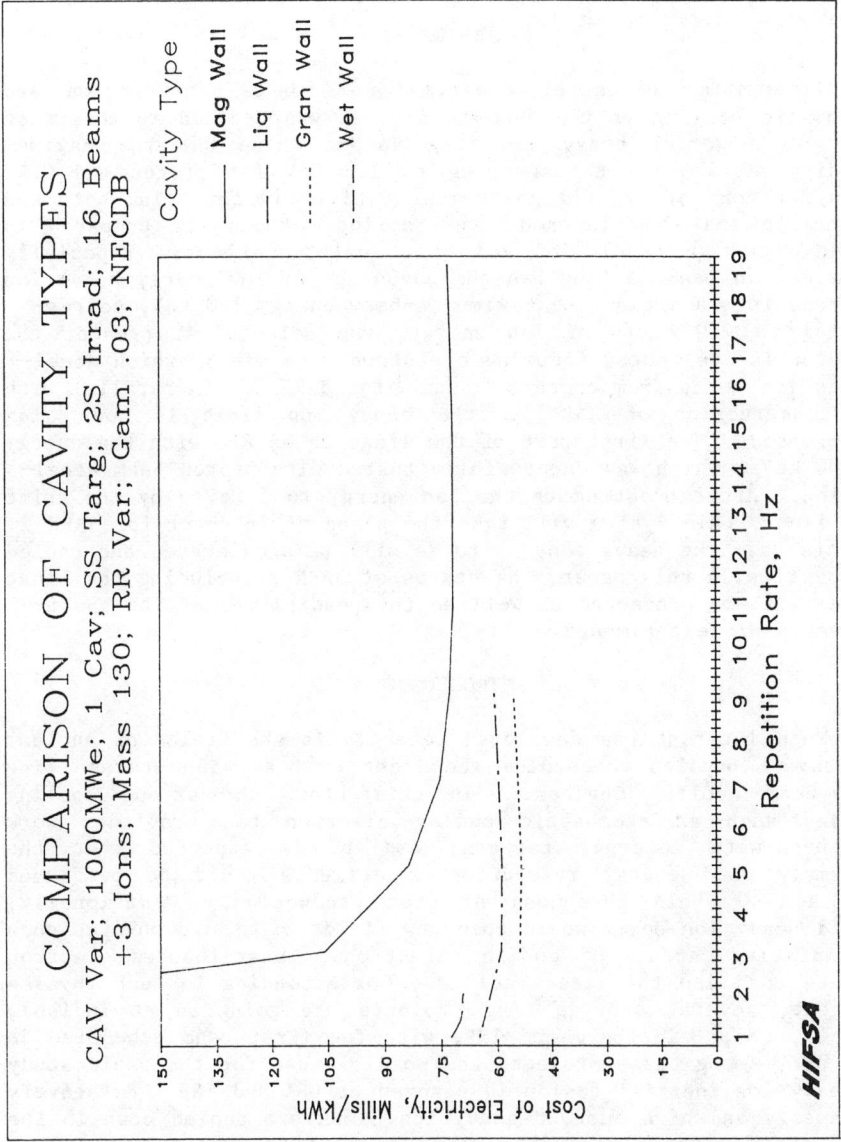

Fig. 4. The COE is compared for different cavity types. The conclusion is that optimum repetition rates lie in the range 3–9 pps, and that more work should be done on reactors capable of such rates.

HEAVY ION ACCELERATION AND COOLING RING TARN 2

Takeshi Katayama
Institute for Nuclear Study, University of Tokyo
3-2-1 Midoricho, Tanashi, Tokyo 188, Japan

ABSTRACT

After the successful experiments of beam accumulation and stochastic cooling at the TARN-1 ring, it was decided to construct the more powerful heavy ion ring TARN-2 which has the maximum rigidity of 7 T-m, corresponding to 1.3 GeV for proton and 0.45 GeV/u for ions of 1/2 charge-to-mass ratio. In this ring both the stochastic and the electron beam cooling methods are prepared to obtain the high resolution and small emittance beam. Especially the electron beam cooling has the advantage of low energy heavy ion beam cooling, and then the maximum e-beam energy 120 keV, corresponding to 200 MeV/u of ion energy, was selected. At present the injector is the sector focusing cyclotron with K = 67 which accelerates ion beams from protons to neon for TARN 2. In parallel with the construction of TARN 2, the heavy ion linac is now being constructed. The first part of the linac is an RFQ with the energy of 800 keV/u which was successfully tested with proton beam acceleration. After boosting up the ion energy to 5 MeV/u by the drift tube linac, this system will take the place of injector for TARN 2. In this case the heavy ions up to Xe will be accelerated and cooled in the ring. In this paper, the status of TARN 2 including the linac system will be presented as well as the possibility of its application to HIF experiments.

INTRODUCTION

Much interest has developed recently in the fields of nuclear and atomic physics concerning the light or heavy ion storage ring with beam cooling devices. In this ring, the strong cooling devices such as stochastic and/or electron beam cooling, work together with internal targets, and it is expected that the extremely good energy resolution experiments would be performed such as threshold phenomena of pion production. Additionally, cooled heavy ion beams would open new fields of atomic physics such as radiative capture of cooling electrons, laser-induced electron capture and also the laser cooling. Corresponding to such physics interest, several storage ring projects are going on at Indiana, Uppsala, GSI, Heidelberg and INS, with the first beam scheduled in 1986-89. Among these projects the possible use for the basic study of heavy ion inertial fusion is planned at GSI and INS. Relatively low energy and high current heavy ion beams are cooled down in the ring, their emittance and momentum spread being reduced, and then extracted with short pulse length. Extracted heavy ions will hit

The target and a high-temperature, 10eV plasma will be created. The behavior of ion beams in such a high-temperature medium, is a key issue which should be studied in the inertial fusion program with heavy ions.

TARN 2 is currently under construction and is scheduled for completion in 1987. The goals of the project are, firstly, to boost up the maximum beam energy to several hundred MeV/u, secondly, to cool down the beam temperature in three phase spaces and, thirdly, to perform the nuclear and atomic physics as well as the application to inertial fusion.

INJECTOR

In the present scheme, the injector of TARN 2 is a SF cyclotron with a K number of 67. This cyclotron can accelerate various kinds of ions from light ions like p, α, to heavy ions like Fe^{8+}. However, due to the restriction of the internal ion source, the charge state of heavy ions is low and the output energy is correspondingly quite low. Among these heavy ions, Ne^{4+} will be the heaviest with a reasonable current of 1 μA and an energy of 2.6 MeV/u, which is adequate for acceleration in TARN 2. In Table 1, the heavier ions are listed with the available intensities and energies from the cyclotron.

On the other hand, we are now constructing the heavy ion linear accelerator system which has a final output energy of 5 MeV/u; the first part of this system is now completed with an energy of 0.8 MeV/u.[1] It is the four-vane type RFQ linac with the name of TALL. The RFQ linac focuses the charged particles by RF quadrupole fields excited with four electrodes (or vanes) and accelerates them by the axial component generated with the scalloped modulation of the vane tip. This structure is very effective for the acceleration of high intensity beams in the low energy region.

Design parameters of TALL are given in Table 2. The RF system operates at the resonant frequency of 100 MHz and is driven with a single-loop coupler. The beam injection energy is 8 keV/u and the output energy is 800 keV/u, the total length being 725 cm. So far H^+ and H_2^+ beams were successfully accelerated, the measured output energy being 825 keV.[2] The energy spread ΔT/T was measured at 1.6% (FWHM) and a transmission efficiency exceeding 90% was obtained for the H^+ beam of 10 μA. A view of the linac is given in Fig. 1.

TARN 2 is installed in the new accelerator hall which became available by clearing up the old experimental hall of the FM cyclotron. (Fig. 2) Ions from the SF cyclotron are transported through the beam line, and at the stripper section located just prior to the analyzer magnet in the line, the orbital electrons of

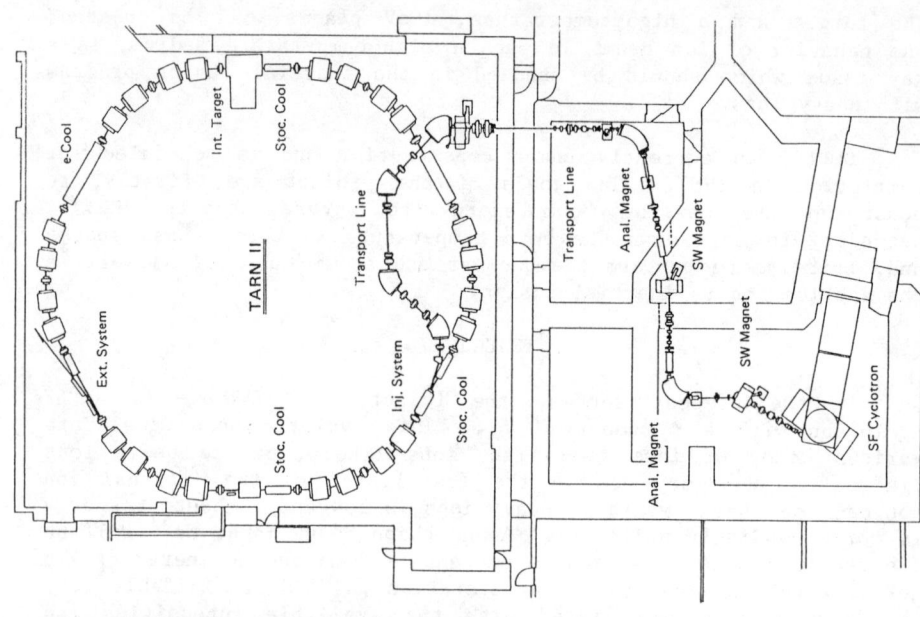

Fig. 2. Layout of TARN2 ring and beam transport line from SF cyclotron.

Fig. 1. View of heavy ion RFQ linac TALL

partially stripped ions are completely taken off. Then ions are injected in TARN 2 by multiturn injection. For heavier ions the injection energy is different for each ion species as is given in Table 1, whereas the proton injection energy will be 20 MeV. During the process of passing through the thin carbon foil with a density of 50 µg/cm^2 at the stripper section, the beam quality will be degraded; for example, the emittance will increase due to the multiple scattering, and the energy spread will be enlarged by the straggling effects. As a most serious case, the Ne^{4+} beam of 2.6 MeV/u is examined which shows that the emittance of 20π mm-mrad of the beam from the cyclotron will be increased to 35π mm-mrad, and the energy straggling will be around 1×10^{-3}. The fraction of fully stripped ions is estimated at one third of the beam.

The cyclotron is essentially a CW machine, and not having a high peak current it does not fit as an injector for a pulsed accelerator such as a synchotron. The peak current at the injection point is around 1 pµA (p, α) and 0.1 pµA (heavier ions) after passing through the beam transport line with magnetic analyzer system. The momentum spread of the injected beam is around ±0.1% which comes from the narrow phase spread of 2 degrees in the RF acceleration field of the AVF cyclotron. The horizontal and vertical emittances are 10π mm-mrad for p, d, α and 35 π mm-mrad for heavy ions. The acceptance of TARN 2 is designed at 400π mm-mrad, and the dilution factor during the process of multiturn injection is assumed at 2.5. The expected beam intensities are around 1×10^8 for p, α, d and 1×10^6 for heavy ions. However, if one uses both the horizontal and longitudinal phase spaces, the expected intensity will be increased by the order of two. This is mainly due to the fact that the AVF cyclotron beam has a very small longitudinal emittance of $\varepsilon_\ell = \Delta\phi(\Delta T/T) \approx 5 \times 10^{-4}$ rad.

Beam life time at the injection energy is mainly determined, for heavy ions by the charge capturing process of fully stripped ions through collisions with the residual gas and for the light ions, such as protons, by Rutherford scattering. Assuming the vacuum pressure in the ring as 1×10^{-10} Torr, the beam life is estimated as follows: 3300 sec for protons (20 MeV), 760 sec for C^{6+} (7.6 MeV/u), and 12 sec for Ne^{10+} (2.7 MeV/u). From this estimate of beam life at the injection energy, it is expected that there is enough time for beam manipulation, such as RF stacking or fast stochastic cooling, even at the flat-base injection period.

GENERAL DESCRIPTION OF THE RING

The ring will be used in three modes of operation: 1) normal synchrotron operation, 2) long spill operation, stretcher mode, and 3) cooling-ring mode. In the synchrotron mode, the repetition cycle is 0.5 Hz with the acceleration period of 0.75 sec, the flat-top of 0.5 sec and the falling period of 0.75 sec. This repetition rate is determined mainly due to the available power at the present

INS electric station. In the stretcher mode the acceleration period will be around 10 sec, whereas the flat-top will be long enough, say 1 hour for 500 MeV protons, which should be a good compromise between the beam life and the ultra-slow ejection method such as stochastic extraction. In the cooling-ring mode, the operation scheme will be the same for the stretcher cycle while the strong beam cooling devices will be operated as well as the internal gas jet target.

A set of lattice parameters for the stretcher and cooling modes is given in Table 3, and the lattice functions are shown in Fig. 3. The circumference of 69.908 m is the maximum ring size that fits into the new accelerator hall. A symmetric, three-period lattice with the six long straight sections, each 4 m long, was adopted. Hence there are three dispersion-free, straight sections and three large dispersion sections which is adequate for beam cooling and for the internal target experiments. In the dispersion-free, straight section, an electron-cooling device, stochastic cooling kickers and an RF accelerating cavity will be installed, while the large dispersion sections are prepared for stochastic cooling pickups, the internal target system and the electric inflector for the beam injection.

In the renovation process of TARN 1, all of the dipole magnets are rebuilt, while the quadrupole magnets being used in the TARN 1 ring will be used again for TARN 2. The magnetic structure of the new ring is made up of 24 dipole magnets and 18 quadrupole magnets. Each dipole magnet is an H type structure with a straight core length of 1m. The edge shape at the end of the yoke is approximately a Rogowski curve cutting the yoke in four steps. The designed field region is 200 mm in width and 60 mm in vertical direction. In order to realize the large good field region for the wide excitation range up to 18 kG, the side pole edges are shaped to give a constant B curve. Also, small shims are attached to suppress the decrease of the magnetic field at the pole ends until the field is saturated.

An RF system will accelerate the ions from the injection energy to the desired working energy. The lowest injection energy among the various ions from the SF cyclotron, is 2.58 MeV/u for Ne^{4+} corresponding to the revolution frequency of 0.307 MHz. At the top energy of 500 MeV/u, the revolution frequency is 3MHz and the RF frequency ratio of the initial and final stages is ten. The harmonic number is chosen as two and the designed acceleration frequency changes from 0.6 MHz to 6 MHz. An RF voltage of 6 kV seems adequate for the acceleration of the beam with the momentum spread of 0.5% within the acceleration period of 0.75 sec. This RF voltage is produced using a cavity loaded with ferrite 2.5 m long (Fig. 4). In this cavity, the resonance frequency has been successfully varied by a factor of 13 with the change of ferrite bias current from 0 to 750A.

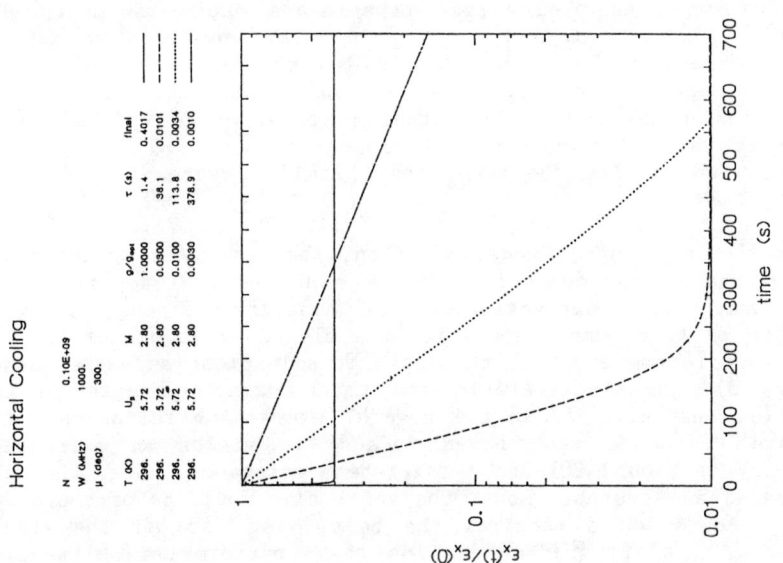

Fig. 5. Calculated beam emittance during the stochastic cooling.

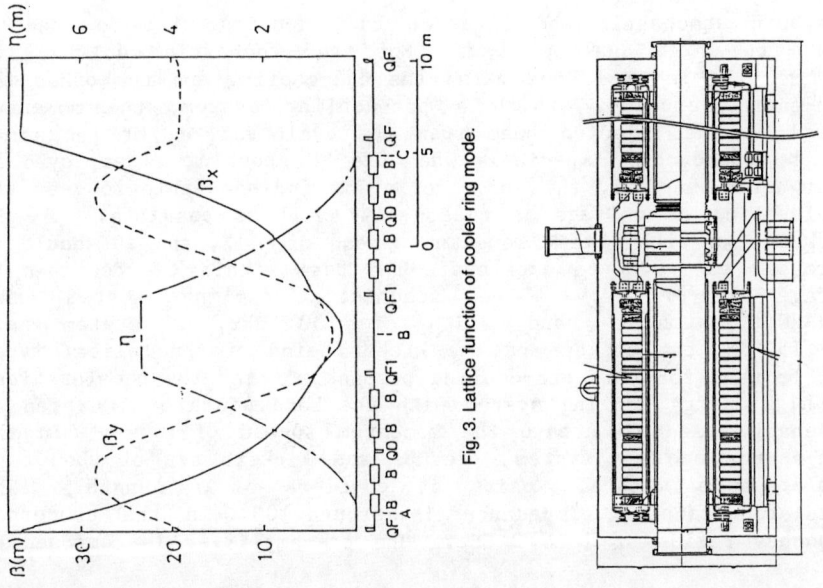

Fig. 3. Lattice function of cooler ring mode.

The residual gas has several effects on beams circulating in TARN 2, namely, 1) the charge-capturing process of fully stripped heavy ions leads to beam loss, 2) multiple Coulomb scattering of light ions determines the beam life in the ring, 3) contribution to background when the jet target experiments are performed. The estimation of the beam life at the injection energy shows that a vacuum pressure of 10^{-11} Torr should be achieved. In TARN 1, an all-metal vacuum system bakable at $200°C$ was used with eight sputter ion pumps and eight titanium getter pumps. The normal operating vacuum pressure of better than 1×10^{-10} Torr was achieved with storing the beam, and a similar system will be used also for TARN 2.

In the synchrotron mode operation, the beam will be extracted with the use of the one-third resonance at the flat-top period of 0.5 second. The extraction system consists of the following elements: 1) four bump magnets for the closed orbit distortion, 2) four sextupole magnets as a chromaticity adjustment and a resonance excitor, 3) one electrostatic and three magnetic septa in the extraction channel. Among the several extraction resonances, the one-third resonance was chosen because it yields an extraction efficiency of around 90% and a small beam emittance. When the ring is operated as stretcher mode, the spill time would be of the order of 100 seconds which requires the beam being far off the linear resonances to avoid sudden beam loss. To perform this ultra-slow ejection, the stochastic extraction system practiced at the LEAR facility at CERN will be used in combination with the stochastic cooling system and the normal extraction equipment.

BEAM COOLING

Both stochastic and electron beam cooling will be used to obtain the high quality beam. For the stochastic cooling, two systems will be used, one being the pre-cooling and the other the high-energy cooling. With a pre-cooling system, the momentum spread of the injected beam from the cyclotron and/or the linac will be improved. Especially when the RF stacking is employed as an injection method, this pre-cooling is indispensable to keep the accelerating RF voltage as reasonably small as possible. The RF stacked beam will have a momentum spread of $\sim 1\%$, and it should be decreased to $\sim 0.2\%$ to accelerate the beam within the designed RF voltage of 6 kV. The stochastic cooling system used at TARN 1, with a band width of 100 MHz, a system gain of ~ 150 dB, the pickup and the kicker being of the helical type, will be used for this pre-cooling purpose. After the acceleration, the high-energy cooling system with the band width of 1GHz can be reasonably used to attain the momentum spread of 10^{-4}. In the high-energy cooling system, pickups and kickers are of the loop-coupler type with 16 pairs of couplers of $\lambda/4$ length. The calculated coupling impedance is around 100 Ω in the concerned frequency region of 1 to 2 GHz. The pre-amplifier is composed of

Ga As, the field effect transistor is cooled down to the temperature of liquid nitrogen. The expected noise figure (NF) is around 0.5 dB. With these parameters the optimum or fastest cooling time is 33 msec when the system gain should be 197 dB with the unrealistic power of 2.5 GWatt. With the reduction of gain to 137 dB, the cooling time increases up to 33 seconds, with the power of 2.5 kW which would be a good compromise between the cooling time and needed RF power. In this case, the rms momentum spread is expected to be 5×10^{-5}.

The variation of horizontal beam emittance $\varepsilon(t)$ is calculated and is given in Fig. 5 where the normalized beam emittance $\varepsilon(t)/\varepsilon(0)$, $\varepsilon(0)$ being the initial emittance, is given with time for various system gains. With the reduction of system gain from the fastest cooling gain, g_{opt}, by 30 dB, the final $\varepsilon(t)/\varepsilon(0)$ is 10^{-3} in spite of the long cooling time of 380 seconds.

Electron cooling is most effective at lower energies, say 100 MeV/u, and for beams which are already relatively cool. It can thus complement the stochastic system, especially in the experiments with the internal circulating beam where the momentum spread is 10^{-5} and the beam size smaller than one cm resulting from an equilibrium between the cooling and the heating through the internal target.

The main parameters of the electron cooling system are listed in Table 4. The system is designed to cool down the ions from H^+ to Ne^{10+} up to 200 MeV/u, limited by the maximum electron energy of 120 keV. The electron energy is variable from 12 to 120 keV, and the maximum current density 0.5 A/cm^2 is available at voltages higher than 60 keV whereas the current at the lower collector voltage is determind by the perveance. The length of the interaction region is 1.8 m which is limited by the length of the straight section (4 m). As the beam size at the cooling section after the acceleration is less than 50 mm, the cathode diameter is designed to be 50 mm, which is a type of flat cathode rather than a spherical one. The cathode is immersed in the uniform solenoidal field having a maximum field strength of 1 kG. An example of electron trajectories in the region of the electron gun is given in Fig. 6, where the collector voltage is 110 keV, the electron current is 10 A, and the perveance is 0.688 $\mu A/V^{3/2}$.
The transversal electron temperature is assumed in this case to be less than 1eV. A layout of the electron cooling device, the so-called U scheme, is shown in Fig. 7, where the electron beam is injected and ejected over the beam line of the ring.[4]

The electron cooling process has been simulated[5] with the help of the CERN program SPEC. A typical example of the time evolution of the ion distribution in the (X,Y) space and (X, $\Delta p/p$) space is given in Fig. 8. As can be seen, a drastic reduction of beam size

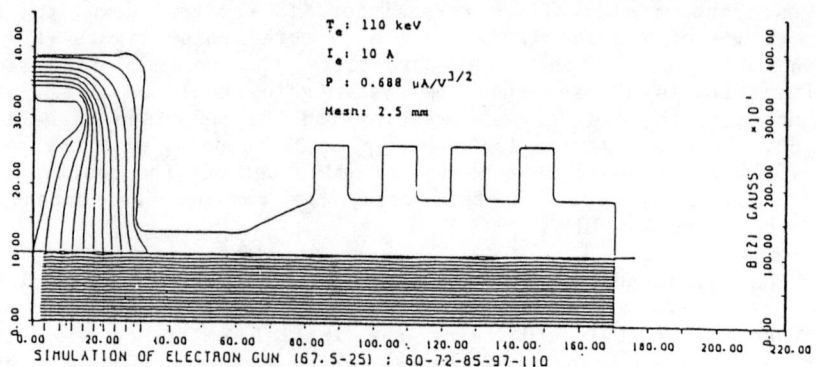

Fig. 6. Calculated electron trajectories in the gun.

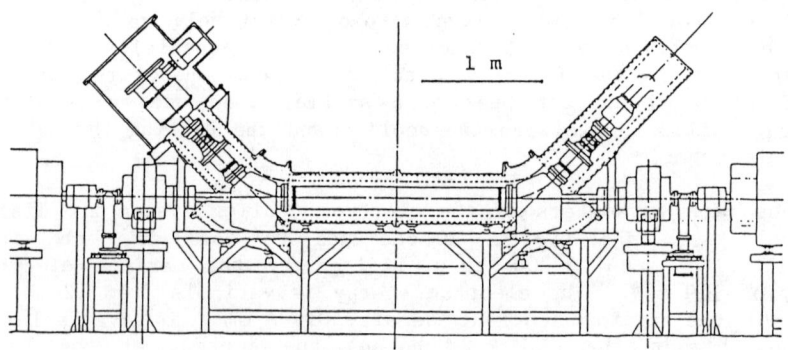

Fig. 7. Layout of the electron cooling device.

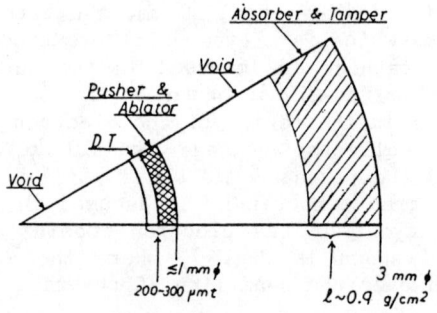

Fig. 9. An example of target for HIF high temperature experiment.

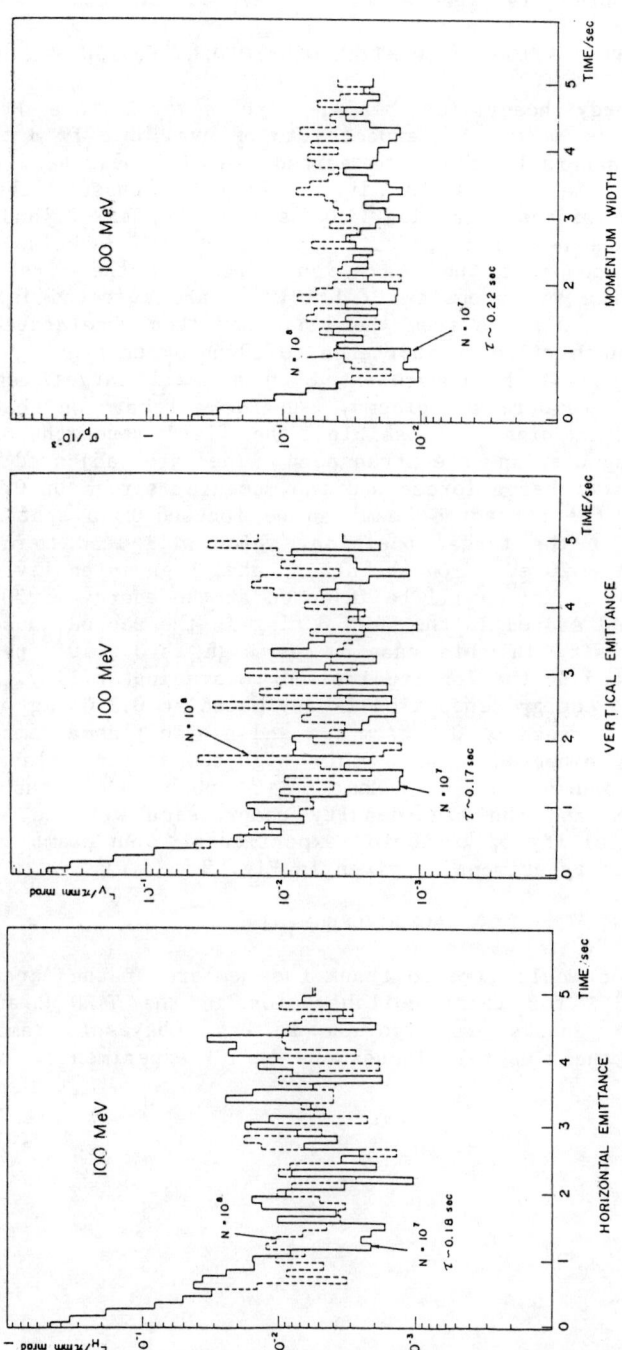

Fig. 8. Time dependence of emittance and momentum spread for different number of stored ions.(ref. 5)

and momentum spread is expected within several seconds.

APPLICATION TO INERTIAL CONFINEMENT FUSION

High energy heavy ion beams, with several tens MeV/u to several hundreds MeV/u, are expected to be available from the TARN 2 ring, and naturally they are suited for ICF near-term physics experiments. The project is divided into two phases, the first being composed of the heavy ion linac system and TARN 2, and in the second phase the new ring, tentatively called TARN 3, being planned to store and compress the heavy ion beam. In the first phase, Ar^{18+} ions with an intensity of 1×10^{11} are injected in TARN 2 at the energy of 5 MeV/u from the linac and then accelerated to 75 MeV/u. The bunch width is shortened to 10 ns at that energy and is fastly extracted to be concentrated on a small target region to produce a high temperature plasma. To focus a beam on the target with as small a size as possible, the field strengths of four quadrupole magnets in the transport line are adjusted taking account of space charge forces and the momentum spread of 0.1%. It is found that the extracted beam can be focused to a spot of 0.15 mm in radius at the target position, which will lead to a plasma temperature of ~ 20 eV. On the other hand, high intensity heavier ion beams such as Kr^{36+} will be injected at the energy of 300 MeV/u from TARN 2 and stored in the TARN 3 ring in the second phase. The expected intensity in this case is as high as 3×10^{13} particles which is limited by the longitudinal microwave instability. In the Au target, the energy deposition is estimated at 0.5 TW/mg with the assumption of a range of 0.9 g/cm^2, and a high plasma temperature of 100 eV is expected. With this high-temperature plasma, the ablation pressure of 10 Mbar can push the pusher of mass 10^{-2} g/cm^2 and the high-density compression will be achieved with the possibility of implosion experiments.[6] An example of the target for this experiment is given in Fig. 9.

ACKNOWLEDGMENT

The author would like to thank the members of the Accelerator Division at INS for their collaboration in the TARN 2 and TALL construction. Thanks are also due to Drs. Obayashi, Yamaki and Nishihara for their useful discussions on ICF experiments.

REFERENCES

1. N. Ueda et al., IEEE Trans. on Nuclear Sci., Vol. NS-32, No. 5, 3178 (1985).
2. N. Ueda et al., elsewhere in this Proc..
3. N. Tokuda et al., IEEE Trans. on Nuclear Sci., NS-32, Vol. 5, 2415 (1985).
4. T. Tanabe et al., Proc. of the Workshop on Electron Cooling and Related Applications, KfK, Karlsruhe, 343 (1984).
5. T. Tanabe et al., INS-T-454, 1986.
6. K. Nishihara, private communication.

Experiments on the ISIS Synchrotron

G H Rees

Rutherford Appleton Laboratory, Chilton, Didcot, Oxon, OX11 0QX, UK.

ABSTRACT

Injection and acceleration studies on the ISIS synchrotron are described. Though HIF related experiments have not been undertaken on ISIS, aspects of machine performance relevant for HIF are discussed, such as coasting beam instabilities and bunch formation. Recently, some rapid growth of beam has been observed shortly after injection which might be due to the longitudinal microwave instability. It is planned to study this topic as a HIF related experiment in the Fall of 1986, while operating the synchrotron as a 70.4 MeV proton storage ring.

INTRODUCTION

The latest performance of the ISIS synchrotron is 40 μA average proton current on the neutron target, corresponding to 5×10^{12} protons per pulse at 50 Hz, which is approximately 25% of the design intensity. The present limits are set both by the injector performance and by the synchrotron beam loading. At lower repetition frequencies, beams of 10^{13} protons have been injected and beams of 6×10^{12} accelerated. These intensities are well above the threshold levels for longitudinal and transverse instabilities but, until the recent observation of some rapid beam growth shortly after injection, no instabilities have been observed.

The initial continuous circulating beam is bunched by appropriate accelerating fields of harmonic number two. The RF voltage per turn is low at the start of acceleration, and it is necessary to control the heavy beam loading by the use of beam feed-forward techniques. The system is adequate for beams of 3×10^{12} protons per bunch, but modifications are required for higher intensities. Due to the rapid-cycling of the synchrotron guide field, the acceleration is non-adiabatic with filamentation present and the development of non-equilibrium bunch distributions. Complex bunch shapes are formed which are intensity dependent due to the effects of the longitudinal space charge forces. A one dimensional longitudinal space charge code has been used to compare the observed and predicted bunch formation.

As the intensity is increased, the bunch shapes become smoother, though the characteristic double-humped shape remains, with more complexity at parts of the synchrotron cycle. In the future it will be necessary to program the RF voltage waveform carefully to improve the beam bunching factor and thus reduce the transverse space charge forces. No space charge induced betatron resonance effects have been observed yet.

INJECTION STUDIES

Typical performance figures for the 70.5 MeV H$^-$ linac injector have been output currents of 5 mA for pulse lengths up to 200 μs at 25 and 50 Hz and currents of 3.5 mA for pulse lengths up to 450 μs at lower repetition frequencies. These figures are well below the design specification and are set by the H$^-$ ion source output and by the fact that, as the average current is raised, there is increased frequency of breakdown of the 665 kV accelerating column. The mechanism of breakdown is under study and the present performance has been achieved only after improved pumping at the high voltage end of the column and after installing inter-electrode shields in the column to intercept the ions before they reach the glass insulators. A new ion source is also under development.

The output beam emittances from the 4-tank, 202.5 MHz, Alvarez linac have been as expected with 95% of the beam within transverse, un-normalised emittances of 20 π μrad-m. Momentum spread measurements in the injection beam line have indicated 95% of the beam within $\Delta p/p$ values of $\pm 1.2 \times 10^{-3}$ and a debuncher cavity is used to reduce the spread to approximately $\pm 5 \times 10^{-4}$. At beam currents of 5 mA, the spread then increases to $\pm 7 \times 10^{-4}$ due to the longitudinal space charge forces within the microbunches as they debunch to a continuous beam. The value for $\Delta p/p$ of $\pm 7 \times 10^{-4}$ is a prediction and it has not yet been measured within the synchrotron.

A schematic diagram of the injection straight of the synchrotron is shown in Figure 1. It is approximately 5 m in length and houses four septum-type dipole magnets for creating a localised bump of the closed orbit. The first of the bump magnets lies adjacent to a H$^-$ injection septum magnet, and the region between the two central magnets is used to house the foil which strips H$^-$ ions to protons. Large aluminium oxide stripping foils have been developed within the laboratory and have proved highly satisfactory in operation. They have a thickness of 0.25 microns.

Up to 300 turns have been injected into the synchrotron with high efficiency by filling horizontal and vertical phase space after H$^-$ stripping. Over 98% of the input H$^-$ ions are stripped to protons and over 1½% to H° particles. There is a separation of the H° beam from the protons after passage through the injection bump magnet just downstream of the foil. A non-destructive monitor of the injected beam is obtained by using an internal scintillator with an external TV camera to view the separated H° beam. Fluctuations of the injected beam are readily seen on this monitor. A second scintillator is used for vertical alignment of the injected beam. It may be lowered to half-fill the aperture so that it may intercept protons on injection and again after one revolution of the ring.

High injection efficiencies have been obtained routinely, and it is only at the end of the last run that detailed studies have been made of the small injection loss. Signals have been compared from beam current toroid monitors at the end of the injection beam line and in the synchrotron ring. In addition, signals have been studied from all the ionisation-chamber radiation monitors that are located inside the ring. Steering of the injected beam may be adjusted to obtain injection efficiencies up to 96.5%, where approximately 2% of the loss is the stripping loss. By viewing the radiation loss monitors, all the loss may be identified as occurring in the injection line or in the injection straight section.

In Figure 2 some features of the injection loss may be seen. The time base is 200 μs cm^{-1}, the injection interval is 90 μs and the injected beam is 3 x 10^{12} protons. The upper trace is for the radiation monitor adjacent to the injection straight, the centre trace for the monitor near the end of the injection line and the lower trace is the upper trace repeated but with the injection orbit bump switched off 150 μs earlier. The initial loss is mainly the stripping loss, but loss continues after injection, decreases and then increases again. This late loss occurs even though the equilibrium orbit in the machine is moving away from the injection septum magnet. The loss ceases once the orbit bump is reduced. For the upper trace, the total loss corresponds to 4.5%, with 2% stripping loss and 2.5% elsewhere in the injection straight. For the lower trace, the loss figure is reduced to about 3.5% by the early switch-off of the bump magnets. The late loss is thought to correspond to horizontal growth of $\sim$ 5 mm in a time of $\sim$ 100 μs.

The injection studies have highlighted a mechanical design error at the point where the injector beam line vacuum pipe merges with the synchrotron vacuum chamber, creating the equivalent of an enhanced septum thickness. This has led to the small reduction in injection efficiency but, more importantly, it has led to the requirement of an associated mis-steering of the H$^-$ beam ahead of the injection septum magnet. This feature is to be corrected later in the year but, in the meantime, it allows observations to continue of the fast growth of beam just after injection.

Further studies are needed to identify the growth mechanism. Envisaged are studies of the growth as a function of beam intensity, Δp/p, Q-values and injection timing. By varying the timing, the rate at which the equilibrium orbit spirals away from the septum magnet may be adjusted. It is important to establish whether growth is due to a betatron resonance or to the microwave instability.

LONGITUDINAL MICROWAVE INSTABILITY

At injection in ISIS, the space charge component of the longitudinal coupling impedance is high, with $Z/n > |-j\ 700|\ \Omega$, and the momentum spread of the circulating beam is $\Delta p/p \sim \pm 7 \times 10^{-4}$. Beams of a few times 10^{12} protons are then above the threshold for the microwave instability. The predicted growth times for modes near 200 MHz are of order 100 μs if the resistive component of Z/n at 200 MHz is assumed to be $\sim$ 10 Ω. The theory of the effect includes a number of approximations and, though it predicts the onset of the instability, it does not predict how the instability develops. Computations[1] have indicated that a tail develops in the momentum distribution which inhibits further growth. Thus it is of much interest to establish whether the microwave instability is the cause of the fast radial growth in ISIS after injection.

The observed radial growth of 5 mm, if entirely due to a

negative momentum tail, corresponds to a $\Delta p/p$ tail of $- 2.5 \times 10^{-3}$, which may be compared with the initial $\Delta p/p$ value of $\pm 7 \times 10^{-4}$. Measurements of the subsequent trapping in ISIS indicate that the core of the beam momentum distribution is not significantly enhanced. However, this aspect needs to be confirmed as the intensity is increased above the present levels. The unstable mode most likely to be observed is the growth of the residual linac bunch structure at 202.5 MHz; this has not been confirmed due to direct pick-up of the linac frequency on the monitors. It is planned to repeat the experiment while operating ISIS as a 70.4 MeV proton storage ring when the measurements should be easier and the effect of varying the mean proton revolution frequency may be studied.

BUNCH FORMATION

The RF system is switched on 145 μs before the guide field minimum (T = 0) and is held at constant frequency and constant volts/turn until T = 0. Subsequently, the frequency is raised to keep the beam centred in the aperture and the volts/turn are rapidly increased from 3 kV to 80 kV by T = 1 ms and to 112 kV by T = 5 ms. Protons undergo a quarter of a synchrotron oscillation by T = 0 at which time two smooth bunch shapes have developed (harmonic number 2). Later motion is non-adiabatic with filamentation present and the development of non-equilibrium bunch distributions.

The shape of both bunches is double humped by T = 100 μs and periodically returns to this form, but with more complex forms at intermediate times. At increased intensity, the shapes become smoother due to enhanced longitudinal space charge forces. A one dimensional longitudinal space charge tracking code[2] has been developed to study the bunch development. There is binning of a large number of representative particles and then an approximation of the longitudinal space charge forces from the bin distributions.

Typical predictions of the code are shown in Figures 3a and 3b for the energy distribution, the phase space distribution and the bunch shape at a defined time in the acceleration cycle, together with a digitized output of an actual bunch signal at the equivalent time in ISIS. The Figures are for an accelerated beam of 5.3×10^{12} protons. Even though the bunch shapes are complex there is reasonable agreement between the predicted and observed shapes. It is of interest to see if the code continues to predict the motion at increased intensity. The code has been used to track 10,000 particles so that the computations are extensive: care has been taken with economy in computation.

The non-equilibrium bunch distributions lead to low bunching factors at times in the acceleration cycle and hence enlarged transverse space charge forces. In future, after replacing the injection beam pipe, it is planned to inject with a larger beam momentum spread and then adjust the accelerating field amplitude to obtain a more uniform distribution in the bunches.

A successful use of the tracking code on ISIS gives some confidence in the use of the code for studying the bunch rotation in the HIF RF linac-storage ring scenarios.

REFERENCES

1. I. Hofmann, Non-linear aspects of Landau Damping in Computer Simulation of the Microwave Instability. Proceedings of 'Computing in Accelerator Design and Operation', Berlin 1983.
2. S. Koscielniak, RAL, Computer Simulation Program.

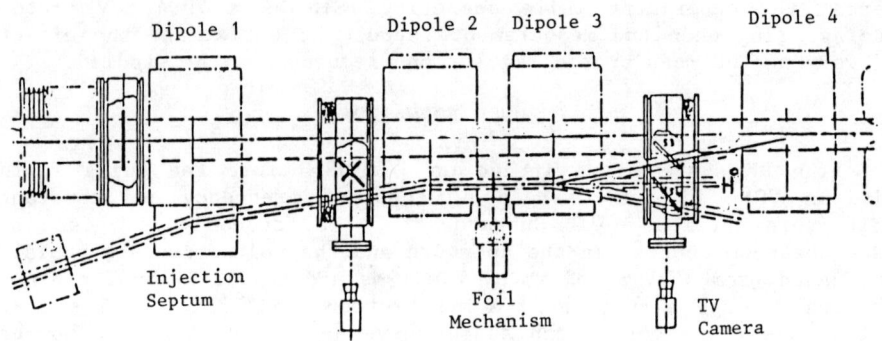

Figure 1   Injection Straight Section

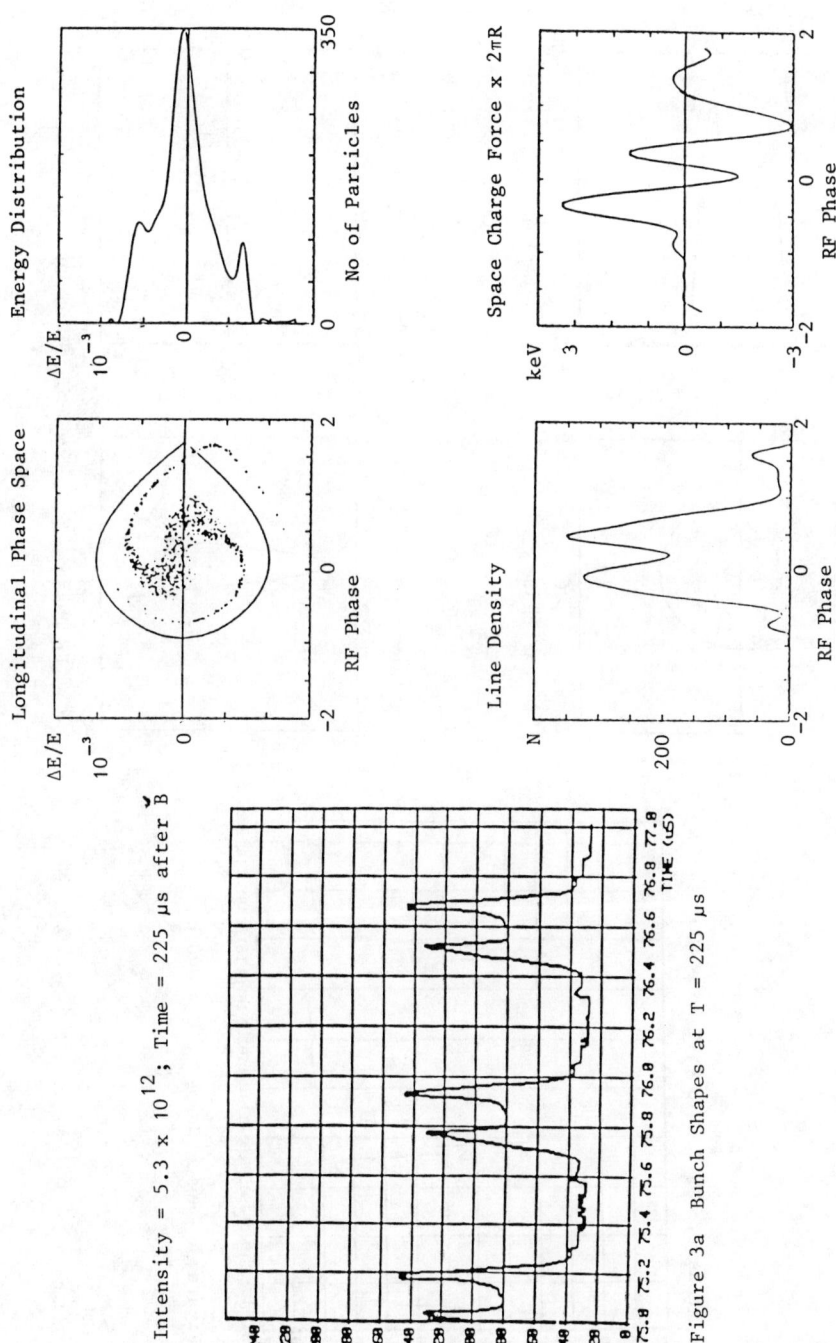

Figure 3a Bunch Shapes at T = 225 μs

144

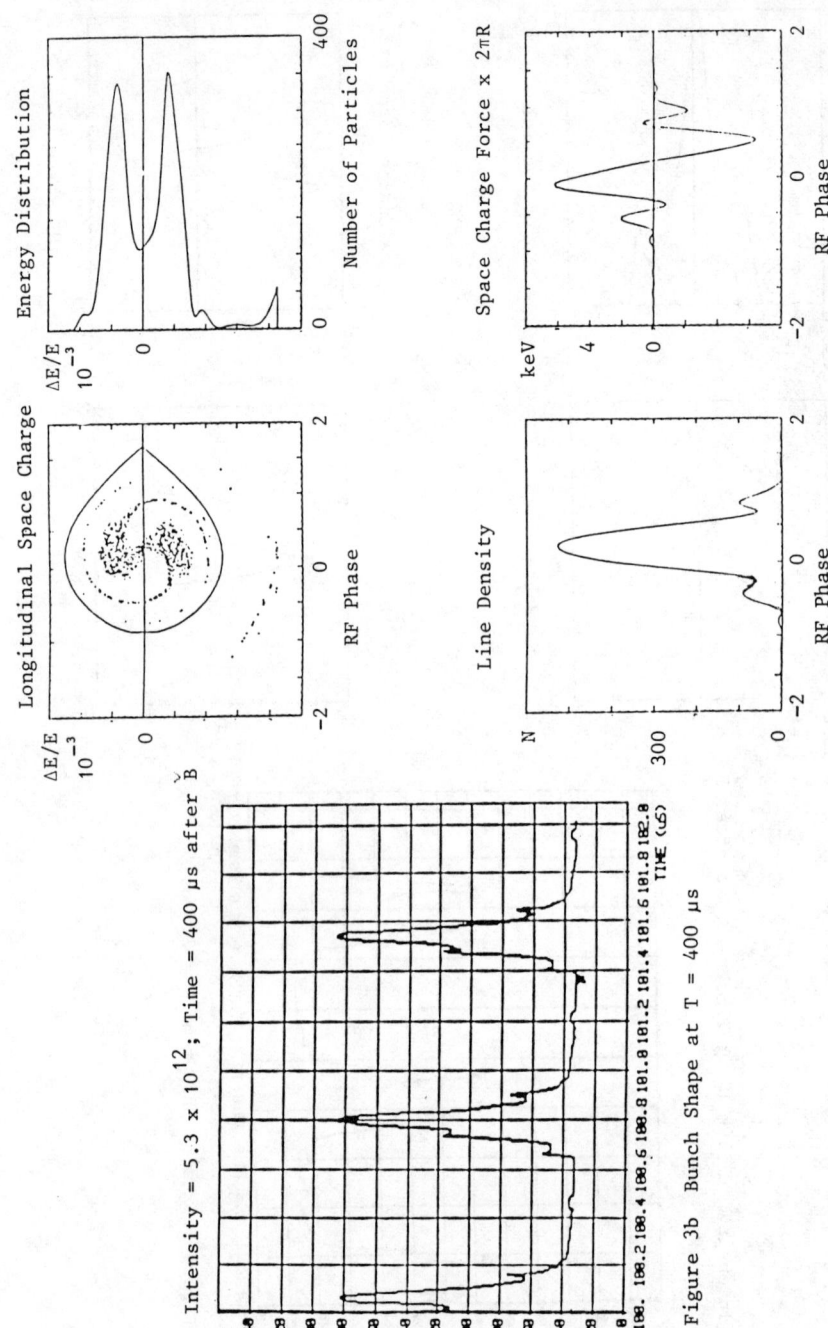

Figure 3b Bunch Shape at T = 400 μs

PRELIMINARY RESULTS FROM MBE-4: A FOUR BEAM INDUCTION LINAC FOR HEAVY ION FUSION RESEARCH*

T.J. Fessenden, D.L. Judd, D. Keefe, C. Kim, L.J. Laslett
L. Smith, and A.I. Warwick

Lawrence Berkeley Laboratory
University of California, Berkeley, CA 94720

Presented by A.I. Warwick

ABSTRACT

Preliminary results are presented from a scaled experimental multiple beam induction linac. This experiment is part of a program of accelerator research for heavy ion fusion. It is shown that multiple beams can be accelerated without significant mutual interaction. Measurements of the longitudinal dynamics of a current-amplifying induction linac are presented and compared to calculations. Coupling of transverse and longitudinal dynamics is discussed.

INTRODUCTION

MBE-4 is an experimental multiple beam ion induction linac which will model on a small scale some of the beam dynamics of a much larger fusion driver. About half the linac is constructed and operational at this time and we present preliminary analyses of the behaviour of the beams.

Four space-charge-dominated beams of Cs^+ ions, initially at 200 keV and with initial currents of 11 mA each, are individually focussed by electrostatic quadrupoles and accelerated, with current amplification, by fields induced when high voltage pulsers discharge through carefully designed circuits looping the induction cores. Figure 1 shows the layout of the linac; at present sections A and B are operational. Overall design details and the performance of the 4-beam injector have been presented at the 1985 Particle Accelerator Conference.[1] The beam envelopes are matched into the quadrupole lattice during transport through the conditioning section.[2] In this paper we will discuss acceleration through sections A and B.

An induction linac fusion driver is likely to have a large number of beams which will be accelerated and amplified in current by shaped induced fields. A number of issues arise, of which the following will be addressed by MBE-4:

a) MBE-4 can demonstrate the principle of a multiple beam linac.

b) Acceleration and current amplification by means of shaped voltage pulses can be demonstrated, requiring successful design and operation of the high voltage pulsers.

*This work was supported by the Office of Energy Research, Office of Basic Energy Sciences, U.S. Department of Energy under Contract No. DE-AC03-76SF00098.

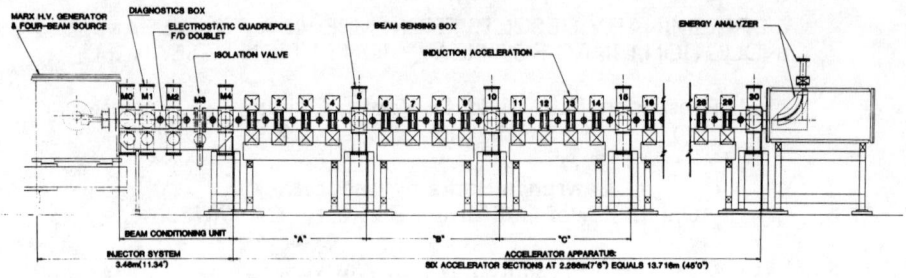

Fig. 1. Complete layout of the MBE-4 accelerator.

c) Because of the low ion velocity, current amplification schedules can be implemented which are more ambitious than any contemplated for a driver. In the 17 m length of MBE-4 it should be possible to model the longitudinal dynamics of a significantly longer portion of a large accelerator.

d) The ends of the bunch must be held together against the effects of longitudinal space charge forces.

e) The effect of longitudinal manipulations can be investigated. They should cause acceptably small growth of the longitudinal emittance.

f) Transverse emittance can be measured to see if it is conserved over the length of the experiment during acceleration.

g) Coupling of the transverse and longitudinal dynamics can be investigated in terms of the envelope oscillations induced during acceleration.

MULTIPLE BEAM EFFECTS

The four beams of MBE-4 are not expected to interact in any significant way during transport and acceleration. The only location where they deflect one another is at the emitter surface of the thermionic ion source. Here the ions are moving slowly and the charge density in the beams is high. The design computations indicated a deflection between beams of a few milliradians which was approximately compensated by shaping the anode and eliminated during initial operation by shimming.

Figure 2 shows current density profiles for a drifting beam at gap 5, both in the presence and absence of the other three beams. There is no appreciable difference. These profiles are measured by moving a small hole (.5 mm x .5 mm) across the center of the beam and sensing the transmitted current.

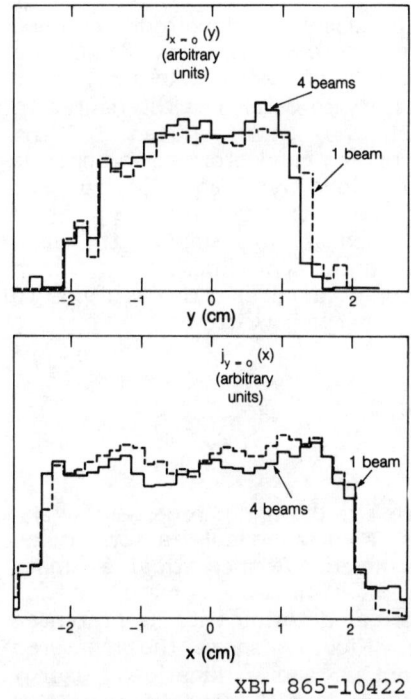

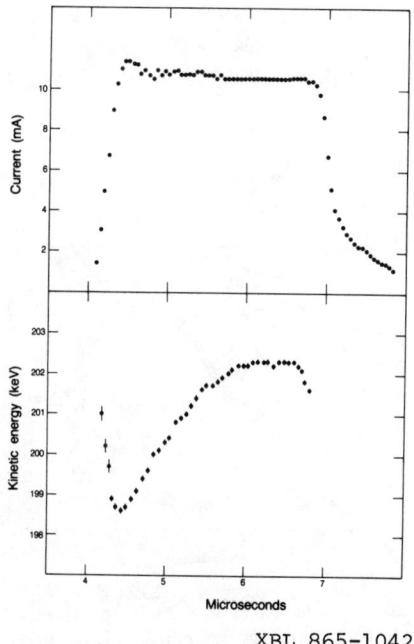

Fig. 2. Horizontal and vertical current density profiles of the bottom beam at the end of section A (gap 5). The profiles measured under normal operation (4 beams) are compared with those measured when the other three beams are blocked at gap 0 (1 beam)

Fig. 3. Initial conditions for acceleration and current amplification; current and kinetic energy measured at the beginning of section A (gap 0)

LONGITUDINAL DYNAMICS

Figure 3 shows the measured initial current and kinetic energy waveforms of one of the beam bunches at the diagnostic location immediately before accelerator section A (gap 0). These data are the starting point for the design of the accelerating schedule. Of immediate concern is the non-uniform kinetic energy waveform produced by the Marx pulse on the diode. Ions take 500 ns to cross the diode gap at 200 kV. There is a transient during the first and last 500 ns of the pulse as the diode fills with, and empties of, charge. It is necessary to precisely shape the rising edge of the 200 kV Marx pulse to minimize current and kinetic energy fluctuations during the initial transient phase.[3] After optimizing the rise time we obtained the energy and current waveforms of Fig. 3. Corrections to the non-uniform kinetic energy have been built into the accelerating schedule, and we will see below that by gap 8 these errors have been removed.

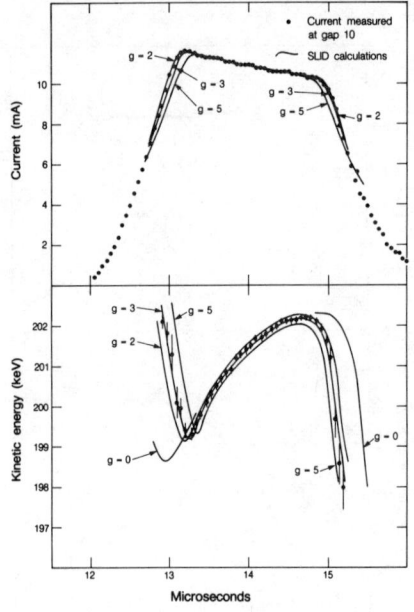

Fig. 4. Empirical determination of the g-factor. Computations of the longitudinal dynamics of a drifting bunch through sections A and B, beginning with the initial conditions of Fig. 3 and using various values of g, are compared to the measured kinetic energy and current waveforms at gap 10.

Because the longitudinal space charge forces are crucial to the operation of the accelerator, it is necessary to determine the degree to which they are reduced by the surrounding conductors. The bunch is 1.4 m long in an electrostatic quadrupole structure with half period nine inches. We employ the long wavelength approximation in which the longitudinal electric field due to space charge is written

$$E = \frac{-g \frac{\partial \lambda}{\partial z}}{4\pi\varepsilon_0 \gamma^2}$$

where λ is the line charge density and g is a geometrical factor to be determined. A theoretical estimate of g = 3.4 has been made and an empirical determination is presented here. Figure 4 shows the measured current and kinetic energy waveforms at the end of section B (gap 10) for a drifting beam. Calculations, which begin with the measured initial waveforms at gap 0 (Fig. 3), are also shown using various values of g. Comparison with the data gives g = 2.8 (± 0.6).

MBE-4 can be tuned to various current amplification and acceleration schedules. Here we will confine the discussion to one schedule which begins with an 11 mA current pulse of 2.5 microseconds duration at 200 keV and which accelerates at the limiting rate set by the amount of induction core on each accelerating gap. The accelerating voltage, which is proportional to the rate of change of flux in the core, appears across the gap until the core material saturates, so that the amount of core can be measured in volt-seconds of acceleration. Table I shows the core and pulsing circuits installed at present. Typically between 60% and 80% of these volt-seconds are used to accelerate beam (see Fig. 5).

The design calculations which specify the accelerating waveforms required at each gap to achieve a desired acceleration and amplification schedule have been described elsewhere.[4] Briefly, they take the form of a one dimensional simulation using fifty particles, accelerated at discrete gaps and interacting amongst themselves according to the charge they carry. A computer program called SLID has been developed by C. Kim. At each accelerating gap the computer takes the current and kinetic energy waveforms before acceleration and generates the accelerating waveform

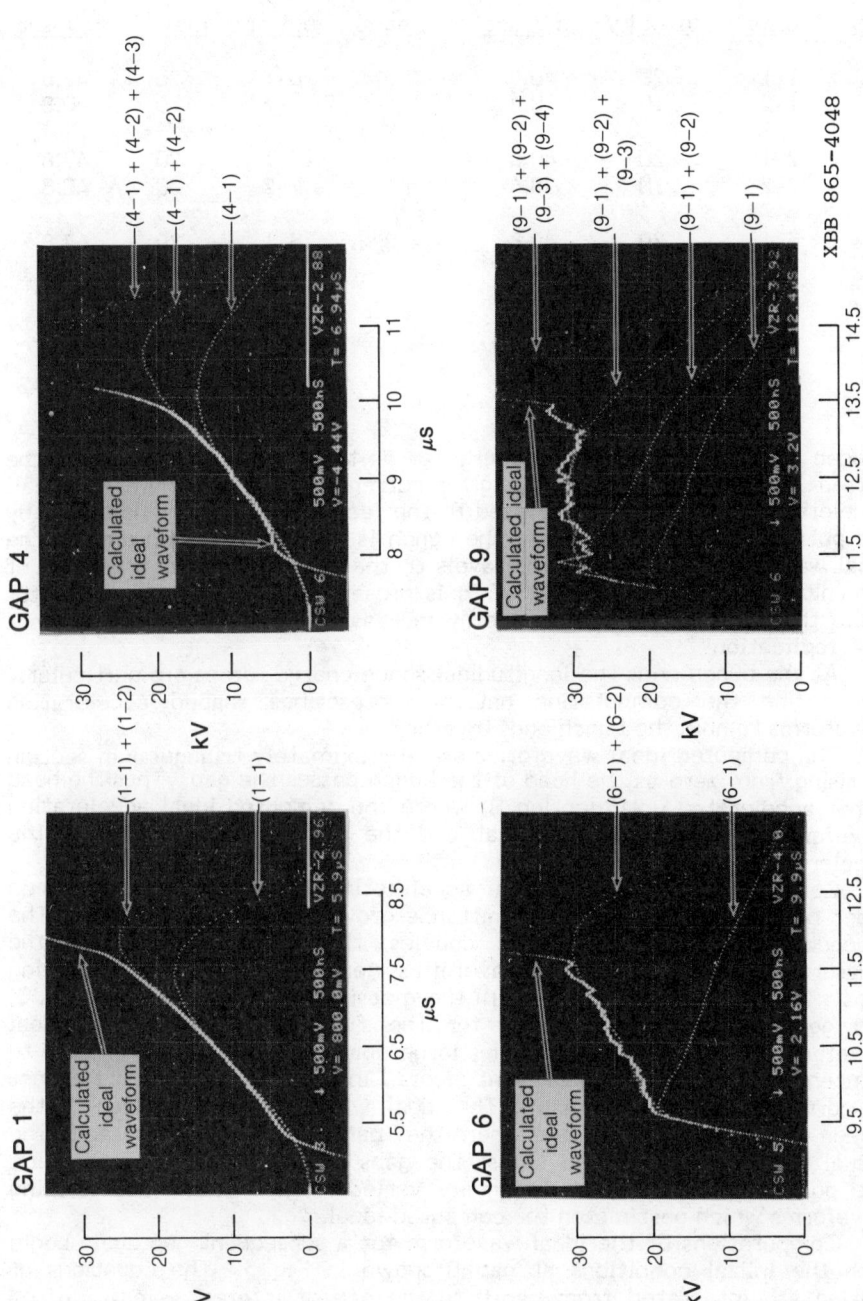

Fig. 5. Actual and ideal waveforms on some of the accelerating gaps. Note the tail corrections at gap 4 and 9 and the tailored risetime to correct the head at gap 6.

Table 1. Core and Pulsers

gap	pulser	max kV	mV-sec	gap	pulser	max kV	mV-sec
1	1-1	20	40.8	6	6-1	20	47.8
	1-2	10	20.4		6-2	20	40.8
2	2-1	20	40.8	7	7-1	20	47.8
	2-2	10	20.4		7-2	20	40.8
3	3-1	20	40.8	8	8-1	20	47.8
	3-2	10	20.4		8-2	20	40.8
4	4-1	20	40.8	9	9-1	20	47.8
	4-2	10	20.4		9-2	10	23.9
	4-3	10	13.6		9-3	10	23.9
					9-4	10	20.4

needed to change the kinetic energy so as to preserve the shape of the current waveform while amplifying the current. These are known as 'ideal' accelerating waveforms as opposed to the 'actual' waveforms delivered by the pulsers. In the computation the bunch is accelerated according to the ideal waveform at the gap and travels to the next gap. Thus the shape of the inital current waveform (Fig. 3) is preserved through the accelerator while the current and kinetic energy increase. This is known as 'current self-replication'.

At the bunch ends the longitudinal space charge forces are particularly strong and the computation naturally prescribes shaped accelerating waveforms to hold the bunch ends together.

The computed ideal waveforms are approximately triangular in section A, rising from zero as the head of the bunch passes the gap. Thus the head is not accelerated until section B, where the computed ideal accelerating waveforms are approximately flat and the tail of the bunch is in the accelerator.

Careful tuning of the actual accelerating waveforms is required in order to minimise longitudinal emittance growth during acceleration. The procedure we have developed couples the computations with the measurements of the actual waveforms generated at the accelerating gaps. The computation begins with the measured initial conditions (Fig. 3) and generates ideal waveforms for the first three gaps. (The ideal waveform for the first gap includes large corrections to the bunch ends to compensate for the diode transient errors. In practice these errors cannot be corrected at a single gap). The ideal waveforms are written to the screen of a digital oscilloscope where they can be directly compared to the actual waveforms produced across the gaps by the high voltage pulsers. The pulser amplitudes and timing are varied to give actual accelerating waveforms which best match the computed ideal.

Computations of the ideal waveform for a subsequent gap again begin with the initial conditions at gap 0 shown in Fig. 3. The equations of motion are integrated from gap 0 to the gap of interest using digitised measurements of the actual accelerating waveforms applied across the intervening gaps, then the ideal waveform for the gap of interest is computed. For example, the computation of the ideal waveform on gap 9 involves integration of the equations of motion from gap 0 to gap 9 using the actual accelerating waveforms applied across gaps 1 - 4, 6 - 8. At the last of the four gaps of each accelerator section (gaps 4 and 9 for that part of MBE-4 already constructed) we install additional pulsers to control the space charge at the bunch ends and compensate as best we can for errors in the previous gaps (Table I). Figure 5 shows the results for gaps 1, 4, 6 and 9.

The shaping of the accelerating waveforms to control the bunch ends is dominated at first by the corrections for the transient effects in the diode. The transient error at the tail of the bunch is corrected at gap 4 (see Fig. 5 where the actual waveform on gap 4 matches the ideal all the way to the tail of the bunch).

The transient error at the head of the bunch calls for deceleration relative to the rest of the ions. Deceleration of the head of the bunch by means of negative voltage pulses with tailored trailing edges has not proved possible. The trailing edge occurs when the core material is beginning to saturate, the impedance is changing and the circuit design is very difficult. Instead, the bunch head is corrected by refraining from accelerating it on gaps 6, 7 and 8, by means of a tailored rising edge on the main accelerating waveform. Figure 5 shows such a waveform on gap 6, here the ideal is computed using the actual waveforms on gaps 1 - 4. The required fast rise-times at gaps 6, 7 and 8 cause some oscillation of the actual accelerating waveforms so that the match to the ideal in section B is not as good as that achieved in section A.

Since the ideal waveform for gap 9 is calculated using the actual waveforms applied to gaps 1 - 4, 6 - 8, this computation calls for corrections for all the errors on the previous gaps and represents a measure of the cumulative error. The frequency of the cumulative error is too high for corrections to actually be

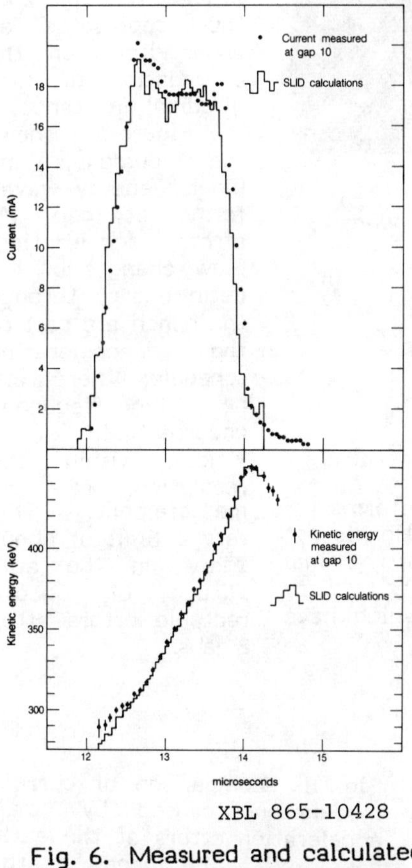

XBL 865-10428

Fig. 6. Measured and calculated current and kinetic energy waveforms for the right beam at the end of section B, under the accelerating schedule described here. Current is amplified by a factor of 1.6. All the errors caused by the diode transients are corrected.

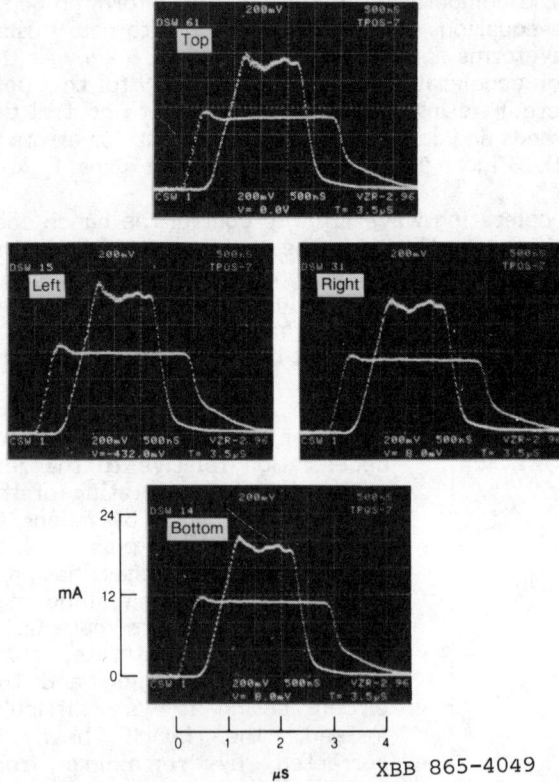

Fig. 7. Initial and final current waveforms for all four beams. Current self-replication is not working perfectly. The modulations of the final current waveform are caused primarily by very small voltage errors on gaps 1-4, which have excited space charge waves.

made and the amplitude, after 7 gaps, is about 2 kV. However the beam is not sensitive to frequency components of the errors above about 2 MHz, integrated over the transit time through the gap (300 ns). The cumulative error of 2 kV thus represents an upper limit on the degradation of longitudinal emittance.

Figure 6 shows the current and kinetic energy waveforms at gap 10, after acceleration. Slow changes of kinetic energy through the bunch are part of the acceleration schedule; there are no high-frequency uncorrectable errors visible within the resolution of the measurement. This sets a limit of about 2 keV on the amplitude of uncorrectable errors after 8 gaps.

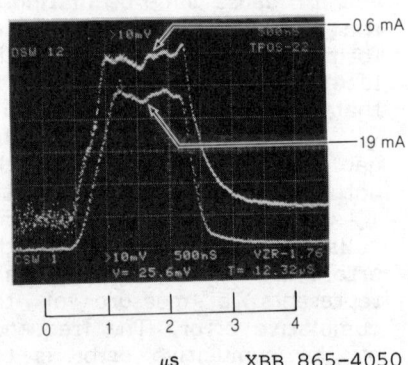

Fig. 8. Comparison of current modulations caused by small acceleration errors at the early gaps on a) space charge dominated 10 mA beam and b) 0.6 mA beam with negligible self-forces.

In the context of a heavy ion fusion driver the acceptable longitudinal emittance is determined by the final lens system and the focal spot requirements. If one imagines random errors on each gap which, through a long accelerator, accumulate as the square root of the number of gaps, one could tolerate uncorrectable voltage errors at about one percent of the accelerating voltage.[5] Our measurements to date indicate that MBE-4 operates just within this limit. The situation will be clearer when the experiment is longer.

Pulser voltage errors soon become beam current fluctuations. In Fig. 7 the initial and final current waveforms are compared for all four beams. Current is amplified by a factor of 1.6. The final current waveforms show modulation which is caused by the voltage errors in the first few gaps. In Fig. 8 we compare the modulations on the standard 11-18 mA beam with those on a low space charge beam starting at 0.6 mA. The modulations are clearly less for the space-charge-dominated beam.

LONGITUDINAL/TRANSVERSE COUPLING

The r.m.s. transverse emittance is derived from measurements of the transverse phase space distributions. A pair of slits, one defining x, the other defining x', are moved across the beam from pulse to pulse and the transmitted current is measured. We find $\epsilon = 1.5 \pm 0.3 \times 10^{-7}$ m-rad where ϵ is defined by the expression.

$$\epsilon \equiv 4 \beta\gamma \left[<x^2><x'^2> - <xx'>^2 \right]^{1/2}$$

The K-V envelope equations[6] describe the beam size during acceleration quite well. We integrate the envelope equations beginning with measured initial r.m.s. radii and divergences at gap 0. At each accelerating gap the kinetic energy, current and beam envelopes are changed discretely according to the accelerating schedule of the head or tail of the bunch. For the head we take a point in the bunch which arrives at gap 10 12.8 μs after the Marx fires (see Fig. 6); for the tail we take 13.7 μs. The quadrupole strength is set to preserve the centre of the bunch at $\sigma_0 = 60°$ as the kinetic energy increases. This calculation proceeds through the ten quadrupole periods of section A and B and is compared in Fig. 9 to the phase space distributions measured at gap 10. The agreement is good and the K-V calculations reproduce the different optics of the head and tail.

The beam size is not strongly dependent on kinetic energy if the current is constant; as the kinetic energy increases from head to tail of the bunch at constant current the increased stiffness is approximately compensated by the decreased perveance. In the acceleration schedule implemented here the current has the same magnitude throughout the bunch as it passes a given point in the accelerator (Figs. 3 and 6), thus the equilibrium size of the head is close to the equilibrium size of the tail, both staying close to their initial values during acceleration. The upper part of Fig. 9 shows the calculated beam envelopes during acceleration through the ten quadrupole doublets of sections A and B. Mismatch oscillations, which can be seen as changes in the maximum beam size, are not a significant problem.

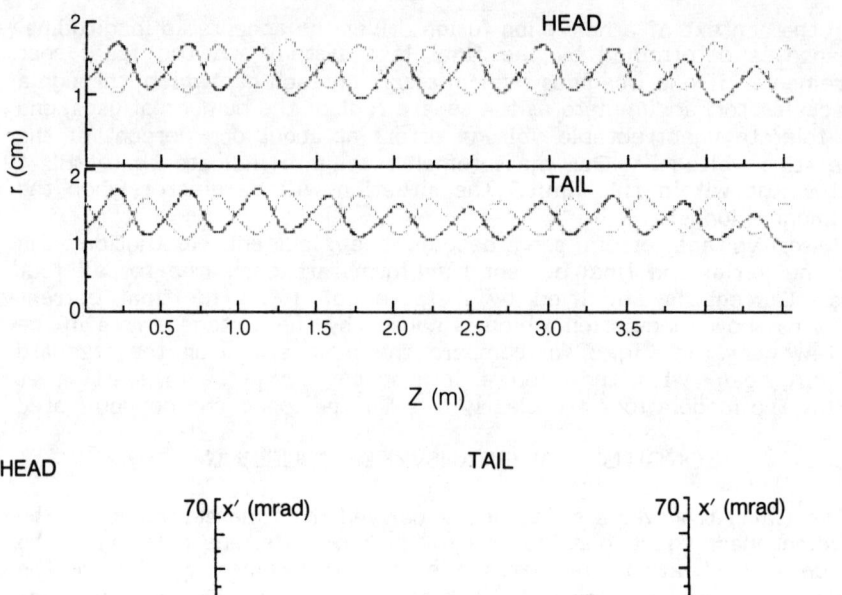
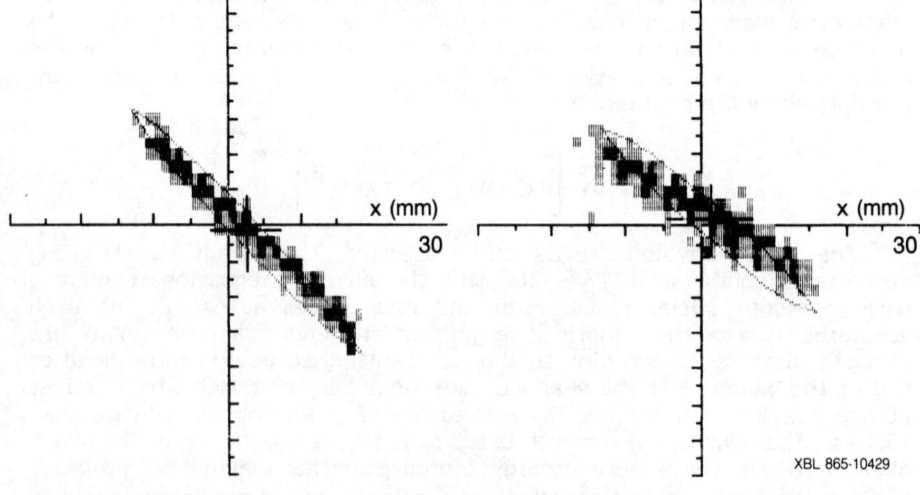

Fig. 9. Behaviour of the envelopes during acceleration. Integration of the K-V envelope equations through the ten periods of section A and B, beginning with measured initial conditions, are shown for the head and tail of the bunch. In the lower part of the Fig. the final K-V ellipses are compared to phase space distributions measured at gap 10.

SUMMARY

MBE-4 is an experiment which has just begun. The achievement to date has been to implement an acceleration schedule and observe that the transverse and longitudinal dynamics behave as calculated. The beams have been shown not to interact with one another, the current has been amplified by a factor of 1.6 with tolerable longitudinal energy errors, bunch end control has been implemented and mismatch oscillations have been investigated and found to be small.

Important questions have still to be answered regarding the effects of misalignments, preservation of transverse emittance and the possibility of making steering corrections during acceleration. We need to refine our techniques to follow the transverse emittance as the angles become smaller during acceleration. We also need to improve the resolution of the energy measurements to detect longitudinal energy errors in the beam itself. The development of longitudinal errors will be particularly interesting as the accelerator gets longer. Current fluctuations are of great interest, but MBE-4 will not operate in the regime of high beam currents where fluctuations may grow due to coupling through the impedance of the structure.

REFERENCES

1. R.T Avery et al., IEEE Trans. Nucl. Sci., NS-32 (1985) 3187.
   A.I. Warwick, D. Vanecek and O. Fredriksson, IEEE Trans. Nucl. Sci., NS-32 (1985) 3196.

2. HIFAR Year End Report, LBL-20310 (1985).

3. M.G. Tiefenback and M. Lampel, Appl. Phys. Lett., 143-1 (1983), 57.

4. C. Kim and L. Smith, Particle Accelerators, 18 (1985) 101.

5. D. L. Judd, Proc. Heavy Ion Fusion Workshop, LBL-10301, (1979).
   E.P. Lee, private communication.

6. F. Sacherer, IEEE Trans. Nucl. Sci., NS-18 (1971) 1105.

STATUS OF THE HEAVY-ION RFQ LINAC "MAXILAC"

R.W. Müller, U. Kopf, R. Keller, P. Spädtke, J. Bolle
GSI, Planckstraße 1, D-6100 Darmstadt, W. Germany

ABSTRACT

An RFQ linac for heavy ions such as $^{127}I^+$ has been constructed at GSI. It is a prototype of a fusion driver, and it will serve to bring up the intensity of the existing heavy ion linac, UNILAC, by several orders of magnitude. Some beam test results are given. Ions are accelerated from 2.3 to 45 keV/amu.

REQUIREMENTS AND SOLUTIONS

Among the different types of machines, a linac is the best choice for injecting heavy-ion beams into a synchrotron or storage ring. Intensities in existing heavy-ion linacs are low, compared with their space-charge transport limit, because high charge states have to be filtered out of the ion-source output, e.g. U^{10+} in case of UNILAC. Fig. 1 shows the injection beam current, as measured in the UNILAC pre-stripper section, for which the synchrotron would be space charge saturated, as a function of the nuclear order number Z. Also shown are ion beam intensities which can be achieved in UNILAC with existing sources. For ions beyond Ar (Z = 18) two or three decades of higher intensities would be desirable for certain classes of experiments.

Improvements can be expected from advanced ion-source schemes like EBIS or ECR. The most efficient (maybe not the cheapest) way, however, is starting with single-charged ions extracted out of high-brillance sources, accelerating up to 120 ... 140 keV/amu, and stripping. Then the critical phase of beam formation has been done with the minimum of space charge, and hence the best chance of beam survival. However, these ions are slow at the beginning of their life and require new techniques of acceleration.

The trend into this new technique has been set by studying inertial confinement of plasmas driven by heavy ions, with the goal of producing electric energy from D-T fusion. In studies like HIBALL (Fig. 2), after

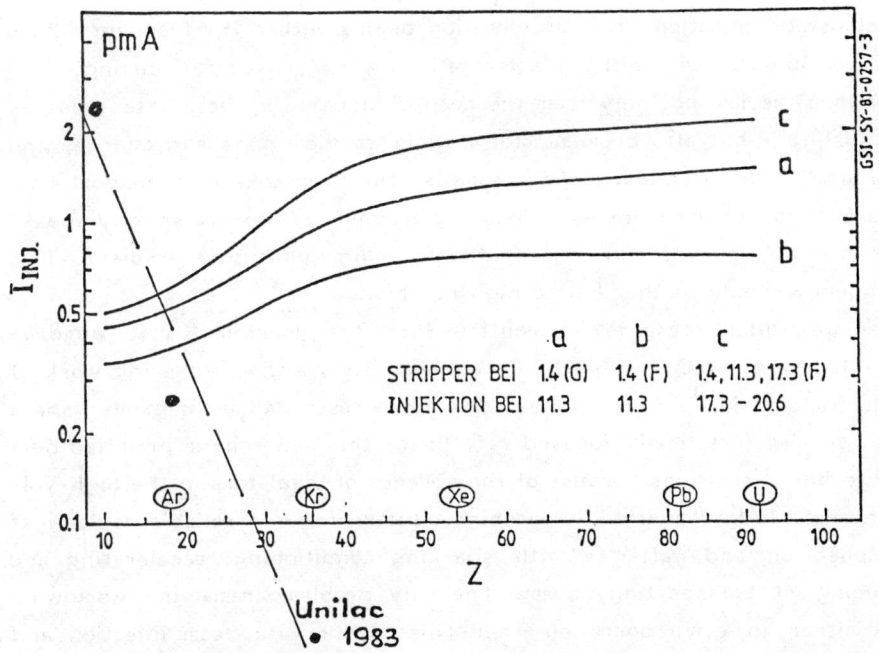

Fig. 1: Space charge limit ("Laslett" limit) of SIS 18 for various ions and injection conditions; multiturn injection. Beam current figures are referring to the pre-stripper section of UNILAC.

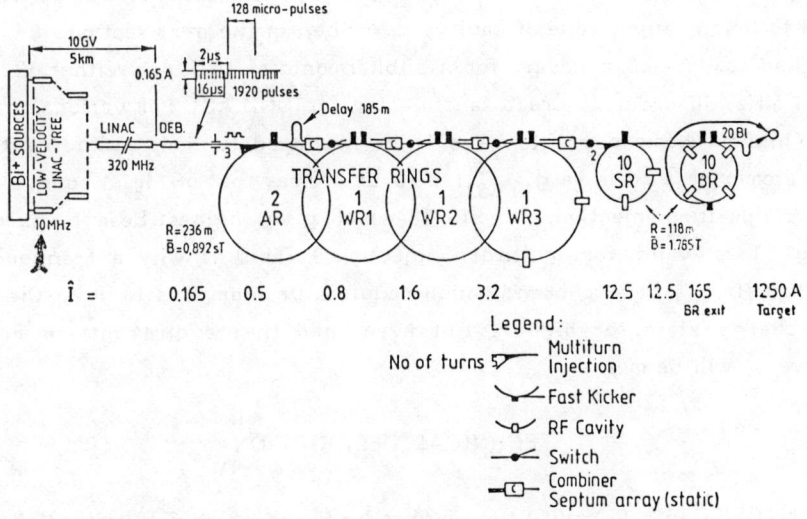

Fig. 2: HIBALL II, schematic ground plan. MAXILAC is a prototype of the linac section indicated by the arrow.

electrostatic injection, r.f. acceleration begins with a low-frequency linac (10 ... 15 MHz) with either electrostatic or r.f. transverse focusing. The frequency is low <u>not only</u> from the need of stacking up beam intensities by "funneling", but also because, for a given particle mass and charge, and technically limited quadrupole gradients, the space-charge transport limit is a function of the transverse focusing frequency, or more strictly speaking, the spatial rate of reversing the focusing quadrupole gradient. This frequency should neither be too high nor too low.

When we studied these developments in 1977, the principle of r.f. quadrupole focusing (RFQ) matured into applicability by the pioneering work of I.M. Kapchinskij [1]. Though a little more restricted in parameter space than an electrostatically focused r.f. linac, this new scheme promised better technical solutions because of the absence of insulators in the high-voltage area. Indeed every RFQ machine built so far is a reliable device just switched on and, after a little sparking conditioning, accelerating and focusing ($\equiv$ transporting) beam. The only problems remaining worldwide are either in c.w. operation (non-pulsed), or with beam injection and matching.

Most RFQ devices, designed for protons or mass-to-charge ratios of up to 7, use the H_{211} "cloverleaf" oscillation mode, at frequencies of 100 to 440 MHz. For 13.5 MHz this mode is no more adequate, the tanks either become too big (>5 m of diameter), or the shunt impedance will be low. We had to invent a new type of cavity, described in the next section.

We had to develop a design for a subharmonic of UNILAC, either 13.55 or 9.04 MHz, in order to create an injector for UNILAC. For various reasons the highest beam currents have to be produced for ions at the centre of the atomic mass scale, e.g. $^{127}_{53}I^{+}$, about 4 times that of Fig. 1, curve a, for single-turn injection into SIS 18 which gives highest beam brightness. (Fig. 1 is valid for multiturn injection.) This is why a frequency of 13.55 MHz has been chosen, which requires uranium ions to be in the double-charge state, or higher. But even then the requirements of Fig. 1, curve b, will be met.

TECHNICAL DESCRIPTION

First of all, an r.f. cavity type had to be found which fulfills the following requirements:

a. Compact dimensions at f = 13.5 MHz;
b. Good resonant resistances, or power economy, or low conduction losses, while feeding a very high system capacity of $C' = 20...30 \cdot \varepsilon_o =$ 180...270 pF/m;
c. Flat voltage distribution;
d. Good mechanical strength, easy adjustment;
e. Modularity, i.e. many cavities making up a system of 10...30 m length are to be tightly coupled without voltage unflatness problems.

Neither the GSI-type Wideröe structure, nor the IH structure, which both were candidates, could fulfill these requirements. A solution was given by the split coaxial (SC) cavity, a TM mode where the magnetic flux inducing the r.f. voltage, toroidally includes the accelerating system which is contained within an inner conductor. This inner conductor is slotted in such a way that four pieces are created whose width is increasing linearly from zero to a quarter of the circumference. At this latter broad end the pieces are fixed to the cylinder end walls.

An easy way to understand this novel type of cavity is a mental evolution from a double $\lambda/4$ cavity oscillating in the push-pull mode (**Fig. 3**, left) into the SC cavity (right).

Fig. 4 shows a longitudinal section through the machine actually built, module no. 1. The machine now has 5 modules coupled together with their "hot ends" and driven by one amplifier. Though the "load" C' is continually varying along the machine from 30 ε_o down to 14 ε_o, the voltage along the whole machine is flat within 10% (i.e., 10% higher at the end where C' is lowest). The individual tuning of the modules is uncritical and scatters by ±2% around the average of 13.25 MHz. This slightly too low a frequency has been aimed at because it is easier to bring the frequency up by magnetically displacing tuners than bringing it down by added capacitors.

The focusing and accelerating system is made up of cylindrical electrodes of stepwise varying diameter; see **Fig. 5**. The bigger diameters and the spacings of opposite poles are constant. Every pair of electrodes is held in a ring of a standardized shape, and fixed to the SC inner conductor pieces by solid copper stems. These stems are to conduct currents and heat (e.g. of lost particles) to the water-cooled inner conductor. The electrodes do not have own water cooling.

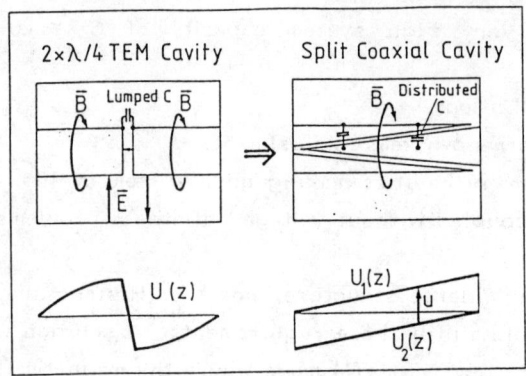

Fig. 3 : Mental evolution of a 2 x λ/4 TEM cavity into an SC cavity.

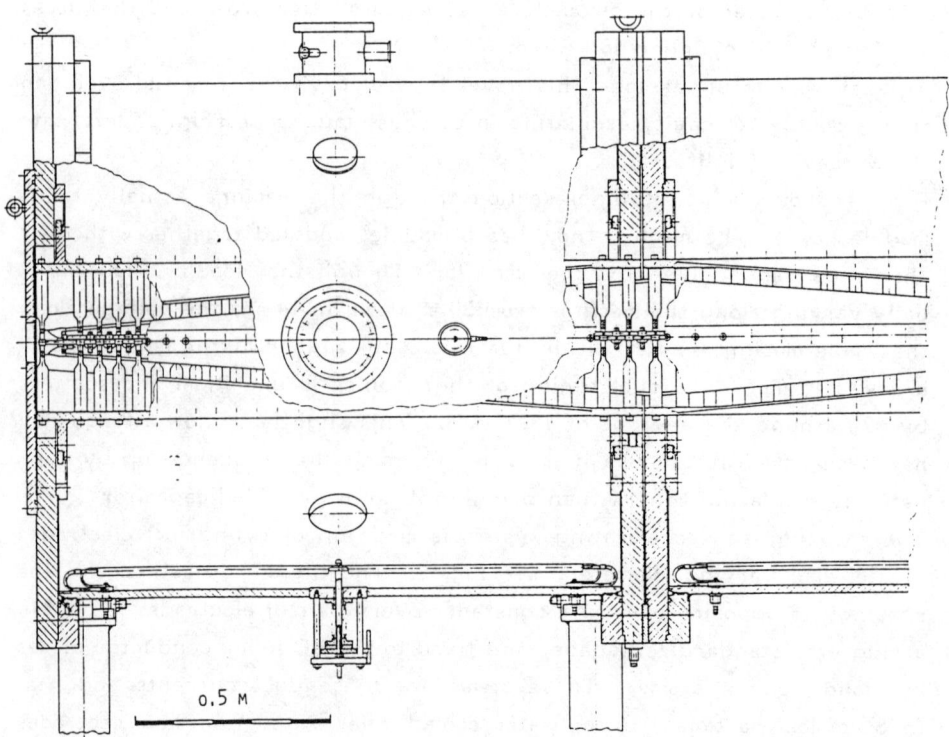

Fig. 4: Longitudinal section through MAXILAC module 1, and a part of module 2.

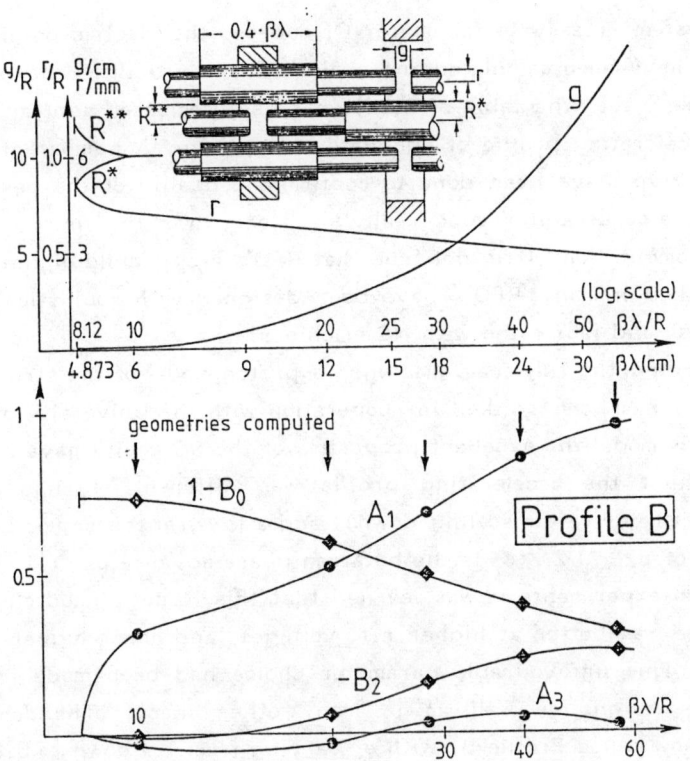

Fig. 5: Quadrupole electrode shaping (top) and parameters.
Bottom: Field coefficients; A_1 and A_3 are the Fourier coefficients of the potential on axis, $1-B_0$ and B_2 of the transverse gradient.

TABLE I: Beam Test results (preliminary)

	$^{40}Ar^+$	$^{84}Kr^+$	$^{132}Xe^+$	$^{238}U^{4+}$	$^{238}U^{3+}$	Duration of Experiments
1 Module 4.5 keV/u	10 mA	20 mA	10 ma	–	–	1 year
3 Modules 20 keV/u	–	10 mA	–	4 mA	5 mA	3 months
5 Modules 45 keV/u	4 mA	–	–	2 mA	–	4 months

Such a system is easy to fabricate. Of course, the electric potential distributions have been calculated and evaluated by finite difference methods. The "profile", i.e. the table of $\beta\lambda(z)$, has been composed semi-analytically into an accelerating profile of adequate bucket size. A number of particle simulation runs have been done to confirm the quality of the design, but the only major computer code really necessary was the finite-difference potential computation. It is not true that RFQ's are a "child of the computer age", though many RFQ's have been designed with sophisticated computer codes, and fabricated with NC machines.

Before building the full-scale machine, a proton model of the structure, at a 1:4 scale, has been studied in cooperation with the University of Frankfurt. In this model the excellent properties of the SC cavity have been confirmed, whilst the accelerating profile was different. It had a higher aperture diameter (after scaling-down), and a low transverse, or betatron, frequency of $\sigma_t \sim 0.2$. (σ_t is the betatron phase advance per r.f. period). In the beam experiments it was learned that this is not a good choice, the performance was better at higher r.f. voltages, and hence higher σ_t, than designed. This unfavourable parameter choice had been made in all the early RFQ designs as well as in many other linacs. The design was changed, now called Profile B, with σ_t varying from 1.4 down to 0.8, and a free aperture diameter of 12 mm (full scale).

The expected space-charge limit, occuring when the transverse oscillation frequency is lowered by space-charge forces by 30%, is at a beam current of 0.2 mA • (A/q), e.g. 25 mA for I^+. If the r.f. conductivity losses are as designed, 60% of the r.f. power are to go into the beam. At present the 5-module RFQ (length 9.5 m) injection energy 2.3 keV/amu, final energy 45 keV/amu, is powered by an 80 kW amplifier (20 kW max. average). This is sufficient only for A/q ≤ 75. For higher masses (design limit A/q = 133, $^{133}Cs^+$) a new amplifier with a 1 MW tetrode tube (RS 2042) is under construction; this amplifier is powerful enough to drive the full beam load even through a structure which is three times the present length.

BEAM RESULTS

Since (at least at present) we cannot diagnose the beam behaviour, especially beam losses, within a linac structure, the assembly of the 5 cavities

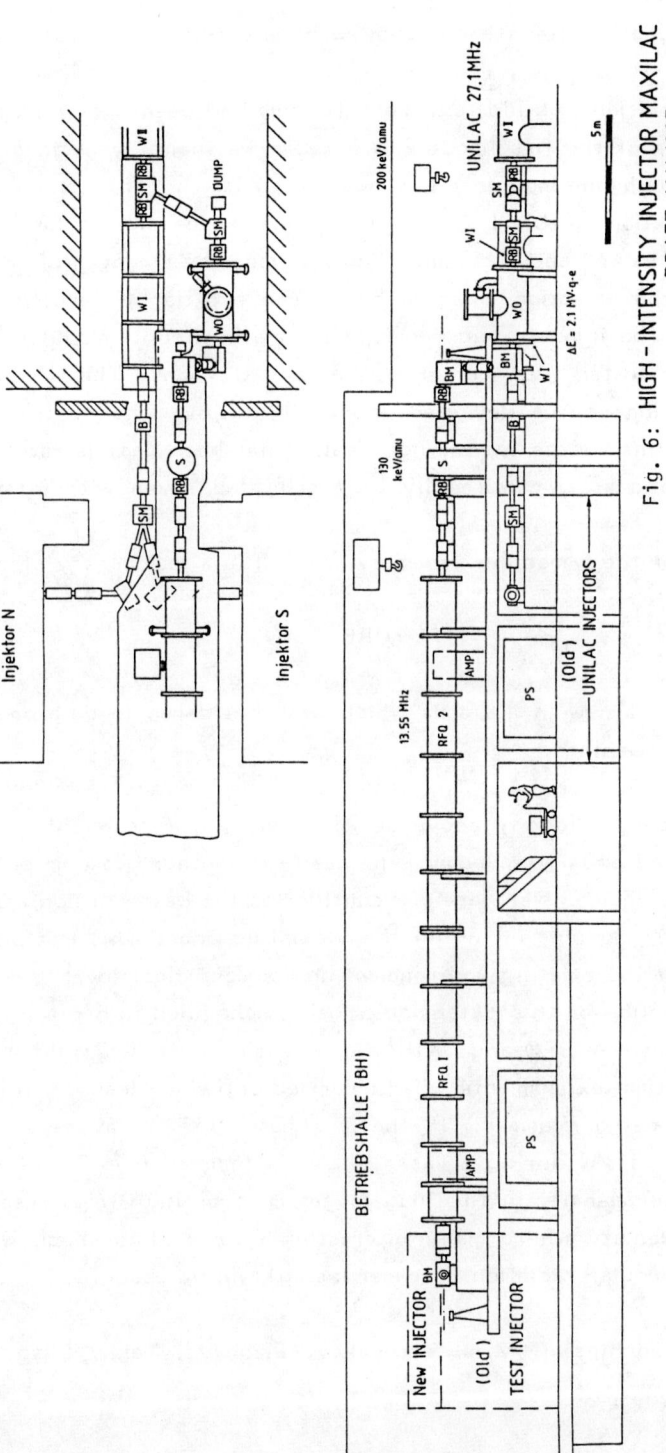

Fig. 6: HIGH-INTENSITY INJECTOR MAXILAC
DRAFT, JAN. 1985

had to be done in steps each followed by a careful beam test: 1, 3 and 5 modules.

Since the decision to build four more modules had been taken after successful operation of the first module, we had more than 1 year to study beam tuning through one module. It is not a surprise that the results were best in this situation (Table I). The theoretical space charge limit has been surpassed for Ar^+ and Kr^+ ions. For Xe^+ ions the strength of the focusing lenses in the injection channel have been a little bit insufficient, only 10 mA could be injected and received in the final cup. Modules 1 through 3 have been tested with Ar^+, Kr^+, U^{4+} and U^{3+} ions. The U ions came out of a spark ion source designed and built at LBL. Though the discharge stability of this source type is not ideal, it has been demonstrated that also ions as exotic as U can readily be accelerated, even with intensities of >1 pmA.

The tests for the 5-modules assembly are still in a preliminary stage.

FUTURE PLANS

The machine is now in a status which was considered to be a prototype of the HIBALL machines.

For injection into UNILAC, the beam has to be brought up to 130 keV/amu to be stripped. Though, above 50 keV/amu, a more conventional type of linac might be equally economic, the need for another beam matching would exclude this choice. We therefore consider as the best solution a 26 m long RFQ, with 12 modules including the 5 existing ones. This machine can be installed into the existing environment on a second floor level; see Fig. 6.

The construction of this extension is being scheduled to begin in 1988, the beam path between RFQ and UNILAC-W2 is being designed right now.

In the meantime experiments will be carried out which lead into the field of heavy-ion heated plasmas. If the I^+ beam (150 kW averaged over the macro-pulse, 1 MW during a micro-pulse) is focused to a 2 mm^2 spot, the specific power density in the 10µg target is 15 or 100 GW/g, respectively. This is a density where many interesting effects, like shock waves and coherent molecular radiation (excimer lasing) will be observed.

1. I.M. Kapchinskij, V.A. Teplyakov; Pribory i Tekh. Eksp. (1970), 19-22.

TABLE II. RFQ ACCELERATOR MAXILAC
PARAMETER TABLE

MASS/CHARGE OF IONS, A/Q	≤ 135
R.F. FREQUENCY	13.55 MHz
INJECTION ENERGY	2.3 keV/AMU
FINAL ENERGY (1986, 5 MODULES)	45 keV/AMU
(PLANNED, 10 ... 12 MODULES)	> 100 keV/AMU)
TRANSVERSE ACCEPTANCE, NORMALIZED	$\pi \cdot 0.6 \ \mu M$
LONG. EMITTANCE	$\pi \cdot 80$ DEG keV/AMU
SPACE CHARGE CURRENT LIMIT	(A/Q) · 0.2 mA
(CALCULATED FOR REL. TRANSV. TUNE DEPR.	- 30 %)
R.F. VOLTAGE AMPLITUDE	(A/Q) · 1.2 kV
MIN. ELECTRODE SPACING	7 mm
KILPATRICK NUMBER (A/Q = 135)	2.0
AVERAGE ACCELERATION GRADIENT, $\bar{E}_{ACC}$	(A/Q) · 5.0 kV M
EFFECTIVE SHUNT IMPEDANCE, $\bar{E}_{ACC}^2 \cdot L/P_{RF}$	25 MΩ/M
R.F. POWER EFFICIENCY, $P_{BEAM}/(P_{BEAM} + P_{RF})$	60 %
R.F. MODE	SPLIT COAXIAL

Fig. 7
5-modules MAXILAC seen from the top of the injector

EMITTANCE GROWTH FROM CHARGE DENSITY CHANGES IN HIGH-CURRENT BEAMS*

T. P. Wangler, K. R. Crandall, and R. S. Mills
Los Alamos National Laboratory, MS H817 Los Alamos, NM 87545

ABSTRACT

We use the relation between field energy and rms emittance, together with the property of charge-density homogenization for intense nonuniform beams in linear focusing systems, to derive equations for emittance growth and minimum final emittance. We discuss three problems in which this charge redistribution mechanism is isolated: the 1-D continuous sheet beam, the 2-D continuous round beam, and the 3-D spherical bunch. For each of the three problems, we identify and compare scaling parameters that determine the emittance growth and minimum final emittance as a function of beam current, emittance, and external focusing strength. Numerical simulations are used to test the equations, to show that the charge redistribution mechanism results in very rapid emittance growth, and to study the detailed time evolution of the beams.

INTRODUCTION

The promise of heavy ion fusion rests on the requirement of a charged particle accelerator system that can accelerate to high energy, and transport to the target, a high beam current within a small emittance.[1] Thus, it is important to understand the causes of rms emittance growth from all sources and to account for these effects in the design of the accelerator. It has been known for many years that the rms emittance growth associated with nonlinear space-charge forces in high-current, low-emittance beams occurs at the low-velocity end of a linear accelerator, where focusing tends to be weaker and where, for a given beam current, the charge density is highest. In spite of much effort to understand the causes of the rms emittance growth in high-current accelerators, many phenomena, observed in numerical studies and in experimental data, have remained unexplained.

Recently, we presented a differential equation[2,3] for 2-D continuous beams with azimuthal symmetry and continuous linear focusing, which expresses a relationship between the rate of change of rms emittance and the rate of change of nonlinear field energy. The nonlinear field energy corresponds to a residual field energy of beams with nonuniform charge distributions. It depends only on the shape of the charge-density distribution and corresponds to the field energy available for emittance growth. Using approximations

---

* Work supported by the US Department of Energy.

valid for a space-charge-dominated beam: (a) constant rms beam size
and (b) homogenization (uniformity) of the final charge density, the
integrated differential equation yielded expressions for emittance
growth and minimum final emittance, which agreed well with the numerical simulations presented. The emittance growth formula also
agreed with a formula proposed earlier[4] to explain numerical simulation results for a periodic quadrupole transport channel After our
initial work, we learned that equivalent forms of the emittance
differential equation had been discovered earlier,[5-7] but it appears
that the utility of this result for obtaining a better understanding
of emittance growth effects in linacs and transport lines had not
been recognized. For the round symmetric beam, a single emittance-growth mechanism was isolated, which consists of the very rapid
charge-density redistribution, as the charged particles, behaving
like a plasma, adjust their positions to shield the external field
from the interior of the beam. The growth of rms emittance arises
physically from the nonlinear space-charge fields that are present
when the beam is not uniform.

The basic relationship between field energy and rms emittance
is not restricted to the 2-D round continuous beam. The relationship for a 1-D sheet beam was derived earlier,[8] and recently we
learned of an independent derivation for the 1-D case.[9] Furthermore, Hofmann has generalized the differential equation to include
2-D continuous asymmetric (elliptical) beams and 3-D bunched beams.[10]
This generalization has allowed us to derive more general formulas
for space-charge-induced emittance growth that include both the
charge-density redistribution effect and a second mechanism associated with kinetic energy exchange towards equipartitioning.[10-12]

With this perspective, we return to the specific problem of
emittance growth associated with charge-density redistribution for
nonstationary beams in linear focusing systems. We identify three
problems in which this mechanism is isolated, characterized in terms
of the number of active degrees of freedom: (a) the 1-D continuous
sheet beam, (b) the 2-D continuous round beam, and (c) the 3-D
spherical bunch. In the 2-D and 3-D examples, the beams are symmetrical in the active degrees of freedom in both position and velocity
space. Each of these problems is described in terms of a single
coordinate, the distance from the beam center. We will present
equations for emittance growth and minimum final emittance, and we
will identify a dimensionless scaling parameter, a function of the
properties of the beam and the beam channel, that characterizes
emittance growth caused by the single mechanism of charge-density
redistribution.

For heavy ion fusion accelerator systems, the 2-D problem may
correspond to a smooth approximation representation of periodic quadrupole focusing for long beams, as for an induction linac, and for
unbunched beams in storage rings. For the latter case, tune depressions may not be low enough for significant emittance growth from
charge-density redistribution. The 3-D problem may correspond to a
smooth approximation of well-bunched beams in an rf linac and may

be important for high-current bunched beams with nonuniform charge
density. If a significant asymmetry exists between longitudinal and
transverse degrees of freedom, the effects of kinetic energy exchange towards equipartitioning should also be taken into
account.[10-12] However, further study will be required to evaluate
these results for periodic quadrupole focusing systems.

EMITTANCE GROWTH FROM CHARGE REDISTRIBUTION

We consider the problem of a paraxial beam, propagating in the
+z direction at constant velocity v, which can be described by a
single coordinate, the distance from the beam center in the rest
frame of the beam. We assume that the beam particles are acted
upon by linear external forces and space-charge forces, both of
which are symmetric in the active degrees of freedom. We will include the continuous 1-D sheet beam that is infinitely extended in
a transverse coordinate y and longitudinal coordinate z; the continuous 2-D round beam, infinitely extended in z; and the 3-D spherical
bunch. The active degrees of freedom are the x-plane for the 1-D
beam, the x- and y-planes for the 2-D beam; and the x-, y-, and
z-planes for the 3-D beam. In these active planes, we will also
assume that the velocity distributions are symmetric. We consider
initial phase-space distributions that generally are not stationary,
so that the beam distribution in the active degrees of freedom will
evolve from an initial state at z = 0 to a final state, which is
stationary or approximately so.

It has already been shown that a differential equation relates
rms emittance to a quantity we call nonlinear field energy.[2,3,8,10]
For the examples of single coordinate beams with n active degrees of
freedom, we can express this result for each active plane as

$$\frac{d\varepsilon^2}{dz} = -\frac{32}{n} \frac{a^2}{Nmv^2\gamma^3} \frac{dU}{dz} \,, \qquad n = 1, 2, 3 \,, \qquad (1)$$

where the rms emittance ε is given in terms of the second moments
of the distribution in x, x' phase space as

$$\varepsilon = 4 \left(\overline{x^2}\, \overline{x'^2} - \overline{xx'}^2 \right)^{1/2} \,, \qquad (2)$$

the rms beam size is $a = \sqrt{\overline{x^2}}$, m is the mass, v is the velocity, and
γ is the relativistic mass factor. The quantity N is the number of
beam particles in appropriate units for each problem; N is the number per unit area in y and z for n = 1, the number per unit length
in z for n = 2, and the number per bunch for n = 3. The quantity U
is the difference between the self-electric field energies of the
actual beam and of the equivalent uniform beam (same second moments

as the actual beam). The quantity U has units of energy per unit area for n = 1, energy per unit length for n = 2, and energy per bunch for n = 3. It represents the residual field energy possessed by nonuniform charge distributions and corresponds to the field energy available for emittance growth.[2,3] The self-magnetic field contribution is included in Eq. (1) by using the relativistic γ factor.

For this class of problems, the rms envelope equation[13] can be written as

$$\frac{d^2a}{dz^2} + k_o^2 a - \frac{\varepsilon^2}{16a^3} - \frac{K_n}{a^{n-1}} = 0 , \qquad n = 1,2,3 \qquad (3)$$

where k_o is the zero-current single-particle wave-number that characterizes the external focusing, and K_n is the space-charge parameter given in Table I for different values of n. The quantity K_2 equals $K/4$, where K is the generalized perveance used in Refs. 2 and 3. For an rms matched beam $d^2a/dz^2 = 0$ and in the extreme space-charge limit (when $\varepsilon = 0$), we obtain a result from Eq. (3) that will be useful later:

$$a^n = K_n/k_o^2 . \qquad (4)$$

It is convenient to introduce the single-particle equation of motion for the equivalent uniform beam. We find that

$$\frac{d^2x}{dz^2} + \left(k_o^2 - \frac{K_n}{a^n}\right)x = 0 . \qquad (5)$$

Thus, the single-particle wave-number k for the equivalent uniform beam (including space charge) is identified as

$$k^2 = k_o^2 - \frac{K_n}{a^n} . \qquad (6)$$

It is easily shown that the plasma wave number (defined by $k_p^2 = e\rho/\varepsilon_o m\gamma v^2$, where e is the charge per beam particle, ρ is the charge density, and ε_o is the usual permittivity of free space) is related to K_n by $k_p^2 = nK_n\gamma^2/a^n$.

It can also be shown that the space-charge electric-field energy w_n, within the beam boundary of the equivalent uniform beam (a useful normalization quantity), is given by the values shown in Table I. The quantity w_2 is identical with w_o in Refs. 2 and 3.

Table I Space-charge and field energy normalization parameters

n	K_n	w_n
1	$\dfrac{e^2 N}{2\sqrt{3}\varepsilon_o mv^2\gamma^3}$	$\dfrac{e^2 N^2 a}{4\sqrt{3}\varepsilon_o}$
2	$\dfrac{e^2 N}{8\pi\varepsilon_o mv^2\gamma^3}$	$\dfrac{e^2 N^2}{16\pi\varepsilon_o}$
3	$\dfrac{e^2 N}{20\sqrt{5}\pi\varepsilon_o mv^2\gamma^3}$	$\dfrac{e^2 N^2}{40\sqrt{5}\pi\varepsilon_o a}$

Then, using the Table I results, Eq. (1) can be rewritten as

$$\frac{d\varepsilon^2}{dz} = -\frac{16 a^{4-n}}{n} \frac{K_n}{w_n} \frac{dU}{dz}, \qquad (7)$$

where U has the same units as w_n.

Equation (7) can be integrated, assuming an rms matched beam with constant a, which is a good approximation when space-charge forces are large. Then we find that ε and U are related by

$$\frac{\varepsilon}{\varepsilon_i} = \left[1 - \frac{1}{n} \frac{(U - U_i)}{w_n} \left(\frac{k_o^2}{k_i^2} - 1 \right) \right]^{1/2} \qquad (8)$$

where ε_i and U_i are initial values of rms emittance and nonlinear field energy. We have used a result that $16 K_n a^{4-n}/\varepsilon_i^2 = k_o^2/k_i^2 - 1$.

The minimum value is $U/w_n = 0.0$ which occurs for a uniform density beam. The quantity U/w_n is positive for both peaked and hollow distributions, increasing as the distribution becomes more nonuniform.[2,3] Furthermore, U/w_n is independent of both beam current and rms beam size and is a function only of the shape of the charge-density distribution. Examples of U/w_n values are given in Table II for different charge-density profiles and different values of n.

Table II The quantity U/w_n for some common spatial distributions

Profile n	Uniform 1	Parabolic $1 - x^2/X^2$	Gaussian $e^{-x^2/2\sigma^2}$
1	0	0.00818	0.0456
2	0	0.0224	0.154
3	0	0.0368	0.308

To predict the final rms emittance growth from Eq. (8), we must know the final value of U/w_n. Fortunately, we are able to propose a good theoretical model based on a reasonable physical assumption. We will assume homogenization (uniformity) of the final charge density ($U_f/w_n = 0$). This is an approximation that we expect to be valid for space-charge-dominated beams in linear focusing systems. This follows because the beam particles, behaving like a plasma, adjust their positions to shield the external field from the interior of the beam, and a uniform charge density produces the required linear space-charge field for exact shielding in the extreme space-charge (zero-emittance) limit. As we will see from the numerical simulation studies, this assumption is not exactly correct even in the extreme space-charge limit because of the formation of a halo that may contain several percent of the particles. Furthermore, beams with a finite emittance tend toward a matched charge-density distribution with a central uniform core and a finite thickness boundary approximately equal to the Debye length.[14] The ratio of Debye length λ_D to rms beam size, which can be written in the nonrelativistic limit as $\lambda_D/a = [n(k_o^2/k_i^2 - 1)]^{-1/2}$, becomes small for space-charge-dominated beams, and $U/w_n = 0$ becomes a good approximation. For emittance-dominated beams, where λ_D is large and $U_i/w_n \neq 0$, the emittance growth predicted from Eq. (8) will be small (because $k_i/k_o \approx 1$), in agreement with simulation. In all cases, this approximation will result in an overestimate of the emittance growth.

Then, the final emittance growth for an rms-matched beam can be written directly from Eq. (8) as

$$\frac{\varepsilon_f}{\varepsilon_i} \approx \left[1 + \frac{U_i}{nw_n}\left(\frac{k_o^2}{k_i^2} - 1\right)\right]^{1/2} = \left[1 + \frac{U_i}{n^2 w_n}\left(\frac{a}{\lambda_D}\right)^2\right]^{1/2} \quad (9)$$

For a space-charge-dominated beam, where Eq. (4) is valid, we can re-write Eq. (9) in the form

$$(k_o \varepsilon_f)^2 = (k_o \varepsilon_i)^2 + \frac{16}{n} \left(\frac{K_n}{k_o^{2-n}} \right)^{4/n} \left(\frac{U_i}{w_n} \right) . \tag{10}$$

Equation (10) implies that as ε_i approaches zero, the final emittance approaches a minimum value, given by

$$k_o \varepsilon_{f,min} = \frac{4}{\sqrt{n}} \left(\frac{K_n}{k_o^{2-n}} \right)^{2/n} \left(\frac{U_i}{w_n} \right)^{1/2} . \tag{11}$$

Equation (11) implies that the minimum final emittance decreases with increased focusing force (larger k_o), increases with beam intensity (larger K_n), and increases with the initial nonuniformity as measured by U_i/w_n. The minimum final emittance effect was first reported in numerical studies of linac injectors.[15-18]

We predict from Eq. (11) that the minimum final emittance depends upon the beam intensity through the parameter K_n as $\varepsilon_{f,min} \propto K_n^{2/n}$. Thus, $\varepsilon_{f,min}$ would vary as I^2/k_o^3 for the 1-D case, as I/k_o for the 2-D case, and as $I^{2/3}/k_o^{1/3}$ for the 3-D case, where I is the beam current. Although the results for nonsymmetric 2-D and 3-D beams are modified to account for kinetic-energy-exchange effects, the dependence of the minimum final emittance on beam current is unchanged from the results presented here for symmetric beams.[11]

Experimental evidence has been reported in support of Eqs. (9) and (10) with n = 2 for unneutralized beams in a real quadrupole transport channel.[19,20] Experimental results[15,16] for a linac beam suggest that the measured dependence of minimum final emittance on beam current is in the range $I^{1/3}$ to $I^{1/2}$. If the result is confirmed, it would be in disagreement with the $I^{2/3}$ dependence for n = 3 from Eq. (11) and would suggest that additional emittance growth mechanisms are present in real linac beams.

In Eqs. (8) and (9), the rms emittance growth is expressed as a product of two factors, one dependent on U_i/w_n, related to the change in the shape of the distribution, and one related to the initial tune depression k_i/k_o, or equivalently to the ratio of rms beam size to Debye length. The tune depression can be shown, by using Eq. (3) for an rms matched beam, to depend on a dimensionless parameter

$$u_n = \frac{2^{n-1} K_n}{\varepsilon^{n/2} k_o^{2-n/2}} . \tag{12}$$

For n=2 this definition reduces to the parameter u defined by Reiser[21] for continuous beams in periodic channels. The expression, relating the parameter u_n to k_i/k_o, is

$$u_n = (1 - k_i^2/k_o^2)/2(k_i/k_o)^{n/2} \qquad (13)$$

Therefore, for given initial beam profile (fixed U_i/w_n), Eqs. (8) and (9) predict that the emittance growth depends on the beam and channel variables K_n, ε, and k_o, through the dimensionless scaling parameter u_n. Then for constant u_n, the emittance growth is constant, and an increase in u_n implies an increase in emittance growth. Because $K_n \propto I$, the emittance growth depends on $I/\varepsilon^{1/2}$ for the 1-D case, I/ε for the 2-D case, and $I/\varepsilon^{3/2}$ for the 3-D case.

NUMERICAL STUDIES

Numerical simulation studies for the 2-D case have been published.[2,3] From the 2-D simulations for different initial charge-density profiles, it has been confirmed that the emittance differential equation [Eqs. (1) or (7)] is accurately satisfied, that for initially nonuniform beams, the emittance grows rapidly in one quarter of a plasma period, and that the final charge density approaches a uniform profile with a low-density halo as the beam becomes space-charge dominated. Furthermore, the emittance growth formulas [Eqs. (9) and (10)], based on the assumption of final state homogenization, agree well with the simulation results over the full range of tune depressions.

The computer code[2,3,22] has been modified for more general use, and we have extended the initial simulation work to include studies of the 1-D and 3-D problems. We have used 4000 particles with different rms-matched initial distributions. In general, we find that our conclusions for the 1-D and 3-D problems are nearly identical to those already presented for the 2-D case. First, we present results for a space-charge-dominated spherical bunch (n = 3) with an initial Gaussian distribution, both in position and divergence, truncated at four standard deviations and with an initial tune ratio, $k_i/k_o = 0.02$. Figures 1a and 1b show U/w_3 and $\varepsilon/\varepsilon_i$ as a function of distance z/λ_p along the beamline, where a plasma length $\lambda_p = 2\pi/k_p$ is the distance the beam travels during one plasma period. The quantity U/w_3, shown in Fig. 1a, decreases to zero during the first quarter plasma period, then oscillates and finally settles down to a steady value, somewhat larger than zero. Figure 1b shows the emittance ratio $\varepsilon/\varepsilon_i$ as a function of z/λ_p, which grows rapidly during the first quarter plasma period, and is in agreement with the emittance differential equation [Eqs. (1) or (7)] and the U/w_3 profile shown in Fig. 1a.

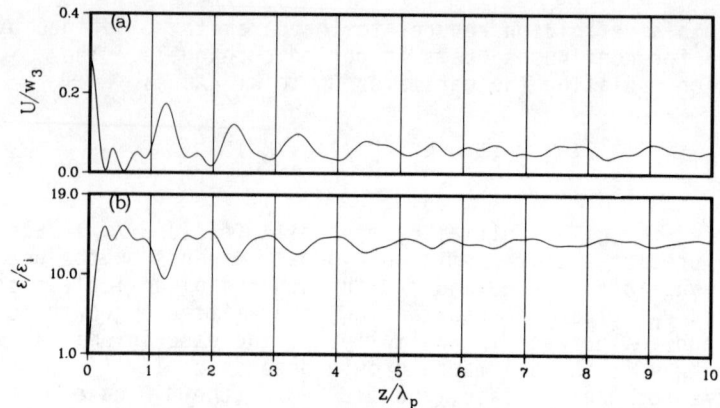

Fig. 1. Results of numerical simulation for 10 plasma periods using an initial Gaussian spherical bunch, truncated at four standard deviations, and initial tune ratio $k_i/k_o = 0.02$. The abscissa is the distance z/λ_p along the beamline, where λ_p is the distance the beam travels in one plasma period.
(a) Dimensionless nonlinear field energy U/w_3.
(b) Emittance growth ratio $\varepsilon/\varepsilon_i$.

Figure 2 shows x,x' phase-space and the radial charge density at $z/\lambda_p = 0.0$, 0.25, 0.5, and 10 for an initial Gaussian spherical bunch with $k_i/k_o = 0.02$. The variation of the phase-space distribution, seen in Figs. 2a through 2c ($z/\lambda_p = 0.0$, 0.25 and 0.5), is qualitatively consistent with the rms emittance curve shown in Fig. 1b. The phase-space distribution at $z/\lambda_p = 10$, shown in Fig. 2d, has evolved to a nearly stationary configuration composed of a central core and an extended diffuse background and corresponds to an emittance growth ratio of about $\varepsilon_f/\varepsilon_i = 16$. Figures 2e through 2h show, as a result of radial plasma oscillations, that the charge density changes from an initial Gaussian distribution at $z/\lambda_p = 0.0$ to a uniform distribution near $z/\lambda_p = 0.25$ and to a hollow beam at $z/\lambda_p = 0.50$. The charge density at $z/\lambda_p = 10$, after the plasma oscillations have damped, can be seen in Fig. 2h, and within statistical uncertainties, is consistent with a uniform distribution having a soft edge or finite thickness boundary. A low-density tail or halo remains outside the central core and contributes to the non-zero final value of U/w_3. The phase-space characteristics and the uniform charge density, observed at $z/\lambda_p = 10$, change very little at larger z/λ_p values, indicating the formation (at least approximately) of a final stationary spherical bunch.

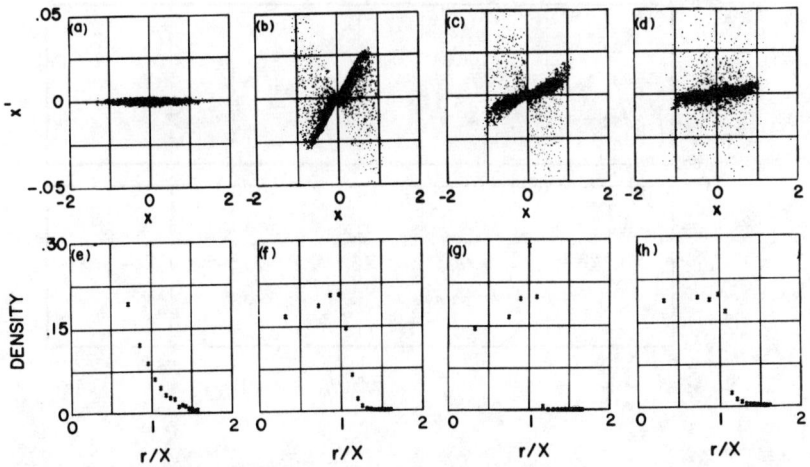

Fig. 2. Phase-space (x,x') and charge density versus normalized radius r/X, where $X = \sqrt{5}a$, at $z/\lambda_p = 0.00, 0.25, 0.50$, and 10.0 from numerical simulation for an initial Gaussian spherical bunch, truncated at four standard deviations, and initial tune ratio $k_i/k_o = 0.02$. The quantities x, x' and charge density are shown in relative units.

Figures 3a and 3b show U/w_3 and $\varepsilon/\varepsilon_i$ versus z/λ_p for an initial semi-Gaussian, or thermal, spherical bunch with $k_i/k_o = 0.25$, corresponding to uniform charge density and a Gaussian distribution in divergence (velocity) space. The quantity U/w_3 increases rapidly, which would imply an emittance decrease according to Eq. (8). Figure 3b, the rms emittance growth ratio $\varepsilon/\varepsilon_i$, shows a correspondingly small decrease in rms emittance. Examination of the charge density, at this and at other tune depressions, shows that the initially uniform beam acquires a tail or soft edge, whose width (consistent with the Debye length) increases with emittance, which leads to an increase in U/w_3. Even so, the assumption of a final uniform beam for this case will lead to only a small error in emittance growth because the emittance decrease is so small.

Figure 4 shows numerical simulation results of $\varepsilon_f/\varepsilon_i$ at $z/\lambda_p = 10$, plotted versus k_i/k_o for an initial Gaussian spherical bunch. The curve obtained from Eq. (9) with $n = 3$, using the value $U_i/w_3 = 0.308$ for a Gaussian beam, is in good agreement with the simulations and shows the steep rise in emittance growth at low tune depressions (high space charge). Figure 5 shows the same plot for an initial thermal spherical bunch, which has an initial uniform charge density and $U_i/w_3 = 0.0$. The rms emittance decrease shown in Fig. 3 also can be seen in some plotted points on Fig. 5, but is a small effect. Again, the curve from Eq. (9) is in close agreement

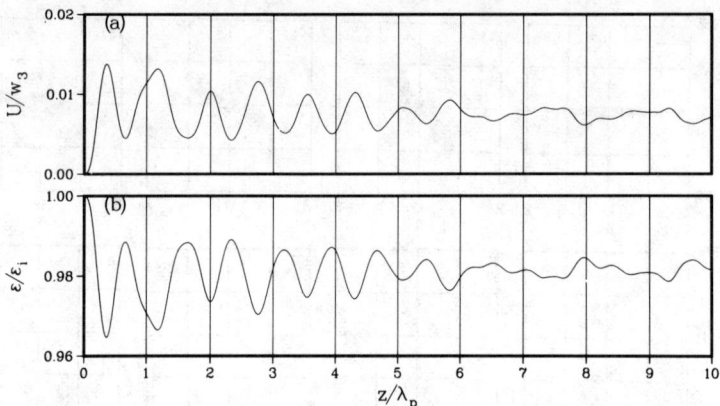

Fig. 3. Results of numerical simulations for an initial semi-Gaussian or thermal spherical bunch, truncated at four standard deviations in velocity space, and initial tune ratio $k_i/k_o = 0.25$. The quantities plotted are the same as in Fig. 1.

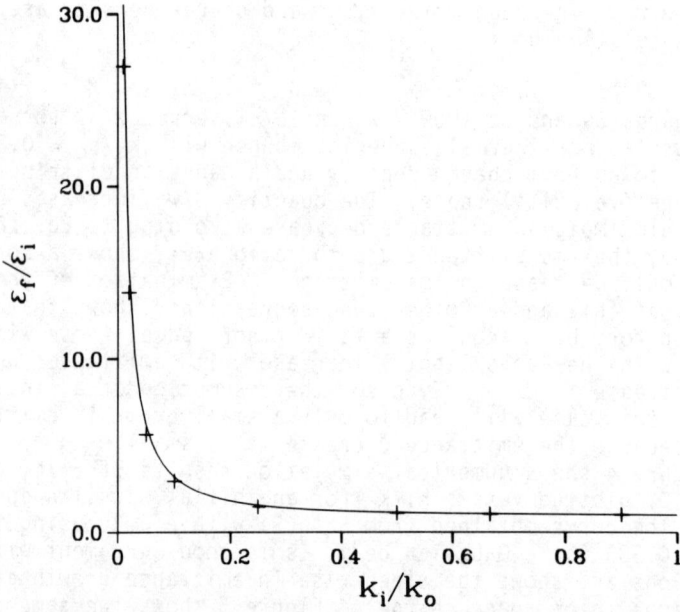

Fig. 4. Final emittance growth ratio versus initial tune ratio for an initial Gaussian spherical bunch, truncated at four standard deviations. The curve is generated from Eq. (9) with n = 3, and the plus symbols show the results of the particle simulations after 10 plasma periods.

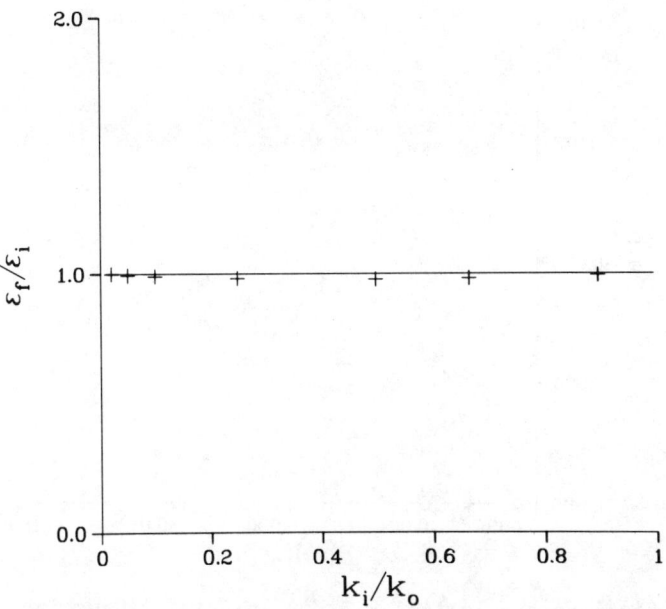

Fig. 5. Final emittance growth ratio versus initial tune ratio for an initial thermal spherical bunch, truncated at four standard deviations. The curve is generated from Eq. (9) with n = 3, and the plus symbols show the results of the particle simulations after 10 plasma periods.

with the numerical simulation results. Comparison of Figs. 4 and 5 shows that the strong emittance increase observed at low tune depressions for the initial Gaussian bunch is absent for the initial thermal bunch, which illustrates the advantage of initially uniform beams for controlling emittance growth from charge redistribution. Furthermore, from the good agreement between the curves and the numerical simulation results (shown in Figs. 4 and 5), we conclude that the final emittance growth can be closely predicted from Eq. (9), using only the initial tune ratio and the initial value of U/w_n. A comparison of Eq. (10) with particle simulations for a spherical bunch is made in Figs. 6 and 7, where $k_0\varepsilon_f$ versus $k_0\varepsilon_i$ is plotted for initial Gaussian and thermal distributions, respectively. The results from the simulations and the curves from Eq. (10) with n = 3 are, again, in close agreement. The nonzero value of the minimum final emittance is evident for the initial Gaussian bunch in Fig. 6, in contrast to the result for the initial thermal (uniform charge density) bunch in Fig. 7, where the minimum final emittance is zero.

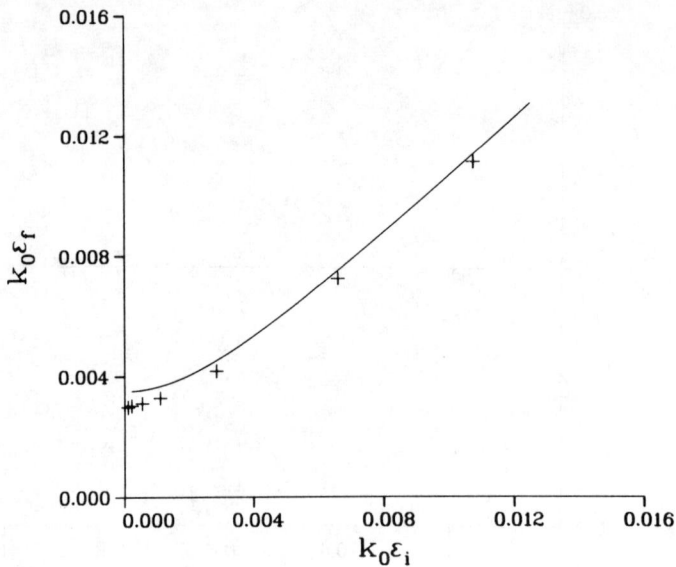

Fig. 6. Final versus initial $k_0\varepsilon$ for an initial Gaussian spherical bunch, truncated at four standard deviations. The curve is generated from Eq. (10) with n = 3, and the plus symbols show the results of particle simulations after 10 plasma periods.

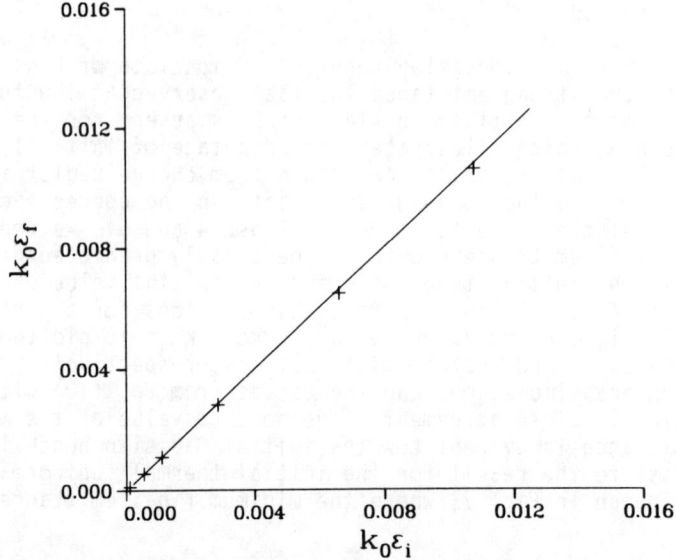

Fig. 7. Final versus initial $k_0\varepsilon$ for an initial thermal spherical bunch, truncated at four standard deviations. The curve is generated from Eq. (10) with n = 3, and the plus symbols show the results of particle simulations after 10 plasma periods.

The results for the 1-D sheet beams are qualitatively similar to the 2-D and 3-D results. In Figs. 8 through 11, we show the results of 1-D simulations for comparison with Eqs. (9) and (10) with n = 1, and (again) the agreement is good for both the Gaussian and thermal beams. The dependence of minimum final emittance on beam intensity predicted by Eq. (11) for the different degrees of freedom n, was also tested by the numerical simulation studies. The results are shown in Figs. 12a, 12b, and 12c for the 1-D, 2-D, and 3-D beams, respectively, where we have plotted minimum $k_0\varepsilon_f$ versus the dimensionless quantity K_n/k_0^{2-n} for simulations of initial Gaussian beams. The curves are obtained from Eq. (11) using the values of U_i/w_n from Table II for Gaussian beams. We observe a close agreement between the simulation results and Eq. (11). We believe that the small discrepancies are caused primarily by the halo that was discussed earlier.

The prediction that the emittance growth from charge redistribution should depend on the beam and beam-channel variables K_n, ε, and k_0, through the dimensionless scaling parameter, u_n, has also been tested by numerical simulation. We show the results in Fig. 13 for the Gaussian spherical bunch at a tune depression $k_i/k_0 = 0.10$, where we have plotted the emittance growth $\varepsilon_f/\varepsilon_i$ versus the dimensionless quantity k_0K_3, at fixed u_3. It is evident from Fig. 13 that the emittance growth is indeed constant at constant u_3.

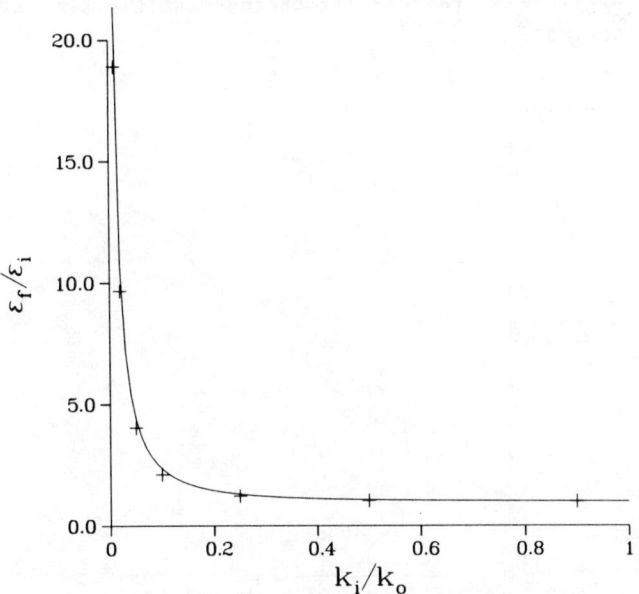

Fig. 8. Final emittance growth ratio versus initial tune ratio for an initial Gaussian 1-D sheet beam, truncated at four standard deviations. The curve is generated from Eq. (9) with n = 1, and the plus symbols show the results of the particle simulations after 10 plasma periods.

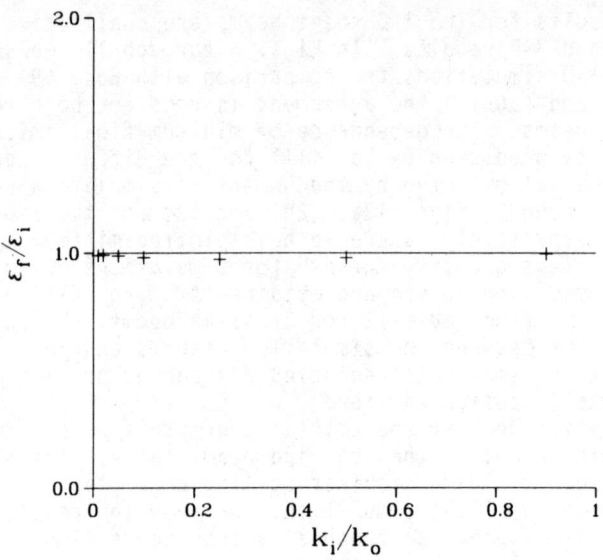

Fig. 9. Final emittance growth ratio versus initial tune ratio for an initial thermal 1-D sheet beam, truncated at four standard deviations. The curve is generated from Eq. (9) with n = 1, and the plus symbols show the results of the particle simulations after 10 plasma periods.

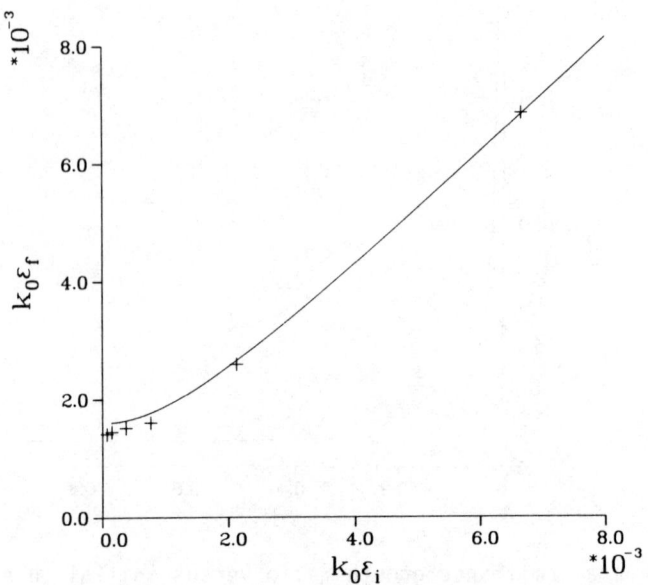

Fig. 10. Final versus initial $k_0\varepsilon$ for an initial Gaussian 1-D sheet beam, truncated at four standard deviations. The curve is generated from Eq. (10) with n = 1, and the plus symbols show the results of particle simulations after 10 plasma periods.

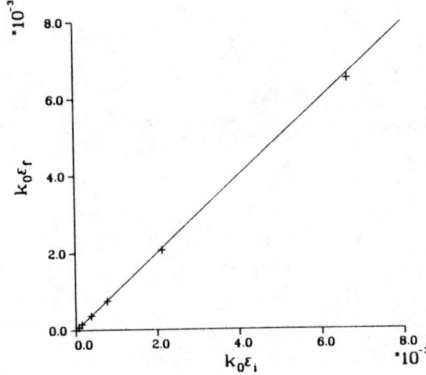

Fig. 11. Final versus initial $k_0\varepsilon$ for an initial thermal 1-D sheet beam, truncated at four standard deviations. The curve is generated from Eq. (10) with $n = 1$, and the plus symbols show the results of particle simulations after 10 plasma periods.

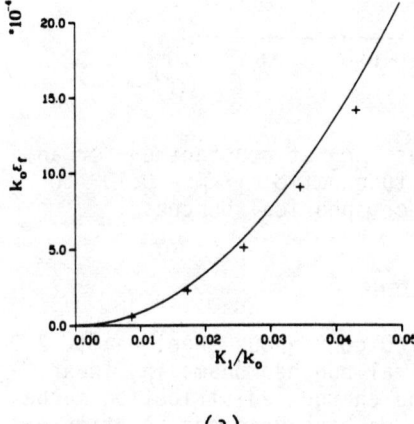

(a)

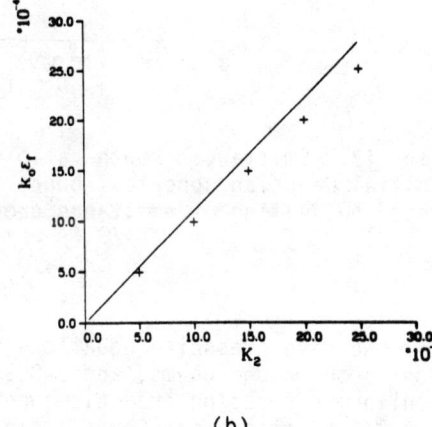

(b)

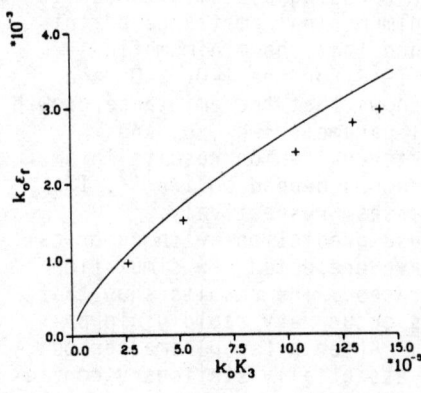

Fig. 12. A test of the dependence of minimum final emittance on beam current for initial Gaussian distributions and for (a) 1-D sheet beams, (b) 2-D round beams, and (c) 3-D spherical bunches. The curves are generated from Eq. (11), and the plus symbols show the results of particle simulations after 10 plasma periods.

(c)

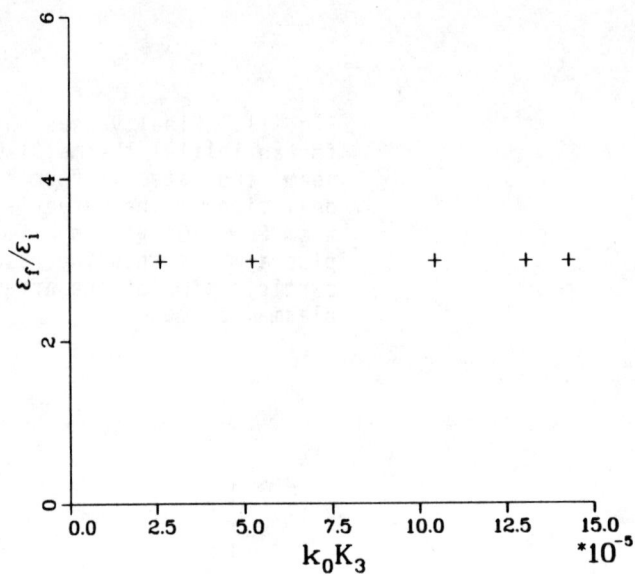

Fig. 13. Emittance growth ratio versus k_0K_3 at constant u_3 for an initial Gaussian spherical bunch at a tune ratio $k_i/k_0 = 0.10$, to test the scaling of emittance growth for spherical bunches.

CONCLUSIONS

We have presented equations for 1-D continuous sheet beams, 2-D continuous round beams, and 3-D spherical bunched beams in linear continuous focusing channels, where the charge redistribution mechanism of emittance growth is isolated. We have used the relation between field energy and rms emittance, together with the approximate property of final charge-density homogenization (uniformity) for intense nonuniform beams in linear focusing systems, to derive equations for emittance growth and minimum final emittance of initial nonstationary beams. We have found that the minimum final emittance should depend on I^2, I, and $I^{2/3}$ for the 1-D, 2-D, and 3-D problems, respectively. We have shown that the emittance growth depends upon the beam and beam-channel parameters K_n, ε, and k_0, through a dimensionless scaling parameter u_n, which results in a prediction that the emittance growth should depend on $I/\varepsilon^{1/2}$, I/ε and $I/\varepsilon^{3/2}$ for the 1-D, 2-D, and 3-D cases, respectively.

We had already tested some of these predictions with numerical simulation for the 2-D problem. We have presented new simulation results that include the 1-D and 3-D cases. The results show that the field energy and emittance changes occur very rapidly, in approximately one-quarter plasma period. After a few plasma periods, the beam distributions settle down to essentially stationary configurations, characterized by a nearly uniform charge density, together with a diffuse halo, as space charge becomes more dominant. For all cases, we find good agreement between the predictions from the equations and the numerical simulation results, thus substantiating the validity of the emittance growth formulas, the dependence of minimum final emittance on beam current, and the scaling of emittance growth. Finally, if these results can be extended to periodic quadrupole systems, they suggest that the challenge for minimizing the emittance growth caused by the charge-density redistribution mechanism in high-current, low-emittance beams is to learn how to produce the beams, whether continuous or bunched, with a uniform charge-density profile.

ACKNOWLEDGMENTS

We wish to thank R. A. Jameson for his encouragement and P. M. Lapostolle for helpful suggestions.

REFERENCES

1. Thomas P. Wangler, "Design Constraints For an RF-Linac/Storage Ring Driver," Laser and Particle Beams $\underline{2}$, Part 4, 413 (1984); and in Proc. of the 1984 INS Int. Symp. on Heavy Ion Accelerators and Their Applications to Initial Fusion, Institute for Nuclear Study, Tokyo, January 23-27, 1984, p. 222.

2. T. P. Wangler, K. R. Crandall, R. S. Mills, and M. Reiser, "Relationship Between Field Energy and RMS Emittance in Intense Particle Beams," IEEE Trans. Nucl. Sci. $\underline{32}$, 2196 (1985).

3. T. P. Wangler, K. R. Crandall, R. S. Mills, and M. Reiser, "Field Energy and RMS Emittance in Intense Particle Beams," Proc. of Workshop on High Current, High Brightness, High Duty Factor Ion Injectors, San Diego, May 21-23, 1985, AIP Conf. Proc. No. 139, 133 (1986).

4. J. Struckmeier, J. Klabunde, and M. Reiser, "On the Stability and Emittance Growth of Different Particle Phase-Space Distributions in a Long Magnetic Quadrupole Channel," Particle Accelerators $\underline{15}$, 47 (1984).

5. P. M. Lapostolle, "Energy Relationships in Continuous Beams," Los Alamos National Laboratory translation LA-TR-80-8 or CERN-ISR-DI/71-6 (1971).

6. P. M. Lapostolle, "Possible Emittance Increase Through Filamentation Due to Space Charge in Continuous Beams," IEEE Trans. Nucl. Sci. $\underline{18}$ (3), 1101 (1971).

7. E. P. Lee, S. S. Yu, and W. A. Barletta, "Phase Space Distortion of a Heavy-Ion Beam Propagating Through a Vacuum Reactor Vessel," Nuclear Fusion $\underline{21}$, 961 (1981).

8. T. P. Wangler, "Energy and Emittance Relations from RMS Equations of Motion in 1-D," Los Alamos National Laboratory memorandum AT-1:84-297, August 27, 1984.

9. O. A. Anderson, "Internal Dynamics and Emittance Growth in Nonuniform Beams," these proceedings.

10. I. Hofmann, "Emittance Growth," presented at the 1986 Linear Accelerator Conference, Stanford Linac Accelerator Center, June 1986.

11. T. P. Wangler, F. W. Guy, and I. Hofmann, "The Influence of Equipartitioning on the Emittance of Intense Charged Particle Beams," presented at the 1986 Linear Accelerator Conference, Stanford Linac Accelerator Center, June 1986.

12. F. W. Guy and T. P. Wangler, "Numerical Studies of Emittance Exchange in 2-D Charged-Particle Beams," presented at the 1986 Linear Accelerator Conference, Stanford Linac Accelerator Center, June 1986.

13. F. J. Sacherer, "RMS Envelope Equations with Space-Charge," IEEE Trans. Nucl. Sci. $\underline{18}$ (3), 1105 (1971).

14. T. P. Wangler and K. R. Crandall, "Characteristics of the Stationary Waterbag Distribution," Los Alamos National Laboratory memorandum AT-1:86-4, January 1986.

15. P. Lapostolle, "Round Table Discussion on Space-Charge and Related Effects," Proc. of 1968 Proton Linear Accelerator Conf., Brookhaven National Laboratory report BNL 50120 (1968), 433.

16. P. M. Lapostolle, C. Taylor, P. Tetu, and L. Thorndahl, "Intensity Dependent Effects and Space Charge Limit Investigations at CERN Injector and Synchrotron," CERN 68-35 (1968).

17. M. Prome, "Effects of Space Charge in Proton Linear Accelerators," Thesis presented at Universite Paris Sud, Centre d'Orsay (1971), Los Alamos translation LA-TR-79-33, 90-92.

18. R. Chasman, "Numerical Calculations on Transverse Emittance Growth in Bright Linac Beams," IEEE Trans. Nucl. Sci. $\underline{16}$, 202 (1969).

19. J. Klabunde, P. Spädtke, and A. Schönlein, "High Current Beam Transport Experiments at GSI," IEEE Trans. Nucl. Sci. $\underline{32}$, 2462 (1985).

20. D. Keefe, "Summary for Working Group on High Current Beam Transport," Proc. of Workshop on High Current, High Brightness, High Duty Factor Ion Injectors, San Diego, California, May 21-23, 1985, AIP Conference Proceedings No. 139 (1986).

21. M. Reiser, "Periodic Focusing of Intense Beams" Part. Accel. $\underline{8}$, 167-182 (1978).

22. K. R. Crandall, R. S. Mills, and T. P. Wangler, "Simulation of Continuous Beams Having Azimuthal Symmetry to Check Relation Between Emittance Growth and Nonlinear Energy," AT-1 Group memorandum AT-1:85-218, June 12, 1985.

THE MARYLAND TRANSPORT EXPERIMENT--GENERAL PERSPECTIVES
AND RECENT RESULTS WITH OFF-CENTERED BEAMS*

M. Reiser, J. McAdoo, D. Kehne, and K. Low
Electrical Engineering Department
and
Laboratory for Plasma and Fusion Energy Studies
University of Maryland
College Park, MD 20742

and

J. D. Lawson and C. R. Prior
Rutherford Appleton Laboratory
Chilton Didcot, Oxfordshire OX11 0QX
United Kingdom

ABSTRACT

The experiment at the University of Maryland was designed to study the physics of space-charge-dominated beam transport through a periodic focusing channel. Its major goal is to check theoretical predictions of current limits and instabilities and to investigate the effects of nonlinear lens forces, misalignments, and beam off-centering. In the recently completed second phase, a 5 kV, 180-200 mA electron beam is transported through 36 equally spaced solenoid lenses. The focusing strength of the channel is measured in terms of the phase shift σ_o without space charge. Below $\sigma_o = 50^0$, the focusing is too weak to confine the beam within the pipe. Between 60^0 and 80^0 a "window" of lossless transmission is observed. Above $\sigma_o = 80^0$, the transported current begins to drop. The structure of the transmission curve versus σ_o in this fall-off region is not smooth, and is very sensitive to alignment errors of the gun and of individual magnets. Results are compared with computer simulations. The potential of the experiment and expected roles of instabilities, nonlinearities, and errors will be discussed.

INTRODUCTION

It was at the second Heavy Ion Fusion (HIF) Workshop held nearly nine years ago[1] that plans for beam transport experiments at Maryland began to be made. Analytical work at Berkeley suggested that beams with the Kapchinskij-Vladimirskij (K-V) distribution in phase-space propagating in periodic channels would show

---

*Research supported by the U.S. Department of Energy.

instabilities for various ranges of the parameters σ_o and σ which represent the phase shift (or "tune") per focusing period of the particle oscillations without and with space charge, respectively. Analytical work with more realistic distributions is intractable, and so programs of computational and experimental work were discussed and initiated shortly afterwards. It was argued that if computation agrees with theory for the K-V distribution, it should then give reliable indications for other distributions and provide a useful guide and check for experiments.

Since that time, a great deal of computational work has been done and new insights achieved. The pioneering theory work on K-V instabilities at Berkeley, including simulation results by I. Haber which had been circulated in several technical reports, was published in 1983 in Particle Accelerators.[2] There have also been fruitful experimental programs at Berkeley, Maryland, and Darmstadt. This progress is chronicled both in HIF workshops and the proceedings of accelerator conferences. Besides presenting the latest experimental findings, it is appropriate here to review what has been learned compared with expectations, and inquire what the next stage in the experimental program should be.

First, our understanding of what is meant by the term ´instability´ has been clarified. Although indicated theoretically by the usual criterion of complex frequencies given by dispersion relations, many instabilities for a K-V distribution are now known to represent slight changes in the distribution functions to a new form with almost the same emittance and physical size. The ´growth´ saturates almost immediately. There are good theoretical reasons for believing that practical distributions are more stable, and computations such as the detailed studies of different phase-space distributions by Struckmeier, Klabunde, and Reiser[3] show this to be the case.

Second, the concept of matching distributions to the channel has been made more precise. For a K-V distribution, matching to a uniform or a periodic distribution is straightforward. In a periodic system, the beam radius has the same periodicity as the channel, and the matched solution can be found quite easily using numerical methods. Other distributions in general will not propagate even down a uniform channel without change of form. Here the term ´rms matching´ is used, as defined by Lapostolle and Sacherer[4]. This means that the distribution is adjusted such that its rms emittance is the same as that of a K-V distribution that is matched to the channel. Such a beam changes form as it propagates along the channel since it is not matched in detail, and its size and emittance may increase. It is not yet known whether the distribution ultimately settles down asymptotically to a steady state. So far the only known matched distribution in a periodic channel is the K-V distribution. Struckmeier[5] showed that the generalized waterbag distribution of a continuous channel also behaves like a matched beam in a periodic system if a suitable transformation of the phase-space coordinates is made. But a general, satisfactory proof for distributions other than K-V is still lacking at this time. If a general, nonuniform distribution is matched in an rms sense, but not in detail, it readjusts itself very rapidly on the time scale of a plasma period. This internal redistribution of particles may result in substantial emittance growth due to conversion of field energy to kinetic energy.[3,6] At low values of σ, all nonuniform distributions tend to become uniform in a focusing channel with linear forces. This redistribution is governed by the conservation of total transverse energy. In addition to redistribution of the transverse energy in the manner indicated, there is also the possibility of parametric excitation of instability, similar to that of a K-V beam, when suitable integral relationships exist between the channel periodicity and plasma oscillations arising from the two transverse dipole modes, or other internal degrees of freedom of the beam. This topic is not easy to summarize succinctly. Here we merely note that although there is evidence of such behavior from computational studies,[3,7] it is difficult to identify a specific resonance effect of this type unambiguously in the experiment since several instabilities as well as other effects might be present.

Looking back to the concerns felt about beam transport in 1977, the main practical outcome has been the realization, confirmed both computationally and experimentally, that at least within the range of σ_o below 90^0, the tune depression σ/σ_o can safely be made very small without too unpleasant effects. Since 90^0 is a comfortable practical value, this result is reassuring. Nevertheless, it might be economically advantageous to use even larger values of σ_o, and it is important to understand what happens beyond 90^0. For sufficiently large σ_o, the motion becomes unstable. When there is no space charge, the stability limit occurs at $\sigma_o = 180^0$. In the presence of space charge, the precise value, always below 180^0, depends on the structure and on the beam intensity. The experimental uncertainties are greatest in this region, as will be discussed below.

THEORETICAL BACKGROUND ON INSTABILITIES AND SCALING LAWS

The degree of instability predicted by analytical theory for a K-V beam in a periodic channel[2] is measured in terms of the parameter ("eigenvalue") λ. For $|\lambda| = 1$ the mode is stable, and for $|\lambda| > 1$ it is unstable. In the latter case, e-fold increase of the initial amplitude of the perturbation occurs in $N = 1/\ell n|\lambda|$ focusing periods of the channel lattice. As an example, for $|\lambda| = 1.1$ one finds $N \approx 10.5$. The two cases considered to be most important are the envelope instabilities predicted to occur for $\sigma_o > 90^0$ and the third-order instabilities for $\sigma_o > 60^0$. Results for a solenoid channel obtained from calculations by Laslett[8] and Struckmeier[9] are shown in Fig. 1 (envelope modes) and Fig. 2 (third-order modes). These figures indicate unstable behavior for certain ranges of σ values. Note that decreasing σ implies increasing beam intensity (space charge). As σ_o increases, the instability regions grow in strength, become wider, and move

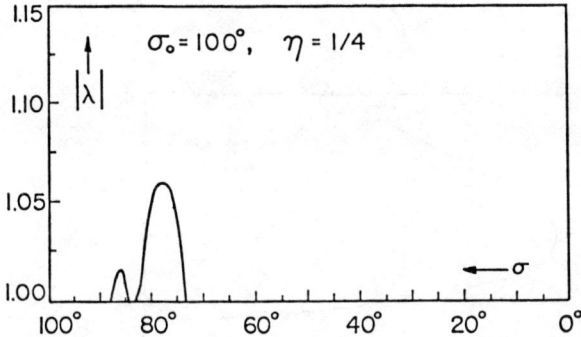

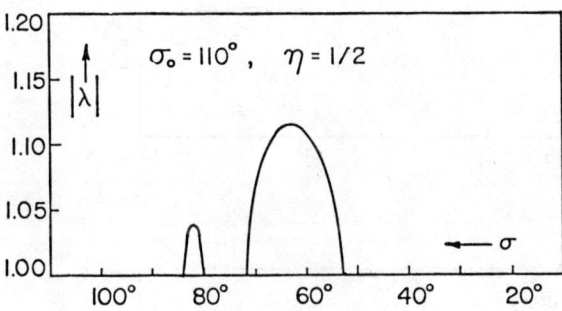

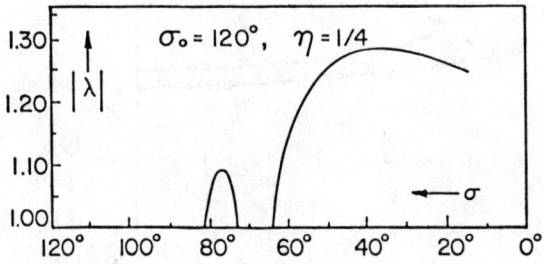

FIG. 1. Growth rate of K-V envelope instability in a periodic solenoid channel for three values of σ_o (100^0, 110^0, 120^0) as a function of the depressed tune σ. (From Laslett[8] and Struckmeier[9])

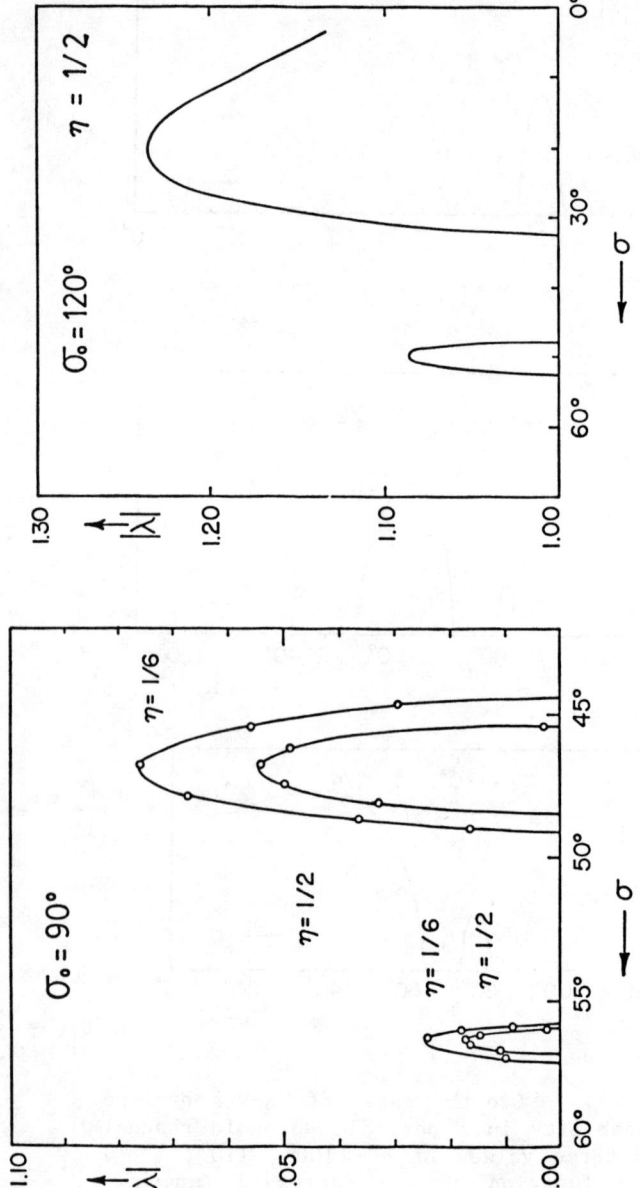

FIG. 2. Growth rate of K-V third-order instability in a periodic solenoid channel for $\sigma_o = 90^0$ and $\sigma_o = 120^0$ as a function of σ. (From Laslett[8]).

towards lower values of σ (higher beam intensities). Note that the scales and occupancy factors (η = ratio of magnet length to cell length) differ in the various graphs. The plot for $\sigma_o = 90^0$ in Fig. 2 indicates that the instability becomes weaker when the occupancy factor η increases. In fact, the third-order instability disappears completely in a continuous solenoid where η = 1. It should also be noted that these instabilities are generally more pronounced in periodic focusing systems using quadrupole lenses rather than solenoids. As mentioned in the Introduction, particle simulation studies with different distributions have shown that some of the instabilities should not occur in a practical beam (such as modes of order higher than three) or are less severe than the K-V theory indicates. The third-order mode becomes weaker the more the distribution deviates from the K-V case, and practically disappears in the case of a Gaussian distribution.[3] This is why it is now believed that beam transport in a system with $\sigma_o = 90^0$ might be safe. On the other hand, envelope instabilities do apparently not decrease in strength when a distribution becomes more nonuniform. Thus, it was found in simulation studies for a solenoid channel with $\sigma_o = 120^0$ and $\sigma = 35^0$ that emittance growth due to envelope instability was more severe in a Gaussian beam than in a K-V beam.[7]

The question as to how large σ_o can become before instability sets in is not only an academic one, but of great practical importance. The current that can be transported through a channel with a given aperture increases with σ_o; in fact below 90^0 it is proportional to $\sigma_o^2[1 - (\sigma/\sigma_o)^2]$ according to the smooth approximation theory.[10] One of the major goals of the Maryland experiment is to explore the upper limits of σ_o for stable transport. An important aspect is to compare experimental data with simulation results and to identify the effects that lead to beam loss and emittance growth. Agreement between experiment and computation would give the confidence in the validity of the codes that is needed for the design of the much larger and more expensive transport systems needed for heavy ion fusion. Such studies must include the nonlinear focusing forces of the lenses as well as beam off-centering and misalignments that will be discussed in the next sections of this paper.

If the aim of the Maryland experiment is to check theory and simulation and if these results are to be applied to a beam transport line such as in an HIF installation, then we must ensure that the scaling laws are understood. Since the instability and emittance growth effects to be investigated had been initially calculated in paraxial approximation, it seemed only necessary to satisfy the paraxial scaling condition that Ka^2/ϵ^2 is conserved, where K is the generalized perveance and a the beam radius.[11] (At fixed energy and fixed radius, this is sometimes also known as I/ϵ^2 scaling.) For complete scaling, to include nonlinear effects it is necessary that K and a/ε be separately conserved; furthermore, focusing elements must be geometrically similar so that end effects and other geometric distortions are reproduced. A characteristic parameter that must be conserved for true scaling, but not for paraxial scaling, is a/λ_o, where λ_o is the betatron wave length for the same paraxial conditions. The parameter a/λ_o characterizes $1/2\pi$ times the typical angle that the trajectories make with the axis. In the Maryland experiment a/λ_o for particle oscillations without space charge could exceed 1%, whereas in a realistic beam line it could be considerably less. For this reason nonlinear effects are expected to be much more serious in the experiment than in a real beam line, and it is desirable that they should have only a small effect on the results. In fact, for large values of σ_o nonlinear effects were found; their effect was to produce classical spherical aberration, giving rise to alternate halos and hollow beams.[12] Their possible effect in the interpretation of experimental results to be presented is considered. Other important effects to be studied in an experimental program are lens imperfections and alignment errors. These are particularly important at the rather low magnetic field levels used in this experiment, where stray fields can be a problem. Although in earlier experiments with twelve lenses[13] alignment of the gun and imperfections of the lenses did not seem to give serious problems, this was not the case when all 36 lenses were in use. Some of the difficulties, and methods used to overcome them are described below. Before considering this important topic further, we proceed to a description of the experimental facility and recent results.

DESIGN CRITERIA AND BASIC PARAMETERS OF THE EXPERIMENT

It is worth re-examining the criteria for a useful beam transport experiment in the light of experience during the past years. This is done with special reference to the work at Maryland. We still feel that a good choice was made for the actual parameters, and repeat below what the main considerations were. Difficulties associated with nonlinear effects and alignment tolerances were both anticipated and found. Especially with focusing stronger than that needed for a beam transport channel, interesting nonlinear effects were found.[12]

First, in planning the experiment several choices had to be made. Solenoids were chosen instead of quadrupoles for reasons of simplicity. Instabilities of the type to be explored occur with solenoids as well as with quadrupoles, though of course the latter would be used in practical systems for HIF. The second choice, of particle beam species and energy, is a matter of experimental convenience. Electron beams of the required perveance can simply and conveniently be produced from Pierce guns in the energy range of a few kV; further, diagnostic equipment is relatively inexpensive. It was originally thought that emittance spoiling grids would be needed to cover the required range of emittance, and such grids were developed.[14] The emphasis on transport near the space-charge limit ($\sigma \rightarrow 0$) and the finding that stable transmission was possible even for very small values of σ/σ_o has, however, meant that the grids have not yet been used.

It turned out that this experiment, of suitable scale for a university laboratory, complemented the larger scale electrostatic quadrupole beam line built for a Cesium beam at Berkeley.[15] The original target of constructing a 36-lens channel was achieved more than a year ago. Ideally, the aim is to scan $\sigma - \sigma_o$ space and correlate these parameters with emittance growth and instability. So far, for reasons to be discussed, it has only been possible to do this in part.

The basic features and parameters of the Maryland facility have been described previously.[13] For convenience, we summarize the values of the most pertinent parameters in Table I. In the fall of 1984, we completed the second phase which brought the facility to its full design length of 36 lenses plus two matching solenoids. During the past year we built alignment systems for the gun and the two matching lenses, and added an aperture plate with holes for different beam diameters. In addition, we made important improvements in the beam diagnostics. A schematic of the facility is shown in Fig. 3.

Table I. The parameters of the Maryland facility.

Electron Gun Voltage	5 kV
Pulse Length	5 μs
Repetition Rate	60 Hz
Beam Current	180–200 mA
Cathode Radius	1.27 cm
Cathode Temperature	$kT \sim 0.13$ eV
Number of Matching Lenses	2
Number of Channel Lenses	36
Length of Channel Period	13.6 cm
Total Length of Transport System	520 cm
Initial (Intrinsic) Beam Emittance	$\varepsilon = 8.8 \times 10^{-5}$ m-rad
Initial Tune Depression in Channel	$\sigma/\sigma_o \approx 0.1$

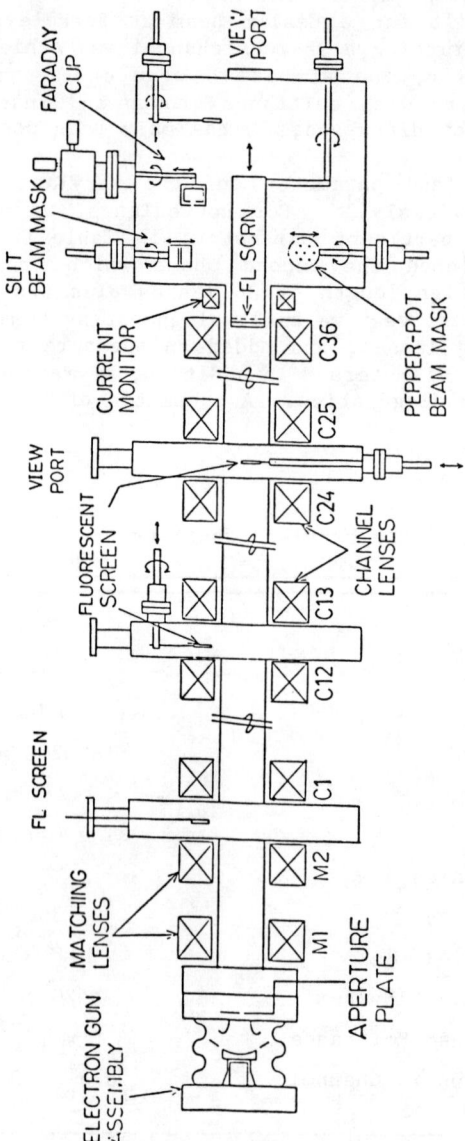

FIG. 3. Schematic of the Maryland electron beam transport experiment. Current monitoring Rogowski coils are located behind the cathode and after coils C1, C9, C14, C20, C26, and C36.

ALIGNMENT AND BEAM DIAGNOSTICS

Alignment of the gun and the lenses was found to be much more critical in the long channel than in our previous studies with 12 lenses. Thus, we have spent considerable time in developing a precise alignment technique which is not yet fully completed and satisfactory.

The alignment of the lenses and electron gun proceeds in two steps. First, a surveyor´s transit is used to locate all the lens centers on a line. Then the drift tube is threaded through the lenses, and the gun and two matching lenses are installed. These latter cannot be aligned with the transit; therefore, a second procedure is necessary. The gun is apertured allowing a fine beam, initially 400 microns in diameter, to drift down the channel. This beam is made to propagate in a straight line with all lenses off, by cancelling residual fields in the vicinity of the beam path. The gun is aimed to make the beam hit the centers of two widely spaced phosphor screens. Then the first lens is energized. If it is off center, the beam will be deflected on the phosphor screens. The lens is moved until all such deflection vanishes. Then the next lens is energized and aligned the same way and so on down the entire channel. This procedure, though simple in theory, is difficult in practice, the main difficulty being the nulling of the residual fields along the beam path. We are currently preparing to follow a procedure used on the Stanford Linear Accelerator in which a combination of Helmholtz coils and magnetic shielding was used to reduce the fields to less than 1 milligauss.[16]

The power supply for the electron gun produces short pulses of typically about 5 µs length (to avoid charge neutralization by the background gas) at a repetition rate of 60 Hz. The pulse amplitude for the experiments described here has been 5 kV.

Four beam parameters are measured in this experiment: 1) beam current, 2) beam position, 3) beam radius, and 4) beam rms emittance. The beam current is monitored with Rogowski coils at six locations along the channel as indicated in Fig. 3. In addition, the cathode current is measured with a Rogowski coil around the cathode heater leads. Beam position is obtained by videographing the image of the beam on phosphor screens at the four locations indicated in the figure. The videographed image is stored on video tape and later digitized for measurement of the distance between the beam center and the channel axis. Due to the combinations of lenses and distances involved, the spatial resolution of the system varies from 0.1 mm at the downstream location nearest the camera to 0.5 mm at the next upstream screen. There is a side port equiped with a mirror at the center station. The camera is moved to this location to view the two most upstream screens. Beam radii are measured with the same videographic system used for the beam position; however, measurement uncertainty is dominated by a qualitative decision about the definition of the beam edge. Generally the image consists of a bright circular core surrounded by an irregular halo. Because the core is usually well defined, we measure its radius rather than the radius of the extended image. Beam emittance is measured by a slit and pinhole combination reported previously.[13] The raw data from the emittance measurement is reduced using an improved technique, described in Ref. 17, which is more accurate than the previous method when the beam profile is nonuniform. The beam may be off-center, but is assumed to have a circular cross section. When the beam is not perfectly round, the method gives only an approximate value of the emittance. However, this usually suffices to judge the degree of beam deterioration that might occur during transport through the channel.

EXPERIMENTAL RESULTS WITH OFF-CENTERED BEAM

As discussed in the previous section, the alignment technique tried so far was not satisfactory since the earth field and stray magnetic fields could not be balanced out adequately with the existing Helmholtz coil. (These fields are only of concern for the alignment procedure; when all of the channel solenoids are energized, their effect on the beam is negligible.) Before proceeding with improving the alignment method, we decided to study the beam transport with imperfect centering. In particular, the electron gun and the two matching lenses were not aligned precisely. As a result, the beam entered the periodic channel (after the two matching lenses) with its center off the axis by an amount of between 1-2 mm and with a small tilt angle that has not been measured.

For a given magnetic field strength corresponding to a calculated value of the parameter σ_o, the amount of beam current transported through the entire channel was maximized by systematic variation of the currents through the two matching lenses (M1 and M2 in Fig. 3). This procedure did not necessarily give good matching of the beam. In fact, the data shows that the beam was usually not well matched. The beam current was measured with Rogowski coils at several stations along the transport system.

Figure 4 shows the fraction of transported current at three positions along the channel as a function of the phase shift σ_o. The decreasing transmission below $\sigma_o = 50^0$ is due to off-centering and the fact that the focusing force is too weak to confine the beam within the pipe. Between 60^0 and 80^0, the transport efficiency is 100%. The most noticeable effects, though, are the rapid loss of current above $\sigma_o = 80^0$ and the occurence of dips at several values of σ_o. These dips are clearly the result of beam off-centering and misalignments and they are enhanced by the longer transport distance. In Fig. 5, we show the beam transmission in the range $80^0 < \sigma_o < 110^0$ in more detail and at additional positions along the channel. From this figure it is clear that beam loss increases with distance over the entire range. The dips, where resonant-like loss occurs, are located at σ_o values of 83^0, 90^0, and 100^0. This behavior cannot be explained by any of the

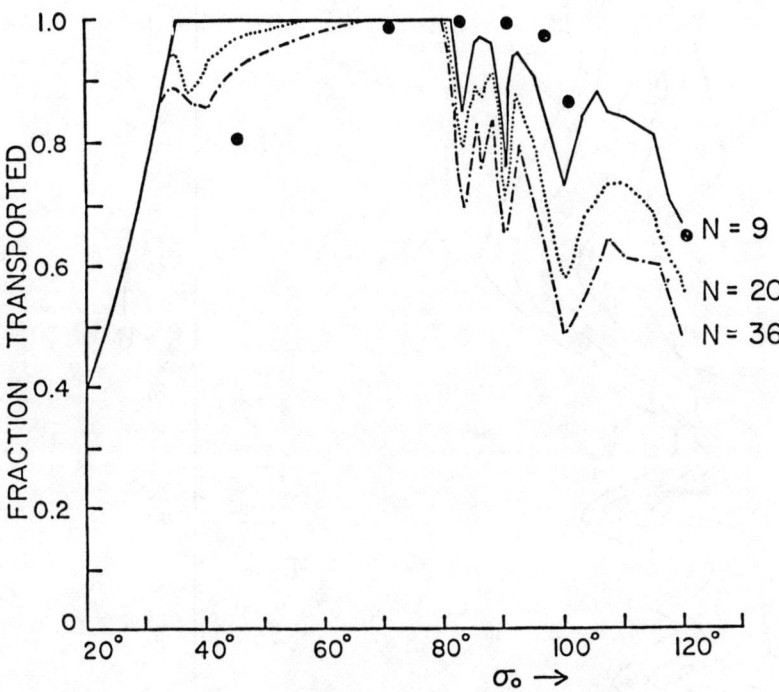

FIG. 4. Fraction of the beam current transported versus σ_o after lens C9, C20, and C36. The dark circles indicate the simulation results at C36 for 1.5 mm initial off-centering of beam.

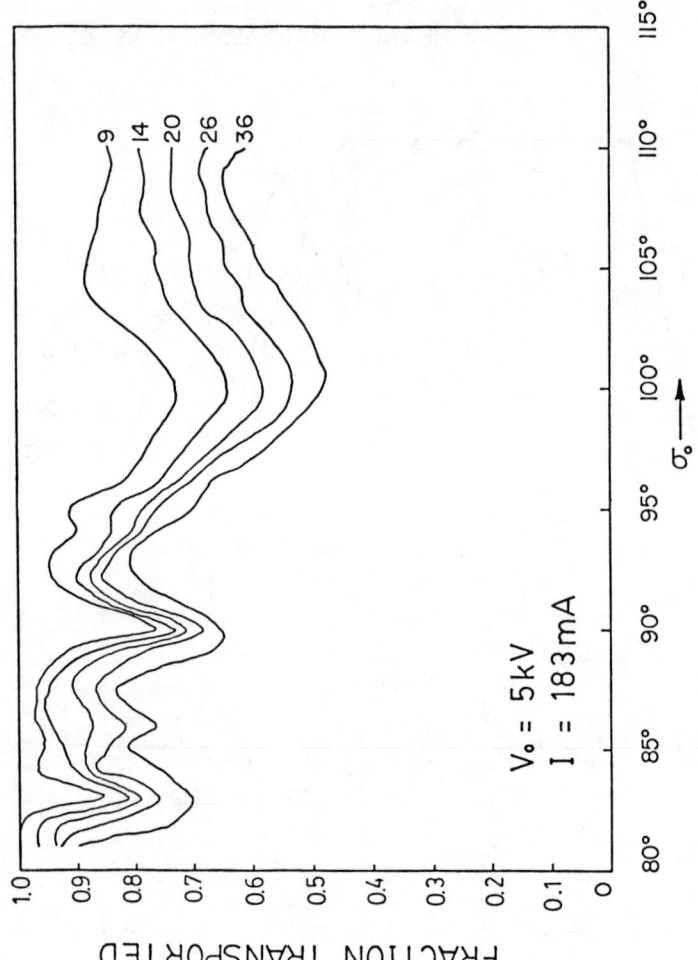

FIG. 5. Transmission curves in the range $80° < \sigma_o < 110°$ showing more details than Fig. 4.

known instabilities discussed above. Indeed, for the parameter regime of our solenoid channel, neither the third-order nor the envelope instability is expected to occur in this region for a depressed tune of $\sigma/\sigma_o \sim 0.1$. Thus, the dips are the result of off-centering and nonlinear forces, which apparently add in a destructive way at the particular values of σ_o. We found that alternating the polarity of the lens fields or placing small permanent magnets at various places along the channel would make the dips disappear. The dark circles in Fig. 4 indicate the results of simulation studies which will be discussed in the next section.

Figure 6 shows the phosphor-screen pictures of the beam for $\sigma_o = 60^0$ (left), where 100% transmission is observed, and for 107^0 (right), where substantial loss occurs, at the channel entrance (N = 0), and three different positions downstream (after lenses C12, C24, and C36). The measured center points of the beam corresponding to these positions are plotted in Fig. 7. Obviously, in both cases the beam is further off center at the end of the channel than at the beginning, which suggests that there might be further lens misalignments down the channel. Although, according to Fig. 4, beam transmission in the $\sigma_o = 60^0$ case is almost 100%, one would expect losses to occur if the transport continued beyond the 36 periods of our system. With regard to the $\sigma_o = 107^0$ case, the original photographs indicate that the substantial beam loss is due to halo formation which is not shown too well in the copy (Fig. 6) except for the last picture frame (N = 36). A well defined beam core remains although its shape (as well as that of the halo) deviates from the circular symmetry of the initial beam. The physics responsible for the halo formation is not yet fully understood. (We think that the effect could be caused by the nonlinear lens forces as discussed in Ref. 12 or by third-order instability.) However, the general behavior of an off-centered beam is consistent with the computational results discussed in the following section.

For comparison with the simulation results we have made a preliminary measurement of the beam emittance at $\sigma_o = 70^0$. The beam current was 185 mA, and the cathode temperature was 1125^0 C which corresponds to an intrinsic thermal emittance of 8.8×10^{-5} m-rad (unnormalized). The measured emittance at the end of the channel was 20×10^{-5} m-rad. This represents an increase by a factor 2.3 from the calculated intrinsic emittance and is somewhat lower than the simulation result to be discussed below.

RESULTS OF NUMERICAL SIMULATION STUDIES

In an effort to obtain some understanding of the experimental behavior, one of us (C. Prior) has started a series of simulation runs with an improved computer code that includes nonlinear forces up to third order, beam rotation, and conducting boundaries. This code, which uses realistic magnetic field profiles, is described in

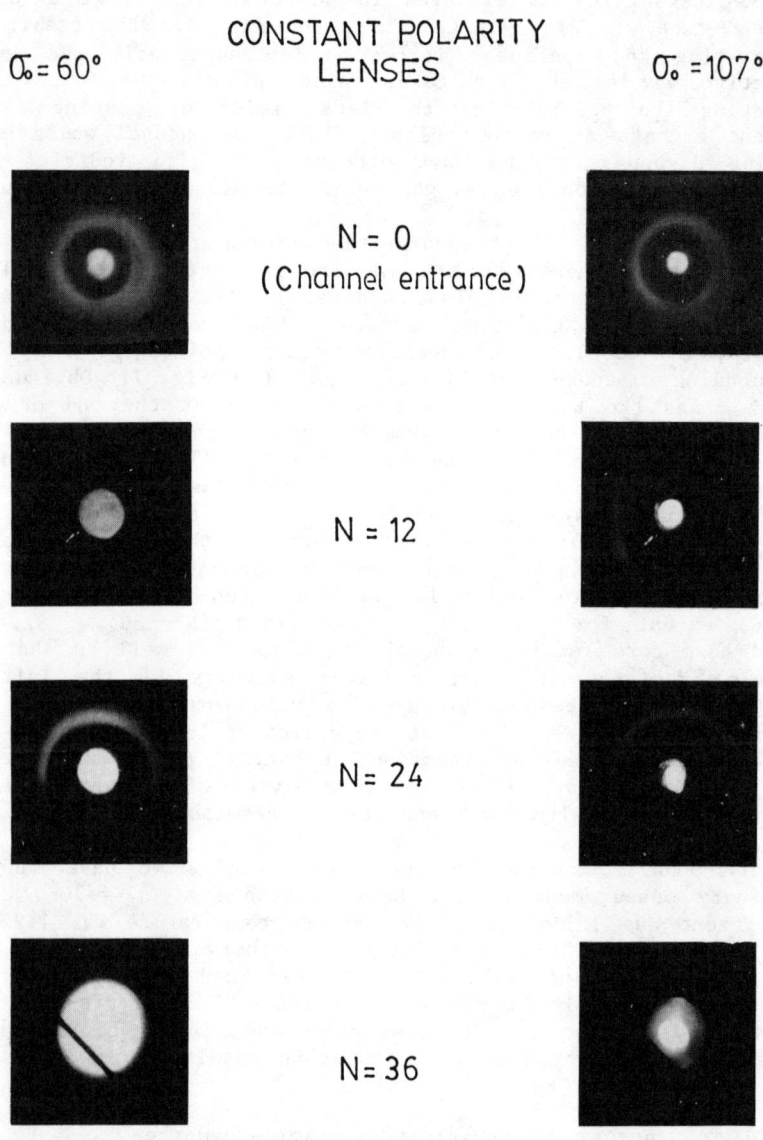

FIG. 6. Phosphor screen image of the beam at the channel entrance (N = 0) between lenses M2 and C1 and after 12, 24, and 36 periods for $\sigma_o = 60^0$ on the left side and $\sigma_o = 107^0$ on the right.

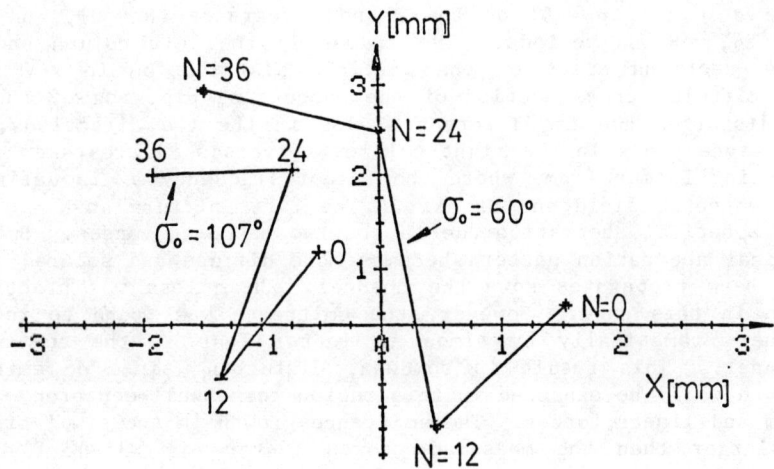

FIG. 7. Beam centers corresponding to the pictures shown in Fig. 6.

a separate paper.[18] Since the lens and gun alignments are not known accurately, the experimental conditions were simulated in an approximate way by launching the beam from the gun with a 1.5 mm displacement from the axis. This axis was defined by the system of lenses which were assumed to be in perfect alignment. Otherwise the parameters for the electron beam and the solenoidal field shape were chosen to be the same as in the experiments, and the particle distribution was taken to be Semi-Gaussian (thermal), i.e., uniform in space and Maxwellian in transverse velocity.

Figure 8 presents the results of the simulation run for the case $\sigma_o = 70^0$, $\sigma = 6^0$ at the channel entrance (N = 0), and after 12, 24, and 36 periods. The frames in the left column show the cross-sectional plot of the particle distribution in x-y space. The circular cross section of the conducting pipe and of the beam is distorted due to different scales in the two directions. The emittance plots in the right column (x´ versus x) are taken in the rotating Larmor frame where the Larmor frequency ω_L is defined by the magnetic field on the axis. The first picture at N = 0 shows the spherical aberration due to the two matching lenses. But this typical aberration pattern becomes more diffuse and smeared out as the beam propagates down the channel. Beam loss was practically zero in this case. However, the emittance was found to increase rather dramatically by almost a factor four at the end of the channel. This result is somewhat disturbing since no emittance growth would be expected in this region for a well-centered matched beam and linear forces. The emittance growth in the simulation run is larger than the measured growth factor of 2.3 which is not surprising since the actual misalignments of lenses or the gun are not exactly known at this time and therefore cannot be modeled precisely in the simulation. Measurements as well as further simulation studies will be carried out to determine how this emittance growth depends on the degree of misalignments and off-centering.

The computer results for the case $\sigma_o = 120^0$, $\sigma = 17.5^0$ are shown in Fig. 9. As expected from the theory discussed in the second section above, the beam is unstable and suffers rapid losses associated with large emittance growth. The role played by the off-centering and nonlinearities is difficult to assess until runs have been made without these effects. From the stability theory of a K-V beam, one would expect that both the envelope and the third-order mode are unstable in this case. Note that the plots are shown at different locations than in the previous figure. The beam becomes hollow after the first channel lens and exhibits a ring after 10 periods where the transmission is down to 83.6% and the emittance has grown by more than a factor eight. At the end of the channel (N = 36), the current has dropped to 64.6% and the emittance of the remaining beam is still a factor 6.4 larger than the emittance of the full beam at injection from the gun. In summary, this particular case exhibits a drastically different behavior than the $\sigma_o = 70^0$ run. This is also observed in the experiment where the region near $\sigma_o = 120^0$ is extremely sensitive

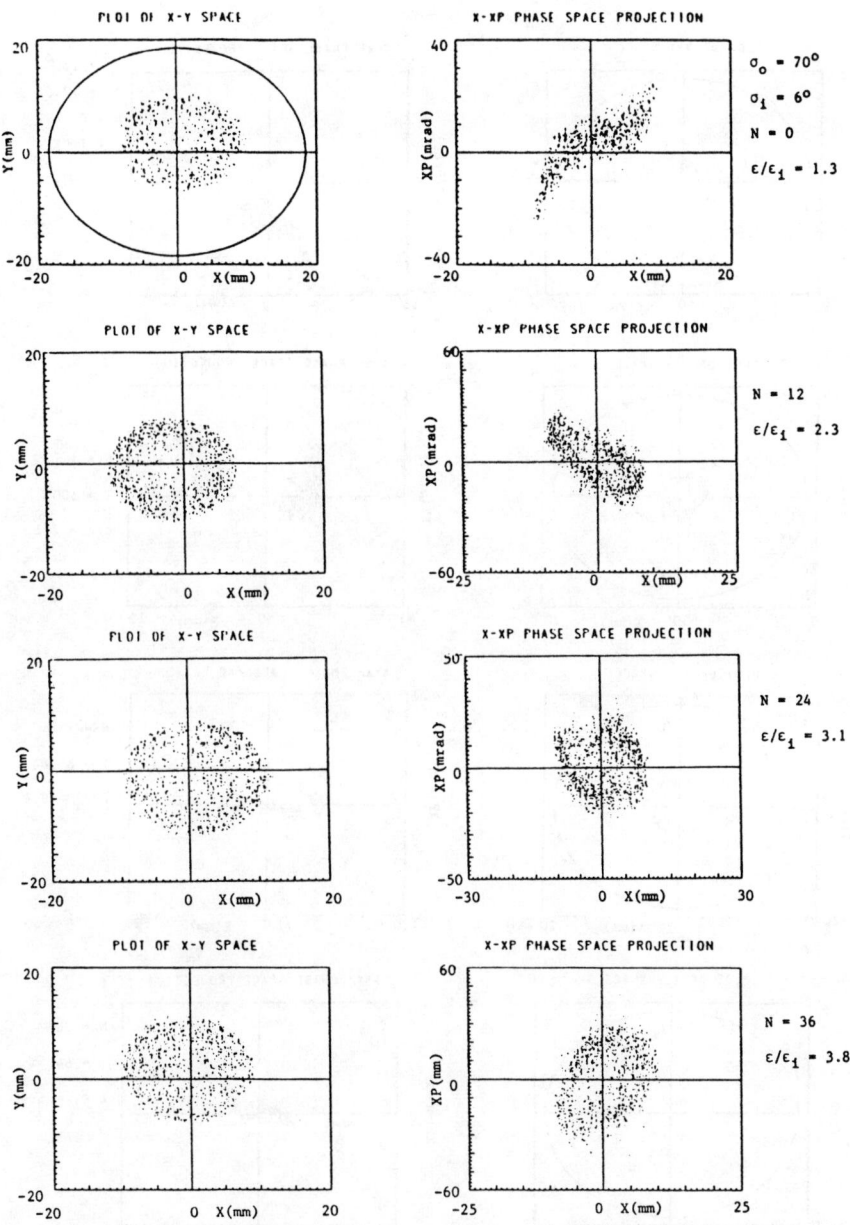

FIG. 8. Simulation plots for semi-Gaussian ("thermal") distribution with 1.5 mm initial displacement at channel entrance (N = 0) and at different locations along the channel for the case $\sigma_o = 70°$. The left side shows x-y space, the right x´-x phase space.

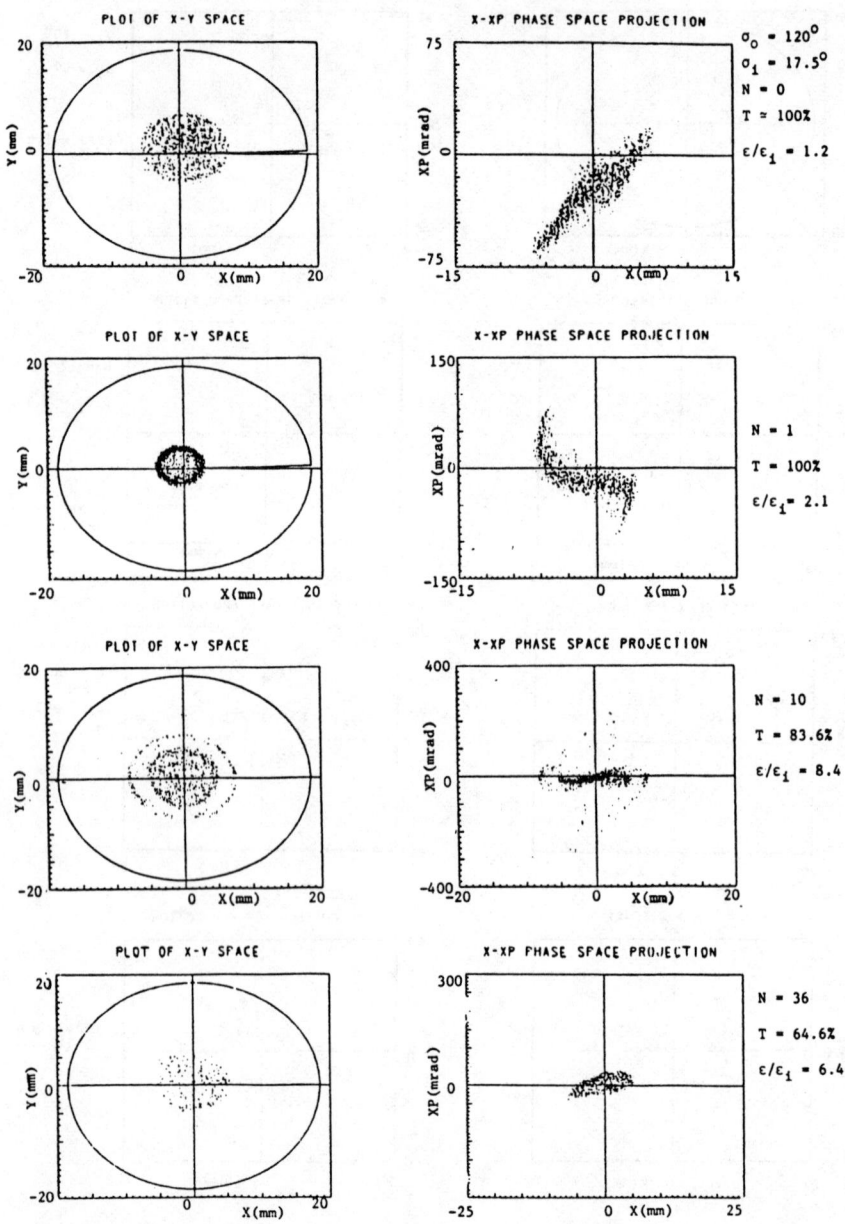

FIG. 9. Simulation results for off-centered beam in the case $\sigma_o = 120°$.

to operating conditions and where catastrophic beam loss occurs, as exhibited in Fig. 4. The transition from the region where the transported current drops off slowly to the rapid onset of instability needs to be explored in further studies.

It is interesting to compare these results with previous simulation studies on the effects of nonlinear forces in the case of a centered beam. For solenoid transport at $\sigma_o = 90^0$ it was found that no emittance growth occurs as long as the beam radius remains small enough such that nonlinear terms are less than 10% of the linear force at the edge of the beam.[19]

CONCLUSIONS

Several important conclusions can be drawn from the experimental and theoretical studies reported above.

First, precise alignment and beam centering is of crucial importance in transport through long channels. If the beam is not well centered, particle losses and emittance growth occur. The latter was found in the simulation studies to be unacceptably large for 1.5 mm off-centering even in the region below $\sigma_o = 90^0$ where no instability is observed and the transport efficiency is almost 100% over the 36 periods of the channel.

Second, beam centering and alignment become extremely critical in the region above $\sigma_o = 90^0$, and unexplained dips in the transmission efficiency occur at several values of σ_o.

Third, the catastrophic beam loss observed for $\sigma_o \gtrsim 120^0$ is consistent with theoretical predictions of envelope and third-order instabilities in this region. However, the onset of instabilities has not been clearly identified yet.

Systematic future studies, both experimental and theoretical, are planned to determine how emittance growth and beam loss depend on beam size, current, initial emittance, off-centering, misalignments, and nonlinear lens forces. The goal is to determine the "linear-aperture" size of the solenoidal channel and the range of σ_o values where stable, loss-free transport occurs. Also, we plan to explore the transition to unstable behavior above $\sigma_o = 90^0$ in more detail than has been possible so far. In the absence of analytical theory concerning the effects of nonlinearities and off-centering, we must rely strongly on systematic numerical simulation studies. Our hope is that these studies will yield better agreement with the experimental observations than has been possible so far. Improvement of the alignment technique and more precise measurements of gun and lens positions will be necessary. If this can be done, we could introduce well-known misalignments of either the beam center or specific lenses and compare experimental observations with computational results.

REFERENCES

1. J. D. Lawson and M. Reiser, Proceedings of the Heavy Ion Fusion Workshop, BNL-50769, (Brookhaven National Laboratory, 1977), p. 104 ff.; M. Reiser, W. Namkung, and M. A. Brennan, IEEE Trans. Nucl. Sci. NS-26, 3026 (1979).
2. I. Hofmann, L. J. Laslett, L. Smith, and I. Haber, Part. Accel. 13, 145 (1983).
3. J. Struckmeier, J. Klabunde, and M. Reiser, Part. Accel. 15, 47 (1984).
4. P. M. Lapostolle, IEEE Trans. Nucl. Sci. NS-18, 1101 (1971); F. J. Sacherer, ibid., p. 1105.
5. J. Struckmeier, IEEE Trans. Nucl. Sci. NS-32, 2516 (1985).
6. T. P. Wangler, K. R. Crandall, R. S. Mills, and M. Reiser, IEEE Trans. Nucl. Sci. NS-32, 2196 (1985).
7. J. Struckmeier and M. Reiser, Part. Accel. 14, 227 (1984).
8. L. J. Laslett (private communication).
9. J. Struckmeier (private communication).
10. M. Reiser, Part. Accel. 8, 167 (1978); J. Appl. Phys. 52, 555 (1981).
11. J. D. Lawson, Use of the Spallation Neutron Source for Heavy Ion Fusion Beam Dynamics Studies, edited by N. M. King, RL-81-080 (Rutherford Appleton Laboratory, 1981), p. 84.
12. P. Loschialpo, W. Namkung, M. Reiser, and J. D. Lawson, J. Appl. Phys. 57, 10 (1985).
13. E. Chojnacki, M.S. thesis, University of Maryland, 1984; M. Reiser, E. Chojnacki, P. Loschialpo, W. Namkung, J. D. Lawson, C. Prior, and G. P. Warner, Proceedings of the 1984 Linear Accelerator Conference, GSI-84-11 (Darmstadt, W. Germany, 1984), p. 309.
14. J. D. Lawson, E. Chojnacki, P. Loschialpo, W. Namkung, C. R. Prior, T. C. Randle, D. H. Reading, and M. Reiser, IEEE Trans. Nucl. Sci. NS-30, 2537 (1983).
15. M. G. Tiefenbach and D. Keefe, IEEE Trans. Nucl. Sci. NS-32, 2483 (1985).
16. R. H. Holm, G. A. Loew, W. K. H. Panofsky, "Beam Dynamics," in The Stanford Two-Mile Accelerator, edited by R. B. Neal (W. A. Benjamin, Inc., 1968), Chap. 7, p. 163.
17. W. Namkung and E. P. Chojnacki, Rev. Sci. Instrum. 57, 341 (1986); M. J. Rhee and R. F. Schneider, Part. Accel. (to be published).
18. C. Prior, "Particle Tracking in Solenoid Channels," paper in this Symposium.
19. H. Dantsker-Rudd, I. Haber, and M. Reiser, IEEE Trans. Nucl. Sci. NS-32, 2620 (1985).

AN INTENSE METAL ION BEAM SOURCE FOR HIF.

Ian G. Brown
Lawrence Berkeley Laboratory, Berkeley, CA 94720

ABSTRACT

We have developed an ion source which can produce high current beams of metal ions. The source uses a metal vapor vacuum arc discharge as the plasma medium from which the ions are extracted, so we have called this source the MEVVA ion source. The metal plasma is created simply and efficiently and no carrier gas is required. Beams have been produced from metallic elements spanning the periodic table from lithium through uranium, at extraction voltages from 10 to 60 kV and with beam currents as high as 1.1 Amperes (electrical current in all charge states).

In this paper a brief description of the source is given and its possible application as an ion source for heavy ion fusion is considered. Beams such as C^+ ($\geq 99\%$ of the beam in this species and charge state), Cr^{2+} (80%), and $Ta^{3+,4+,5+}$ (mixed charge states) have been produced. Beam emittance measurements and ways of increasing the source brightness are discussed.

INTRODUCTION

Heavy ion drivers for the development of inertially confined fusion call for an ion source which can produce high current, high quality, short pulse beams of heavy ions. The required source parameters are severe, and there has not yet evolved an ion source which meets all of the requirements at the same time.

Recently we have developed at LBL a new kind of ion source for the injection of high current beams of metal ions into the Bevalac (linac and synchrotron) particle accelerator facility for heavy ion nuclear physics research - the MEVVA ion source. In this paper we describe the MEVVA source and how it works, and summarize the source and beam performance characteristics that we've measured in the process of developing the source for Bevalac application. Then we report on some work we've done recently with heavy ion fusion application specificlly in mind. The results are very encouraging, and further development of this kind of ion source for the HIF program could be advantageous.

PRINCIPLE AND DESCRIPTION OF THE MEVVA SOURCE

In the MEVVA ion source,[1-3] the plasma from which the ions are to be extracted is created directly from the solid by means of a metal vapor arc discharge between two metallic electrodes in vacuum. A characteristic of the metal vapor vacuum arc is the formation of 'cathode spots' on the surface of the cathode. These are minute regions of intense current concentration (many megamps/cm^2 over a spot of diameter of order microns), and it is at these spots that the metal plasma is generated from the solid surface[4-8]. In general many cathode spots will participate in the arc, and the assemblage of spots constitutes a prolific source of metal plasma produced from the cathode material. This quasi-neutral plasma plumes away from the cathode toward the anode and persists for the duration of the arc current drive. The anode of the discharge is located on axis with respect to the cylindrical cathode and has a central hole in it through which a part of the plasma plume streams; it is this component of the plasma that forms the medium from which the ions are extracted. The plasma plume drifts through the post-anode region to the set of grids that comprise the extractor - a three-grid, accel-decel, multi-aperture design. A small magnetic field, produced by a simple coil surrounding the arc region and of magnitude up to about 100 gauss, serves to help duct the plasma plume in the forward direction, but this is not an essential ingredient to the source.

A schematic of the embodiment of the concept with which we've done most of our work is shown in Figure 1, and the partially disassembled source in Figure 2. This is the device called MEVVA II. The various components and features refered to above can be seen. The extractor diameter is 2 cm, and so also is the initial beam diameter.

We've also made other versions of MEVVA ion source. One of these is the MicroMEVVA; this is a miniature source - about 6 cm long and 1.5 cm diameter - which can produce pulsed low emittance metal ion beams up to the 10 ma level[9]. MEVVA IV is another version which is presently in fabrication and which has an array of 16 cathodes disposed in a Gatling gun -like arrangement so that one can switch between cathodes rapidly and while under vacuum.

The work described here has been done using the MEVVA II ion source.

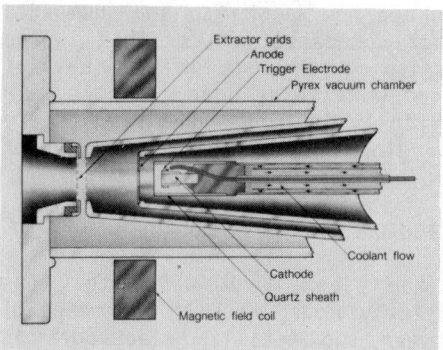 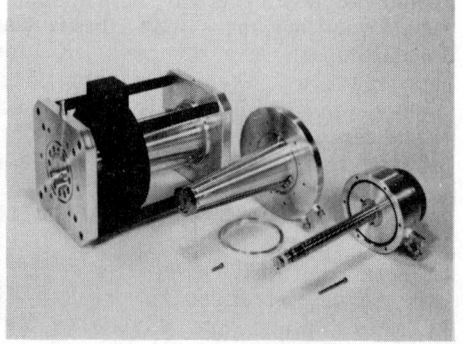

Fig. 1. Schematic of the MEVVA ion source.

Fig. 2. The partially assembled MEVVA II ion source.

SOURCE PERFORMANCE AND BEAM PARAMETERS

For application in the nuclear physics program at LBL, for which use the MEVVA source was developed, the main interest is in producing intense uranium beams in the SuperHILAC and the Bevalac particle accelerators[10,11]. The Bevalac - a heavy ion synchrotron - calls for pulses about 1 millisecond long at a repetition rate of about 1 pulse per second. Thus most of our development work has been done under conditions similar to these.

We've run with a wide range of cathode materials and have produced high current beams of Li, C, Mg, Al, Si, Ti, Cr, Fe, Co, Ni, Cu, Nb, Mo, Sn, La, Gd, Ho, Ta, W, Au, Pb, and U. The cathode may also be made from a conducting compound, and in this case the beam contains a mixture of the component species; we've made beams from FeS, PbS, LaB_6, CdSe, SmCo, SiC, and WC. It is noteworthy that in this way beams can be made which contain non-metallic species - eg, B from LaB_6 and S from FeS and PbS; the requirement seems to be just that the cathode material be a conductor. So far the source has worked with every cathode material we've tried.

The beam current has been measured using several different diagnostics, including Faraday cups with magnetic suppression of secondary electrons, beam current transformers, and beam power calorimeters. The record high beam current that we've measured to-date is 1.1 Amperes (electrical current in all charge states). Beams of a few hundred milliamperes can be produced routinely while operating the source at only moderate power levels. These beam currents can be obtained from all cathode materials/beam species. Operationally, the arc current (charging voltage of the LC pulse line which drives the arc) is varied so as to maximize the beam current measured into the acceptance of the Faraday cup, which occurs when the plasma density is best matched to the extractor parameters[12,13].

The beam emittance has been estimated in a number of ways. One can obtain a reasonable estimate simply from the beam divergence and the geometry of the source and beam-monitoring Faraday cup and other source and beam parameters. We have also obtained some emittance diagrams using a 'pepper-pot' diagnostic[14]. Emittance measurements obtained in this way confirm the simpler, geometry-obtained estimates. We find that typically half the beam current resides within a normalized emittance of from 0.2 to 0.5 π mm. mrad. An emittance plot obtained using the pepper pot diagnostic is shown in Figure 3. We remark parenthetically that the MicroMEVVA source shows an emittance of about 0.1 π mm. mrad. (normalized), as an upper limit[9].

The composition of the ion beam is of paramount importance - both purity and charge state distribution. The diagnostic we've used for these measurements is a time-of-flight device - a submicrosecond sample of the beam pulse is drifted down a field-free region where it separates into its different charge-to-mass components; it is a Q/A diagnostic.

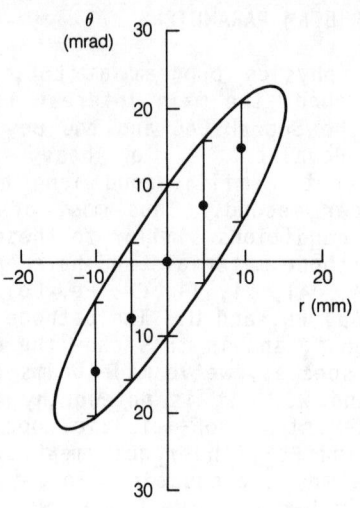

XBL 865-1895

Fig. 3. Typical emittance diagram. Cr beam, V_{ext} = 60 kV, I_{beam} = 200 ma. ε = 0.33π mm. mrad. (normalized).

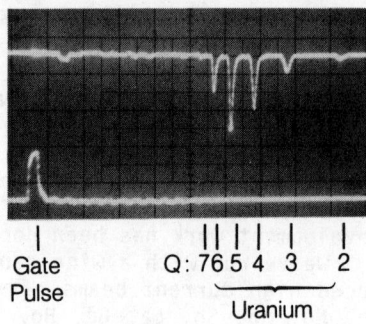

Fig. 4. Charge state distribution for uranium.

The beam spectra obtained in this way have been confirmed for a few special cases by a more conventional magnetic analysis. A typical spectrum, and one of some interest, is that of uranium, shown in Figure 4. The charge state distribution is peaked at U^{5+} and small amounts of U^{2+} and U^{7+} can be seen. The charge state distribution can be varied to a small extent via the arc current, but this effect is small. (As the arc current is increased more cathode spots form to participate in the arc, but the plasma physics of each spot is not greatly changed). The spectrum is fairly clean, showing no contamination from the stainless steel trigger, the alumina trigger/cathode insulator, or other components of the source, presumably reflecting the fact that the origin of the plasma is indeed the cathode spots, which form only on the cathode. A small amount of H^+ can be seen; this is a common metallic contaminant and can be removed by vacuum baking the cathode. We have obtained charge state distribution data such as shown in Figure 4 for all of the cathode materials we've tried, listed above. In general the lower Z materials show lower charge states and the higher Z materials show higher charge states.

RECENT HIF-RELATED RESULTS

We have looked at a number of aspects of the MEVVA source operation that are of particular pertinence to its use within the Heavy Ion Fusion program. These include short pulse operation, high current/low emittance operation, and charge state distribution.

For short pulse operation the LC pulse line that normally drives the arc current was replaced by a simple RC discharge with R = 0.8 Ohms and C = 12 µF, producing an arc current pulse with a width of around 15 microseconds, as shown in the lower trace of Figure 5. A large Faraday cup with a strong transverse magnetic field for secondary electron suppression was located close to the ion source extractor for beam current measurements, and a pulse so

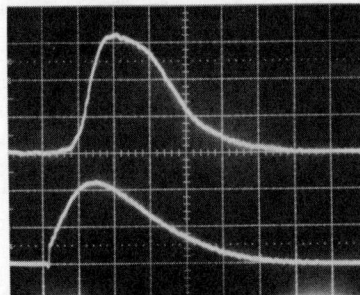

Fig. 5 Beam current (upper trace, 200 ma/cm) and arc current (lower trace, 200 A/cm); 5 μsec/cm; nearby Faraday cup.

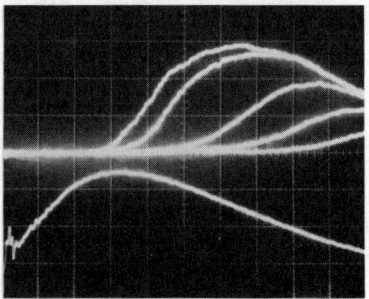

Fig. 6 Beam current (upper trace, 40 ma/cm) at distant Faraday cup (z = 50 cm) for P = 0.5,1,2,5,10x10^{-5} Torr; 2 μsec/cm.

measured is shown in the upper trace of Figure 5. The peak beam current in this case is 650 ma. Here the arc current was 450 Amps and the extractor voltage 35 kV; the cathode material was gadolinium (Z = 64, A = 155 to 160). The risetime of the beam pulse is 2 - 3 microseconds, and one might speculate that a more rectangular arc current pulse, eg from a short LC pulse line, might work fine down to a beam pulse width of perhaps 5 microseconds. However there is a problem in providing adequate space charge neutralization for good beam transport (absence of space charge blowup) for short times. This problem is manifest in Figure 6. Here the beam current measured by a Faraday cup located 50 cm distant from the extractor is shown for several background gas pressures from 5 x 10^{-6} Torr up to 1 x 10^{-4} Torr. At the lowest pressure the beam blowup is so severe that most of the beam is lost during the pulse length of the beam; at the highest pressure there appears to be adequate neutralization occurring in the arc pulse risetime to allow good transport. Some fraction of the neutralizing electrons are provided by the final, grounded extractor grid. From these data it would seem probable that in order to provide high current beam pulses that can propagate with submicrosecond risetimes it will be necessary to provide some form of enhanced beam neutralization.

An upper limit to the emittance of the beam for this case can be obtained from the geometry. The 2.5 cm radius Faraday cup was located 50 cm from the source extractor, giving a beam half-angle divergence of approximately 50 mrad if we neglect the extractor width. For a gadolinium beam with Q = 2 and A = 158, and an extraction voltage of 20 kV, the normalized emittance within which the measured beam current resides is 0.37π mm. mrad. An oscillogram of the beam current pulse measured by the Faraday

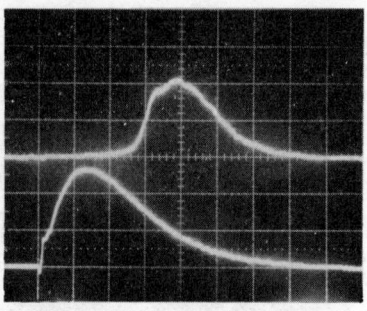

Fig. 7 Beam current (upper trace, 100 ma/cm) and arc current (lower trace, 200 A/cm); 5 μsec/cm; distant Faraday cup.

cup is shown in Figure 7 – 200 mA of beam into 0.37π mm. mrad (normalized), and a significant part of the beam has been lost due to space charge blowup. Note also the quality of the beam current fluctuation level (beam noise) which is evident from this oscillogram – it's a quiet beam.

It is advantageous for the charge state distribution to contain only a single charge state. The distribution that is actually obtained in a given case depends on the particular cathode material and on the arc current. Since for optimum beam divergence the plasma density must be matched to the extractor geometry and voltage, the arc current that one needs to run with in turn depends on the extractor voltage. Thus the charge state distribution that is obtained is difficult to predict, and every case needs to be actually measured. Figure 8 shows the time-of-flight charge state distribution obtained for the case of the gadolinium beam used in the present work. Here we have the serendipitous result that in excess of about 80% of the beam current is in a single charge state, Gd^{2+}. One might thus perhaps be able to operate the source in this case without the complication of a magnetic charge state analysis.

Several other charge state distributions are shown in Figures 9 through 11. The distributions that are shown should be taken only as indicative of what can be obtained. One might envisage an embodiment of MEVVA source designed and fabricated to optimise performance for the short pulse, high current requirements of HIF application; in this case the precise charge state distributions obtained might differ a little from those shown here.

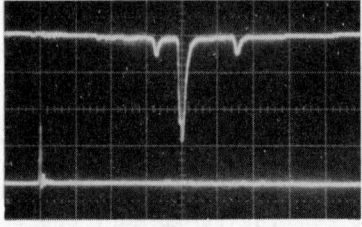
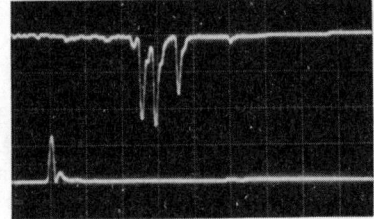

Fig. 8 Charge state distribution for gadolinium.

Fig. 9 Charge state distribution for tantalum.

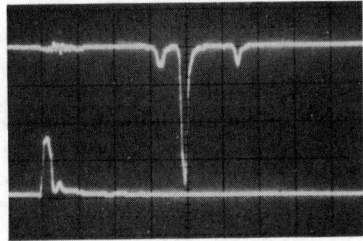
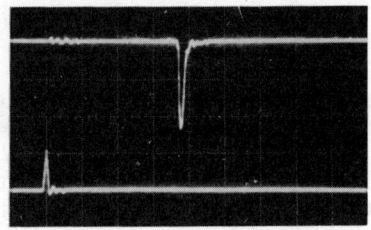

Fig. 10 Charge state distribution for chromium.

Fig. 11 Charge state distribution for carbon.

DISCUSSION

The results described here are characteristic of the kind of MEVVA ion beam that one might obtain in a more extensive R&D program specifically geared to HIF. A beam current of over 500 ma can be obtained, the normalized emittance can be in the range 0.3 - 0.4π mm. mrad. or better, and charge state distributions can be produced containing a high fraction of the beam in a single charge state, singly ionized for light metal species and multiply ionized for heavy metal species. The production of transportable short pulse (of order a few microseconds) beams is intimately connected to the concern of full and rapid beam space charge neutralization, and this area needs work; nevertheless, beams of about 10 μsec width can be produced already.

It is interesting to note that for the case of the gadolinium beam produced here, the linear charge density at 2 MV is 0.06 - 0.24 μC/m, where the lower limit corresponds to the low emittance beam actually measured (200 ma at 20 kV extraction) and the upper limit corresponds to what would be obtained for the case of better space charge neutralization and less beam loss (ie, the measured 650 ma at 35 kV extraction).

Related work is being carried out elsewhere. Humphries and coworkers[15-18] at the University of New Mexico have developed a very nice method for extracting well focused beams from a metal vapor vacuum arc source, using a grid to gate the electron flow to the extractor gap. In this manner they have produced extremely quiescent high current metal ion beams. A hybrid device, incorporating this feature into the MEVVA geometry described here, could have characteristics important to HIF application. Keller[19,20], at GSI, Germany, has modified the extraction geometry of a MEVVA II source to incorporate a post-acceleration section immediately after a single-aperture extractor, and has in this way produced a 160 kV uranium beam of very high brightness. The 14 mA beam was of initial diameter 8 mm, normalized emittance

0.029π mm. mrad., and brightness 16.5 Amps/(mm.mrad)^2, (where brightness is defined as the current divided by the square of the normalized emittance). The compound extractor system allows generation of high current beams from the MEVVA ion source at extraction voltages over 150 kV with very high brightness values.

In conclusion, the MEVVA ion source provides a means of creating high current, low emittance, pulsed heavy ion beams that can be of use to the HIF program. Further development of this new kind of source, specifically oriented to HIF application, is called for.

REFERENCES

1. I. G. Brown, J. E. Galvin and R. A. MacGill, Appl. Phys. Lett. 47, 358 (1985).
2. I. G. Brown, IEEE Trans. Nucl. Sci. NS-32, 1723 (1985).
3. I. G. Brown, J. E. Galvin, B. F. Gavin, and R. A. MacGill, Lawrence Berkeley Laboratory Report LBL-19976 (to be published in Rev. Sci. Instrum., June 1986).
4. J. M. Lafferty, ed., "Vacuum Arcs - Theory and Application", John Wiley and Sons, New York, 1980.
5. A. A. Plyutto, V. N. Ryzhhov and A. T. Kapin, Sov. Phys. JETP 20(2), 328 (1965).
6. W. D. Davis and H. C. Miller, J. Appl. Phys. 40, 2212 (1969).
7. IEEE Trans. on Plasma Science, special issue on vacuum discharge plasmas, IEEE PS-11, Sep. 1983.
8. IEEE Trans. on Plasma Science, special issue on vacuum discharge plasmas, IEEE PS-13, Oct. 1985.
9. I. G. Brown, J. E. Galvin, R. A. MacGill and R. T. Wright, Lawrence Berkeley Laboratory Report LBL-21487; to be published.
10. J. R. Alonso, IEEE Trans. Nucl. Sci. NS-30, 1988 (1983).
11. J. R. Alonso, Nucl. Instrum. and Methods A244, 262 (1986).
12. T. S. Green, Reports on Progress in Physics 37, 1257 (1974).
13. T. S. Green, IEEE Trans. Nucl. Sci. NS-23, 918 (1976).
14. R. Keller, GSI 1980 Annual Report, GSI Report 81-2, p.263.
15. S. Coffey, G. Cooper, S. Humphries, Jr., M. Savage and D. Woodall, Bull. Am. Phys. Soc. 29, 1314 (1984).
16. C. Burkhart, S. Coffey, G. Cooper, S. Humphries, Jr., L. K. Len, A. D. Logan, M. Savage and D. M. Woodall, Nucl. Instrum. and Methods B10/11, 792 (1985).
17. S. Humphries, Jr., C. Burkhart, S. Coffey, G. Cooper, L. K. Len, M. Savage, D. M. Woodall, H. Rutkowski, H. Oona, ,and R. Shurter, J. Appl. Phys. 59, 1790 (1986).
18. L. K. Len et al, paper presented at this Symposium, paper P12.
19. R. Keller, private communication.
20. R. Keller, paper to be presented at the 1986 Linear Accelerator Conference, June 2-6, 1986, Stanford Linear Accelerator Center, Stanford University.

HIF TRANSPORT ISSUES FOR $p > 10^{-3}$ TORR AND $Z > 1$

C. L. Olson
Sandia National Laboratories, Albuquerque, New Mexico 87185

ABSTRACT

Final transport schemes for HIF are examined, with emphasis on transport for $p > 10^{-3}$ Torr and $Z > 1$ since this should simplify the reactor design and reduce the length of the accelerator. Specifically the question of charge neutralization is addressed. We find (1) the fractional neutralization f_i needed scales as $f_i = (1 - Z^{-2})$ which means $f_i > 0.89$ is needed for $Z > 3$; (2) axially-trapped electrons limit the net beam potential to $e\phi_{min} = \alpha (1/2\, m_e v_i^2)$ with $1 \leq \alpha \leq 4$; (3) radially-expelled plasma ions increase f_i, especially near the pellet; (4) radially-oscillating plasma electrons have an adiabatic limit of $f_i \approx 0.5$; and (5) as f_i approaches unity, plasma particle trajectories may involve drift motions along and radially away from the ion beam. Also, criteria are given for the maximum Z/A allowed for transporting very large currents. For the HIF parameters used, it appears that neutralization will probably be adequate for $Z \leq 3$.

INTRODUCTION

Final transport forms an important portion of a heavy ion fusion (HIF) reactor system. From a physics standpoint it has been shown that it is advantageous to have high β_i (~0.4), high A (~200), low Z (~1), and low to moderate vacuum ($p \leq 10^{-3}$ Torr) for transport considerations.[1,2] Here $\beta_i c$ is the ion velocity, c is the velocity of light, A is the ion atomic number, Z is the ion charge state, and p is the pressure. However, for economic considerations, it may be advantageous to alter the values of these parameters. It is in this context that we examine the neutralization and transport of HIF beams for $p \geq 10^{-3}$ Torr and $Z > 1$.

Ion beam propagation modes are summarized in Fig. 1. As these have been discussed earlier,[2] we will comment only briefly. HIF may use any of the seven modes listed. Specifically, HIF has a unique advantage in that ballistic transport with a bare beam is possible, i.e., no neutralization is needed at all. However, for certain HIF systems, it may be advantageous to utilize some neutralization as in modes (2)-(5). In this paper we will examine such neutralization.

Transport schemes for HIF reactors are summarized in Table I. As these have been examined earlier,[1,2] we only note here that as a baseline case the moderate vacuum regime ($10^{-4} - 10^{-3}$ Torr) with Z = 1 offers transport with relatively high confidence and permits Li temperatures up to 450°C in the reactor chamber. However, recent HIF systems studies indicate advantages in pushing the upper limits of this regime to higher pressures and higher charge states. In this context, we proceed to investigate the regime $p > 10^{-3}$ Torr and $Z > 1$.

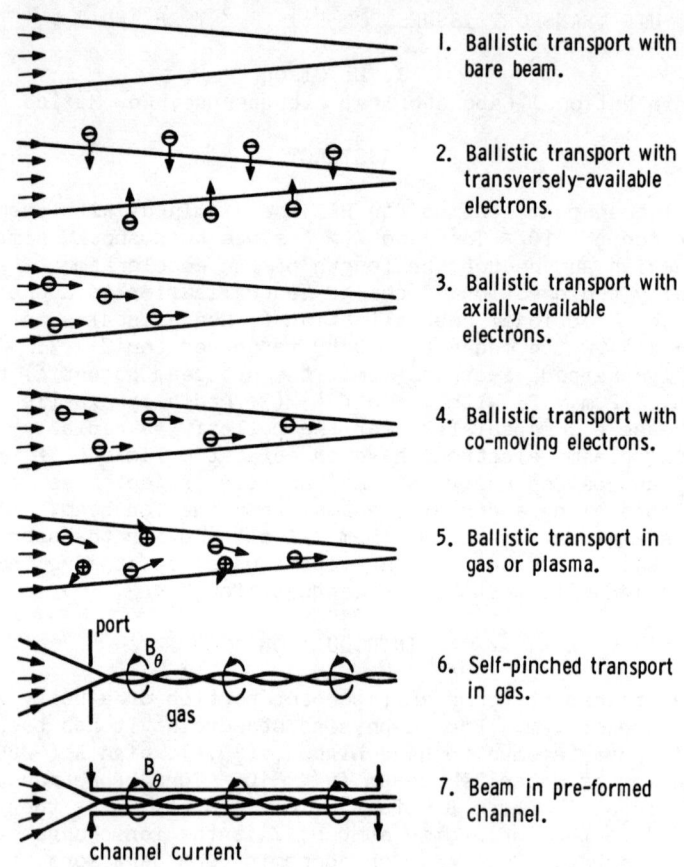

Fig. 1. Ion beam propagation modes.

NEUTRALIZATION AND TRANSPORT FOR $p > 10^{-3}$ TORR and $Z > 1$

A brief list of potential problems with the regime of $p > 10^{-3}$ Torr and $Z > 1$ are as follows:

1. For $Z > 1$, more space charge is present. This will require some neutralization to reduce beam spreading to an acceptable level.
2. For $p > 10^{-3}$ Torr, the background gas strips the beam ions to higher charge states, and produces a spread in charge states. Stripping above the injected charge state level means even more neutralization is needed.
3. For $p > 10^{-3}$ Torr, the ion beam ionizes some of the background gas and produces plasma, which should help to neutralize the beam's space charge. However, as will be evident below, it cannot be argued that the plasma will simply and instantaneously neutralize the ion beam. This is especially true for HIF reactor

TABLE I. Transport Schemes for HIF Reactors.

Transport Scheme	trajectories	Z	accelerator length	reactor chamber	comment
hard vacuum	ballistic	1	long	large, dry wall	high confidence
moderate vacuum	ballistic	1	long	Li to 450° C	high confidence
co-moving electrons	ballistic	>1	shorter	Li to 450° C	merits study
"1 Torr window"	ballistic	sizeable	shorter	"1 Torr"	questionable
self-pinched	oscillating	large	short	10's of Torr	complication of counter-streaming electron beam
pre-formed channel	oscillating	large	short	10's of Torr	complication of channel-forming method

transport because then the entire beam propagates in free space, away from all boundaries.

4. The e-i two-stream instability was previously examined for $p \lesssim 10^{-3}$ Torr, and saturation effects were estimated to be negligible.[1,2] At higher pressures (~ 1 Torr) this instability is collisionally quenched. Its effect at intermediate pressures (10^{-3} Torr to 1 Torr) was recently examined, assuming charge neutrality, and it was concluded that because of the large range of k values excited as the focused beam converges toward the pellet that the net fields produced should not be disruptive.[3]

5. The filamentation instability, knock-on electron problem, and multiple scattering all become more severe as 1 Torr is approached.

Presently the key problem appears to be how to adequately neutralize a $Z > 1$ beam in the $p > 10^{-3}$ Torr regime. First we comment that neutralization of an ion beam is more complicated than neutralization of an electron beam. For electron beam propagation in low pressure gas,[4] the beam produces plasma, the mobile plasma electrons are expelled radially, and an ion background remains to provide charge neutrality. For ion beam propagation in low pressure gas, the beam produces plasma, but to achieve neutralization requires that, e.g., relatively immobile plasma ions be expelled, or electrons be trapped axially. After estimating the neutralization required for $Z > 1$ beams, we proceed to assess several methods for producing the required neutralization.

Neutralization required for $Z > 1$

The radial force equation for an ion at the edge of the beam is[2]

$$\gamma_i A M_p \ddot{r} = (eZ)(2ZI_p)(\beta_i cr)^{-1}[(1-f_i)-\beta_i^2(1-f_m)] \qquad (1)$$

where the ion velocity is $\beta_i c$, $\gamma_i = (1-\beta_i^{-2})^{1/2}$, c is the speed of light, M_p is the proton mass, e is the magnitude of the electron charge, I_p is the particle current, f_i is the fractional space charge neutralization, and f_m is the fractional current neutralization. If we consider an ion beam with straight trajectories that are focused from an injection radius R at $Z = 0$ to a point at $Z = L$, then because of the radial spreading given by (1), the beam radius at $Z = L$ will not be zero, but will have a minimum value r_p. Solving (1) for I_p and setting it equal to the radial space charge spreading current I_p^{RS} gives

$$I_p^{RS} = (1/4)(\beta_i^3 \gamma_i)(A/Z^2)(M_p c^3/e)[(1-f_i)-\beta_i^2(1-f_m)]^{-1}$$
$$\cdot (R^2/L^2)[\ln(R/r_p)]^{-1} \qquad (2)$$

This is the peak particle current that can be transported in the reactor chamber during final focus a distance L to a pellet of radius r_p.

It is useful to rewrite (2) as

$$(1-f_i)-\beta_i^2(1-f_m) = (A/Z^2)(\beta_i^3 \gamma_i/4)[(M_p c^3/e)/I_p^{RS}](R^2/L^2)[\ln(R/r_p)]^{-1} \qquad (3)$$

Now suppose we have a beam with $Z = 1$, $f_i = f_m = 0$, and $I_p = I_p^{RS}$, and we ask if we increase the charge state to $Z > 1$ what neutralization is needed to still transport the same particle current. For $f_i \neq 0$, $f_m = 0$, the required neutralization is

$$f_i = \gamma_i^{-2}(1-Z^{-2}) \tag{4}$$

For $f_i = f_m = f$, the required neutralization is

$$f = 1-Z^{-2} \tag{5}$$

For HIF, $\gamma_i \approx 1$ and results (4) and (5) differ only slightly. Result (5) is plotted in Fig. 2. Note that f rises rapidly as Z increases above unity. For example, if a beam with $Z = 1$ and no neutralization is changed to a beam with $Z = 3$, a neutralization of $f \geq 0.89$ is required for the beam to still hit the pellet.

Note that inclusion of finite beam emittance effects does not alter results (4) and (5). Inclusion of finite emittance lowers the value of I_p^{RS} in (2) to $I_p^{RS+\epsilon}$ given by

$$I_p^{RS+\epsilon} = I_p^{RS}[1-(L^2/R^2)(\epsilon^2/r_p^2)] \tag{6}$$

where ϵ is the transverse emittance. However, if an unneutralized $Z = 1$ beam is at the limit (6), then the neutralization required for $Z > 1$ is still given by (4) or (5).

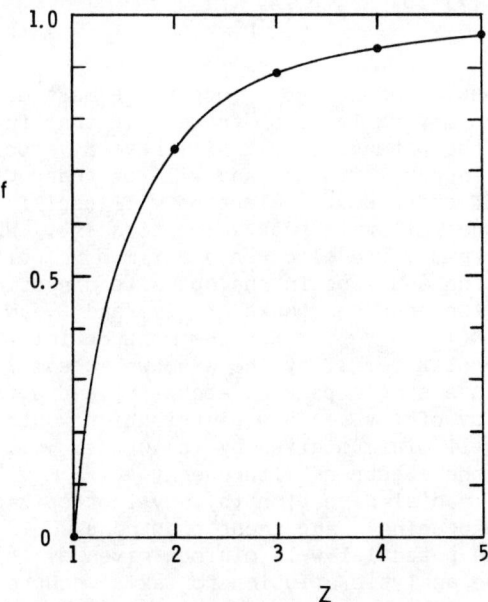

Fig. 2. Fractional neutralization f required for a $Z > 1$ beam to match the radial space charge spreading of an unneutralized $Z = 1$ beam.

Neutralization by Axial Trapping of Plasma Electrons

As the ion beam ionizes the background gas, it creates a plasma of density n_p inside the beam boundaries. A long potential well is created by the ion beam space charge, and plasma electrons created inside the well may become trapped in it as it passes by. If trapping occurs, then the trapped electrons will provide some charge (and some current) neutralization for the ion beam. The potential well depth $\Delta\phi$ between the outer edge of the beam and the inside center of the beam is

$$\Delta\phi_o = ZI_p(\beta_i c)^{-1} \tag{7}$$

where, for the unneutralized case, the maximum allowed value of I_p is I_p^{RS+E}. The potential well will accelerate and trap electrons as the well back passes by until $\Delta\phi$ is reduced to $\Delta\phi_{min}$ where

$$e\Delta\phi_{min} \approx \alpha(1/2 \, m_e v_i^2), \tag{8}$$

$1 \leq \alpha \leq 4$, m_e is the electron mass, and $v_i = \beta_i c$. The maximum neutralization possible is then given by

$$(1 - f_i)\Delta\phi_o = \Delta\phi_{min} \tag{9}$$

or

$$f_i = 1 - (\alpha/2)(m_e c^3/e)\beta_i^3 (ZI_p)^{-1} \tag{10}$$

using (7) and (8).

A physical picture of how $\Delta\phi_{min}$ occurs is most easily seen by transforming to the moving ion beam frame. In that frame, plasma electrons born in the potential well will have a velocity $-v_i$, and therefore kinetic energy $1/2 m_e v_i^2$, and will be trapped only if $e\phi \geq 1/2 \, m_e v_i^2$. If trapped, the electron will oscillate back and forth axially in the well with peak velocities $\pm v_i$. When transformed to the laboratory frame, the electron's maximum velocity will be $+2v_i$ in the beam direction and zero in the opposite direction. The velocity distribution then has peaks at $+2v_i$ and 0, and the peak kinetic energy is $4(1/2 \, m_e v_i^2)$. For the most optimistic case, this distribution will relax (e.g., by the e-e two stream instability) to a distribution with a single peak at about $+v_i$ and a thermal temperature velocity of $\leq v_i$. This distribution would be contained by the potential well minimum given by (8) with $\alpha = 1$. For the most pessimistic case, the electrons with energy $4(1/2 \, m_e v_i^2)$ are scattered into the radial direction (by any isotropization mechanism such as elastic scattering), and these electrons would be lost radially unless the potential well minimum given by (8) has $\alpha = 4$. In 1-D, the analytic solution for axial neutralization oscillates with a peak potential of $\alpha = 4$. In 1-D numerical simulations of axial neutralization, $\alpha \approx 2-4$ is observed.[5,6] Thus $\alpha = 1$ represents the absolute minimum potential and $\alpha = 4$ represents a realistic boundary for the maximum potential.

Result (10) is plotted in Fig. 3a for $I_p = I_p^{RS}$, with $\alpha = 1$ and $\alpha = 4$, and for some characteristic HIF parameters. Comparing with Fig. 2, we conclude that $1 \leq Z \leq 3$ is probably permissible, but that higher Z's should not be used.

Another perhaps more important conclusion is that for Z/A below some value (specified shortly), very large currents should be transportable in a single beam. This result occurs by using f_i given by (10) in (1) and ignoring the pinch term $\beta_i^2(1 - f_m)$ for now. This gives

$$\ddot{r} = [(\alpha/\gamma_i)(Z/A)(m_e/M_p)(\beta_i^2 c^2)](1/r) \qquad (11)$$

whereas substituting the known space charge spreading current (2) into (1) gives

$$\ddot{r} = [(1/2)(R^2/L^2)(\ln R/r_p)^{-1}(\beta_i^2 c^2)](1/r) \qquad (12)$$

Comparing the bracketed expressions in (11) and (12) we find the beam will propagate with very large currents (i.e., there is no dependence on I_p) provided

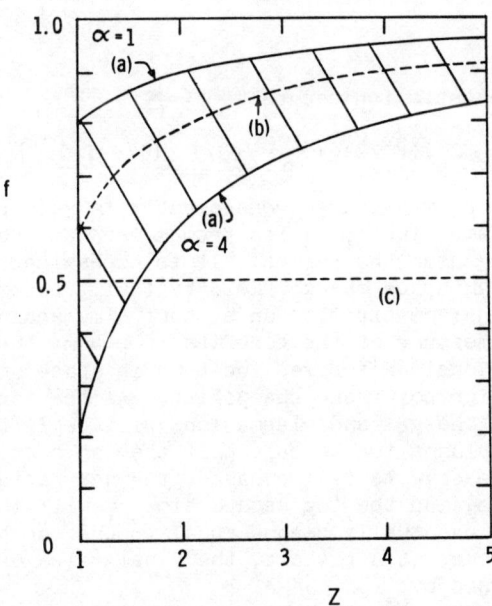

Fig. 3. Neutralization provided by various processes.
(a) Neutralization by axial trapping of plasma electrons.
(b) Neutralization by plasma ion expulsion.
(c) Neutralization by plasma electron radial oscillations.
[$\mathcal{E}_i$ = 15 GeV, A = 200, R = 6 cm, L = 600 cm, r_p = 0.15 cm, (β_i = 0.38, I_p^{RS} = 2.3 kA). In (b), $r_b^p(z)$ = 3 cm, t_b^i = 12 ns, A_g^p = 7, Z_g = 3.]

$$Z/A \leq [\gamma_i/(2\alpha)](M_p/m_e)(R^2/L^2)[\ln(R/r_p)]^{-1} \qquad (13)$$

For the parameters of Fig. 3a (R/L = 0.01, R = 6 cm, r_p = 0.15 cm) this is $Z/A \leq 0.027/\alpha$. For A = 200, this is $Z \leq 1.3$ (using $\alpha = 4$) or $Z \leq 5.4$ (using $\alpha = 1$).

Neutralization by Radial Expulsion of Plasma Ions

Note again, that as the ion beam ionizes the background gas, it creates a plasma of density n_p inside the beam boundaries. For an ion at the beam edge, there is effectively no neutralization because the plasma ion and plasma electron space charge fields cancel each other. Some neutralization occurs as the plasma ions are expelled outward and the plasma electrons are drawn inward.

We examine the plasma ion motion first. A plasma ion born at radius r has the equation of motion for $0 \leq r \leq r_b$,

$$A_g M_p \ddot{r} = Z_g e(2\Delta\phi_o/r_b)(r/r_b)(1 - f_i) \qquad (14)$$

where the subscript g refers to the background gas which is the source of the plasma ions, and $\Delta\phi_o$ is given by (7). The solution of (14) is

$$r = r_o e^{t/\tau_i} \qquad (15)$$

where the characteristic ion escape time τ_i is

$$\tau_i = \{(1/2)(A_g/Z_g)(\beta_i/Z)[(M_p c^3/e)/I_p](1 - f_i)^{-1}\}^{1/2}(r_b/c) \qquad (16)$$

During final focus r_b may change considerably (e.g., from 10 cm to 0.1 cm). At the same time τ_i varies from $\tau_i \gg t_b$ at the beginning of the focus to $\tau_i \ll t_b$ near the pellet. It is emphasized that near the pellet, the ion beam space charge field typically becomes so large that the plasma ions are expelled on a short time scale.

A very rough measure of the benefits of plasma ion escape can be made as follows. Consider a fixed location in space somewhere between the injection point and the pellet. As the ion beam passes by, it will ionize the gas and plasma ions will begin to escape radially. As the plasma ions escape past the ion beam radius, f_i for the beam will increase. As f_i increases, the net radial electric field will decrease, and the ion escape time τ_i will increase. As f_i increases, τ_i will eventually become equal to the ion beam pulse length t_b. Therefore, at a given z, the final value of f_i that occurs may be defined by

$$\tau_i(f_i, z) = t_b \qquad (17)$$

or using (16)

$$f_i = 1 - (1/2)(M_p c^3/e)[r_b(z)^2/(\beta_i^2 c^2 t_b^2)](A_g/Z_g)\beta_i^3(ZI_p)^{-1} \qquad (18)$$

In Fig. 3b, we plot result (18) for $I_p = I_p^{RS}$ and $r_b(z) = [r_b(z = 0)]/2$, i.e., at a point half way between the injection point and the pellet. This means that at this point, as the ion pulse passes by, the plasma ion escape time will just equal the ion pulse length. Closer to the pellet f_i will be higher than that given by (18), and closer to the injection point f_i will be lower than that given by (18). Thus Fig. 3b indicates the final f_i value at the mid point of the final focus.

Very large currents are again transportable provided Z/A is less than some value. Using (18) in (1) we find

$$\ddot{r} = [(1/\gamma_i)(Z/A)(r_b(z)^2/t_b^2)(A_g/Z_g)]1/r \quad (19)$$

Comparing this with (12) gives the criterion

$$Z/A \leq (\gamma_i/2)[\beta_i^2 c^2 t_b^2/r_b(z)^2](Z_g/A_g)(R^2/L^2)[\ln(R/r_b)]^{-1} \quad (20)$$

For typical parameters, we find Z of a few is allowed by evaluating (20) at the point midway between injection and the pellet. However, since f_i is time dependent, and f_i varies as the ion bunch approaches the pellet, the transport should really be evaluated numerically.

Neutralization by Radial Motion of Plasma Electrons

We now examine the plasma electron motion. For gas ionization by the ion beam, initially all plasma electrons will be contained within the ion beam boundaries. As the electrons are attracted in, they may locally provide some neutralization to the core of the beam, but overall at the beam edge, they provide no net neutralization. (If a plasma is created at a radius larger than the beam radius, then neutralization can occur over the whole beam thickness as electrons are drawn in.) In any case, if f_i is fixed in time, an electron will oscillate radially with an amplitude equal to its initial radial value. As more electrons are born and begin to oscillate, f_i should increase. However, as f_i increases, the restoring force on all previous electron oscillations decreases, so their oscillation amplitudes will increase, which in turn will tend to lower f_i. A limiting f_i occurs when these effects balance.

Quantitatively, the radial electron oscillations are described by

$$m_e \ddot{r} = -e(2\Delta\phi_o/r_b)(r/r_b)(1 - f_i) \quad (21)$$

This is an harmonic oscillator equation with a slowly varying parameter f_i. For the adiabatic case, the action angle variable W/ω remains constant where W is the local total energy and ω is the local oscillation frequency. The solution of (21) is

$$r = r_o \frac{1}{(1 - f_i)^{1/4}} \cos \int_o^t [K(1 - f_i)]^{1/2} dt \quad (22)$$

where $K = Ze2I_p(\beta_i c r_b^2 m_e)^{-1}$. This result shows the scaling

$$r \sim (1 - f_i)^{-1/4} \tag{23}$$

To determine the equilibrium value of f_i, we assume we are at equilibrium, and ask what will happen if we try to increase the electron density slightly more to raise f_i. If the equilibrium electron density inside the ion beam is n_o (so $f_i = n_o/n_i$) and we try to increase it by Δn, then all of the previously existing radial oscillations will have the quantity $(1 - f_i)$ change to $[1 - f_i - (\Delta n/n_o)]$. According to (23) the radial amplitude r_j of the j'th electron's oscillation will increase to r_j' where

$$r_j'/r_j = (1 - f_i)^{1/4}/[1 - f_i - (\Delta n/n_o)]^{1/4} \tag{24}$$

The density n_o will accordingly decrease by the ratio $(r_j/r_j')^2$ which is the same for all electrons. Equating the attempted increase in density to the resulting decrease in density gives

$$\Delta n_o = n_o - n_o (r/r')^2 \tag{25}$$

or

$$\Delta n/n_o = 1 - [1 - f_i - (\Delta n/n_o)]^{1/2} [1 - f_i]^{-1/2} \tag{26}$$

Expanding the bracketed expression for the limit $\Delta n/n_o \ll 1 - f_i$, we obtain $(1/2)/(1 - f_i) = 1$ or

$$f_i = 0.5 \tag{27}$$

This is the equilibrium f_i value that the radially oscillating electrons will produce. If more electrons are added, the previous oscillation amplitudes simply increase so as to maintain result (27). There is no Z dependence to result (27) as shown in Fig. 3c.

Result (27) shows rather poor neutralization for the radial motion of the electrons. In simulations with transversely available electrons, rather poor neutralization is also observed, with potentials sometimes reaching the full bare beam potential.[5]

Drift Trajectories as f_i Approaches Unity

Lastly, we comment on the radial motion of plasma electrons and ions for f_i approaching unity (e.g., when f_i is produced by plasma ion expulsion). Initially for the bare ion beam self fields, we have $E_r > B_\theta$ so plasma ions will be expelled radially, and plasma electrons will be drawn in radially and oscillate. However as f_i increases, E_r/B_θ will drop below unity and the particles will assume trajectories that resemble $E_r \times B_\theta$ drift trajectories directed toward the front of the beam. The drift is given by

$$v_z \approx (E_r/B_\theta) c \approx [(1 - f_i)/\beta_i]c \tag{28}$$

and the condition for onset of the drift is $E_r/B_\theta < 1$ or

$$f_i = 1 - \beta_i \tag{29}$$

Similarly as the ion beam converges toward the target, the bulk of the beam will have an E_z pointing opposite the beam direction. For $E_z < B_\theta$, the trajectories will resemble $E_z \times B_\theta$ trajectories in the outward radial direction. Thus the net effect of these drifts is that as f_i approaches unity, (1) particles do not have simple radial motions but may be trapped with <u>axially-directed</u> drifts, and (2) <u>radially-outward</u> drifts may also occur.

Summary Picture

The neutralization effects discussed above are indicated schematically in Fig. 4 for an HIF beam as it converges toward the pellet. At each location, a different mechanism may actually provide the best possible neutralization. Near injection, there is essentially no ion expulsion, axial trapping of electrons is just beginning, and yet radial electron oscillations may provide some core neutralization. Near the center, ion expulsion is becoming significant, and axial trapping of electrons should be important. Near the pellet, ion expulsion is excellent (with the caveats of possible trapped particle drifts and plasma density depletion).

CONCLUSIONS

We have examined several neutralization mechanisms for transport with $Z > 1$ and $p > 10^{-3}$ Torr. Comparing the required neutralization of Fig. 2 with the possible neutralization of Fig. 3, shows that for the HIF parameters used, neutralization will probably be adequate for $Z \leq 3$. In addition, criteria (13) and (20) indicate maximum allowed values of Z/A for transporting very large particle currents.

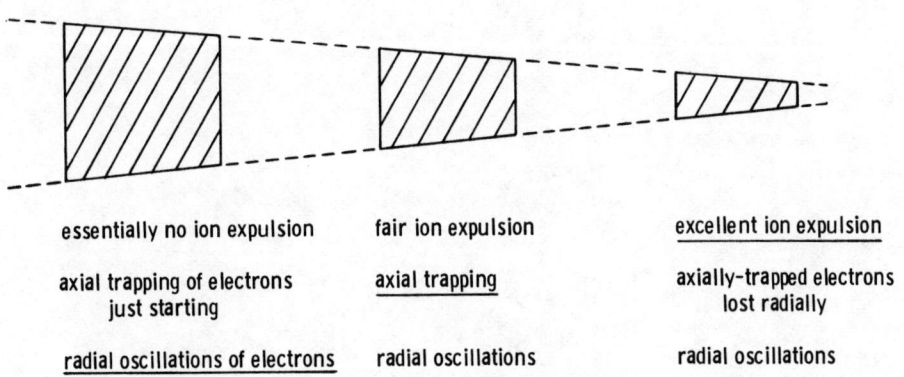

Fig. 4. Characteristic neutralization effects as the HIF beam converges toward the pellet.

ACKNOWLEDGEMENTS

Discussions with members of the Heavy Ion Fusion Systems Assessment Project, especially T. Godlove, W. Herrmannsfeldt, D. Keefe, E. Lee, P. Stroud, D. Wilson, W. Saylor, and D. Dudziak, are gratefully acknowledged.

This work was supported by the U.S. Department of Energy under contract DE AC04-76-DP 00789.

REFERENCES

1. C. L. Olson, Proc. Heavy Ion Fusion Workshop, Berkeley, CA, October 29 - November 9, 1979, LBL-10301, SLAC-PUB-2575, p. 403 (1980).
2. C. L. Olson, J. Fusion Energy $\underline{1}$, 309 (1982).
3. P. Stroud, to be published.
4. C. L. Olson, Phys. Rev. A $\underline{11}$, 288 (1975).
5. J. W. Poukey, Sandia Report SAND80-2545, November 1980.
6. S. Humphries, Jr., T. R. Lockner, J. W. Poukey, and J. P. Quintenz, Phys. Rev. Lett. $\underline{46}$, 995 (1981).

STUDIES ON LONGITUDINAL BEAM COMPRESSION IN INDUCTION ACCELERATOR DRIVERS*

J. W-K. Mark, D. D-M. Ho, S. T. Brandon, C-L. Chang[1],
A. T. Drobot[1], A. Faltens[2], E. P. Lee[2], and G. A. Krafft[3]
Lawrence Livermore National Laboratory, Livermore, CA 94550 USA
[1]Science Applications Inc., McLean, VA 22102 USA
[2]Lawrence Berkeley Laboratory, Berkeley, CA 94720 USA
[3]CEBAF, Newport News, VA 23606 USA

ABSTRACT

Longitudinal beam compression is an integral part of the U.S. induction accelerator development effort for heavy ion fusion. It occurs before final focus and fusion chamber beam transport and is a key process determining initial conditions for final focus hardware. Determining the limits for maximal performance of key accelerator components is an essential element of the effort to reduce driver costs. We outline here studies directed towards defining the limits of final beam compression including considerations such as: maximal available compression, effects of longitudinal dispersion and beam emittance, combining pulse-shaping with beam compression to reduce the total number of beam manipulations, etc. We summarize initial results on these limits gleaned from a survey including already more than 30 runs of 20-30 minutes each on our Cray computers.

In addition, we illustrate one of several possible techniques for utilizing the beam compression process to provide the pulse shapes required by a number of targets[1,2]. Without such capabilities to shape the pulse, an additional factor of two or so of beam energy would be required by the targets.

The use[3,4] of higher ion charge state $Z \geq 3$ is likely to test the limits of the previously envisaged hardware for beam compression and final focus. A more conservative approach is to use additional beamlets in final compression and focus. On the other end of the spectrum of choices, alternate approaches might consider new final focus with greater tolerances for systematic momentum and current variations.

INTRODUCTION

For applications as an ICF reactor driver, a heavy-ion beam pulse must be longitudinally compressed by almost one order of magnitude to achieve the peak power required to ignite a target. This process of beam compression is a crucial element of an accelerator for heavy-ion fusion; it occurs primarily after the main phase of acceleration and before final focus onto target. Space-charge forces play a vital role in halting the compression before the final-focus lens system is reached. Earlier simulations are summarized in Refs. 5-7.

---

*Work performed under the auspices of the U. S. Department of Energy by the Lawrence Livermore National Laboratory under contract number W-7405-ENG-48.

Here we examine the compression of a drifting heavy-ion pulse with the aid of particle simulations using our 2-1/2-D particle code Condor guided by simple models and physical estimates. We have made an initial survey of beam compression including effects of magnitudes of initial velocity tilt Δv_z, ion charge state Z, beam emittances and radial dependance in the g-factor relating geometry to longitudinal self electric fields, realistic boundary conditions of an isolated bunch as well as focusing forces that increase with beam compression. This includes more than 30 runs on our Cray computers of 20-30 minutes each. One example is illustrated later in Figs. 1-2. Summary results are given in Fig. 3 for a third of the runs. We hope that this survey can be used for normalizing simple models of these processes.

We have also explored some processes for pulse shaping during beam compression (see Figs. 4-5). We can obtain pulse shapes like those required in typical targets[1,2].

In addition, we are engaged in a series of efforts to address various realistic effects of beam compression. Of particular interest here is the possibility that space charge forces cause modifications of the compressed pulse-shape, particularly for beam ion charge state $Z \geq 3$. Additional effects such as more realism in transverse focusing forces and non-rectilinear beams are also being considered for further simulations.

The use of higher charge states is one major initiative[3,4] towards driver cost reduction in HIF. When accompanied by the $\geq$300-400 TW peak-power requirements in many targets, this initiative is likely to test the limits of beam compression, final focus and reactor beam transport as outlined in previous investigations.

The achievable peak power is the product of the achievable beam compression factor and the output power of the main accelerator. Attendant to the desire to push the limits of beam compression, we felt compelled to emphasize the maximum acceptable beam compression. In earlier studies acceptable compression depends mainly on the final momentum dispersion ($\delta p_z/p_z$) after compression. However the desire for high peak power with higher charge state ions ($Z \geq 3$) can result in rapid systematic pulse-shape and tilt changes in a compressed pulse. This is not acceptable for some final focus designs. It is this additional limit which we use to invoke a conservative choice such as more beamlets (say 30-50 in specific cases). With this choice we can expect the requisite peak power and consider advantages such as certain targets with higher target gain or reduced energy and power requirements as well as pulse-shape control before final focus (Refs. 1, 8, 9; note that higher charge state ions are also easier to bend).

Alternatively, innovative final focus can be developed which can accommodate considerable systematic variations in momentum and currents in the pulse. Developing innovative possibilities such as Robertson lenses allow additional advantages, for example, combining beam compression and pulse-shaping. The pulse length would then continue to change even within and after final focus. Apart from continued beam pulse compression in length, the space-charge forces causing internal changes in pulse-shape are largely eliminated by beam neutralization. But nonlinear residual tilts cause easily predictable residual pulse-shaping. Control over the pulse Δv_z variations for focusing purposes is expected to be ameliorated by the low field requirements (less than kilogauss) of this type of lens system and the fact that only a small fraction of this field needs to be rapidly pulsed. With the beam being neutralized during the final phase of compression, the initial tilts Δv_z could also be smaller in magnitude. A spectrum of choices is of course to be expected in any major new initiative with consequences in cost reductions.

BEAM COMPRESSION AND 2-1/2-D PARTICLE-CODE SIMULATIONS

An accelerated beam pulse has finite extent in space and time (in what follows, we take the z-coordinate as lying along the direction of beam propagation). Beam compression is initiated by pulsing the accelerating modules so that the tail end of the beam moves faster than the head. Especially when the variation of velocity is a linear function of distance (measured from the center of the pulse), this differential velocity can be viewed as a velocity "tilt" in longitudinal (z, v_z) phase space (see Fig. 1a).

We used a 2-1/2-dimensional LLNL version of the particle code Condor to produce a survey of beam compression. These simulations of the survey use a 18 x Z-μC beamlet of Pb ions with Z = 1, 3, 5 allowed. The beamlet initial rms dispersions in velocity δv_z approximate (1, 3, 10) x 10^4 m/s (values change slightly after particle loading). Since we use an electrostatic field solver in the beam frame, relativistic and self-current effects are ignored (a good first approximation). Thus the actual beam velocity can vary. If the beamlet ions were at ~ 10 GeV kinetic energy, the above z-velocity dispersion amounts roughly to $(\delta p_z/p_z)_0$ = (1, 3 and 10) x 10^{-4}. Also for similar kinetic energy ions, initial velocity tilts <20% are surveyed. The beam radius has the value a ~ 3 cm, and the perfectly conducting pipe has radius b = 6 cm. The boundary conditions at z → ±∞ are approximated accurately by doubling the periodicity length of the simulation region in z, because the Green's function falls off exponentially. The transverse quadrupole focusing forces are approximated by an axially symmetric radial electric field with simple time history $E_r = E_0 (1 + \alpha t)$ r/b. The constants E_0 and α are chosen for each simulation so that the beam radius approximates half the pipe radius. In the survey typically 40,000-60,000 particles are used on a 30 x 512 grid, representing a region 6 cm x 5 m in the r and z directions, respectively. The initial line density $\lambda(z,0)$ is made nearly parabolic in this survey because the pulse shape remains relatively invariant in the simulations despite the high perveance of the Z = 3 - 5 beamlets. We will discuss pulse-shape modifications later.

We illustrate one of these simulations in Figs. 1-2, where for convenience of the graphics in making this figure, only a representative 10% of the particles are plotted. As the beam is compressed, a conservative factor is that we initiated the simulation with a near equilibrium rather than a true equilibrium, so that extra dispersions arise. Transverse beam spread need not be an impediment to the longitudinal beam compression needed in heavy-ion fusion to increase peak power by almost one order of magnitude.

Figure 3a illustrates the available compression as functions of tilt velocity Δv_z, ion charge state Z and for initial particle dispersion $(\delta v_z)_0 = 3 \times 10^4$ m/s; Fig. 3b illustrates final δv_z (this number is at present less accurately determined than the simulation itself). Acceptable compression depends on the type of final focus and the considerations of the next section.

CHARGE STATES $Z \geq 3$ AND PULSE-SHAPE MODIFICATIONS

According to the simple model[7] of a "cold" 1D beam bunch, the quantity $(Ze g \lambda/m)$ acts like the "enthalpy" of ideal adiabatic gas dynamics and a related equivalent "sound speed" $C_s = (Ze g \lambda/m)^{1/2}$ of the space-charge waves can be defined. Here (Ze) and m are charge and mass of beam ions; λ is line charge density of beam and $g \partial \lambda / \partial z$ is proportional to longitudinal electric field. Particularly in bunch shapes with steep $\partial \lambda / \partial z$ (such as square pulses) our simulations confirm the approximate applicability of this concept even in the presence of transverse motions in the beam.

Let us apply this sound speed concept to a beamlet pulse of 10 GeV Pb ions and a total 18 x Z-µC of charge ($Z \geq 3$) distributed in a constant line charge density $\lambda(z)$. The final compressed beam length is $\simeq 1$ m. These numbers correspond roughly to scenarios where with 16 beamlets we achieve ~3 MJ on target at ~300 TW. Use of multiply-charged ions $Z \geq 3$ shows that: if the bunch shape after compression is a square pulse with near zero velocity tilt, then it stays near that particular pulse shape for only an additional 20 m of beam transport. After that short distance (or time), "expansion wave-like" phenomena eat into the square-pulse and start modifying the entire pulse shape and velocity tilt in a major way. We have confirmed these estimates with Condor simulations.

Since 20 meters or so is certainly shorter than the distance within standard final focus arrays, strong beam current and $\Delta p_z / p_z$ tilt variations develop within the final focus lens system even if they were eliminated beforehand during pulse-compression. More beamlets than 16 is certainly the obvious available solution.

An alternative with attractive features involves application of innovative final focus for which concepts and experiments exist but where some research and development efforts are required. The final focus of interest must allow beamlets with systematic current and momentum variations to be focused onto the same spot. A possibility in this category is discussed in the final section. Here we point out further that if we were to develop such final focus, an additional advantage would be that it allows the combining of longitudinal pulse-compression and pulse-shaping into a more compact hardware system. Although we concentrate in the next section on charge state Z = 3, <u>similar pulse-shaping processes apply to all relevant Z</u>.

A SIMPLE TECHNIQUE FOR COMBINING PULSE-SHAPING WITH BEAM COMPRESSION

The initial shape of the charge bunch after leaving the accelerator and before bunch compression is likely to be a square pulse in longitudinal distance z (perhaps with slightly rounded ends). Consider the time evolution in beam compression of two pulses with nearly square initial shape, but with different amounts of initial velocity tilts; the amount of tilt or dv_z/dz is directly related to the steepness of the final pulse shape $\lambda(z)$.

The most obvious variation of this concept is to give the beam head a tilt that is less steep than that for the beam tail. This is done by imposing an initial tilt, shown in Fig. 4a, with two different slopes but with continuous v_z. The initial square pulse evolves to the pulse shape of Fig. 4b at the time that line-charge density (or power) is increased by a factor of 7. This type of pulse shape is one that we could effectively use to drive high gain targets. We can, for example, compare our Fig. 4b with the pulse required by the target of Ref. 1 (Fig. 2a). To understand this comparison we must note in particular how the target sees the pulse reversed because the beam head reaches the target before the beam tail.

The initial low power front porch of Ref. 1, Fig. 2, does not require pulse compression. Its effect could be included by the simple device of creating an initial velocity tilt which approximates three or more connected straight lines in (z, p_z) phase-space.

The intriguing fact is that pulse-shapes of great interest to other types of targets could also be obtained. Another effect of such control by just three linear tilts in (z, p_z) space is best seen by Figs. 5a and 5b. These figures show the initial tilt and final pulse shape attained at 4.5-fold power amplification, given the accompanying initial velocity tilt of the beam from head to tail shown in Fig. 5a. We see that we can produce different rates of rise to the front porch of the pulse seen by the target. Similar techniques might reproduce two ramped pulses, perhaps even approaching the pulse shapes shown in Ref. 2 if we used multiple pulses of this type.

Although our summary does not elucidate all the physics or all possible shapes of final pulses, many pulse shapes are technically possible. More complicated shapes require more linear components to $v_z(z)$. The use of 3 or more linear pieces of velocity tilt is sufficient to give flexibility to the pulse-shape for target requirements alone.

CAN ROBERTSON LENS BE USED FOR FOCUSING HIGH CHARGE STATE HEAVY IONS?

The collective focusing lens proposed by Robertson[10] has been developed further by Krafft[11]. This lens is applied in a nearly space-charge neutralized environment while at the same time retaining part of the property of performance improving with increased values of beam ion charge Z. The low field requirement facilitates pulsed operation. These characteristics seem ideally suited to address the

needs of our previous section. Furthermore, since even Z>3 might
be possible for the main accelerator and collective lenses are not
averse to higher Z, we have retained Z=5 in the bulk of our paper
(note that more work might also be needed on the accelerator front
end to accommodate such high Z-values).

Estimates based on re-examining results of Krafft[11] are
centered here on the use of 10 GeV Pb ions in beamlets with $\sim 18 \times Z$-µC
of charge and unnormalized emittance $\epsilon \sim 21$ µm-rad. The
conclusion is that even with this lens working as expected
theoretically, it is preferable to have either (i) smaller emittance
(by about a factor of 2) and/or (ii) more beamlets. Based on the
above numbers the focal spot radii are somewhat large from the
target physics point of view. A lower limit on focal-spot radius
that is a factor of two smaller might be preferred.

We illustrate with three plausible cases we found. Here
$\ell = 25.3$ m is the size of the lens and L = 5 m is the assumed
standoff distance from lens to target. For charge state Z = 3 we
can at the moment only establish that a focal spot of 4 mm can be
obtained with a magnetic field in lens of B = 208 gauss. With Z = 5
and B = 161 gauss we can attain a 3 mm spot. Size of magnetic field
is obviously not the limiting factor. The electrons will be
coinjected at the beam velocity. Unless further studies are made,
we simply assumed that the azimuthal electron velocity cannot be
allowed to become relativistic. Also we have confirmed that
diamagnetic effects are unimportant.

There are many uncertainties in these alternate focusing systems
which would require corroboration by extensive numerical simulations
and experiments. There are also however strong incentives such as
the promise of a much more compact final beam handling system which
matches the more compact higher charge-state accelerator.

SUMMARY

Using our 2-1/2-D particle code CONDOR, we have made a series of
computer simulations of the beam compression process similar to
those illustrated in Figs. 1-2. Some global characteristics of one
third of the runs are summarized in Figs. 3a,b. Each point in these
latter figures represent a separate simulation.

The use of higher ion charge state $Z \geq 3$ is one method[3,4]
towards driver cost reduction in HIF. Consideration of typical
$Z \geq 3$ beam bunches after pulse compression suggests that the
subsequent behavior are likely to test the maximal limits of the
final focus process. We might choose the conservative viewpoint of
using $\geq$ 30-50 beamlets in specific cases or might consider the
other extreme of new final focus designs with greater tolerances for
systematic momentum and current variations. Development of such
final focus concepts would also allow more compact (and hopefully
cheaper) hardware packages where the previously separate processes
of beam compression, pulse-shaping and final focus occur as
partially combined and nearly concurrent beam manipulations.

We have outlined one of several of these methods which allow the
combining of pulse shaping and beam compression operations to reduce

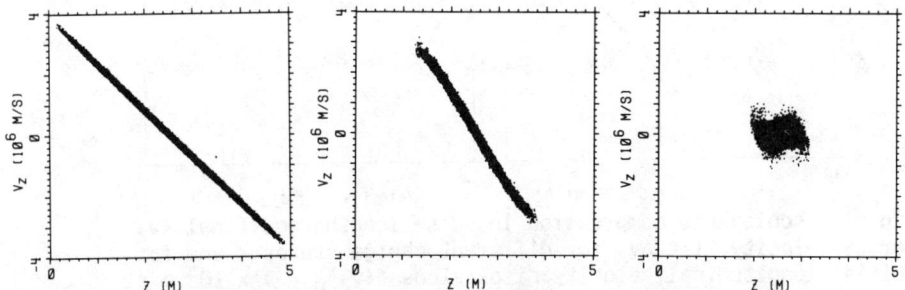

Fig. 1. The (z, v_z) phase space of a 2-1/2-D particle-code simulation (chosen as typical of our survey, see text): (a) at time t=0 (initial velocity tilt), (b) t = 0.36 μs, (c) t = 0.72 μs (final compressed state). Here charge state Z=3 and initial $(\delta v_z)_0 = 3 \times 10^4$ m/s. Note that Δv_z is the full height of frame (a) while δv_z refer to half heights for example of frame (c).

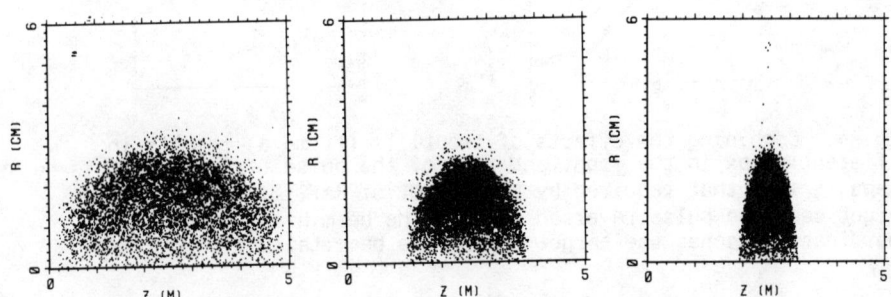

Fig. 2. The spatial positions in (r, z) coordinates of a representative sample of the particles in the 2-1/2-D simulation of Fig. 1 at corresponding times. (Plotting more points would result in burn-out of the photographic images). Note different horizontal and vertical length scales.

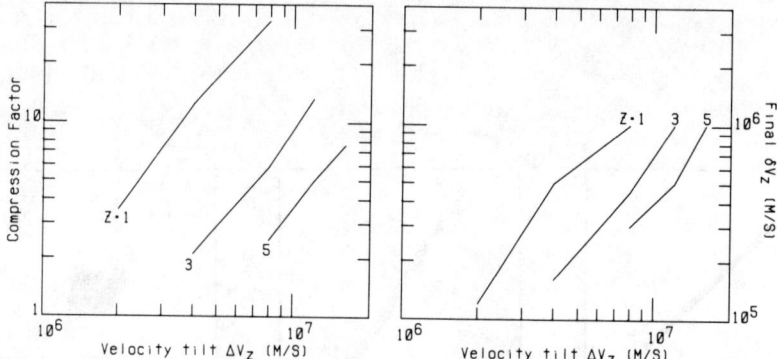

Fig. 3. Achievable compression in pulse lengths and final δv_z versus velocity tilt Δv_z for different charge states Z and for initial longitudinal velocity dispersions $(\delta v_z)_0 = 3 \times 10^4$ m/s. Each point in this curve is gleaned from a simulation such as illustrated in Figs. 1-2. Only a third of these simulations are included here because of space limitations. The smaller (or larger) δv_{z0} values of the rest of our survey has some corresponding effect on final Δv_z.

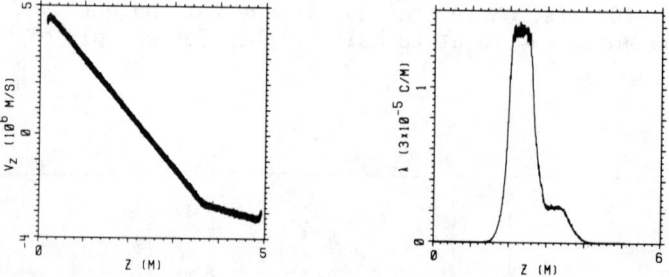

Fig. 4. Combining the effects of two tilts (frame a) results in different ramps in the front and rear of the pulse (frame b). This pulse shape is like that required by the target of Ref. 1. Note that the target sees the pulse reversed because the beam head (larger z coordinate) reaches the target before the beam tail.

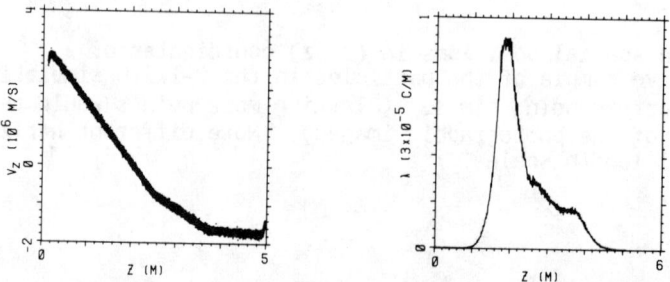

Fig. 5. Pulse shape at a 4.5-fold increase of peak power when a particular 3-component initial tilt is chosen. This illustrates that we can generate a more moderate rate of rise to the front porch of the pulse as seen by the target.

the number of final beam manipulations. This process uses the simple technique of an initial longitudinal velocity tilt in (z, v_z) consisting of several linear pieces as illustrated in Figs. 4-5. We can produce pulse shapes similar to those required by a number of targets[1,2].

REFERENCES

1. J. W-K. Mark and J. D. Lindl, "Symmetry Issues in a Class of Ion Beam Targets Using Direct Drive Pulses," Lawrence Livermore National Laboratory report UCRL-94346 (1986); and elsewhere in this proc.
2. J. W-K. Mark, "Charged Particle Fusion: Introduction," Lawrence Livermore National Laboratory Laser Program Annual Report for 1982, UCRL-50021-82, p. 3-17 (1983).
3. A. Faltens, E. Hoyer and D. Keefe, High-Power Beams 81, Eds. H. J. Doucet and J. M. Buzzi (Ecole Polytechnique - Palaiseau, France, 1981), p. 751.
4. J. Hovingh, V. O. Brady, A. Faltens, D. Keefe and E. P. Lee, "The Cost of Induction Linac Drivers for Inertial Fusion for Various Target Yields," Lawrence Berkeley Laboratory Report HIFAN-318 (1986).
5. I. Hofmann and I. Boszik, Symposium on Accelerator Aspects of Heavy Ion Fusion (Gesellschaft fuer Schwerionenforschung, Darmstadt W. Germany, 1982), p. 181.
6. I. Haber, Symposium on Accelerator Aspects of Heavy Ion Fusion (Gesellschaft fuer Schwerionenforschung, Darmstadt W. Germany, 1982), p. 372.
7. J. Bisognano, E. P. Lee and J. W-K. Mark, IEEE Trans. Nucl. Sci, NS-32, 2477 (1985).
8. J. W-K Mark and Y. L Pan, "Factors of Two Reductions of Beam Energy and Peak Power Requirements in Targets for Heavy Ion Fusion," Lawrence Livermore National Laboratory report UCRL-93046 (1986); and elsewhere in this proc.
9. D. D-M. Ho, R. O. Bangerter, J. W-K. Mark, S. T. Brandon, and E. P. Lee, "Longitudinal Compression of Heavy-Ion Beams with Minimum Requirements on Final Focus," Lawrence Livermore National Laboratory report UCRL-94347 (1986); and elsewhere in this proc.
10. S. Robertson, Phys. Rev. Letters 48, 149 (1982).
11. G. A. Krafft, "Collective Focusing of Intense Nonrelativistic Ion Beams by Co-moving Electrons," doctoral thesis, Lawrence Berkeley Laboratory and Physics Department, University of California Berkeley Campus (1985).

THRESHOLD FOR EMITTANCE GROWTH OF A-G FOCUSED INTENSE BEAMS

I. Haber
Naval Research Laboratory, Washington, DC 20375

ABSTRACT

The behavior of an intense beam propagating in a long alternate-gradient focusing channel is examined using computer simulations to determine, for phase advances between $60°$ and $90°$, the emittance growth which might be expected to result from space charge nonlinearities and instabilities.

While strong evidence of space charge driven instabilities is found at a phase advance of $90°$, for all beam distribution functions tested, these instabilities rapidly diminish as the phase advance is lowered. For an initial semi-Gaussian distribution at a phase advance of $80°$, for example, no instability-caused emittance growth is observed down to the numerical limits of the simulations. This behavior appears to be relatively insensitive to lens fill and small initial mismatch, but the thresholds for emittance growth do appear to shift as the initial beam distribution is varied, so that a similar $80°$ K-V system does show evidence of instability-caused emittance growth.

INTRODUCTION

The cost advantage which results from the use of multiply-charged ions in an induction linac driver has increased the importance of propagating beams with very low depressed phase advances while simultaneously avoiding any emittance growth during the propagation. It is also advantageous to employ, for at least part of the accelerator system, the largest phase advance which is consistent with the constraint on emittance growth.

A number of simulations, at various values of phase advance and beam intensity, for both initial K-V and semi-Gaussian distributions, have therefore been run to aid in optimizing the choice of transport system operating parameters. Sensitivity of these results to the details of the initial distribution function as well as to a small initial mismatch and to differences in lens occupancy factor were also examined.

Emphasis has been placed on the parameter regime appropriate to a transport system several hundred cells long in which the emittance growth must be limited to several percent. It is in the simulation of such long systems with intense beams that numerical parameters must be chosen carefully to obtain reliable results and yet stay within the computer resources available. In fact, in some cases, reliable extrapolation to the operating regime desired will require the use of more particles than was possible for this report. It is therefore necessary to regard some of the conclusions presented here as preliminary, and the emittance growths observed as an upper bound to the emittance growth likely in an actual system. Further work is necessary to refine the actual limiting values for emittance growth.

DEPENDENCE OF EMITTANCE GROWTH ON PHASE ADVANCE

It has been known for some time that alternating gradient (A-G) focusing systems with 90° phase advance per cell are subject to space charge driven instabilities which can cause rms emittance growth[1-4], while 60° systems, even with very low depressed phase advances[3-5], are not subject to the same emittance growth even though such systems can be shown to be linearly unstable for a beam with a Kapchinskij-Vladimirskij (K-V) distribution[3].

The Single Beam Transport Experiment (SBTE) at LBL[6,7] while providing corroboration that instability is observed at 90° and not 60°, shows a sharp drop in emittance growth as the phase advance is decreased below 90°. Since driver design is facilitated if the stronger focusing at higher phase advances is used, it is advantageous to operate as close to 90° as possible provided that any emittance growth is within acceptable bounds. Applicability of the SBTE results is, however, limited both by the short length of the transport system in the experiment and by the difficulty of obtaining very low beam emittances and of measuring small growths in emittance. It is therefore difficult to predict, from the experiment alone, the emittance growth expected when a very low emittance beam propagates down a transport system which is several hundred cells long.

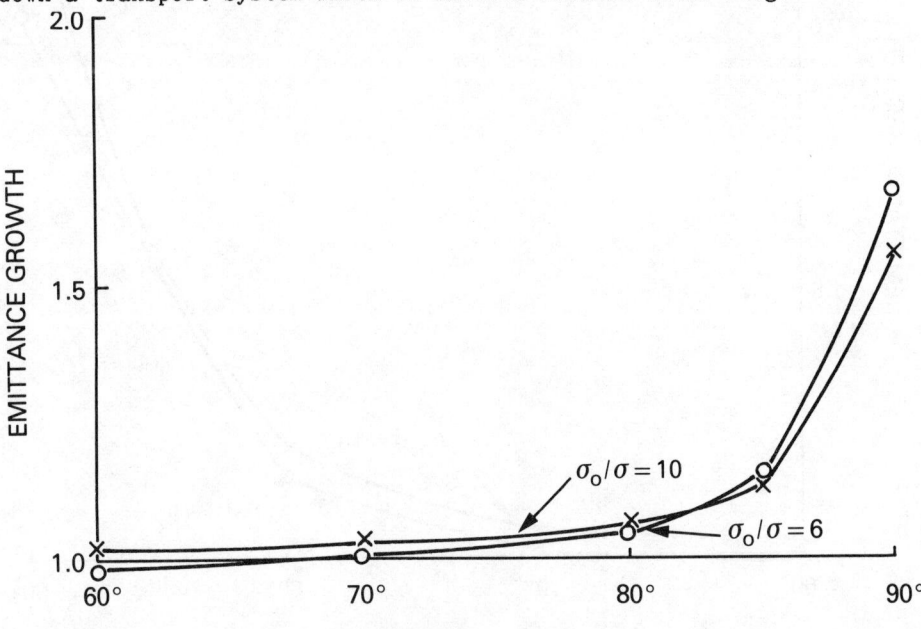

Fig 1. Emittance growth after 100 periods vs. phase advance (σ_o) for an initially semi-Gaussian distribution with phase advance depressed by factors of six and ten. The factor-of-ten curve was run with twice as many particles (32K) as the factor-of-six case to reduce the systematic differences due to numerical collisions.

A series of simulations has been run both to test the agreement between simulation and experiment, and to extend the experimental results to the lower emittances and the lower growth rates which might be important in long systems, and yet not be observable in SBTE. The results of these simulations are summarized in Figs. 1 and 2, which are plots of the rms emittance growth, after propagation down a 100 period channel, of initially semi-Gaussian (Gaussian in velocity and uniform in configuration space) and K-V distributions respectively. The two curves on each figure correspond to depressions in the phase advance by ratios, σ_o/σ, of 6 and 10. Both the semi-Gaussian and K-V distributions can be seen to reproduce the sharp drop in emittance growth which is observed experimentally as the phase advance is reduced below 90°.

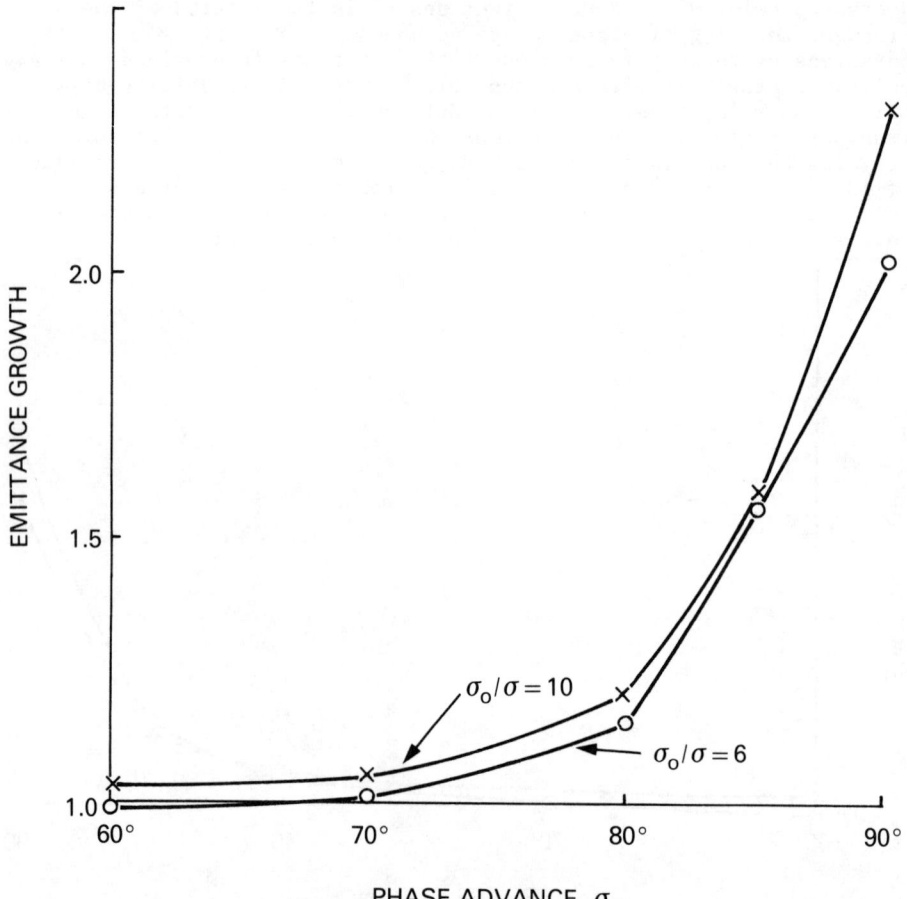

Fig. 2. Emittance growth after 100 periods vs. phase advance (σ_o) for an initially K-V distribution and depressions in the phase advance by factors of six and 10.

Because of the finite number of particles in the simulation, there is an anomalously high fluctuation level which causes collisional emittance growth[8]. While it appears difficult to predict the parametric scaling of collisional emittance growth in general, at a particular phase advance, and in the limit of low emittances, the beam size divided by the area of a Debye "ellipse", and therefore the expected collision frequency, scales as the square of the depression in phase advance, when the number of particles in the simulation is constant. The semi-Gaussian curve with depressed phase advance of $\sigma_o/\sigma = 10$ was therefore run with twice as many particles (32K) as the $\sigma_o/\sigma = 6$ curve to reduce systematic differences due to collisions. A limited number of runs have also been performed using 64K particles. At a phase advance of $80°$, doubling the number of particles to 64K reduced the emittance growth after 100 periods from 7% to 4% in 100 periods, so that much of this growth appears collisional. At $85°$, on the other hand, this doubling of particles reduces the emittance growth only from 15% to 12% and therefore does not appear to be primarily collisional. At $90°$ collisions appear to be even less significant, since a decrease in emittance growth only from 57% to 51% is observed as the number of particles is quadrupled from 16K to 64K. Though it is possible that collisions can stabilize an instability, and the growth in emittance might then increase if the number of particles were further increased, previous and more detailed studies[8] of collisional emittance growth have not found this to occur.

The conclusions drawn from the above are, therefore, that an $80°$ transport system will not suffer more than a 4% emittance growth in 100 magnet periods and the actual growth rate might be somewhat less, and that the simulations, in agreement with the SBTE experiment, do not show evidence of instability-caused emittance growth at least down to the level of 4% in 100 periods. An $85°$ transport system, on the other hand, does not seem as well behaved.

Some caution is in order, however, because of the much greater emittance growth observed in the $80°$ system with a K-V initial distribution. And while a K-V system is not usually encountered physically, there are many instances, as will be discussed below, where non K-V distributions approximate the behavior of their K-V counterparts.

OTHER FACTORS AFFECTING EMITTANCE GROWTH

Since an actual accelerator will likely have a beam distribution which is neither precisely semi-Gaussian nor K-V, it is desirable to know something about the sensitivity of emittance growth to distribution function in general and, in particular, in view of the differences in emittance growth observed between K-V and semi-Gaussian distributions, whether it is possible to stabilize the instabilities which cause emittance growth by a proper choice of distribution. Examination of the sensitivity of emittance behavior to such factors as the occurrence of a small initial mismatch or differences in lens occupancy factor is also useful in choosing an optimum operating regime. Particularly if, for example, a lens system with $80°$ phase advance is more sensitive to initial mismatch than one at $70°$.

The influence of initial distribution on instability growth rates has been examined previously for a 90° transport system.[9,10] The method employed was to gradually increase the current in the simulation so as to simulate the center of a long beam being longitudinally bunched. By starting the simulations at low currents, it is easily possible to construct equilibria as a superposition of non-interacting K-V distributions with densities chosen to give any desired total density. Though instability thresholds can be modified by changing the initial distribution, the basic parametric behavior of emittance growth remains largely unchanged. The typical behavior of a K-V distribution bunched in this way is a period of little change in emittance as the current is increased from a low value until instability threshold is reached. Any unstable eigenfunctions then begin to grow, but since the unstable modes preserve rms emittance at low amplitude, only after several periods of growth is the emittance affected. The emittance then grows rapidly until saturation, which, for a system with 90° phase advance, bunched over 100 periods, with the current increasing by a factor of 200 until reaching a phase advance depressed by a factor of six, occurs after a growth of about a factor of 2.6 in emittance.

A 90° system bunched similarly, and having the same rms values, but with an initial distribution consisting of a parabolic density in phase space (water-bag), instead of the flat phase space density of the K-V distribution, behaves similarly, but the onset of rapid emittance growth is delayed by approximately 10 periods corresponding to a 10% increase in the current at onset. The final emittance growth in this case is about a factor of 2.1. An initial distribution with a triangular density distribution, also exhibits almost the same behavior. Rapid emittance growth in this case is delayed by 14 periods compared with the K-V case, and the final emittance grows a factor of of 2.0.

Another distribution which has been used in these "bunching" simulations is generated by "debunching", i.e. reducing the current to a low value, after the instability has saturated. This distribution which can be closely approximated by a composite distribution, consisting of a 75% parabolic component added to a 25% triangular distribution extending to twice the phase space radius. When such an initial distribution is bunched the emittance behaves somewhat differently than the ones above. The onset of emittance growth is much more gradual, and the very rapid growth which characterizes the other cases is absent. However, no saturation of the growth occurs, so that when the final current has been reached, the total observed emittance growth of about a factor of 2.0 is similar to the other cases. Furthermore, if the rate of current growth is increased, the rate of emittance growth also increases, leaving the ratio I/ε relatively unchanged. Since the composite distribution is an approximation to the distribution which results after an instability has saturated, this value of I/ε, corresponding to an rms-equivalent depressed phase advance of approximately 30°, is a limit to the current intensity which can be transported in a 90° transport system. However, since the emittance is still growing at the end of the simulation, the actual limit suggested by this type of simulation is actually a somewhat lower intensity.

Similar bunching simulations have been performed for transport systems with phase advances of $60°$, $70°$, $80°$ and $85°$, using K-V and composite distributions. While the composite distribution appears to reduce the emittance growth resulting from instabilities, the space charge nonlinearities near the beam edge can cause emittance growth so that at $70°$ and below the composite distribution suffers more emittance growth when bunched than the rms equivalent K-V. Note that even though a $60°$ K-V system which is not bunched will not exhibit any emittance growth, it does go unstable during bunching and the resulting non K-V distribution does have some small nonlinearities at the beam edge which can cause emittance growth as the beam is bunched.

In summary, though description of the detailed behavior of these bunching runs is fairly complex and therefore beyond the scope of this paper, a sharp decrease in the emittance growth again for these types of simulations as the phase advance is lowered from $90°$. At $85°$ the emittance growth for a K-V distribution, when bunched to a factor of six rms-equivalent depressed phase advance, goes from the approximate factor of two observed at $90°$ to 1.5. This emittance growth decreases further, to about 1.2, at $80°$ phase advance.

In addition to the simulations described above it was found that changing the lens occupancy from the thin lenses used above to a 50% fill factor resulted in almost no noticeable change either to emittance growth or instability threshold in the small number of test cases run. Typical of the effect of an initial mismatch was to increase the emittance growth in an $80°$ K-V system, with the phase advance depressed a factor of six, to 16% from 15% and for a semi-Gaussian to 6% from 5%. But here too, since no radically unexpected behavior was observed, the number of tests was small.

CONCLUSIONS

A large number of simulations have been performed in an attempt to find optimum operating conditions for transporting a very low emittance high-current beam in a long alternating gradient channel. Assuming an initial semi-Gaussian distribution, an $80°$ transport system appears capable of transporting a current with the phase advance depressed by a factor of ten with only a modest amount of emittance growth. However, since an actual fusion driver system will require some changes in the parameters of the transport system as the beam is accelerated, and these changes can be accompanied by emittance growth which increases as the phase advance of the transport system increases, a more detailed investigation will probably be required to further optimize any trade-offs. In addition to further refinements to the numerics in some of the simulations described here, more work is also required to examine such effects as image and external field nonlinearities in any proposed operating regime. It does appear likely, however, that phase advances well above $60°$ can be employed for at least part of the transport system without any unacceptable emittance growth, with an $80°$ phase advance showing promise as an acceptable compromise.

ACKNOWLEDGEMENT

This work was supported by the United States Department of Energy under contract DE-AI05-84ER-40156.

REFERENCES

1. P.M. Lapostolle, "Quelques Proprietes des Distributions de Charges dans une Systeme de Focalisation," CERN Int. Report ISR/300LIN/69-63 (1969).
2. I. Haber and A. W. Maschke, "Steady State Transport of High Current Beams in a Focused Channel," Phys. Rev. Lett. $\underline{42}$, 1479 (1979).
3. I. Hofmann, L.J. Laslett, L. Smith and I. Haber, "Stability of the Kapchinskij Vladimirskij (K-V) Distribution in Long Periodic Transport Systems," Part. Accel. $\underline{13}$, 145 (1983).
4. I. Hofmann, "Transport and Focusing of High-Intensity Unneutralized Beams," in Applied Charged Particle Optics, Part C: Very-High-Density Beams, ed. by A. Septier, (Academic Press, New York, 1983).
5. I. Haber, "Simulation of Low Emittance Transport," Proc. of the INS Symposium on Heavy Ion Accelerators and their Application to Inertial Fusion, Jan. 23-27, 1984, Institute for Nuclear Study, University of Tokyo, 451 (1984).
6. D. Keefe, "Experiments and Prospects for Induction Linac Drivers," elsewhere in this PROC.
7. M. G. Tiefenback and D. Keefe, "Measurements of Stability Limits for a Space-Charge-Dominated Ion Beam in a Long A.G. Transport Channel," IEEE Trans. Nucl. Sci. NS-32, 2483 (1985).
8. I. Haber, "High-Current Simulation Codes," in High Current, High Brightness, and High Duty Factor Ion Injectors, ed. by George H. Gillespie, Yu-Yun Kuo, Denis Keefe and Thomas Wangler, AIP Conf. Proc. $\underline{139}$, (American Inst. of Physics, New York, 1986).
9. I. Haber, "Space Charge Limited Transport of Non K-V Beams," IEEE Trans. Nucl. Sci. NS-26, 3090, (1979).
10. J. Struckmeier, J. Klabunde and M. Reiser, "On the Stability and Emittance Growth of Different Particle Phase-Space Distributions in a Long Magnetic Quadrupole Channel," Par. Accel. $\underline{15}$, 47 (1984)

LONGITUDINAL COMPRESSION OF HEAVY-ION BEAMS WITH MINIMUM REQUIREMENTS ON FINAL FOCUS*

D. D.-M. Ho, R. O. Bangerter, J. W.-K. Mark, and S. T. Brandon
Lawrence Livermore National Laboratory, Livermore, CA 94550

E. P. Lee
Lawrence Berkeley Laboratory, Berkeley, CA 94720

ABSTRACT

A method is developed to compress a heavy-ion beam longitudinally in such a way that the compressed pulse has a constant line-charge density profile and uniform longitudinal momentum. These conditions may be important from the standpoint of final focusing. By realizing the similarity of the equations that describe the 1-D charged-particle motion to the equations that describe 1-D ideal gas flow, the evolution of λ and the velocity tilt can be calculated using the method of characteristics developed for unsteady supersonic gasdynamics. Particle simulations confirm the theory. Various schemes for pulse shaping have been investigated.

INTRODUCTION

In induction linac for heavy-ion fusion, the heavy-ion beam must be compressed longitudinally in order to provide the required power for target implosion. Various authors have attempted to investigate this problem previously. Haber[1] studied the compression of a rectangularly-shaped pulse with a linear velocity tilt (variation of longitudinal particle velocity from beam-head to beam-tail). Bisogano, Lee, and Mark[2] investigated the compression of a parabolic-shaped pulse again using a linear velocity tilt. These studies showed that the compressed beam does not have constant line-charge density λ and/or uniform particle longitudinal momentum. However, the requirements for the compressed beam to have uniform λ and velocity may be important from the standpoint of final focusing. The purpose of this paper is to show that it is possible for the beam to satisfy these requirements at the end of the compression process if the beam is given proper λ and velocity-tilt profiles while the beam is traveling in the accelerator before the compression starts.

The first part of this paper describes the compression scenario and the theoretical basis behind it. Fluid equations and the method of characteristics developed for unsteady supersonic gasdynamics are applied to this problem. The second part presents the particle simulation which confirms the theory. Possible schemes for pulse shaping and other considerations are discussed in the final part of this paper.

COMPRESSION SCENARIO

The procedure for beam compression consists of the following steps.

(1) When the beam is still in the accelerator, the space-charge effect (the longitudinal electrostatic force resulting from the gradient of λ) is relatively low since the beam is still long ($\sim$ 10-15m). The line-charge density can be assumed to be constant along the beam. Under these circumstances, it is possible to change the constant λ profile to any reasonably shaped λ profile by giving the beam a proper velocity tilt. This can be understood by first writing the continuity equation (fluid equations can be used if the longitudinal temperature is negligible) in the Lagrangian coordinate, i.e.

$$\frac{\partial \Gamma}{\partial t} = \frac{\partial V_z}{\partial m} , \qquad (1)$$

where $\Gamma = 1/\lambda$, V_z is the longitudinal fluid velocity in the center of mass frame, and $dm = \lambda\, dz$. If space charge is not important, then V_z for each particle is almost a constant after the required velocity tilt is imposed on the beam at the beginning of the process. If the desired pulse shape is achieved in time T, then after integrating Eq. (1) trivially in time and converting the equation back to the Eulerian coordinate gives

$$\left. \frac{\partial V_z(Z)}{\partial Z} \right|_T = \left(1 - \frac{\lambda(T,Z)}{\lambda(0,Z)} \right) \frac{1}{T} . \qquad (2)$$

Since $\lambda(0,Z)$ is independent of Z and $\lambda(T,Z)$ is given, the profile of the velocity tilt can be obtained from Eq. (2).

Figure 1(a) shows that the beam in the accelerator does not have a velocity tilt initially. A velocity tilt is gradually imposed on the beam. This tilt reaches a maximum in Fig. 1(b) and is removed gradually afterwards. The removal is complete when the desired λ profile is achieved in Fig. 1(c).

(2) A linear velocity tilt is now gradually given to the beam until the peaks of the tilt reach $\pm\, 2\, C_s$ (see Fig. 1(d)). Here, the ion sound velocity is

$$C_s \equiv \sqrt{(Ze/m)g\lambda} ,$$

where Ze and m are ion charge and mass, and g is a constant. Due to the compression resulting from the linear velocity tilt, the

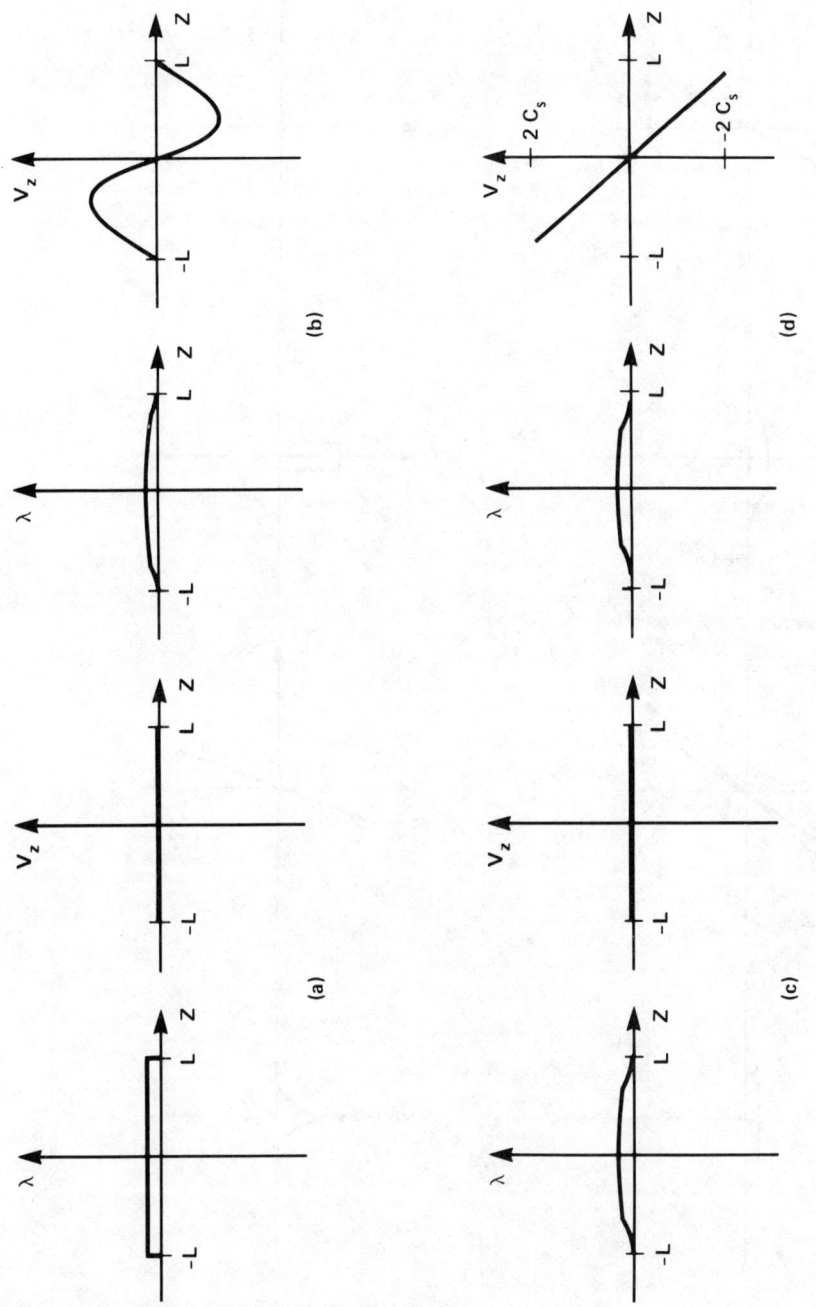

Fig. 1 Compression scenario

246

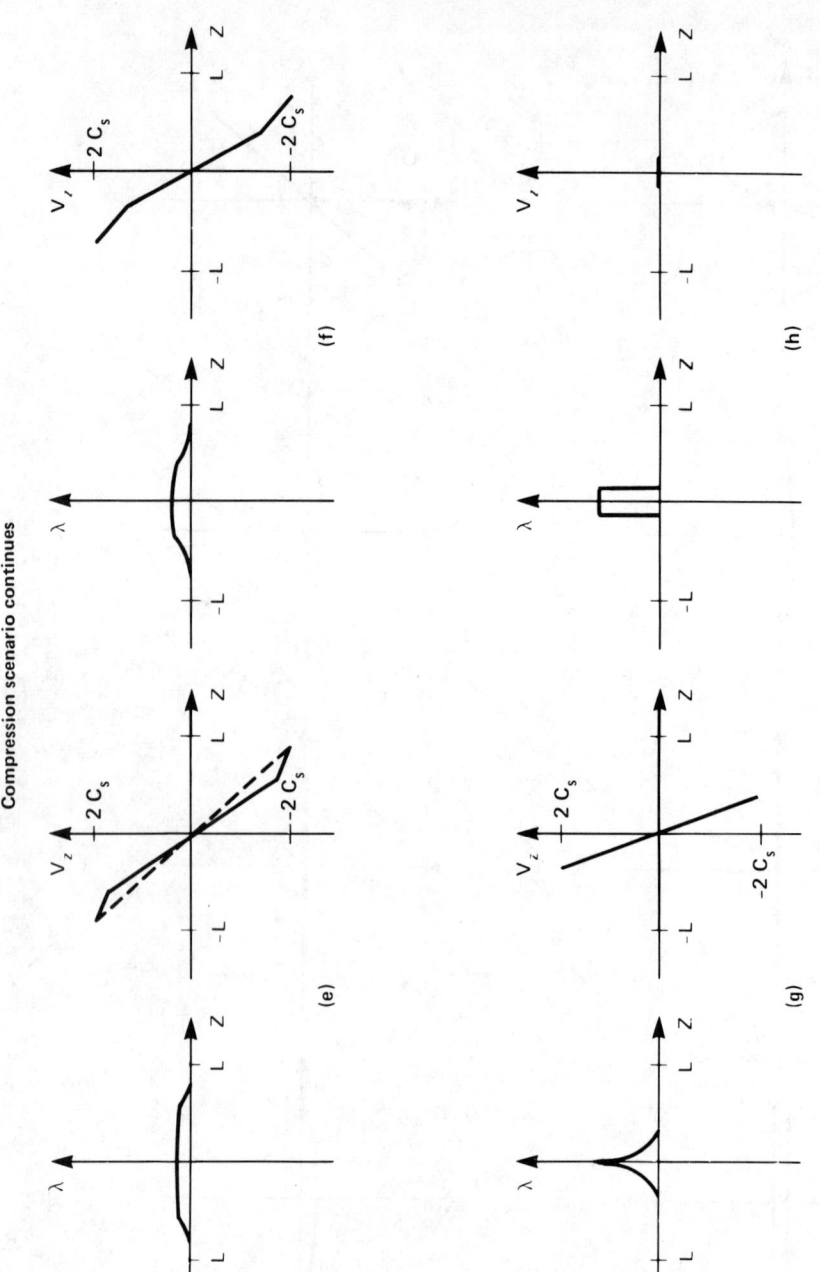

beam length (as shown in Fig. 1(d)) is now shorter than it was in Fig. 1(c). However, compressing the beam by a linear velocity tilt has preserved the functional form of λ.

(3) At this point, a "mid-course" correction is given to the beam in order to change the linear velocity profile to the shape given in Fig. 1(e). Since the maximum difference in velocity between this velocity profile and the linear profile is small, the "mid-course" correction can be carried out in a time scale so short that the functional form of λ in Fig. 1(e) remains essentially the same as that shown in Fig. 1(d).

(4) No more external manipulation of the beam is required beyond Fig. 1(e). (The λ and velocity profiles of the beam at the instant when the beam leaves the accelerator are shown in Fig. 1(f).) The velocity tilt shown in Fig. 1(e) compresses the beam in such a way that the electrostatic force generated by the gradient of λ removes all the velocity tilt at end of the compression process (see Fig. 1(h)).

In order to determine the λ and velocity-tilt profiles in Fig. 1(e), one realizes that the equations for 1-D charged-particle motion (with negligible longitudinal temperature) are identical to those for 1-D ideal gas motion except that the pressure force ∇P in the fluid equation is now replaced by the electrostatic force $(Ze/m)g\partial\lambda/\partial Z$. Therefore, the longitudinal compression process (between Figs. 1(e) and 1(h)) can be treated exactly as the time-reversal process for the free expansion of a slab of gas with initial constant pressure and density profiles[3,4]. The time-reversibility of this problem was first suggested by Faltens and Lee[5]. Thus, the λ and velocity profiles in Fig. 1(e) can be obtained by using the method of characteristics developed for unsteady supersonic gasdynamics[4]. The discussion of this method is involved and hence is deferred to a forthcoming paper.

PARTICLE SIMULATION

A beam with parameters listed in Table I is used for the simulation using the 2-1/2-D Condor particle code.

TABLE I
BEAM PARAMETERS

Beam Length (at the end of acceleration)	10 m
Beam Radius	3 cm
Total Beam Current	715 amp/beam
Total Charge/Beam	69.4 micro coulomb
Ion Energy (at the end of acceleration)	11.5 GeV
Ion Mass	210 amu
Charge State of Individual Ion	+ 3
Ion Sound Velocity	4×10^5 m/sec
λ	6.9×10^{-6} coulomb/m
After Compression	
Final Beam Length	1 m
λ	6.9×10^{-5} coulomb/m

In the simulation, the beam particles are represented by 9×10^4 test particles. The radial boundary condition in the simulation represents a cylindrical perfect conductor at a radius of 6 cm. The quadrupole magnetic field is represented by an azimuthally symmetric radial electric field which is constant in the longitudinal direction and the field strength is proportional to the radius.

The simulation starts when the beam is 5m long. The λ and velocity-tilt profiles shown in Fig. 2(a) are obtained using the method of characteristics. Initially, the beam has a radial temperature of 200 eV and has no longitudinal temperature. After a time of 4.2×10^{-7} sec, the beam is compressed to a length of approximately 3 m as shown in Fig. 2(b). Figure 2(c) is the λ and velocity-tilt profiles obtained from analytical calculations corresponding to the same instant of time as that of Fig. 2(b). Comparison of Fig. 2(b) and 2(c) shows that the particle simulation agrees very well with the theoretical calculations.

Since the radial electric field used in the simulation is constant in time, the simulation is restarted when the beam length reaches 3m. The input data used are those shown in Fig. 2(c). Also, a new equilibrium radial electric field is used. After a time of 4.2×10^{-7} sec, the compression process is completed and the compressed beam now has a length of 1m. As shown in Fig. 2(d), the final λ profile is almost flat and the velocity tilt is essentially removed. Thus, particle simulation again confirms the theory. In addition, no substantial growth in longitudinal or radial emittance is observed in the simulation.

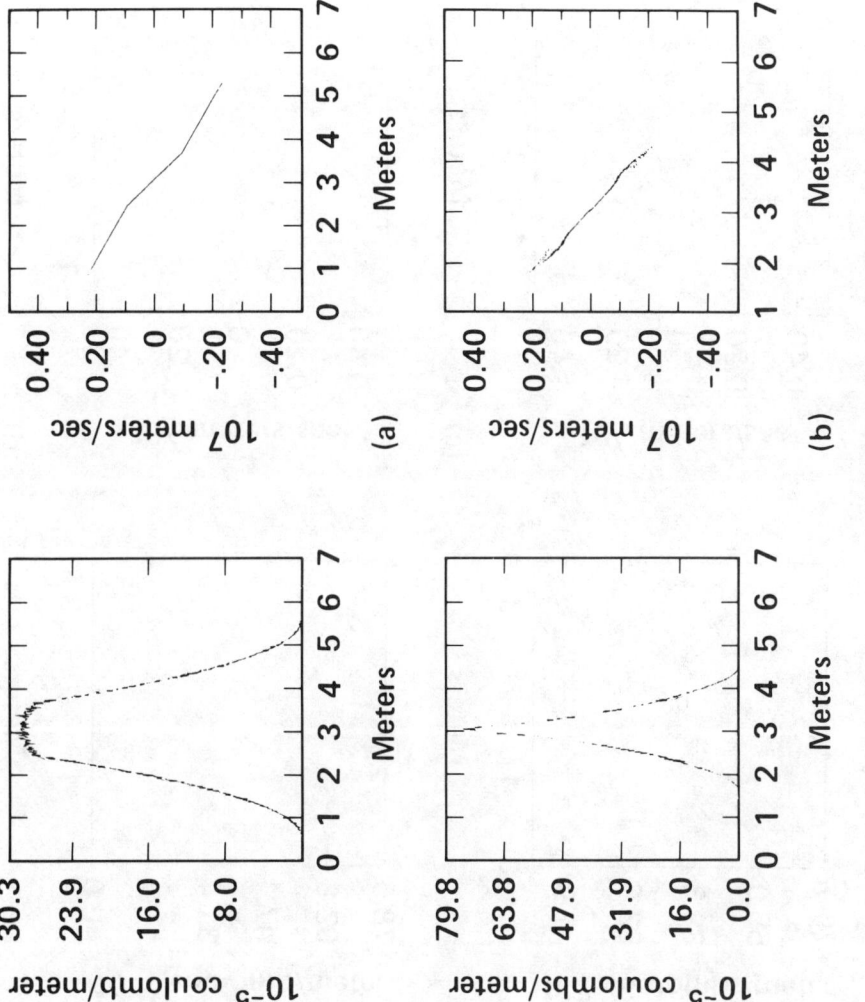

Fig. 2 Line - charge density and velocity - tilt profiles from particle simulation

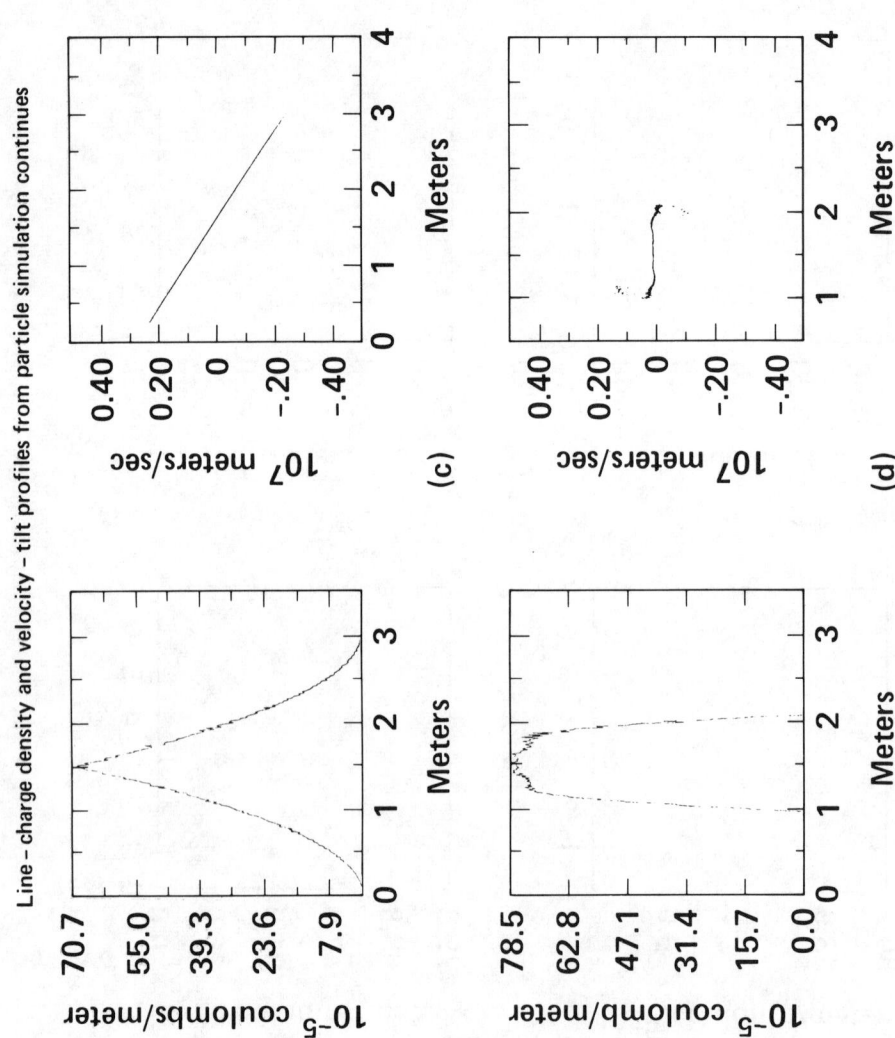

Line – charge density and velocity – tilt profiles from particle simulation continues

PULSE SHAPING AND OTHER CONSIDERATIONS

Pulse shaping can be obtained easily by delaying the relative arrival time of different pulses on the target. The delaying of some pulses can be obtained by making the beam lines for some of the pulses a few meters longer than those for the other pulses. Another possible way of achieving pulse shaping is to give each pulse a non-symmetrical (observed from center of the beam) λ and velocity-tilt profiles. The profiles for the non-symmetric case are shown in Fig. 3(a). Figure 3(b) shows the λ and velocity-tilt profiles at the end of compression. This method may not require delaying beams; however, the focusing system would have to accommodate the variation in current and momentum.

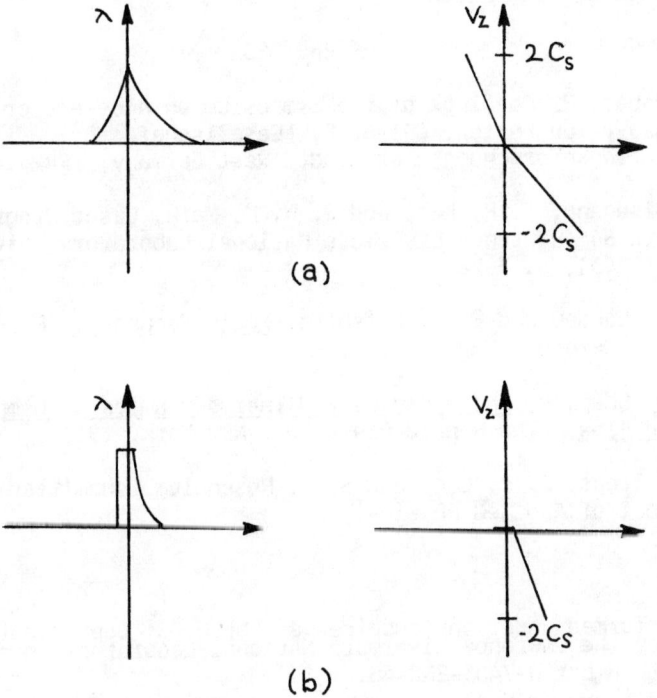

Fig. 3. Non-symmetric λ and Velocity-tilt Profiles

Using the beam parameters given in Table I and taking engineering restrictions into account, calculations indicate that the initial manipulation of λ and velocity-tilt profiles (between Figs. 1(a) and 1(e)) can be carried out rapidly enough so that the space-charge effect does not cause appreciable pulse shape deterioration.

CONCLUSIONS

A method is developed to compress a heavy-ion beam longitudinally in such a way that the compressed pulse has a uniform λ profile and uniform longitudinal momentum. These conditions may be important from the standpoint of final focusing. The initial λ and velocity-tilt profiles for the pulse is calculated by invoking the method of characteristics developed for unsteady supersonic gasdynamics. Particle simulations confirm the theory. Various schemes are available for achieving the desired pulse shape. No substantial growth in either longitudinal or transverse emittance is observed. Estimates show that the initial manipulation of λ and velocity-tilt profiles can be carried out rapidly enough so that space-charge effect does not cause serious pulse shape deterioration.

REFERENCES

1. I. Haber, Proceedings of the Symposium on Accelerator Aspects of Heavy Ion Fusion, GSI-82-8, (Gesellschaft für Schwerionenforschung, Darmstadt, West Germany, 1982), p. 372.

2. J. Bisogano, E. P. Lee, and J. W.-K. Mark, Laser Program Annual Report 84 (Lawrence Livermore National Laboratory, Livermore, CA., 1985), p. 3-28.

3. L. D. Landau and E. M. Lifshitz, <u>Fluid Mechanics</u> (Pergamon Press, Oxford, 1959), Ch. X.

4. A. H. Shapiro, <u>The Dynamics and Thermodynamics of Compressible Fluid Flow</u>, (The Ronald Press Co., New York, 1953), Ch. 24.

5. A. Faltens, E. P. Lee, and S. S. Rosenblum, submitted to Journal of Applied Physics.

Work performed under the auspices of the U. S. Department of Energy by the Lawrence Livermore National Laboratory under contract number W-7405-ENG-48.

SOME MECHANISMS AND TIME SCALES FOR EMITTANCE GROWTH

O. A. Anderson
Lawrence Berkeley Laboratory, Berkeley, CA 94720

ABSTRACT

We discuss several processes which affect transverse rms emittance and analyze the growth rate for two of them: (1) a fast process by which electrostatic self-field energy is converted into fluid motion, and (2) a slow process by which a nonlinear focusing force converts beam-size mismatch energy into additional emittance growth. The analysis for these two cases is done by solving the particle equation of motion. For the latter case we find that the final emittance depends primarily on the beam-size mismatch and that the time scale is inversely proportional to the nonlinearity parameter. We also discuss, in a more general way, another process by which wave-breaking in phase space causes partial conversion of fluid motion to quasi-thermal energy. In all cases, insight is gained by splitting the mean-square emittance into fluid-motion and quasi-thermal parts. Among other benefits, this allows a clearer understanding of the wave-breaking process.

INTRODUCTION

It is well-known that changes in rms emittance can be related to changes in self-field energy [1-4] and that this process occurs rapidly [4,5]. However, little attention has been paid to a different type of emittance growth that can result from an initial mismatch in rms beam size, or to the slow time scale involved.

We begin by introducing an emittance-splitting technique which is illuminating in all cases. Then we illustrate how to find the time dependence of emittance from particle dynamics by reviewing our fast-scale analysis [5], and we discuss the subsequent wave-breaking and mixing process. We use energy conservation to discuss this second process and also to estimate the maximum emittance growth that can result from beam-size mismatch. Finally, we use the particle-dynamics method to analyze the time development of the latter process for the case of a nonlinear external force.

In this short report we concentrate primarily on the relatively simple case of sheet beams. Results for round beams are summarized at the end of the paper.

DECOMPOSITION OF EMITTANCE INTO THERMAL AND FLUID PARTS

We use Sacherer's definition of mean-square emittance

$$\epsilon^2 = \langle x^2 \rangle \langle x'^2 \rangle - \langle x x' \rangle^2. \quad (1)$$

To divide this into thermal and fluid-flow parts we use the local dimensionless fluid velocity and specific pressure tensor (n is the density):

$$u(x,z) = n(x,z)^{-1} \int_{-\infty}^{\infty} dx' \, x' \, f(x,x',z) \quad (2)$$

and
$$T(x,z) = n^{-1} \int_{-\infty}^{\infty} dx' \, (x' - u)^2 \, f(x,x',z). \quad (3)$$

$T(x,z)$ represents the local temperature if the velocity distribution is Gaussian. The local mean-square velocity is $\langle x'^2 \rangle_{av} = T(x,z) + u^2(x,z)$. Its x-average is

$$\langle x'^2 \rangle(z) = \langle T \rangle(z) + \langle u^2 \rangle(z) \quad (4)$$

where

$$\langle T \rangle = N^{-1} \int_0^{\infty} dx \int_{-\infty}^{\infty} dx' \, (x' - u)^2 \, f(x,x',z) ; \quad (5)$$

N is the half-beam line density. It is easy to show that $\langle x x' \rangle = \langle x u \rangle$ so that Eq. (1) becomes

$$\epsilon^2 = \langle x^2 \rangle \langle T \rangle + \left[\langle x^2 \rangle \langle u^2 \rangle - \langle x u \rangle^2 \right]. \quad (6)$$

The first term is proportional to the mean quasi-thermal energy; the term in brackets comes from fluid motion. Eq. (6) can be written

$$\epsilon^2_{total} = \epsilon^2_{th} + \epsilon^2_{fluid} \quad (7)$$

where the subscript "th" stands for quasi-thermal or thermal, depending on circumstances.

We will show that on the fast time scale, electrostatic energy goes into the fluid term in about one quarter of a plasma period. Later on, there is partial conversion to the $\langle T \rangle$ term.

SHEET BEAMS, FAST TIME SCALE

In this section we excerpt some results from Ref. [5]. Our simple model deals with non-relativistic, singly charged particles with a longitudinal velocity v and transverse velocities « v. The beam is symmetric in the direction of transverse motion, x, and only the upper half ($x \geq 0$) is considered. Before giving the results, we set up the notation.

The density is $n(x,z)$, and the particle number per cm² within x is

$$N_x(x,z) = \int_0^x n(x_1,z) \, dx_1. \quad (8)$$

If the beam is reasonably thin, the transverse self field is $E_s = 4\pi e N_x$. In a uniform channel with linear external focusing represented by k^2,

$$x'' = -k^2 x + P N_x(x,z)/N \quad (9)$$

where $x'' = d^2x/dz^2$ and $P = 4\pi N e^2/mv^2$ is the normalized line perveance, representing total space charge. For a cold beam the particle motion is laminar on the fast time scale so that N_x is preserved for each particle at its position $x(z)$. If we write ξ for the initial position of the particle, then $N_x(x,z) = N_x(\xi,0)$ for all x in the laminar range of z, and the space-charge term in Eq. (9) is constant. The solution, with initial condition $x'(\xi) = 0$ and with the definition $x_e(\xi) = (P/k^2) N_x(\xi,0)/N$, is

$$x(\xi,z) = x_e(\xi) + (\xi - x_e(\xi)) \cos kz \quad (10)$$

so that

$$dx/d\xi = n(\xi)/n_u + (1 - n(\xi)/n_u) \cos kz \quad (11)$$

where $n_u = N k^2/P$. Laminar motion ceases at the distance z_c where the left side of Eq. (11) vanishes, i.e., where

$$\cos kz_c = -(n_u/n(\xi) - 1)^{-1}. \qquad (12)$$

<u>Laminarity Criterion</u>. If $n(\xi) > n_u/2$ for all $\xi < h$, then z_c does not exist and the motion is laminar for all z. If z_c exists, then it lies between $\lambda/4$ and $\lambda/2$; $\lambda \equiv 2\pi/k$ and is also the plasma period for a cold beam.

Laminar motion implies $n(x,z)dx = n(\xi)d\xi$, so Eq. (11) gives the beam density z dependence (in this paper we often refer to z dependence as <u>time</u> dependence since $z = vt$ and with our approximations v is constant):

$$n(x,z)/n_u = [1 + (n_u/n(\xi) - 1)\cos kz]^{-1} \qquad (13)$$

where the x dependence is found by simultaneous use of Eq. (10). Note that $n(x,z)$ becomes uniform at $z = \lambda/4$ and then a density reversal occurs: particles originating in underdense regions find themselves in overdense regions and vice versa.

If the laminarity criterion is violated, then $n \to \infty$ as $z \to z_c$ for particles originating at the point of minimum density; this marks the beginning of the wave-breaking process discussed in the next section.

If there are gaps in the density profile, then Eq. (11) needs some modification. We do not consider such cases here.

<u>Emittance Growth</u>. The solution, Eq. (10), is used to calculate the moments in the emittance equation (1). With the notation $X^2(z) = \langle x^2 \rangle$ we find [5]

$$\epsilon^2(z) = \frac{p^2 X_0^2}{3k^2} U_{no}(1 - \frac{U_{no}}{4})\sin^2 kz. \qquad (14)$$

We have not assumed that the beam is matched. X_0 is the initial value of X and U_{no} is the initial value of the shape factor

$$U_n(z) = 2 - 2\sqrt{3}\, W/PX \qquad (15)$$

where
$$W(z) = P\langle x N_x \rangle / N \qquad (16)$$

is the virial moment [3]. Ref. [5] gives a table of U_n values for selected profiles; U_n is the normalized free self-field energy for a sheet beam.

If the laminarity criterion is satisfied, then $\epsilon(z) \sim |\sin kz|$ so that the emittance periodically returns to zero. However, when the criterion is violated, ϵ never returns to zero. The two cases are illustrated by the numerical simulations [6] shown in Fig. 1.

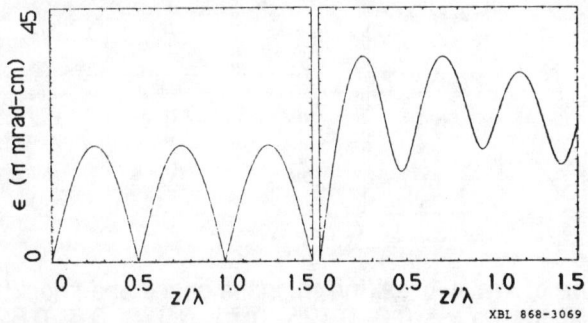

Fig. 1. Time dependence of emittance for different initial profiles: (a) $n/n_u = 1 - 0.45\cos(\pi\xi/h)$, $\xi < h$; (b) $n/n_u = 1 - 0.7\cos(\pi\xi/h)$, $\xi < h$. The latter violates the laminarity criterion.

Warm Sheet Beams. For a cold beam, during the period of laminar particle motion, all the emittance is due to the fluid term in Eq. (6). For a warm space-charge dominated beam the fluid motion on the fast time scale is the same as for the corresponding cold beam, except for a slight change in timing. This was discussed in Ref. [5], where we showed that if the beam is matched (X nearly constant), the thermal term in Eq. (6) is essentially constant, and to good accuracy

$$\epsilon_{tot}(z) = \left[\epsilon_0^2 + \frac{p}{\sqrt{3}} X_0^3 U_{no} \sin^2 k_1 z\right]^{1/2} \quad (17)$$

where k_1 is the slightly corrected wave number.

Eq. (17) shows that the upper limit mentioned in [3] and [4] is actually reached and shows that it is reached at about $z \approx \lambda/4$.

WAVE-BREAKING; SHOCKS

In this section we discuss cold sheet beams which violate the laminarity criterion. A transition occurs at the point z_c given by Eq. (12). Before this point, the distribution on the phase plane is a well-behaved curve with slope

$$\frac{dx'}{dx} = \frac{dx'/d\xi}{dx/d\xi} = \frac{-(1 - n(\xi)/n_u) k \sin kz}{n(\xi)/n_u + (1 - n(\xi)/n_u) \cos(kz)}. \quad (18)$$

Comparison with Eq. (12) shows that this slope is finite for $z < z_c$ so that $x'(x)$ is single valued. As $z \to z_c$ the curve steepens like the shape of a wave about to break in phase space. In real space the cold beam equation (13) shows that the charge density becomes singular as $z \to z_c$.

We illustrate these effects in Fig. 2 with a simulation [6], using the same hollow initial profile as for Fig. 1b. Immediately after the point z_c, $x'(x)$ becomes triple valued so that the specific pressure tensor $T(x)$ jumps from zero to a finite value. The density singularity spreads out from the center like a pair of shock, or pressure, fronts.

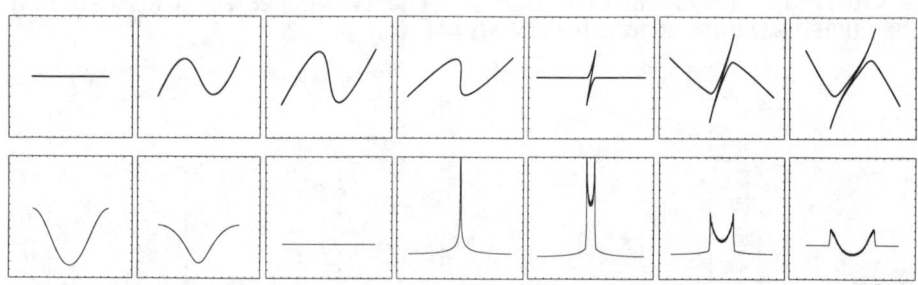

Fig. 2. Example of wave-breaking in phase space and shock formation in configuration space; z/λ = 0.0, 0.125, 0.25, 0.375, 0.5, 0.625, and 0.75. Upper row, phase plots; lower row, charge densities. Results agree with Eqs. (18) and (13) for $z < z_c$. Wave-breaking starts at $z_c = 0.32\lambda$ (Eq. 12). Cold beam behavior is illustrated. Space-charge dominated warm beam results are similar but the shock-like structures are smoother.

SHEET BEAM ENERGY INVARIANT

After wave-breaking occurs, energy is transferred from the $\langle u^2 \rangle$ term in Eq. (4) to the $\langle T \rangle$ term. Since we are considering a uniform focusing channel, we can use energy conservation to gain more insight. Integrating Eq. (9) over x, averaging, and using Eq. (4) we obtain

$$\langle T \rangle + \langle u^2 \rangle + k^2 \langle x^2 \rangle - 2W(z) = \text{Const.} \qquad (19)$$

where the virial moment $W(z)$ is given by Eq. (16). From Eq. (15), W can also be written $W = XP(1 - U_n/2)/\sqrt{3}$; if $U_n(z)$ decreases, W increases. As the beam becomes more uniform, $\langle u^2 \rangle$ or $\langle T \rangle$ must increase if $X \approx$ constant (matched beam); this is the basis of conventional emittance growth from release of self-field energy.

In the next section we analyze emittance growth arising from beam size mismatch and find the maximum value and the growth rate for a slightly nonlinear focusing force. But we can estimate the maximum emittance using energy conservation. We take the limiting case in which $X' \to 0$ and $U_n \to 0$, and we use "∞" subscripts to indicate this idealized limit. Eqs. (1), (4) and (19) then yield

$$\epsilon_\infty = \frac{X_\infty}{X_0}\left[\epsilon_0^2 + \frac{P}{\sqrt{3}}X_0^3 U_{no} + k^2 X_0^2(X_0^2 - X_\infty^2) + \frac{2P}{\sqrt{3}}X_0^2(X_\infty - X_0)\right]^{1/2}. \qquad (20)$$

This result includes both the effect of non-uniform initial beam shape and of beam-size mismatch. If $X_\infty = X_0$, then ϵ_∞ equals the initial peak value given by Eq. (17).

If the beam size is not well matched, however, we see that there can be additional emittance growth. To find X_∞ we use the rms envelope equation for sheet beams [5]

$$X'' + k^2 X = \epsilon^2/X^3 + W/X. \qquad (21)$$

For our model
$$k^2 X_\infty^2 = \epsilon_\infty^2/X_\infty^2 + PX_\infty/\sqrt{3}. \qquad (22)$$

It is straightforward to eliminate X_∞ from Eqs. (20) and (22) with the help of Eq. (6). The resulting expression for ϵ_∞ is unwieldy but can be simplified by expanding in powers of small parameters. One such parameter is U_{no}. Another is $(\epsilon_0/kX_u^2)^2$, which is essentially the square of the so-called tune depression; we define $X_u = P/k^2\sqrt{3}$. We also define the mismatch parameter $\mu = (X_0/X_u) - 1$. Then, omitting a negligible term $\sim U_{no}^2$ and all third-order terms, we obtain

$$\epsilon_\infty = \left[\epsilon_0^2 + \frac{P^4}{9k^6}(U_{no} + \mu^2 + \mu U_{no}) - 2\mu\epsilon_0^2 + \frac{9k^6}{P^4}\epsilon_0^4 + 2\epsilon_0^2 U_{no} + \cdots\right]^{1/2}. \qquad (23)$$

Some of these terms can be dropped for a strongly space-charge dominated beam. For example, for a parabolic profile and an initial tune depression of 0.1, we have [5] $U_{no} = 0.008$ and $(\epsilon_0/kX_u^2)^2 = 0.01$. Then for $\mu = 0.1$ the dominant terms are

$$\epsilon_\infty = \left[\epsilon_0^2 + \frac{P^4}{9k^6}(U_{no} + \mu^2) + \cdots\right]^{1/2} \qquad (23')$$

and the (squared) mismatch parameter μ^2 affects the final emittance on an equal footing with the shape factor U_{no}.

We will find that the above result based on an idealized limit gives a good approximation to a result based on particle dynamics.

EMITTANCE GROWTH RATE FOR MISMATCHED BEAM

If the initial beam size is mismatched, there will be envelope oscillations. A small nonlinearity in the field can slowly couple the energy of these oscillations into emittance growth. For example, such nonlinearity arises from beam profiles which are non-uniform (the generic case). In this section we analyze a comparable but simpler case where the initial beam profile is uniform but the channel focusing force is slightly nonlinear. We replace the equation of motion (9) by

$$x'' = -k_0^2 x(1 + \sigma x^2/q^2) + PN_x(x,z)/N \qquad (24)$$

where $q \equiv P/k_0^2$. The nonlinearity parameter σ is assumed to be $\ll 1$; we will show that emittance grows on a time scale $\sim 1/\sigma$. We will also derive a condition for laminar motion to persist over the same time scale (cold beam model), so that the last term in Eq. (24) remains constant for each particle. We use again the notation $x(0) = \xi$ for initial position and h for the initial beam edge. Since the beam is assumed initially uniform,

$$N_x/N = \xi/h \equiv \chi \qquad (25)$$

with $0 \leq \chi \leq 1$. We define the equilibrium position $x_{eq}(\chi)$ and the normalized deviation from equilibrium $y = (x - x_{eq})/q$ and get

$$y''/k^2 + (1 + 3\sigma x_{eq}^2/q^2)y + 3\sigma(x_{eq}/q)y^2 + \sigma y^3 = 0; \qquad (26)$$

$$x_{eq}/q = \chi - (x_{eq}/q)^3 = \chi - \sigma\chi^3 + O(\sigma^2). \qquad (27)$$

Eq. (26) can be solved by standard perturbation techniques if the initial beam size mismatch is small. We define the mismatch parameter

$$\mu = (h/q) - 1. \qquad (28)$$

To first order in μ and σ

$$x(\chi,z) = q\chi[1 - \sigma\chi^2 + (\mu + \sigma\chi^2)\cos kz]; \qquad (29a)$$

$$k = k_0(1 + \tfrac{3}{2}\sigma\chi^2). \qquad (29b)$$

<u>Laminarity</u>. The distance between two particles relative to their small initial separation is $dx/d\xi$; trajectories cross when this ratio vanishes.

$$dx/d\xi = (q/h)[1 - 3\sigma\chi^2 + (\mu + 3\sigma\chi^2)\cos kz - 3\sigma k_0 z(\mu\chi^2 + \sigma\chi^4)\sin kz] \qquad (30)$$

The beam density is found by taking the reciprocal of this; it becomes singular when trajectory crossing first occurs. A study of Eq. (30) shows that this happens at the beam edge where $\chi = 1$ and at a point where $\cos kz = 0$, $\sin kz = 1$. Then $dx/d\xi = 0$ gives the critical distance

$$3\sigma k_0 z_{crit} = (1 - 3\sigma)/(\mu + \sigma) \approx (\mu + \sigma)^{-1}. \qquad (31)$$

We will see that most of the emittance growth can occur for $z < z_{crit}$.

Mean square beam size. To lowest order in μ and σ

$$x^2/q^2 = \chi^2 - 2\sigma\chi^4 + 2(\mu\chi^2 + \sigma\chi^4)\cos[(1 + \frac{3}{2}\sigma\chi^2)k_0 z] \quad (32)$$

so that

$$\langle x^2 \rangle = \int_0^1 x^2(\chi) d\chi = q^2[1/3 - 2\sigma/5 + 2\mu I_1 + 2\sigma I_2] \quad (33)$$

where I_1 and I_2 are combinations of trigonometric functions and Fresnel integrals which we will not write out here. Instead we give the expansions for small z and for large z. It is convenient to define the normalized distance

$$\zeta = 3\sigma k_0 z. \quad (34)$$

For small ζ, to order ζ^2

$$\frac{\langle x^2 \rangle}{q^2} = \frac{1}{3} - \frac{2\sigma}{5} + \frac{2}{3}\left[\mu + \frac{3\sigma}{5} - \frac{\zeta^2}{5 \cdot 7}(\mu + \frac{\sigma}{3})\right]\cos k_0 z + \frac{2\zeta}{5}(\mu + \frac{3\sigma}{7})\sin k_0 z.$$

For large ζ the leading terms are

$$\langle x^2 \rangle = q^2\left[\frac{1}{3} - \frac{2\sigma}{5} + \frac{2}{\zeta}(\mu + \sigma)\sin k_0 z\right]. \quad (35)$$

The coefficient of $\sin k_0 z$ is $2(\mu + \sigma)^2$ at the point where laminarity breaks down [Eqs. (31) and (34)]. This coefficient is very small for reasonably small nonlinearity and mismatch parameters. This shows that the envelope oscillations are largely damped before the point z_{crit} is reached. We will see that the oscillation energy is mostly transferred to emittance growth.

Emittance. Eq. (29) gives, to lowest order,

$$x'(\chi, z) = -k_0 q \chi(\mu + \sigma\chi^2)(1 + \frac{3}{2}\sigma\chi^2) \sin[(1 + \frac{3}{2}\sigma\chi^2)k_0 z] \quad (36)$$

which, together with Eq. (29) gives the phase space configuration for any $z < z_{crit}$. We see that particles near the edge of the beam oscillate at a different frequency from those near the center and become increasingly out of step. The phase-space configuration becomes increasingly irregular, and this is the cause of the emittance growth.

To get ϵ^2 [Eq. (1)] we evaluate $\langle x'^2 \rangle$ and $\langle xx' \rangle$ in terms of Fresnel integrals and then expand for small and large z as we did for $\langle x^2 \rangle$. The small z result, to lowest order in μ and σ, is

$$\epsilon^2 = \frac{p^4}{6 \cdot 7 \cdot 25 k_0^6}[4\sigma^2 + A_1 \zeta^2 - \cdots] \quad (37a)$$

$$A_1 = (\mu^2 + \frac{10}{9}\mu\sigma + \frac{5}{21}\sigma^2). \quad (37b)$$

In Eq. (37) we have omitted a term $-4\sigma^2 \cos 2k_0 z$ and other oscillatory terms proportional to $\zeta^2 \cos 2k_0 z$, $\zeta \sin 2k_0 z$, $\zeta^2 \cos k_0 z$, and $\zeta \sin k_0 z$, because we are interested here in the mean growth rate only. During the main part of the emittance growth, the A_1 term dominates and the growth rate, using Eq. (34), is approximately

$$\frac{d\epsilon}{dz} = \left[\frac{3A_1}{2\cdot 175}\right]^{1/2} \frac{\sigma p^2}{k_0^2}. \qquad (38)$$

For large z, to lowest order in ζ^{-1}, and again omitting oscillatory terms,

$$\epsilon(\zeta) = \frac{p^2}{k_0^3 \sqrt{18}} \left[(\mu^2 + \frac{6}{5}\mu\sigma + \frac{3}{7}\sigma^2) - \frac{9}{\zeta^2}(\mu + \sigma)^2\right]^{1/2}. \qquad (39)$$

If we neglect the last term and divide by Eq. (38) we find the nominal emittance growth distance

$$k_0 z_{growth} = \frac{\sqrt{175}}{3^{3/2}\sigma}\left[1 + \frac{2}{45}\frac{\sigma}{\mu} + \cdots\right]. \qquad (40)$$

Usually we consider cases where the nonlinear parameter is at least as small as the mismatch parameter so that we can neglect terms in (σ/μ) and ignore the bracketed term, as we do below.

At this point we can write the condition for the growth to be essentially complete before $z = z_{crit}$. Dividing by Eq. (31) we have

$$z_{growth}/z_{crit} = (175/3)^{1/2}(\mu + \sigma). \qquad (41)$$

In the case of a bad mismatch, laminarity will end before the growth is complete. Nonlinear damping of the oscillation energy may be expected to continue, however, and the emittance may be expected to reach a final value given approximately by Eq. (23).

We can use Eqs. (37) and (39) to find the time dependence of emittance to first order in the mismatch and nonlinearity parameters. Fig. 3 shows the analytic result for the case $\mu = 0.03$, $\sigma = 0.02$.

These first-order analytic results were checked by numerical simulation in collaboration with L. Soroka and were found to give adequate accuracy. We also simulated cases with larger μ where wavebreaking occurs before full growth. The maximum emittance agreed well with the energy-invariant prediction of Eq. (23). We also found that for space-charge dominated warm beams the thermal part of ϵ^2 can be added in as discussed in an earlier section.

Even with a perfectly linear external force, badly mismatched beams can exhibit slow but substantial emittance growth. This case is difficult to analyze because the cold-beam laminar model is inadequate. But a non-uniform density profile produces a nonlinear field which should give the

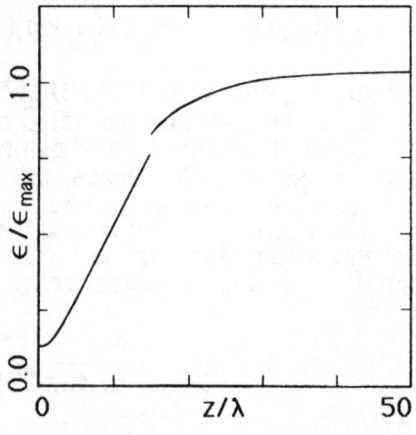

Fig. 3. Emittance growth from mismatch; small z and large z approximations. See text.

effect. In fact, most initially uniform warm beams become non-uniform after rotation in phase space so that mismatching can lead to large emittance growth. An upper limit for emittance can be obtained from energy conservation, but there is no guarantee that it will be reached.

Our 1-D model neglects coupling of energy into the other two degrees of freedom. Some of the nonlinear mechanisms which allow emittance growth due to beam size mismatch could produce such coupling, and this would affect our results.

ROUND BEAMS, FAST TIME SCALE

We introduce some notation from [5]. If the beam density is $n(r,z)$, the number of particles per cm of length within radius r is

$$N_r(r,z) = \int_0^r 2\pi r \, n(r_1,z) \, dr_1. \tag{42}$$

The self field is $E_s = 2eN_r/r$; the total perveance is $K = 2Ne^2/mv^2$; N is the total particle number per cm. Then with a linear external focusing force the equation of motion is

$$r'' = -k^2 r + K N_r(r,z)/(N r). \tag{43}$$

This equation is solvable by perturbation methods; the solution [5] is in powers of the parameter η

$$\eta(\rho) = \rho/r_e(\rho) - 1 \tag{44}$$

where $r_e(\rho)$ is defined by Eq. (45). We note that η vanishes for a uniform matched beam and is small everywhere for most nonuniform profiles if the beam mismatch is not too large. The equilibrium radius is

$$r_e(\rho) = \left[\frac{K}{k^2} \frac{N_r(\rho)}{N}\right]^{1/2} \tag{45}$$

For round beams, we define $n_u = Nk^2/(\pi K)$.

<u>Laminarity Criterion</u>. The motion is laminar for all z if

$$n(\rho) > \frac{n_u}{2}(1 - \frac{4}{3}\eta) + O(\eta^2)$$

for all ρ within the beam. The beam density time dependence, to first order in η, is

$$n(r,z) = \frac{n(\rho)}{\frac{n(\rho)}{n_u} + \left[1 - \frac{n(\rho)}{n_u}\right]\left[\cos\omega_0 z + \frac{2}{3}\eta(1 - \cos\omega_0 z)^2\right]} \tag{46}$$

where $\omega_0^2 = 2k^2$. If the laminarity criterion is violated, then the critical distance $z = z_c$ is defined by the vanishing of the denominator.

<u>Peak Rms Emittance</u>. We show in [5] that essentially all of the free self-field energy is converted into fluid energy at $z = \lambda_0/4$, where $\lambda_0 = 2\pi/\omega_0$. To a good approximation

$$\epsilon^2(\lambda_0/4) = K R_0^2 U_{no}/16 \tag{47}$$

where R_0 is the initial rms radius and U_{no} is the initial value of the normalized free self-field energy U_n. This is

$$U_n(z) = 4\int_0^b dr\, N_r^2(z)/N^2 r - (1 + 4\ln b/R\sqrt{2}) \tag{48}$$

where $R^2 = \langle r^2\rangle$ and b is chosen large enough to include all the beam.

For matched, warm round beams the total emittance at $z \approx \lambda_0/4$ is

$$\epsilon_{peak} = [\epsilon_0^2 + (K/16) R_0^2 U_{no}]^{1/2} \tag{49}$$

as discussed in Ref. [5]. Our prediction that the emittance peaks at $z \approx \lambda_0/4$ agrees with the simulations reported by Wangler et al. [4].

ROUND-BEAM ENERGY INVARIANT AND BEAM SIZE MISMATCH

The invariant corresponding to Eq. (19) is

$$\langle T_r\rangle + \langle u_r^2\rangle + k^2\langle r^2\rangle + KU_n(z)/4 - K\ln R = \text{Const.} \tag{50}$$

where u_r and T_r are radial versions of the quantities in Eqs. (2) and (3), and where $U_n(z)$ is defined by Eq. (48).

If the laminarity criterion is violated, shock-like behavior (cf. Fig. 2) begins at $z = z_c$. Energy is transferred from the fluid-flow term to the quasi-thermal term, as discussed earlier.

In transport of a mismatched space-charge dominated beam there will be slow rms emittance growth if the focusing force is slightly nonlinear. As with sheet beams, the maximum emittance can be estimated using energy conservation. We consider the idealized limit $R' \to 0$, $U_n \to 0$, and use "∞" subscripts to indicate this limit. Then

$$\epsilon_\infty = \frac{R_\infty}{R_0}\left[\epsilon_0^2 + \frac{K}{16}R_0^2 U_{no} + \frac{k^2}{4}R_0^2(R_0^2 - R_\infty^2) + \frac{K}{4}R_0^2 \ln\frac{R_\infty}{R_0}\right]^{1/2} \tag{51}$$

which agrees with Ref. [3] if $R_\infty = R_0$. If the beam size is mismatched, we use the round beam version of Eq. (22) to eliminate R_∞ from the above equation. We find to lowest order

$$\epsilon_\infty = \left[\epsilon_0^2 + \frac{K^2}{32 k^2}(U_{no} + \frac{1}{2}\mu_r^2) + \cdots\right]^{1/2} \tag{52}$$

where we define the radial mismatch parameter $\mu_r = R_0(2k^2/K)^{1/2} - 1$. Eq. (52) is similar to Eq. (23) and we again note that the effect of beam-size mismatch may well exceed the effect of beam non-uniformity in producing emittance growth.

Although there is insufficient space for presentation here, we have also used particle dynamics to analyze emittance growth for a mismatched round beam in a nonlinear channel. All the stages of the analysis are analogous to those for a sheet beam, and the resulting growth rate is similar.

ACKNOWLEDGMENTS

I wish to thank Chris Celata for helpful discussions and advice, Bill Cooper for careful reading and criticism of the manuscript, Lloyd Smith for critical remarks, and John Lawson and Martin Reiser for valuable advice and encouragement. For understanding the longer time scales numerical simulations were essential and the collaboration of Ludmilla Soroka was invaluable.

This work was supported by US DOE Contract DE-AC03-76SF00098.

Note added. — During the editing of this manuscript, M. Reiser pointed out another paper which deals with nonlinear focusing forces: P. Loschialpo, W. Namkung, M. Reiser, and J. D. Lawson, "Effects of Space Charge and Lens Aberrations in the Focusing of an Electron Beam by a Solenoid Lens," J. Appl. Phys., 57, 10 (1985).

Although these authors do not discuss rms emittance growth, they do point out that nonlinearity can destroy laminar flow. They treat a case quite different from ours. Whereas we analyze transport in a uniform channel of a slightly mismatched beam (small change in beam size), they treat focusing by a nonlinear lens to a small spot (large change in size). Their simulations and experiments show effects in phase space and in density profiles that are analogous to our results (Fig. 1) for beams with strongly non-uniform initial profiles.

REFERENCES

[1] P.M. Lapostolle, CERN-ISR-DI/71-6 (1971).

P.M. Lapostolle, IEEE Trans. Nucl. Sci., 18, 1101 (1971).

[2] E.P. Lee, S.S. Yu, and W.A. Barletta, Nuclear Fusion, 21, 961 (1981).

[3] J. Struckmeier, J. Klabunde, and M. Reiser, Particle Accel., 15, 47 (1984).

[4] T.P. Wangler, K.R. Crandall, R.S. Mills, and M. Reiser, IEEE Trans. Nucl. Sci., 32, 2196 (1985). This paper lists many other important references.

[5] O.A. Anderson, "Internal Dynamics and Emittance Growth in Space-Charge Dominated Beams," Lawrence Berkeley Laboratory report LBL-19996; to be published in Particle Accel.

There is an abridged version of this paper: O.A. Anderson, "Internal Dynamics and Emittance Growth in Non-Uniform Beams," LBL-21633, published in the AIP Proceedings of the 1986 Linear Accelerator Conference, Stanford Linear Accelerator Center, June 2-6, 1986.

[6] O.A. Anderson, D.G. Kane, and L. Soroka, Bull. Am. Phys. Soc., 30, 1595 (1985).

HEAD-TO-TAIL VELOCITY TILT IN AN ION INDUCTION LINAC*

C.H. Kim, D.L. Judd, L.J. Laslett, L. Smith, and A.I. Warwick
Lawrence Berkeley Laboratory
University of California, Berkeley, CA 94720

ABSTRACT

In the earlier stages of acceleration in a heavy-ion-induction linac, acceleration and bunching rates are constrained by the allowable value of head-to-tail velocity tilt at a given location. If focusing parameters at a given location are fixed, the velocity tilt should be less than a certain upper bound to avoid too much envelope variation and consequent beam losses. For space charge dominated beams, we found some favorable particle distributions in longitudinal phase space for which the maximum-matched-beam envelope at a given location is almost constant with respect to time, in spite of the presence of a large velocity tilt. Mismatch oscillations can be reduced by slow variation of the velocity tilt and slow current amplification. Under these circumstances, the velocity tilt can be as large as allowed by the usable range of σ_0. Behavior of Cs ion beams with very large velocity tilts (up to 40%) are studied experimentally in MBE-4 and the results are presented.

INTRODUCTION

In cost optimizing the design of an induction-linac-based HIF driver, design tradeoffs are necessary between the required core material and the cost of focusing structures; a shorter pulse requires less core material but the resulting higher current requires greater focusing strength. As a result, specific acceleration and current-amplification schedules are imposed along the accelerator.[1] The acceleration rate and the current-amplification rate are kinematically interrelated with each other through the head-to-tail velocity tilt; for instance, a larger velocity tilt permits a higher acceleration rate for a required current-amplification rate.

In the earlier stages of acceleration when the kinetic energy is low and the bunch length is large, the acceleration rate is severely limited by this constraint. Since the focusing strength at a given location is fixed in time, any variation of the kinetic energy for different longitudinal positions along the bunch may result in a variation of the beam radius and possibly in beam losses. A large velocity tilt and the consequent brisk acceleration and bunching can cause mismatch oscillations. The purpose of the present investigation is to find a practical upper bound for the permissible values of the velocity tilt.

A kinematic relationship between velocity tilt, acceleration rate and the bunching rate is derived in Section II. Some favorable longitudinal particle distributions are discussed in Section III for electric and magnetic focusing systems. These distributions can tolerate higher velocity tilts because the maximum-matched-beam radius at a given location is quite

---

*This work was supported by the Office of Energy Research, Office of Basic Energy Sciences, U.S. Department of Energy under Contract No. DE-AC03-76SF00098.

insensitive to the kinetic energy. An experimental demonstration of a high velocity tilt in an electric focusing system is described in Section IV. A discussion on mismatch oscillations is included in Section V.

A KINEMATIC RELATION

A bunch of ions moves along an induction linac under the influence of a series of accelerating kicks. Its progress may be approximately represented by a set of smooth curves in z-t space traversed by particles at the bunch head (H), tail (T), and center (C) as sketched in Fig. 1. Also assume that the bunch center is located at z where the pulse duration is Δt and at time t when the bunch length is L. The slope of a particle's curve is its velocity (v) and the rate of change of the slope is its acceleration (A). The velocity tilt is defined as:

$$\Delta\beta/\beta = (v_T - v_H)/v_0 \qquad (1)$$

We expand v_{tail} and v_{head} in a Taylor series about the respective points T_0 and H_0 in terms of $(t_T - t_{T_0})$ and $(t_H - t_{H_0})$, and keep only the first order terms. Using the relations,

$$v_{H_0} - v_{T_0} = dL/dt ,$$

$$t_T - t_H \cong L/v_0$$

$$d/dt = v_0\, d/dz ,$$

and

$$A_0 = (qe/m)\, dV_0/dz ,$$

one can rewrite equation (1) in a more convenient form:

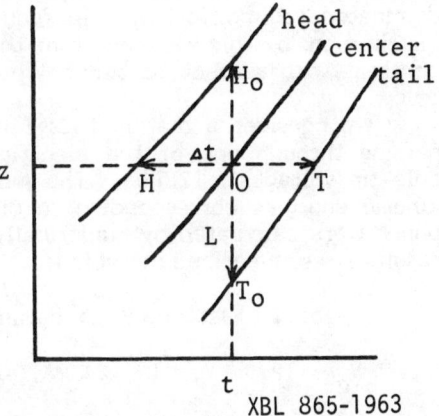

Fig. 1. Trajectories of particles at the beam head (H), center (C), and tail (T).

$$(1/L)\,(\Delta\beta/\beta) = (1/2V_0)\,(dV_0/dz) - (1/L)\,(dL/dz) , \qquad (2)$$

where V_0 is the kinetic energy (E) for particles at the bunch center divided by the charge state (q).

Accordingly, if it is desirable to impose a restriction on the magnitude of $\Delta\beta/\beta$ for reasons which will be discussed in the following sections, then a limitation is necessarily imposed on the acceleration rate, dV/dz. This restriction, unfortunately, acts to lengthen the accelerator and is particularly restrictive if the bunch length is large and the kinetic energy is low.

The average current is defined as $I(z) = Qv_0/L$ where Q is the total charge. Differentiating I(z) with respect to z and using equation (2), we have an expression for the rate of current amplification:

$$\left(1/I(z)\right) dI(z)/dz = (1/L) \Delta\beta/\beta .\qquad(3)$$

MATCHED BEAM CONSIDERATIONS

The presence of a velocity tilt can have several important consequences. With the focusing field gradient of an individual quadrupole lens constant with respect to time, the focusing spring constant, K, becomes significantly different for ions of different kinetic energy. ($K \propto 1/V$ for electrostatic focusing and $K \propto 1/v$ for magnetic focusing). The value of σ_0 thus becomes quite different for particles at various locations within the bunch. The presence of a velocity tilt also will lead to differences in the size of the matched envelope at various portions of the beam.

We are interested in finding longitudinal particle distributions for which the maximum-matched-beam-radius at a given location is constant with respect to time in spite of the presence of the velocity tilt. A distribution which closely satisfies this condition in the space charge dominated regime for electric focusing systems is the constant-current distribution; by this we mean that the beam current at a given location is constant with respect to time. We illustrate this point with an example below.

We consider a matched Cs+1 beam of a constant current of 24 mA passing through one of the electric quadrupole lenses of MBE-4[2] with pole-tip voltages ± 12.5 kV. The matched beam parameters for different kinetic energies corresponding to different longitudinal segments of the bunch were calculated by numerically solving the envelope equations. The results are summarized in Table I.

Table I - Matched Beam parameters at a given lens (MBE-4)

V_q = 12.5 kV,   I = 24 mA,   and ϵ_N = 1.5x10^{-7} π.m.rad

K.E.	(keV)	200	250	300	350	400	450
A_{max}	(mm)	21.1	21.2	21.4	21.7	22.0	22.3
A_{avr}	(mm)	15.4	16.4	17.4	18.2	18.9	19.5
σ_0	(deg)	80.0	61.5	50.7	43.1	38.4	35.0
σ	(deg)	8.8	6.8	5.4	4.5	3.8	3.4

From the tabular material, one notes the very marked variation of the energy and σ_0. In contrast, one notes the virtually negligible variation of the maximum-matched-beam radius (A_{max}). The constant-current distribution can be obtained (at least for the flat part of the bunch) by designing the accelerating voltage wave-forms in a current-self-replicating fashion.[3]

For magnetic focusing, the focusing strength diminishes less rapidly with higher beam velocity than electric focusing; thus the spread of σ_0 is much narrower for the same range of energy. We found that, for the space charge dominated regime, the desirable distribution for which the maximum-matched-beam radius varies the least is the constant-density distribution; by this we mean that the line-charge density at a given location is constant with respect to time. Notice that this distribution is somewhat different from the one considered in a constant-density acceleration scenario[4] where the density at a given time is assumed to be constant with respect to the longitudinal position along the bunch.

Significantly higher velocity tilts can be tolerated in magnetic focusing systems than in electric focusing systems.

THE EXPERIMENT

An experiment was performed in MBE-4 to see whether beams with very high velocity tilt would pass through the transport channel. Four individually focused Cs+1 beams, 200 keV, 10 mA each with 3 μsec pulse durations are injected into the multiple-beam induction linac which was operated with 8 accelerating gaps. These experimental conditions are similar to the ones described in reference (2) except that the last four accelerating wave-forms were modified to triangular shapes by rearranging the timing sequence of the individual pulsers. These wave-forms produced a final velocity tilt of up to 29% over the flat-current portion and up to 38% over the total length of the bunch.

Injected current, I(t), and injected kinetic energy, E(t), were measured as shown in Fig. 2a. These data and the measured accelerating voltage wave forms were used in digitized forms to calculate the expected values of E(t) and I(t) at the end of the linac (Fig. 2b). The

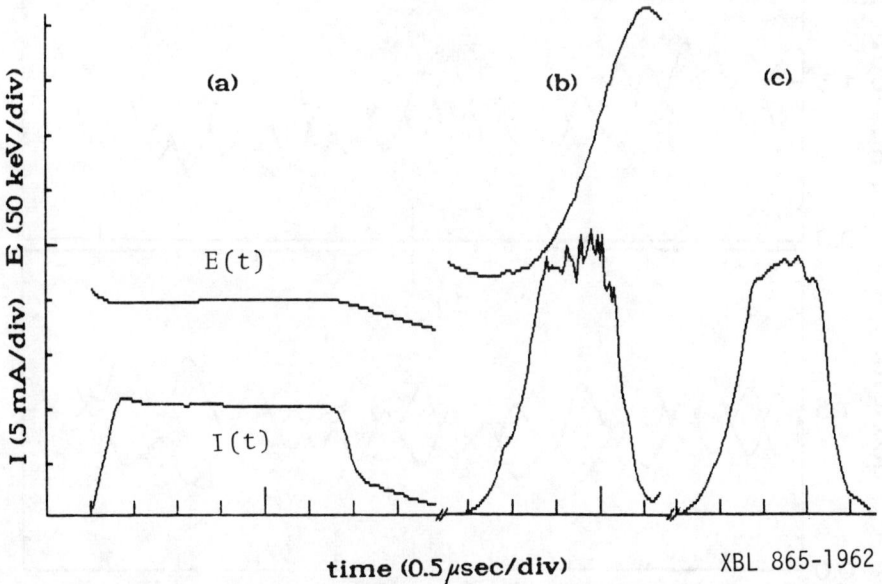

Fig. 2. (a) Measured beam energy, E(t), and beam current, I(t), at injection. (b) Calculated E(t) and I(t) at the end of MBE-4. (c) Measured I(t) at the end of MBE-4.

calculation was done numerically with a 1-dimensional code including the space charge effects (SLID).[3] Measured current at the end of the linac (Fig. 2c) shows a good agreement with the calculation. No beam losses were observed within the experimental accuracy.

The calculated I(t) showed more high frequency noise than the measured I(t) because of the presence of high frequency noise in the accelerating voltages and of the fact that infinitely narrow acceleration gaps were assumed in the calculation. For a finite ion-transit time (τ), a fluid analysis showed that the frequency response decreases by the factor, $\sin(\omega\tau/2)/(\omega\tau/2)$.

Beam emittances were measured at the end of the linac for two representative longitudinal locations of the bunch. No emittance growth was observed within the experimental accuracy.

Beam envelopes for about a dozen representative longitudinal segments of the beam were calculated by numerically integrating the envelope equations. Calculated values of beam currents and kinetic energies for each lattice period were used. These calculations agree very well with the measured values at the end of the fourth gap and at the end of the linac. For the front and central portion of the beam, mismatch oscillations are small as shown in Fig. 3a. The amplitude of mismatch

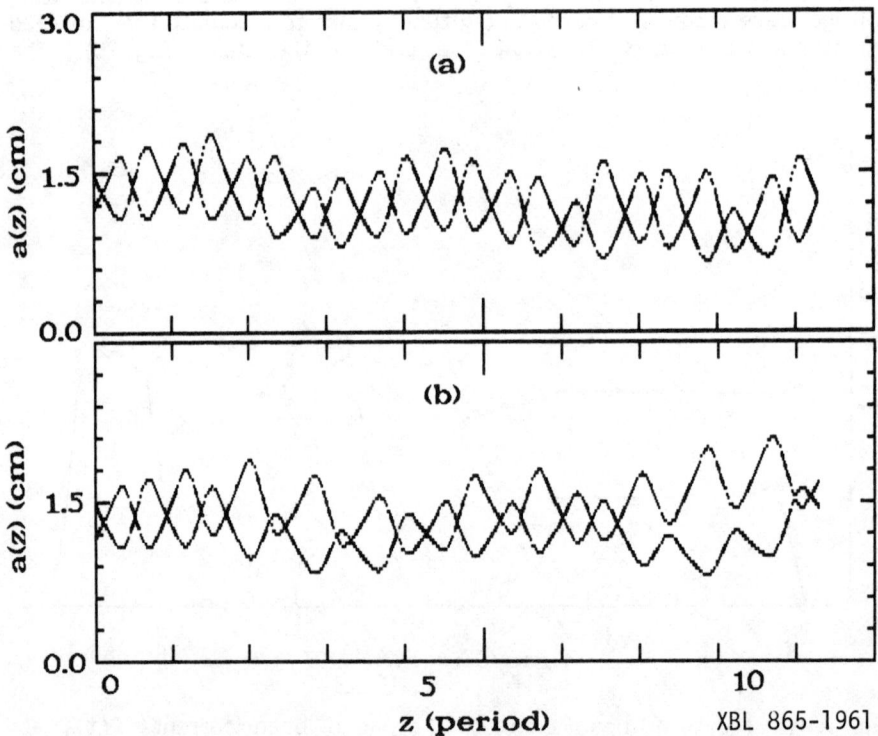

Fig. 3. Calculated beam envelopes (vertical and horizontal) typical for the beam head and center regions (a) and for the beam tail region (b).

oscillations is larger for the tail end of the beam as shown in Fig. 3b, but always well within the quadrupole-bore radius.

The high velocity tilt produced in this experiment may not be sustained in the subsequent acceleration gaps (which will be added later) because of too much current amplification as given in equation (3), unless the acceleration rate is increased from the value we are currently planning to implement.

MISMATCH OSCILLATIONS

We wish to study the qualitative behavior of mismatch oscillations, to which end it is sufficient to consider a round beam in smooth approximation with zero emittance. The envelope equation, taking acceleration into account is:

$$\frac{d}{dz} \beta \frac{da}{dz} = - K\beta a + \frac{qeIZ_o}{4\pi E_o a} \quad (4)$$

where K is the external force constant, E is kinetic energy in eV, and Z_o is 120 πohms. The equation applies separately to each longitudinal segment of the beam; if we adopt a constant current scenario, then I at fixed z is the same for all segments. For a matched solution (at the bunch center, for instance), take a independent of z. This condition determines K_o at the reference energy, E_o:

$$K_o = \frac{qeIZ_o}{4\pi E_o \beta_o a_o^2} \quad (5)$$

Then let $a = a_o(1+\alpha)$ for other segments of the beam at different energies and expand equation (4) to first order in α and $\Delta\beta/\beta$:

$$\frac{d^2\alpha}{d\tau^2} + \omega^2 \alpha = - \epsilon K_o \beta_o^2 \frac{\Delta\beta}{\beta} \quad (6)$$

where $\omega^2 = 2K_o\beta_o^2$ is the free envelope oscillation frequency in the variable $\tau = \int dz/\beta_o$ = (c times time of progress of the beam center), and $\epsilon = 2$ for magnetic focusing and 1 for electric focusing. To keep things simple, consider a scenario in which I is proportional to β_o so that ω is a constant. The solution of Eq. (3) for $\alpha = \dot{\alpha} = 0$ at $\tau = 0$ is then:

$$\alpha = - \frac{\epsilon}{2} \omega \int_0^\tau d\tau' \frac{\Delta\beta}{\beta} \sin \omega(\tau-\tau') \quad (7)$$

As an example, take

$$\frac{\Delta\beta}{\beta} = \Delta_{max} \left[e^{-\lambda_1 t} - e^{-\lambda_2 t} \right] \quad (8)$$

with $\lambda_2 \gg \lambda_1$. This represents a rapid rise at the rate λ_2 to a maximum of Δ_{max} followed by a slow decay at the rate of λ_1. For $\lambda_2 \tau \gg 1$ and $\lambda_1 \ll \omega$, the solution is:

$$\alpha = \frac{\varepsilon}{2} \Delta_{max} \left[\frac{\lambda_2}{\sqrt{\lambda_2^2 + \omega^2}} \sin\left(\omega\tau + \tan^{-1}\frac{\lambda_2}{\omega}\right) - e^{-\lambda_1 \tau} \right] \quad (9)$$

Other examples show similar behavior. Notice that the mismatch does not accumulate as in the case of misalignment errors. Rather, a ringing is set up, the magnitude of which is determined by the maximum value of $\Delta\beta/\beta$ and the rate of increase of $\Delta\beta/\beta$, λ_2, compared to the natural frequency, ω.

If I increases as a higher power of β, the natural oscillations are damped as $(1/\beta)^{1/4}$, but later contributions in (7) are increased. The net effect is a slight decrease in (9).

SUMMARY

We have shown that the acceleration and current amplification rates in a heavy ion induction linac are constrained by the restrictions imposed on the value of velocity tilt at a given location; this restriction is particularly severe when the kinetic energy is low and the bunch is long.

We also have shown that a large velocity tilt can be tolerated if certain particle distributions are used: a constant current distribution for electric focusing and constant-density distribution for magnetic focusing. We have demonstrated experimentally in MBE-4 the electric focusing case.

Mismatch oscillations do not accumulate as the misalignment errors do; rather, a ringing is set up, the magnitude of which is determined by the magnitude and the rate of increase of the velocity tilt.

If any quadrupole misalignments are present, the velocity tilt will cause the frequency of coherent betatron oscillations to be different for different segments of the beam. These effects require further investigation.

REFERENCES

1. J. Hovingh, V.O. Brady, A. Faltens, D. Keefe and E.P. Lee, in these Proceedings
2. A.I. Warwick, T.J. Fessenden, D.L. Judd, D. Keefe, C. Kim, L.J. Laslett, and L. Smith, in these Proceedings
3. C.H. Kim and L. Smith, Particle Accelerators, Vol 18, 101 (1985).
4. L.J. Laslett, Lawrence Berkeley Laboratory Heavy Ion Fusion Group Report, HI-FAN-229 (1983).

BEAM TEST OF THE RFQ LINAC 'TALL'

N. Ueda, A. Mizobuchi, S. Yamada, S. Arai, M. Olivier*
T. Nakanishi, T. Fukushima, S. Tatsumi**, and Y. Hirao

Institute for Nuclear Study, University of Tokyo
Tanashi, Tokyo 188, Japan

ABSTRACT

The RFQ linac 'TALL' is of four vane structure driven with a loop coupler. It is 58 cm in diameter and 730 cm in length. It is designed to accelerate heavy ions with charge to mass ratio of 1 ∼ 1/7. A field uniformity was obtained within an error of ± 2% azimuthally and ± 5% longitudinally by use of side inductive and end capacitive tuners. The TE210 mode was tuned at 101.3 MHz, 0.9 MHz lower than that of the closest mode TE111. Beam test was done by using proton beam. Transmission exceeding 90% was obtained. Energy and its spread of the ouput beam were measured by an analyzer magnet as T = 825 keV and $\Delta T/T$ = 1.6% in FWHM. They agree with calculated values.

INTRODUCTION

The RFQ linac 'TALL' was constructed as the first stage of an injector linac system for a heavy ion synchrotron 'TARN II' which is under construction at INS.[1] The RFQ linac is best for the stage because it can accept low velocity beam and has bunching function. The machine is designed on the basis of the experience of the test RFQ linac LITL.[2] The design parameters of the TALL are given in Table 1. The machine can accept ions with charge to mass ratio of 1 ∼ 1/7. By using an ECR ion source it can accelereate heavy ions up to Xe and W.[3] Injected ions at 8 keV/u are accelerated up to 800 keV/u. For 100 MHz, the $\beta\lambda$ at the output energy is 12.4 cm and long enough to be accepted by the following drift tube linac.

ACCELERATION CAVITY

The acceleration cavity is of four vane structure driven with a loop coupler. The cavity is 58 cm in diameter and 730 cm in length. The cavity is longitudinally separated into four sections, each of which is 1.8 m long.

Each section is assembled and aligned independently. The vane is mounted in a cavity cylinder with three base plugs. The cylinder is made of mild steel, copper plated to a thickness of 100 μm. Each section has 16 holes of 100 mm in diameter for side tuners, pumping ports and rf power feed. It also has one monitor loop in each quadrant.

---

* On leave from Laboratoire National Saturne, CEN, Saclay, France
** Present address: Niihama Works, Sumitomo Heavy Industries, Ltd., Niihama, Ehime Prefecture, 792 Japan.

The cylinders are joined with rf contactors of silver coated metal O-rings. The vanes have no rf contact but narrow gaps of 0.2 mm to tolerate machining errors and unequal thermal elongation at the longitudinal joints. The vane separation effectively reduces L^2-dependence of the voltage variation, and allows vane positioning with realizable accuracy.

VANES

Two sets of vanes are prepared for the TALL. One is for low power operation. It is made of aluminum and has no cooling channel. The other is for high power operation with cooling channel. The field tuning and beam test were done with the aluminum vanes. The vanes and cylinders are electrically contacted with C-shaped contactors made of stainless steel, silver coated to a thickness of 50 μm.

The transverse geometry of the vane tip is approximated by a circular arc with a varying radius, similarly to the LITL. The modulation was machined with a ball end mill of 30 mm dia. in most of the vane length. Mills of 12 and 20 mm dia. were used on the first section where the cell length is short and the modulation factor increases steeply. The modulation machining was checked to be within a tolerance of ± 30 μm by an inspection machine.

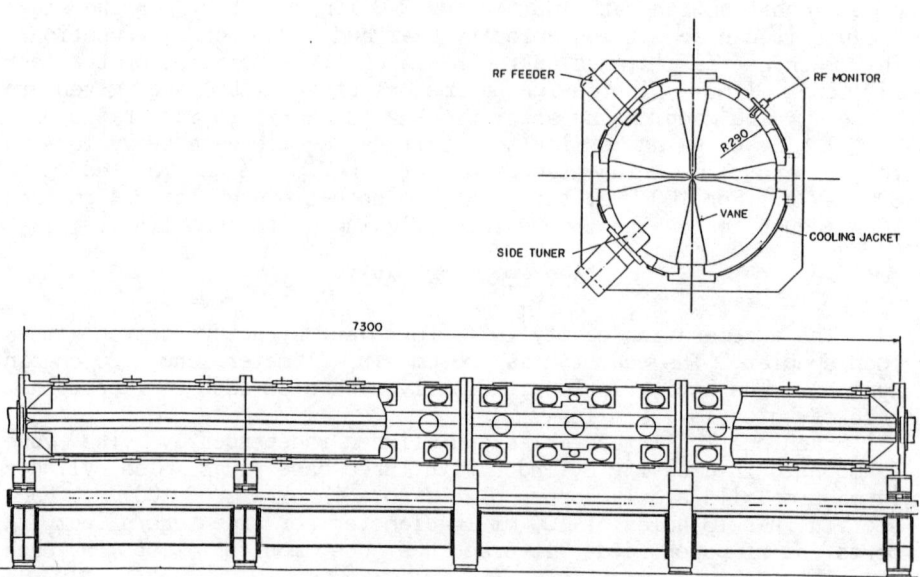

Fig.1. Schematic drawing of the TALL acceleration cavity.

ALIGNMENT

The cylinder has square flanges on both the ends. The vertical and horizontal rims of the flanges are the fiducial planes for the alignment. Both the vane ends have fiducial holes near the vane tip. They were used to measure the radial positions of the vane tips. The side flats of the vanes near the tip and base were used as the azimuthal fiducial planes. The accuracy of the vane positioning was checked with an inspection machine (Fig.2).

The vanes were assembled first with no rf contactor and no vacuum seal. After the vanes were aligned within an error of $\pm$ 50 μm , the positions of the vanes, base plugs and cylinders were fixed with locator pins. Then the cavity was disassembled and cleaned up. Guided with the locator pins, the vanes were assembled with rf contactors and vacuum seals. Again the vane position was measured with the inspection machine and re-aligned when necessary.

The four section was joined on a bed. The bed has five support flats. They were leveled within an error of $\pm$ 20 μm. The square flanges were jointed so that there remained no clearance between the horizontal fiducial planes and the flats, and so that the vertical fiducial planes were in a plane.

The beam axis was aligned within an error of 200 μm over the length of 7.3 m. The steps between the longitudinally adjacent vanes are within 100 μm at the joints. A computer simulation shows that alignment errors of the beam axis of 100 μm at three joints do not decrease the transmission significantly.

Fig.2. Inspection of the vane alignment.

FIELD TUNING

On the input and output ends, a part of each vane end is removable in order to vary the inductance of the cavity ends. The end wall has four movable capacitive end tuners. They are aluminum rods of 25 mm in diameter. Each quadrant has 4, one for each section, movable side tuners. They are water cooled copper cylinder 100 mm in diameter driven with stepping motors in a stroke of 50 mm. They will compensate the resonant frequency shift due to thermal elongation. Aluminum cylindrical blocks of 100 mm in diameter and various thicknesses are inserted through the side holes to obtain uniform field.

Resonant frequencies of various modes were measured with the vane ends shorted to the both end plates. From the dispersion relations the cutoff frequencies were determined at 97.6 and 101.1 MHz for the dipole and quadrupole modes, respectively. The calculated ones by SUPERFISH for the cross sectional geometry at the quadrupole symmetrical plane are 98.3 and 100.6 MHz, respectively.

The field distribution was tuned roughly by varying the shape of the vane ends. Then fine tuning was done by using side inductive and end capacitive tuners. The electric field distribution near the vane tops was measured by use of a dielectric pertubator moving guided by the vanes. A field uniformity within a deviation of ± 2% azimuthally and ± 5% longitudinally was obtained, by using two dozen side tuners of fixed length and three end capacitive tuners. The distribution does not depend on the position of the coupling loop. The separation of 0.93 MHz between the TE210 and the closest TE111 mode is satisfactory.

BEAM TEST

Ions extracted from a microwave ion source at 8 keV are transported to a magnet with two einzel lenses. Protons are separated form other ions with the magnet. They are focused into the RFQ entrance with a triplet of electric quadrupole lenses and an einzel lens.

The accelerated ions are focused with a triplet of quadrupole magnets on an object point of an analyzer magnet.

The energy of the output beam was measured by the magnet as T = 825 keV/u. Considering that the frequency is tuned at 101.3 MHz, the measured energy is reasonable. The energy spread was measured as $\Delta T/T = 1.6\%$ in FWHM. It agrees with a computer simulation by PARMTEQ.

Transmission efficiency was measured for input proton beam of 10 µA. The emittance and intensity of the input beam were measured just in front of the entrance or, downstream of the final einzel lens. The intensity of the output beam was measured at the object point. Transmission exceeding 90% was obtained (Fig.4).

The beam test was done in pulse operation. The duty factor was 16.7% (2 ms duration/12 ms repetition period). The proton acceleration required rf power of 4.6 kW (peak). The unloaded quality factor Q_0 was measured at 7100. It is about 70% of a

Table 1. Design parameters of the TALL.

Ions (q/A)	1 ～ 1/7
Operating frequency (MHz)	100
Input energy (keV/u)	8
Output energy (keV/u)	800
Total number of cells	300
Cell number of radial matching section	40
Vane length (cm)	725
Cavity diameter (cm)	58
Characteristic bore radius, r_o (cm)	0.54
Minimum bore radius, a_{min} (cm)	0.29
Margin of bore radius, a_{min}/a_{beam}	1.15
Maximum modulation, m_{max}	2.5
Focusing strength, B_o	3.8
Maximum defocusing strength, Δ_b	− 0.075
Synchronous phase, ϕ_s (deg)	− 30
Intervane voltage for q/A = 1/7 (kV)	81
Maximum field (kV/cm)	205 (1.8 Kilpat.)
Rf power wall loss for q/A = 1/7 (kW)	180
Transmission for input beam 0 mA	0.94
with a normalized emittance 2 mA	0.91
of 0.6 π mm · mrad for q/A=1/7. 10 mA	0.63

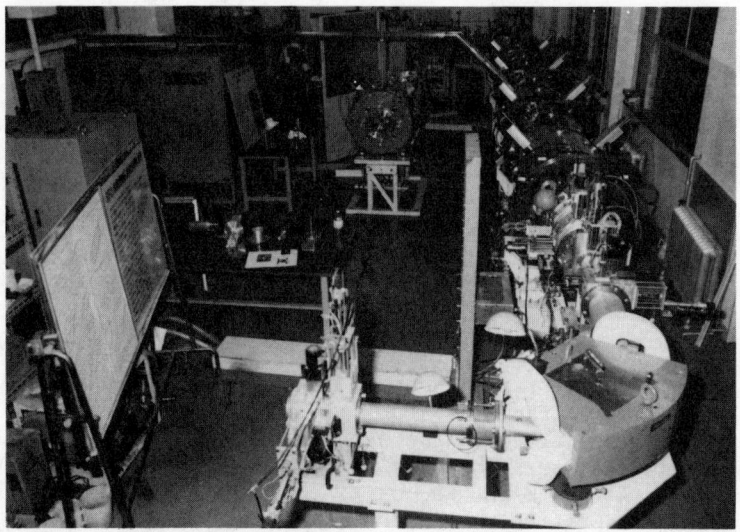

Fig.3. View of the TALL beam test stand.

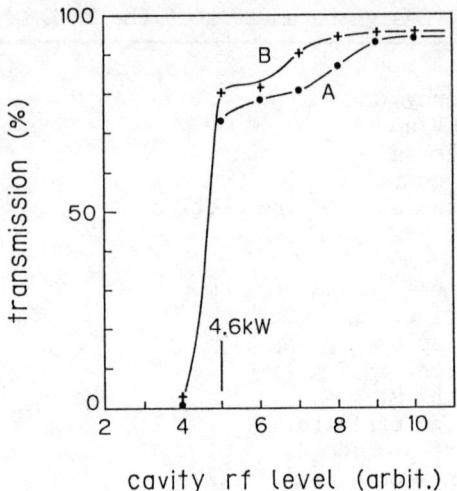

Fig.4. Transmission vs. cavity rf level for input H⁺ beam of 10 μA. Input beam emittance for A and B is given in Fig.5.

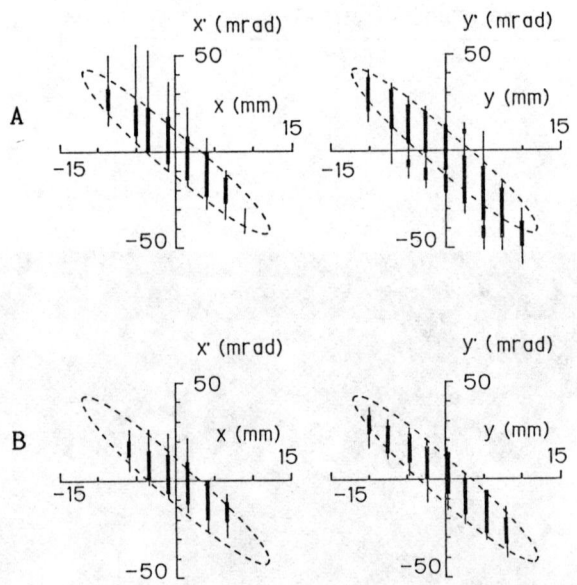

Fig.5. Emittance of the input H⁺ beams of 10 μA measured at 21 cm upstream the RFQ entrance. Thick and thin bars cover 95% and 100% of the beam, respectively. The ellipses are ones for RFQ acceptance at the design voltage. The area is 145π mm·mrad, or 0.6π mm·mrad normalized. A and B are for beams limited by apertures of 13 and 8 mm dia., respectively, inserted between electric quadrupole lenses.

caluculated value for the aluminum vanes. With the loop coupler the cavity was stably operated upto the full power of 25 kW of a power supply now available. Multipactoring was observed in three ranges below the proton acceleration rf level. On the first power test they were easily surmounted after a few hour outgassing.

The cavity is pumped with two turbo-molecular pumps of 500 l/s. The vacuum pressure was $1 \cdot 10^{-6}$ Torr with no rf power. It increased to a range of 10^{-5} Torr on the outgassing process.

CONCLUSION

We obtained a field uniformity within an error of ± 5% both azimuthally and longitudinally with a loop coupler. No vane coupling ring was used. The TE210 mode was tuned at 101.3 MHz. The closest mode TE111 has a suffiecient separation of 0.9 MHz. Transmission exceeding 90% was obtained. The measured energy and its spread agree with the calculation.

On the basis of the experiment the shape of the end part of the high power vanes was determined. The aluminum vanes are now replaced by them. Field tuning and beam test with the new vanes will begin in June.

ACKNOWLEDGMENT

The authors are grateful to the members of the Accelerator Research Division, INS, for their discussions and assistance. They thank the assistance of the staff of the computer room, INS, in beam dynamics design, cavity design and preparing the manuscript with FACOM M380R. The acceleration cavity of the TALL was manufactured by Sumitomo Heavy Industries, Ltd.

REFERENCES

1. T. Katayama et.al., Contribution to the Symposium.
2. N. Ueda et al., IEEE Trans. Nuclear Sci., NS-30, No.4, August, 1983.
3. R. Geller & B. Jacquot, Proc. 1984 Linear Acc. Conf., Seeheim, FRG, May 7-11, 1984, GSI-84-11.

SIMULATION OF TRANSVERSE COMBINING OF
SPACE-CHARGE DOMINATED BEAMS*

C. M. Celata
Lawrence Berkeley Laboratory
University of California, Berkeley, CA 94720

ABSTRACT

Rms emittance growth in the transverse plane due to the transverse combining of four identical elliptical beams of uniform density has been investigated. The emittance growth can be related by conservation of energy to the change in the electrostatic field energy. Its dependence on initial beam positions and radii has been calculated analytically for round beams and by computer simulation for elliptical beams.

INTRODUCTION

Limitations on the charge per unit length which can be focused suggest that the ion beam for heavy ion fusion be transported as separately focused multiple beams at the low energy end of the accelerator. When the energy of the beams is large enough, it may be more cost-effective to combine several of them. In this paper the transverse combining of four such beams is considered, with the goal of selecting initial conditions which will minimize transverse rms emittance growth. Only transverse dynamics are considered -- longitudinal fields are ignored -- and the beam is assumed to be coasting, with no spread in longitudinal energy.

A system for combining beams might consist of a matching section, where the shape of the beams is tailored to minimize emittance growth during the combining process; a bending section, where the beams' centroid trajectories are deflected to decrease the distance between beams; and a merging section, where the beams "see" each other and form a single beam. In this paper we consider only the merging section. The merging is assumed to occur in a focusing lattice which may be alternating-gradient or constant focusing.

We will measure the growth of the one-dimensional rms emittance using the ratio of the final emittance to the total initial phase space area in that dimension, i.e., $\varepsilon_{fx}/(2\varepsilon_{ix})$, where ε_{ix} and ε_{fx} are the initial single beam emittance and the final emittance in the x direction. This quantity can be written as

$$M_x \equiv \frac{\varepsilon_{fx}}{2\varepsilon_{ix}} = \left(\frac{\varepsilon_{ox}}{2\varepsilon_{ix}}\right)\left(\frac{\varepsilon_{fx}}{\varepsilon_{ox}}\right), \qquad (1)$$

where ε_o is the emittance of the composite beam when the initial beams first see each other. The first factor in Eq. (1) is due to the fact that empty spaces between the beams fill with particles, adding to the phase

---

*This work was supported by the Office of Energy Research, Office of Basic Energy Sciences, Department of Energy under Contract No. DE-AC04-76SF00098.

space occupied by the beam. This "geometric" emittance dilution would be the only increase in emittance for low intensity beams. The second factor is the phase space dilution caused by space charge (conversion of electrostatic field energy to transverse kinetic energy) and, if the rms radii of the beam change, by work done by the focusing fields. Simulation has shown that because space charge couples the transverse dimensions, the final state of the beam tends to have equal emittance in the two planes. This has also been observed experimentally in the SBTE at Lawrence Berkeley Laboratory for a single space-charge-dominated beam.[1] Then if the rms beam radii are not changed during the merging, at locations where $<x,x'> = 0$ conservation of energy gives the result:

$$\frac{\varepsilon_{fx}}{\varepsilon_{ox}} = \sqrt{\frac{<x^2>(\varepsilon_{oy}/\varepsilon_{ox})^2 + <y^2> - <y^2>\Delta U/T_{ox}}{<x^2> + <y^2>}} \qquad (2)$$

and similarly for y, where U is the total electrostatic field energy per unit length (beam and surrounding vacuum), $\Delta U = U_f - U_i$, and T_{ox} is the total x kinetic energy per unit length for the four beams. This is a general result for the change in emittance when the spatial configuration of a beam is changed, and reduces to the result of Wangler[2] for the case of azimuthal symmetry. If the rms beam radii change during the process, the work done by the focusing fields can be added to Eq. (2). For constant focusing, for instance, with focusing force $F_x = -k_x x$, ΔU in Eq. (2) would then become $\Delta U + Nk_x \Delta a^2/2$, where "a" is the beam radius and N is the number of particles per unit length, $<x^2>$ and $<y^2>$ in the equation would signify the initial values of these moments, and the right hand side of the equation would be multiplied by $<x^2>_f/<x^2>_0$.

This paper will consider the combining of four identical beams which are upright ellipses in configuration space, and have uniform density and gaussian velocity distributions before the merge. The initial rms x and y momenta for each beam are spatially uniform, and the x and y emittances are equal. Initially (when the beams first "see" each other) the beam centroid positions are $(x,y) = (\delta_x, \delta_y)$, $(-\delta_x, \delta_y)$, $(\delta_x, -\delta_y)$, and $(-\delta_x, -\delta_y)$, with the origin at the vacuum chamber axis and the quadrupoles on the x and y axes in the case of AG focusing (see Fig. 1). The boundary condition is a perfectly conducting pipe of radius R. We consider first the simple case of four round beams of radius "a" with $\delta_x = \delta_y = \delta/\sqrt{2}$.

COMBINING ROUND BEAMS

Wangler[2] has indicated that the equilibrium state for a space-charge-dominated beam is likely to have a uniform density. Anderson[3] has shown this profile to be a minimum of the electrostatic field energy for a one-dimensional (sheet) beam. This neglects the sheath at the beam edge, but the sheath width is of the order of the Debye length, λ_D, and $\lambda_D \ll a$ for space-charge-dominated beams. We will assume then that the beam evolves after the merge to a state with uniform density. Further, since the initial configuration and assumed final beam state both follow the same rms envelope equation, we set the initial rms radius equal

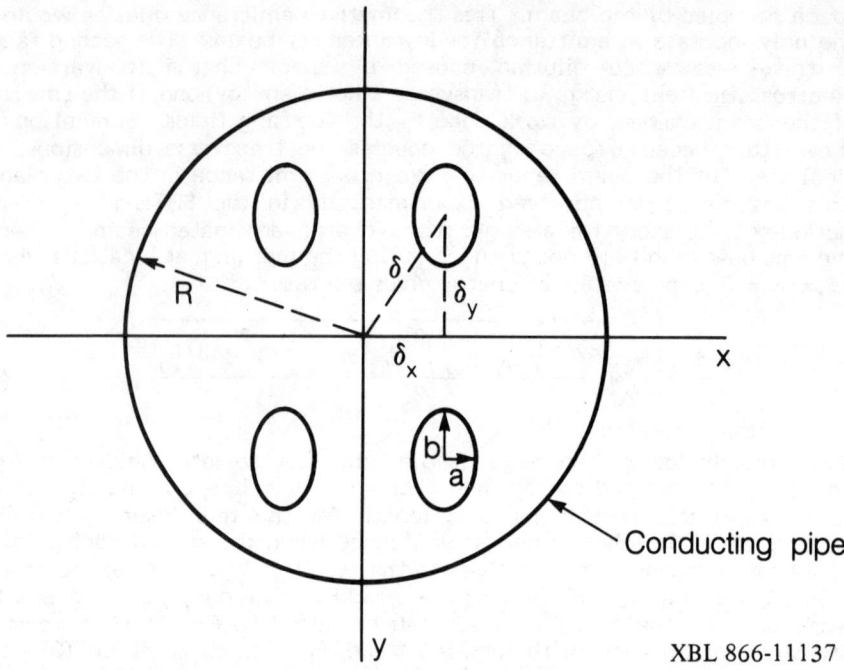

Fig. 1. Transverse plane geometry showing four beams when they first "see" each other.

to its matched value, and assume the final beam will then be matched, with the same rms radius. Use of a particle-in-cell simulation code shows both assumptions to be true only for small δ/a, $\delta/a < 2$. For larger δ/a, the rms radius of the beam oscillates and particle loss occurs. With these assumptions,

$$\Delta U = \frac{N^2 q^2}{16} \left[3 - 4 \ln \frac{1-g^8}{4} \left(\frac{a}{\delta}\right)^3 \left(1 + 2 \frac{\delta^2}{a^2}\right)^2 \right], \qquad (3)$$

where $g = \delta/R$. Note that U_i, U_f, and therefore ΔU, will always be proportional to the total charge per unit length squared, regardless of the beam shape, orientation, or density profile. This is the only dependence of M on Nq. Since g^8 is generally much less than unity, ΔU essentially depends only on δ/a and Nq. M_x can now be calculated from Eqs. (1), (2), and (3), since $\varepsilon_{ox}/(2\varepsilon_{ix}) = 0.5 \sqrt{1+2\delta^2/a^2}$.

We can use these results to determine desirable qualities for beam merging geometries. We will keep constant the emittance of the original beams, so that $T_{ox} \propto a^{-2}$, and also the charge per unit length, in order to study scaling with geometrical factors. We might wish to hold δ/a, $\delta-a$, or δ constant for practical experimental reasons. If δ/a is kept constant, M_x increases monotonically with "a". This reflects the fact that ΔU is unchanging, but the additional transverse kinetic energy it supplies makes a larger relative change in the total when the beam is colder. The results when the clearance between beams is held constant, to allow, for instance, for a septum in the bending section, are shown in Fig. 2

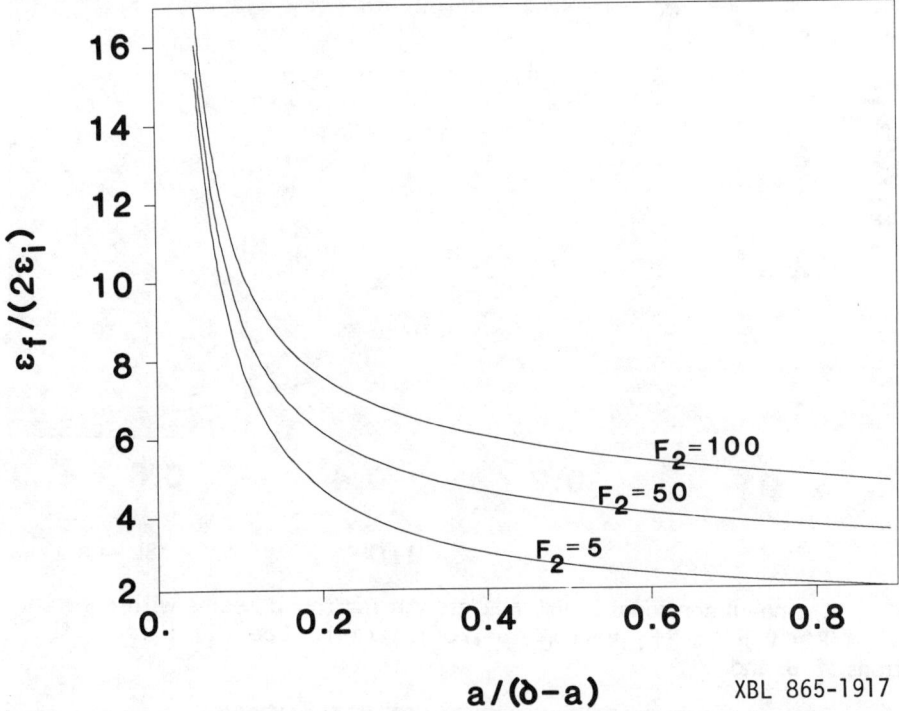

Fig. 2. One-dimensional emittance growth for round beams with constant clearance, δ-a, and varying "a". (δ-a)/R = 0.284. $F_2 = [(\sigma_0^2-\sigma^2)/\sigma^2](\delta-a)^2/a^2$, where σ and σ_0 are the betatron phase advance per lattice period with and without space charge for a uniform beam with the same charge per unit length and x and y kinetic energy as the initial configuration.

for a typical set of parameters. Note that the monotonic decrease of M_x in Fig. 2 is not representative of its behavior at very small values of δ-a, (δ-a < 0.46a), as can be seen from Eqs. (1)-(3). M_x decreases with increasing "a" if δ is held constant and "a" is increased (Fig. 3). In both of these cases, as "a" increases, ΔU and T_{ox} both decrease. Of course for constant beam radius, M_x monotonically increases with δ/a and Nq. Finally, we might wish to fix the radius of the final beam. Then if δ increases, "a" must decrease. In this case M_x increases monotonically with decreasing "a", as the space charge dilution factor decreases, while the area dilution increases.

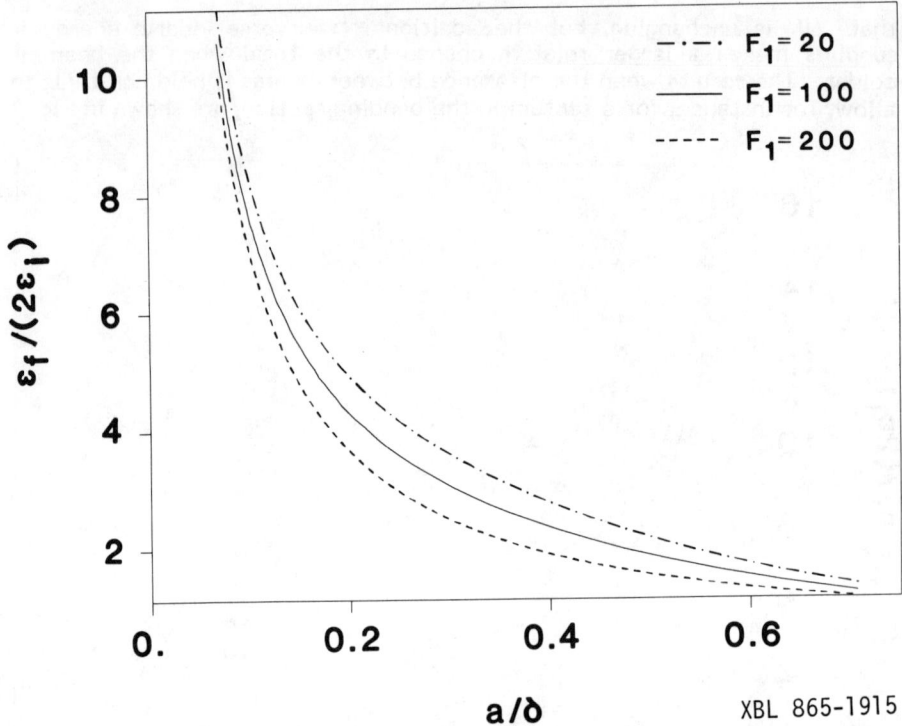

Fig. 3. One-dimensional emittance growth for round beams with constant δ. $\delta/R = 0.506$. $F_1 = [(\sigma_0^2 - \sigma^2)/\sigma^2](\delta^2/a^2)$. See Fig. 1 for definitions of σ and σ_0.

COMBINING FOUR ELLIPTICAL BEAMS

The two-dimensional particle-in-cell simulation code SHIFTXY has been used to compute the quantity $\varepsilon_f/\varepsilon_0$ for the case of four elliptical beams described in the Introduction. Alternating gradient focusing was produced using a thin lens approximation. For all runs a focusing system which provided a single particle phase advance of 60° per lattice period was used. Emittance of the initial single beams was correct to give a space charge depressed phase advance of 20° per lattice period in a 60° lattice upstream of the merge. Again the attempt was made to rms-match the beam. The rms radii of the initial configuration were set equal to those of a matched uniform beam with the same charge per unit length, since this was the assumed final state. Therefore the initial $<x^2>/<y^2>$ was 2.79 for all runs. The initial value of x_{rms}/R was 0.35.

For $\delta_x/a = \delta_y/b$, where a and b are the initial beams' major and minor radii, simulation results for M_x and M_y had the values which would be calculated using the value of ΔU given in Eq. (3) for round beams, with $\delta = \sqrt{2}\,\delta_x$. (Runs were done for $1 \leq \delta_x/a \leq 2.5$.) Therefore the quantity $\zeta = (\delta_x/a)/(\delta_y/b)$ would seem to be of interest, as a "shape factor" characterizing the departure of the behavior of the system from that of round beams ($\zeta = 1$). The geometry of the initial configuration is then completely specified by ζ, δ_x/a, and (x_{rms}/y_{rms}). A brief investigation of the dependence of M_x and M_y on the first two of these parameters was done, but much remains to be studied. The results are shown in Figs. 4-6. The initial rms radii of the configuration, single beam emittances, and focusing system were held constant for all runs. Therefore as δ_x/a increased, "a" decreased, and T_{ox} increased.

As shown in Fig. 4, matching was again a problem for configurations where the beams were not almost touching. For these cases the final beam was also not uniform after ~ 100 lattice periods, even though the emittance growth seemed to have ceased. $<x^4>/<x^2>^2$ was larger than its value for a uniform beam, indicating probably some hollowing of the beam. Beam loss also occurred for $\delta_x/a \gtrsim 2$, and increased with δ_x/a. For the case of Figs. 4 and 5, 5% of the beam was lost in 150 lattice periods. For $\delta_x/a = 2.5$, $\zeta = 1$, 13% was lost in 100 periods. Figure 5 demonstrates that, as shown computationally by Wangler[2] and analytically by Anderson,[3] most of the emittance growth due to the change in the electrostatic field energy occurs within a fraction of a plasma period, though subsequently the x and y emittances equilibrate.

Though the range of parameters spanned was limited by computer time constraints (the small hot beams for large δ_x/a require extremely good spatial and temporal resolution) the results given in Fig. 6 indicate that, as expected, emittance growth increases with δ_x/a -- the initial

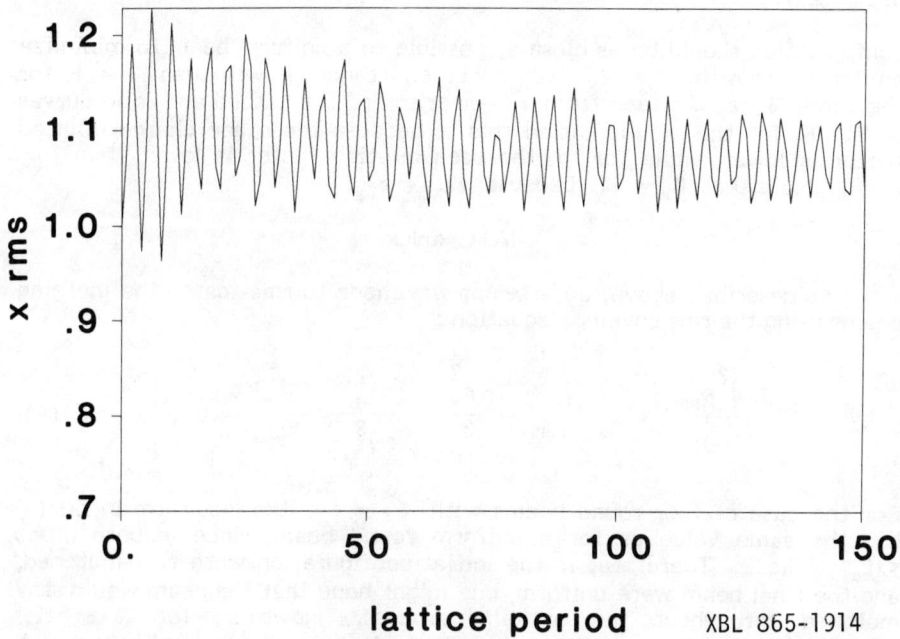

Fig. 4. x_{rms} vs. z for $\zeta = 1$, $\delta_x/a = 2$. Other parameters are given in the text.

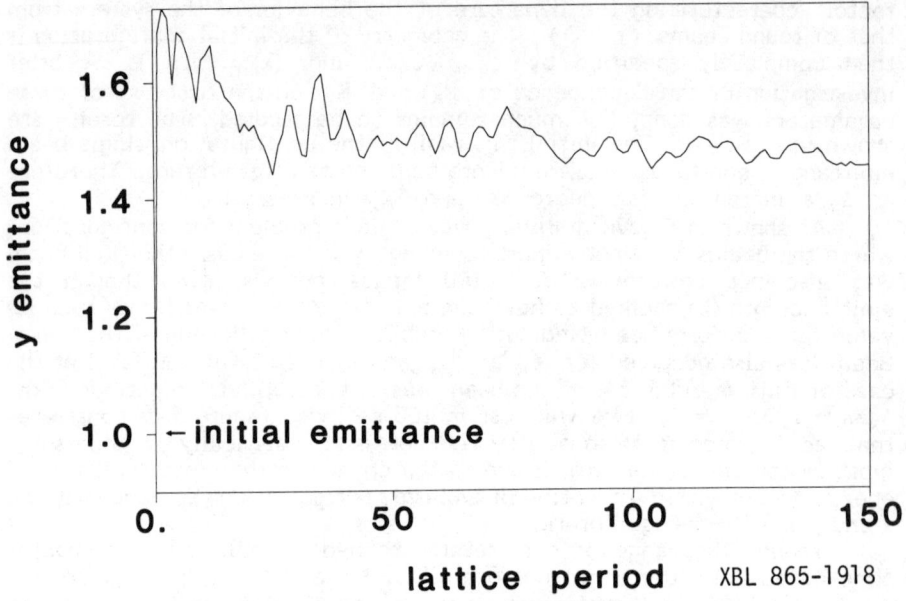

Fig. 5. $\epsilon_{fy}/\epsilon_{oy}$ vs. z for $\zeta = 1$, $\delta_x/a = 2$. Other parameters are given in the text.

configuration should be as close as possible to a uniform beam to minimize emittance growth. $\zeta = 2$ gives lower emittance growth than $\zeta = 1$ for the same δ_x/a. As seen from a comparison of the dashed and solid curves in Fig. 6, for the x dimension this is because the space charge dilution factor is lower, while for y the area dilution factor is lower than for $\zeta = 1$. Note that $M_x = M_y$ because $\epsilon_{fx} = \epsilon_{fy}$.

MATCHING

As described above, an attempt was made to rms-match the merging beams using the rms envelope equation :

$$\frac{d^2 x_{rms}}{dz^2} + k_x x_{rms} - \frac{\epsilon_x^2}{x_{rms}^3} - \frac{q}{m\gamma^3 v_z^2} \frac{\langle xE_x \rangle}{x_{rms}} = 0 . \qquad (4)$$

For the case of four round beams with $\delta_x = \delta_y$, the last term in Eq. (4) has the same value as for a uniform round beam, since in both cases $\langle xE_x \rangle = Nq/2$. Therefore, if the initial configuration were rms-matched, and the final beam were uniform, one might hope that the beam would stay matched throughout the merging process. However, for $\delta/a \gtrsim 2$, oscillations of the rms radii were seen at all z, with an amplitude which decreased with increasing z. This demonstrates a limitation on the use of the rms envelope equation for matching during time development of the

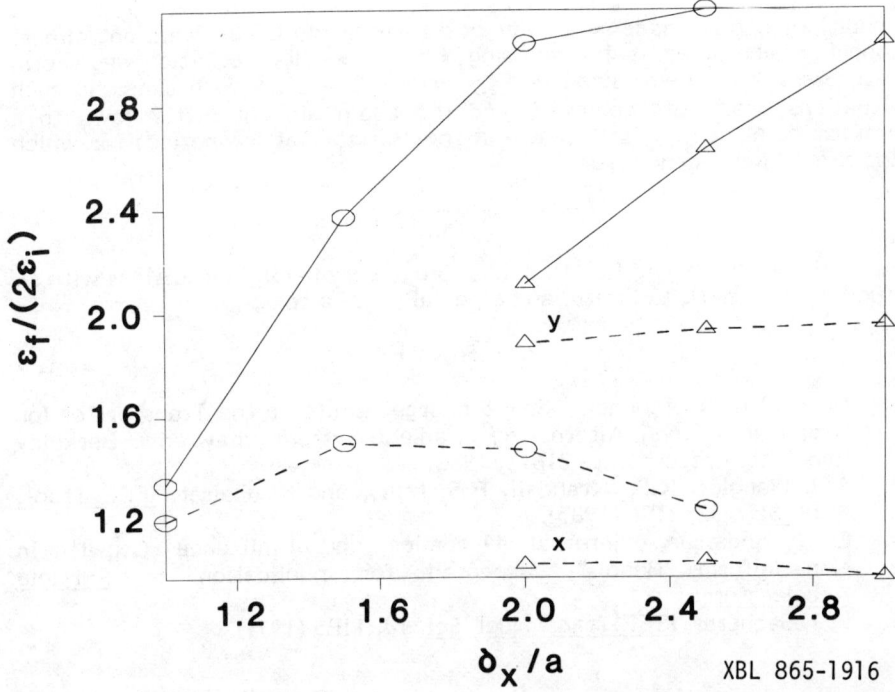

Fig. 6. One-dimensional rms emittance growth vs. δ_x/a for $\zeta = 1$ and $\zeta = 2$, computed by SHIFTXY. Circles and triangles are simulation results for $\zeta = 1$ and $\zeta = 2$ respectively. Solid lines are total emittance growth, $\epsilon_{fx}/(2\epsilon_{ix})$, while dashed lines are the space charge dilution factor only.

distribution function. Note that as the merging progressed, the emittance term in the envelope equation remained negligible compared to the space charge term, so that the changing emittance was not important in matching the beam.

For elliptical beams in an AG focusing system a similar result was found. The matched rms radii appeared to be a function of z, changing as the distribution function evolved.

CONCLUSIONS

Transverse rms emittance growth due to the transverse combining of identical uniform elliptical beams with gaussian velocity distribution and uniform x and y rms momenta has been investigated. For round beams the "space charge dilution factor" depends on δ/a and the initial beam kinetic energy, with negligible dependence on the vacuum chamber radius. For elliptical beams, simulations show that round beam results can be used to calculate emittance growth for $\delta_x/a = \delta_y/b$, i.e., $\zeta = 1$. Some results for other cases ($\zeta = 2$) have been given. The one-dimensional relative emittance growth, $\epsilon_{fx}/(2\epsilon_{ix})$, for experimentally reasonable conditions can be made ≈ 2. For cases where the beams were not almost touching each other, matching using the rms envelope equation was shown to present a problem, since $<xE_x>$ and $<yE_y>$ vary with time. In such situations beam loss occurred, and the beam did not settle down to a uniform final density profile within the 100-150 lattice periods for which the simulation code was run.

ACKNOWLEDGEMENTS

The author would like to acknowledge very helpful discussions with D. Judd, L.J. Laslett, L. Smith, and especially A. Faltens.

REFERENCES

1. Michael G. Tiefenback, "Space-charge Limits on the Transport of Ion Beams in a Long Alternating Gradient System", Lawrence Berkeley Laboratory report LBL-21611, 1986.
2. T.P. Wangler, K.R. Crandall, R.S. Mills, and M. Reiser, IEEE Trans. Nucl. Sci., 32, 2196 (1985).
3. O.A. Anderson, "Internal Dynamics and Emittance Growth in Non-Uniform Beams", accepted for publication in Particle Accelerators.
4. F.J. Sacherer, IEEE Trans. Nucl. Sci. 18, 1105 (1971).

THEORY AND SIMULATION OF EMITTANCE, SPACE CHARGE AND ELECTRON PRESSURE EFFECTS ON FOCUSING OF NEUTRALIZED ION BEAMS

Don S. Lemons and Michael E. Jones
Los Alamos National Laboratory
Los Alamos, New Mexico 87545

ABSTRACT

We investigate the final focus mode characterized by warm comoving electrons and vacuum propagation. In particular, we extend a previous envelope equation analysis of ion focusing in this mode to include the effects of ion emittance as well as ion space charge and initial electron temperature. Our major result is a simple equation relating initial R_o and final R_f beam radii to ion emittance ε and perveance K and electron Debye length λ_D which is supported by one dimensional, electrostatic, particle-in-cell simulations of radial ion focusing. Finally, we use this equation to find the allowed temperature of neutralizing electrons for typical Heavy Ion Fusion reactor and High Temperature Experiment scenarios.

INTRODUCTION

Accelerator requirements for Heavy Ion Fusion reactor can be significantly relaxed, consequently reducing costs, by using beams of multiply rather than singly charged ions as target drivers.[1] However, it may then be necessary to charge-neutralize such beams in order to focus them in the target chamber. The High Temperature Experiment (HTE), proposed to test propagation, focusing, and target physics[2], will also require beam neutralization. Several methods for achieving effective charge neutralization have been discussed including: ballistic propagation in a plasma background, channel guided and magnetically pinched beam propagation.[3,4] Here we continue investigation of a conceptually more simple neutralized beam propagation mode which, unlike the others, avoids beam-plasma instabilities: ballistic propagation of ions and comoving electrons through vacuum.[5,6] Our major result is a simple equation relating the focused beam envelope radius R_f, the initial beam radius R_o, the ion emittance ε and perveance K, and the electron Debye length λ_D. This equation should be useful in future system and design studies.

A single parameter λ_D characterizes the neutralizing electrons. When λ_D is much larger than the beam radius, the neutralization is ineffective; when it is smaller than the beam radius, λ_D^2 acts much like an electron emittance. Since λ_D^2 is proportional to the electron temperature T_e, the equation can, for instance, be used to determine the initial T_e necessary for effective neutralization.

ENVELOPE EQUATION

The envelope equation for an axially symmetric, paraxial, charged particle beam has been derived in many publications.[7] Here we give the equation for an ion beam

$$R\ddot{R} = \varepsilon^2/(4R^2) + (Ze/M\gamma_o^3)\langle Er \rangle_i \qquad (1)$$

where R is the rms radius of ions within a disc of unit length moving with the beam velocity V_z, ε is the ion rms emittance, M and Z are the ion mass and charge state, respectively, E is the self radial electric field, and a dot " · " indicates differention with respect to time. The dependence on the relativistic factor $\gamma_o = (1-V_z^2/C^2)^{-1/2}$ incorporates both relativistic inertia and self azimuthal magnetic field effects. R and ε are defined by $R^2 = \langle r^2 \rangle_i$ and $\varepsilon^2 = 4(\langle r^2 \rangle_i \langle v^2 \rangle_i - \langle rv \rangle_i^2)$ where the bracket notation indicates a quantity averaged over its value for all the ions in the disc. Thus

$$\langle r^2 \rangle_i = \Sigma r_k^2 / N_{Li} \qquad (2)$$

where the sum is over k=1 to N_{Li} and $N_{Li} = \Sigma$ is the number of ions in the disc, or equivalently the ion line density. We assume that ion emittance is conserved.

When the ion beam radial profile is uniform and no electrons are present Gauss' law may be integrated directly and the quantity $\langle Er \rangle_i$ shown to be independent of R, thus closing the system.[8] When thermal electrons are present the closure is not so simple but is still possible. First, we assume there are enough electrons in the moving disc to achieve global charge neutrality. Thus $N_{Le} = Z N_{Li}$. But the local ion and electron densities are not necessarily equal because the species have different "temperatures". Global charge neutrality does, however, lead to the equation

$$\langle Er \rangle_i = \langle Er \rangle_e \qquad (3)$$

which comes from multiplying Gauss' law by (Er) and integrating from $r=0$ to $r=\infty$. Here $\langle Er \rangle_e$ is most conveniently defined in terms of a continuous electron density n_e so that $\langle Er \rangle_e = 2\pi \int dr r n_e (Er)/N_{Le}$. We also assume the electrons are in equilibrium, with pressure P_e balancing the electric field $-en_e E = \partial P_e / \partial r$. Multiplying this last equation by r^2 and integrating we obtain

$$e \langle Er \rangle_e = 2 \langle T_e \rangle_e \qquad (4)$$

where $T_e = P_e/n_e$.

In Refs. 5 and 6 we derived an equation of state relating $\langle T_e \rangle_e$ to the ion radius R from local electron isentropy $P_e n_e^{-\gamma} = $ constant. Here we motivate the equation of state by assuming: 1) global electron entropy conservation for a two dimensional gas $\langle T_e \rangle_e R_e^2 = $ constant and 2) the electron mean squared radius R_e^2 exceeds R_i^2 by a constant proportional to λ_D^2, called the electron Debye length, which in turn is proportional to the constant $\langle T_e \rangle_e R_e^2$ and 3) the envelope equation recovers the uniform bare beam result when $\lambda_D^2 \gg R^2$. Thus the equation has the form

$$\langle T_e \rangle_e = \langle T_e \rangle_{eo} R_{eo}^2 / (R^2 + 4\lambda_D^2) \qquad (5)$$

where $\lambda_D^2 = \langle T_e \rangle_{eo} R_{oe}^2 / (2e^2 N_{Le})$.

Equations (3-5) complete the envelope equation description. Using Eqs. (3-5) Eq. (1) becomes

$$R\ddot{R} = \varepsilon^2/(4R^2) + 2V_z^2 K \lambda_D^2 / (R^2 + 4\lambda_D^2) \quad (6)$$

where K, the ion beam perveance, is given by $K=2Z^2e^2N_{Li}/(MV_z^2\gamma^3)$. Equation (6) can be integrated once between t=0 and $t=t_f$ to obtain the result

$$\frac{R_o^2}{(4\lambda_D^2)} = \frac{[R_o^2/R_f^2 - \exp\{\alpha^2/K - (\varepsilon_{un}^2/(2KR_o^2))(R_o^2/R_f^2 - 1)\}]}{[\exp\{\alpha^2/K - (\varepsilon_{un}^2/(2KR_o^2))(R_o^2/R_f^2 - 1)\} - 1]} \quad (7)$$

where the boundary conditions used are $\dot{R}(t=0)=\alpha V_z/\sqrt{2}$, $\dot{R}(t=t_f)=0$, $R(t=0)=R_o$, $R(t=t_f)=R_f$. Also, ε_{un} is the rms equivalent of the so called unnormalized emittance. So that $\varepsilon_{un}=\varepsilon/V_z$. Also, the parameter α is the focal angle for a uniform profile beam.

Eq. (7) reduces to previous results in various limits. The zero ion emittance limit appears in Ref. 6. When the beam is unneutralized, $\lambda_D^2 \to \infty$, Garren's equation for an unneutralized beam is recovered[9] which in turn reduces to Lawson's equation for an unneutralized, cold beam when $\varepsilon_{un}=0$.[8] Finally, for short but not vanishing Debye lengths focusing is limited only by electron pressure and ion emittance.

The solid lines in the Figure show how, according to Eq. (7), the focusing ratio R_o^2/R_f^2 depends on the parameter $R_o^2/(4\lambda_D^2)$. Remember that the latter is proportional to $N_{Li}/\langle T_e \rangle_{eo}$. Both curves are for $\alpha^2/K=2$ while the upper one has $\varepsilon_{un}=0$ and the lower one $\varepsilon_{un}=10^{-2}\alpha^2 R_o^2$. Also shown are results from a series of one dimensional simulations of the focusing.

ONE DIMENSIONAL PARTICLE SIMULATIONS

Simulations of a focusing ion beam and neutralizing electrons were performed with a one-dimensional, electrostatic, particle-in-cell (PIC) code written especially for this purpose. The code simulates charged particle motion in either (r-v_r) or

$(r-v_r-v_\theta)$ phase space. The particles are pushed by an electric field computed from Gauss's law with the boundary condition $E(r=0)=0$ and the equations of motion are advanced with a leap frog time advance algorithm. Linear weighting is used to construct charge densities on the grid. These and other standard PIC techniques were used in the code.[10]

The ions and electrons were initialized with equal uniform densities from $r=0$ to $r=R_o\sqrt{2}$. They were both given Maxwellian velocity distributions characterized

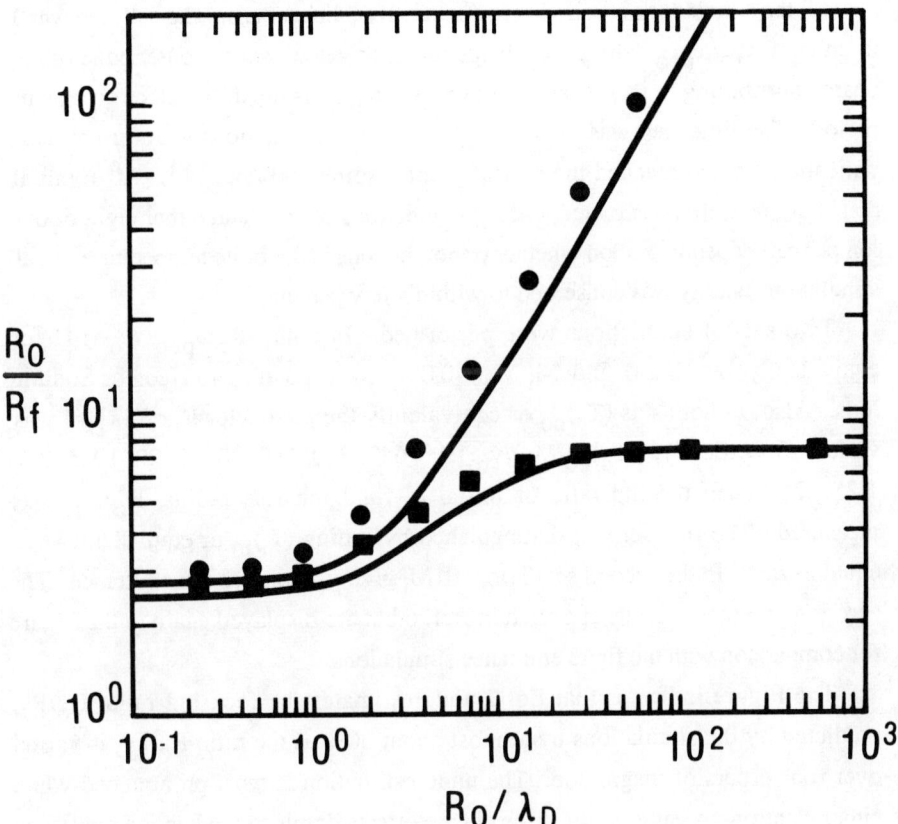

FIG. 1. Focusing ratio R_o/R_f versus R_o/λ_D from Eq. (7) (solid curves) and one dimensional simulations (dots and squares). Upper curve and dots are for zero ion emittance. Lower curve and squares are for finite ion emittance.

by uniform temperatures $\langle T_e \rangle_{eo}$ and $\langle T_i \rangle_{io}$. Thus $\varepsilon_{un}^2 = 4\langle T_i \rangle_{io} R_o^2/(M_i V_z^2)$. In addition, the ions were initialized with a linear velocity profile, $v_r(r) \propto r$, which if undisturbed would eventually focus the ions to a point on the symmetry axis at r=0. The electrons were not given a bulk velocity. Consequently, it took the first few plasma periods of each simulation for the electrons to relax to a quasistatic equilibrium. Global charge neutrality was maintained by reflecting all particles from both the inner symmetry axis at r=0 and an outer wall placed at several times the initial beam radius.

All the simulations reported here contain 1000 linear cells. The cell size was 1 in units of v_{ro}/ω_{peo} where v_{ro} is the ion bulk velocity at the outer edge of the beam contributing to its focusing motion and ω_{peo} is the initial electron plasma period. The time step was .005 ω_{peo}^{-1} and the simulations were run at least until the ions had reached their focal point.: some 200 ω_{peo}^{-1}. Also, initial distributions were constructed with eight electrons per cell and either eight or one ion per cell depending upon whether or not the ions had a finite temperature. In all simulations energy was conserved to within a few percent.

Two sets of simulations were performed. In both, $R_o \omega_{peo}/v_{ro}=141$ or equivalently $\alpha^2/K=2.0$ and $M_i/M_e=40,000$ corresponding to Neon or Sodium ions. Also, in both sets $\langle T_e \rangle_{eo}$ or equivalently the parameter $R_o^2/(4\lambda_D^2)$ was varied. Specifically, $(\langle T_e \rangle_{eo}/M_e)/v_{ro})^{1/2}=640, 320, 160, 80, 40, 20, 10, 5, 2.5, 1.25, .25$. Then the the ratio of initial to final ion rms radius, R_o/R_f, was measured. The first set was distinguished by setting $\langle T_i \rangle_{io}$ or equivalently ε_{un} equal to zero. In the second set $\langle T_i \rangle_{io}=.01 M_i$ giving a nonzero ion emittance. The zero ion emittance results appeared in Ref. 6 but are displayed again in the Figure for comparison with the finite emittance simulations.

The Figure indicates that Eq. 7 underestimates the focusing ratio, R_o/R_f, predicted by the simulations by at most about 30% as the ratio R_o/λ_D is varied over four orders of magnitude. The underestimation is most pronounced when either electron pressure or ion space charge effects dominate; when ion emittance dominates theory and simulation agree quite to within a few percent. This result gives us confidence in Eq. 7 as a conservative estimate of either ion focusing R_o/R_f or, equivalently, necessary initial electron temperature $\langle T_e \rangle_{eo}$.

APPLICATIONS

Since λ_D^2 is proportional to $\langle T_e \rangle_{eo}$ we can rewrite Eq. (7) as a formula for the initial temperature of the neutralizing electrons necessary for adequate focusing. In practical units the formula is given by

$$\langle T_e \rangle_{eo} = 14.98 (ZI_b/\beta) \frac{[\exp\{\alpha^2/K - (\varepsilon_{un}^2/(2KR_o^2))(R_o^2/R_f^2 - 1)\} - 1]}{[R_o^2/R_f^2 - \exp\{\alpha^2/K - (\varepsilon_{un}^2/(2KR_o^2))(R_o^2/R_f^2 - 1)\}]} \quad (8)$$

where $K = 6.387 \; 10^{-8} (Z^2/A)(I_b/(\beta\gamma)^3)$, $\langle T_e \rangle_{eo}$ is measured in ev, I_b in Amperes of particle current per beamlet(i.e. the charge state is assumed to be 1 in calculating I_b) and $A = M_i/M_p$ where M_p is the proton mass.

<u>High Temperature Experiment(HTE)</u>. In the proposed High Temperature Experiment[2] beams of moderate kinetic energy (150 Mev) and mass (Sodium) will be focused on a slab of high Z material in order to test theories of transport, focusing and target interaction. Current design parameters call for a number of 50 Amp beamlets each of which is to be focused from a radius of 2.5 cm down to 1 mm over a distance of 1.78 m. Unnormalized emittance is estimated to be $3.41 \; 10^{-6}$ meter-radians.[2,11] These numbers imply $\alpha = .0142$, $K = 8.43 \; 10^{-5}$, $\alpha^2/K = 2.42$, and according to Eq. (8) $\langle T_e \rangle_{eo} = 91$ ev.

<u>Reactor</u>. Recent reactor studies have suggested using Z=3, A=131, 8 Gev ions as drivers.[12] When such ions are divided into 28 beamlets of 1.68 kA current each and each one has an initial radius of 7 cm focused down to 2.58 mm over a distance of 5.52 m with an unnormalized emittance of $32 \; 10^{-6}$ meter-radians, a high degree of neutralization is required. Specifically, $\alpha = .0127$, $K = 1.50 \; 10^{-4}$, $\alpha^2/K = 1.072$, and from Eq. (8) $\langle T_e \rangle_{eo} = 18$ ev.

REFERENCES

1. J. Hovingh, V. O. Brady, A. Faltens, and E. P. Lee, Lawrence Berkeley Laboratory Report LBL-20979 (1986).

2. R. O. Bangerter, in Proceedings of the Symposium on Accelerator Aspects of Heavy Ion Fusion, (Darmstadt, West Germany, 1982) Gesellschaft fur Schwerionen- forschung Report GSI-82-8, pp. 7-18.

3. C. Olson, in Proceedings of the Heavy Ion Fusion Workshop edited by W. B. Herrmannsfeldt (Berkeley, California, 1980), Lawrence Berkeley Laboratory Report LBL-10301, pp. 403-425.

4. Y. K. Kim and G. Magelssen, Report on the Workshop on Atomic and Plasma Physics Requirements for Heavy Ion Fusion (Argonne, Illinois, 1979) Argonne National Laboratory Report ANL-80-17.

5. D. S. Lemons and L. E. Thode, Nucl. Fusion **21**, 529 (1981).

6. D. S. Lemons and M. E. Jones, Nucl. Sci. **NS-32**, 2474 (1985).

7. See for instance E. P. Lee and R. K. Cooper, Part. Accel. **7**, 83 (1976).

8. J. D. Lawson, J. Electron Cont. **5**, 146 (1958).

9. A. A. Garren, in ERDA Summer Study of Heavy Ions for Inertial Fusion (Berkeley, California, 1976) Lawrence Berkeley Laboratory Report LBL-5543, pp. 102-109.

10. J. Denavit and W. L. Kruer, Comments on Plasma Phys. **6**, 35(1980).

11. D. C. Wilson, Los Alamos National Laboratory, private communication.

12. D. Dudziak, and W.B. Herrmannsfeldt, this proceedings.

TRANSPORT OF INTENSE PROTON BEAMS IN AN INDUCTION LINAC BY SOLENOID LENSES[*]

W. Namkung, J. Y. Choe, and H. S. Uhm
Naval Surface Weapons Center
Silver Spring, Maryland 20903-5000

ABSTRACT

In the proposed proton induction linac at NSWC, a 100 A and 3 μs proton beam is accelerated to 5 MeV through a series of accelerating gaps. This beam can be effectively focused by solenoid lenses in this low energy regime and can be transported by adjusting the focusing strength in each period. For the transport channel design to reduce the number of independently controlled lenses, a theory of matched beams in the space-charge dominated regime has been developed. This study can be applied to cost efficent designs of induction accelerators for heavy ion fusion and free electron lasers.

INTRODUCTION

It has been well known that induction linacs are the most suitable accelerators for high current beams due to their low impedances. Even though there are limited experiences on ion beam accelerations, the induction linac technology is considered to be well established for electron beams[1]. There are renewed interests in both ion and electron linacs for their applications to, for example, heavy ion fusion, free electron lasers, neutron production, and many others. In the proposed heavy ion fusion and free electron laser accelerators[2,3], the number of induction modules is ranging from several hundreds to thousands. Accordingly, as many focusing lenses or transport channel periods are required. These beams can be transported, in principle, by either adjusting the periodicity or focusing strength of the individual lenses. In both cases, we encounter difficulties such as physically rearranging a long linac or employing independently controlled power supplies, i.e., high cost.

Recently, we proposed an intense proton induction linac at Naval Surface Weapons Center as an injector to another type of induction accelerator, a quadrupole betatron[4,5]. This linac is designed to deliver a proton beam of 5 MeV, 100 A, and 3 μs pulse length. Even in this relatively short linac of 36 modules, we want to reduce the number of independently controlled lenses to as few as possible. Since the beam cross section is very large at the low energy end, and beam energy changes in each period, the transport channel design is complicated under various practical limitations.

After briefly describing the NSWC proton induction linac, we present a transport theory of a space-charge dominated particle beam in a solenoid focusing channel. Parameters of the NSWC proton induction linac have been used for an exercise of this theory which can be useful for cost reductions of long induction linacs.

DESCRIPTION OF NSWC PROTON INDUCTION LINAC

The basic parameters of the proposed NSWC proton induction linac is summarized in Table I, and the conceptual design is biefly described in this section.

Table I. Parameters for the 100 A and 3 µs proton linac

	Source	Preaccelerator	Linac
Energy (keV)	50	550	5000
Beam radius (cm)	15	10	6
Number of gaps		1	36
Number of cores per gap		20	5
Total length (m)		2	18

In the past decade, high-quality and high-power duopigatron ion sources have been intensely developed for neutral beam heating of fusion plasmas. A suitable ion source for this high current linac is the one that has been developed at the Oak Ridge National Laboratory (ORNL) and used for neutral beam injectors (NBIs) to tokamaks, PLT and PDX at Princeton, and ISX at ORNL[6].

The ion beam current extracted from duopigatrons for the NBIs is generally so large that it is beyond the limiting current for propagation in free space or in a vacuum drift tube. The limiting current in a vacuum drift tube of radius b is expressed as

$$I_\ell = \frac{I_o(\gamma^{2/3}-1)^{3/2}}{1 + 2 \ln (b/a)}, \qquad (1)$$

where I_o is 31 MA for proton beams, a is the beam radius, and γ is the relativistic mass factor. As a numerical example, the limiting current of a proton beam with a = b, is 99 A for 300 keV and 213 A for 500 keV beams. One notes that this limiting current does not affect the beam extraction in the NBIs since there is a neutral gas cell immediately after the extraction grids. One way to overcome this difficulty due to the limiting current is an increase of the particle energy immediately after the extraction. In other words, one may consider it as a triode operation. This last stage of acceleration is called a preacceleration. Figure 1 shows schematically an ion source and preaccelerator, or ion triode. An axial magnetic field is applied to enhance convergence of the beam

radial stacking method[1] has been adopted. In addition, the radial
stacking of cores provides a voltage step-up between primary and
secondary windings. By this way the high voltage breakdowns in the
pulse forming network are relieved. In this core design, the inner
radius of modules is taken as 0.3 m for the transport system of a
100 A proton beam, and the number of cores for radial stacking is
five for effective use of core materials. Taking axial core
thickness as 0.3 m, the outer radius is approximately 0.82 m. When
the axial packing factor is taken as 0.6 for the gap space, the
average accelerating gradient of the linac yields 250 kV/m with 3 μs
pulse length. The total length of the 5 MeV linac is then 20 m
long and core material weighs approximately 160 tons.

BEAM TRANSPORT IN A PERIODIC FOCUSING CHANNEL

A periodic focusing system is required to transport an intense
proton beam through a series of induction modules. Since the beam
is large and circular at this low energy end, axisymmetric focusing
is more suitable than quadrupole lens focusing. A solenoid
magnetic field is periodiacally interrupted by axial space required
for induction gaps and for the system access. In addition, the
focusing effects of the interrupted solenoids with alternative
polarities are effective not only for the ion focusing but also for
creating a cusp magnetic field in the accelerating gaps. This cusp
field is an effective means of the magnetic insulation for
electrons in the gaps. The subject on how high current is
transportable in this kind of long periodic focusing channels, has
been investigated in connection with heavy ion beam fusion
accelerators. In a recent experiment at University of Maryland, a
space-charge dominated electron beam has been successfully
transported through 40 periods without loss of current[7].

Even though one can easily find parameters of the transport
channel[4] from the smooth approximation theory[8], we present an
alternative theory for space-charge dominated beams in which beam
emittance effects are negligible.

In the paraxial approximation, the envelope equation for an
axisymmetric beam is

$$\frac{d^2R}{dz^2} + \kappa R - \frac{K}{R} - \frac{\epsilon^2}{R^3} = 0 , \qquad (2)$$

where, $\kappa = (qB/2m\gamma\beta c)^2$ is the focusing factor, $K = 2I/(I_o \gamma^3 \beta^3)$ is
the generalized perveance, and ϵ is beam emittance. For space-
charge dominated beams, i.e., $(K/R) \gg (\epsilon^2/R^3)$, the last term in
Eq. (2) can be neglected. Let us consider an ideal field
distribution shown in Fig 2(a) for simplicity. In free space, Eq.
(2) becomes

$$RR'' - K = 0. \qquad (3)$$

to a further reduced size before injection into the main linac column. In addition to the beam focusing, the magnetic field also provides a magnetic insulation of the preacceleration gap to deflect backstreaming electrons. The gap voltage is provided by an inductive module in contrast to the electrostatic voltage applied to the extraction grids.

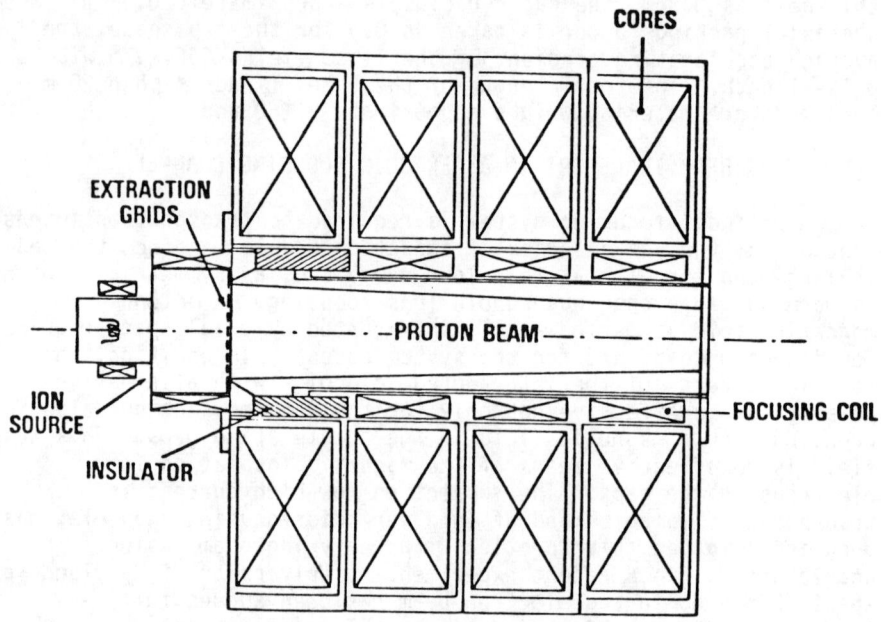

Fig. 1. Schematic of ion source and preaccelerator.

The basic physics of the induction cores is the same for both electron and ion accelerations. Core material commonly used for a long pulse longer than 1 µs is iron alloys. Silicon steel is chosen from a careful comparison between candidates. In an induction linac the beam current and the pulse length in time are conserved unless an axial beam bunching is attempted. When the pulse length is fixed at 3 µs and the final energy of 5 MeV, the total volt-sec of the linac cores is 15 V-S. The cross sectional area of core material is approximately 6.25 m^2. The number of gaps is determined by the details of core geometry and other factors. When the number is taken as 40, each gap voltage is 125 kV, and the cross sectional area per gap becomes 0.156 m^2.

The magnitude of the magnetic induction in a toroidal core is inversely proportional to the layer radius. This results in an earlier saturation to inner layers than the outer layers, thereby distorting the gap voltage wave form from a desired flat-top distribution. In order to maintain proper voltage wave forms, a

On the other hand, in the uniform focusing region it becomes

$$R'' + \kappa R - \frac{K}{R} = 0. \tag{4}$$

When we look for an envelope solution which is periodic shown in Fig. 2(b), we can use well known solutions in each region. The solution to Eq. (3) is the universal envelope curve[9,10],

$$\frac{Z}{R_0} = \sqrt{\frac{2}{K}} \frac{R}{R_0} F\left[\sqrt{\ell n \, (R/R_0)}\right], \tag{5}$$

where

$$F(x) = e^{-x^2} \int_0^x e^{t^2} dt. \tag{6}$$

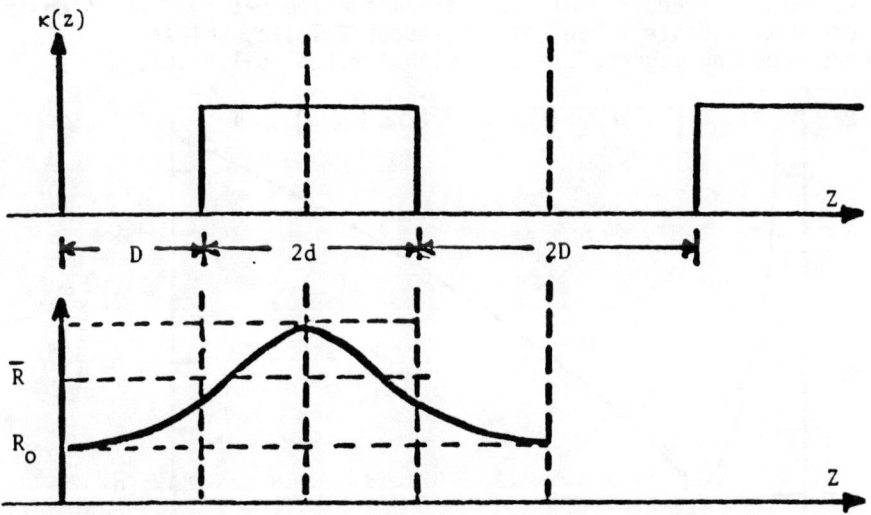

Fig. 2. Channel configuration. (a) Axial magnetic field distribution, and (b) matched beam envelope.

It can be expressed as

$$R = H(z). \tag{7}$$

The equilibrium solution to Eq. (4) with $R'' = 0$ is simply $\bar{R} = \sqrt{K/\kappa}$. With a small envelope perturbation, $R = \bar{R}(1 + \delta)$, the rippled envelope solution can be found as

$$R = \bar{R}\left[1 + \delta_0 \cos\left(\sqrt{2\kappa}\, z\right)\right]. \tag{8}$$

A matched beam condition is easily found by matching R and R' at the boundary. It is shown as

$$H(D) = \bar{R}\left[1 + \frac{H'(D)}{\sqrt{2\kappa}} \cot(\sqrt{2\kappa}\, d)\right]. \qquad (9)$$

When the argument x in Eq. (6) is smaller than 0.3, it can be simplified as

$$1 + g = \frac{R}{R_0}\left[1 + g \cot(\sqrt{2\kappa}\, d)\right], \qquad (10)$$

where $g = (K/2)^{1/2}(D/R_0)$.

The matched beam radius and corresponding magnetic field for a 100 A proton beam versus particle energy are plotted in Fig. 3 with fixed channel parameters of d = 15 cm and D = 10 cm. The matched beam radius decreases rapidly, and the corresponding magnetic field increases as beam energy increases. From this plot, one easily notes that the required magnetic field for the matched condition is beyond a manageable value, namely, about 2 Tesla, unless superconducting magnets are used with limited coil space.

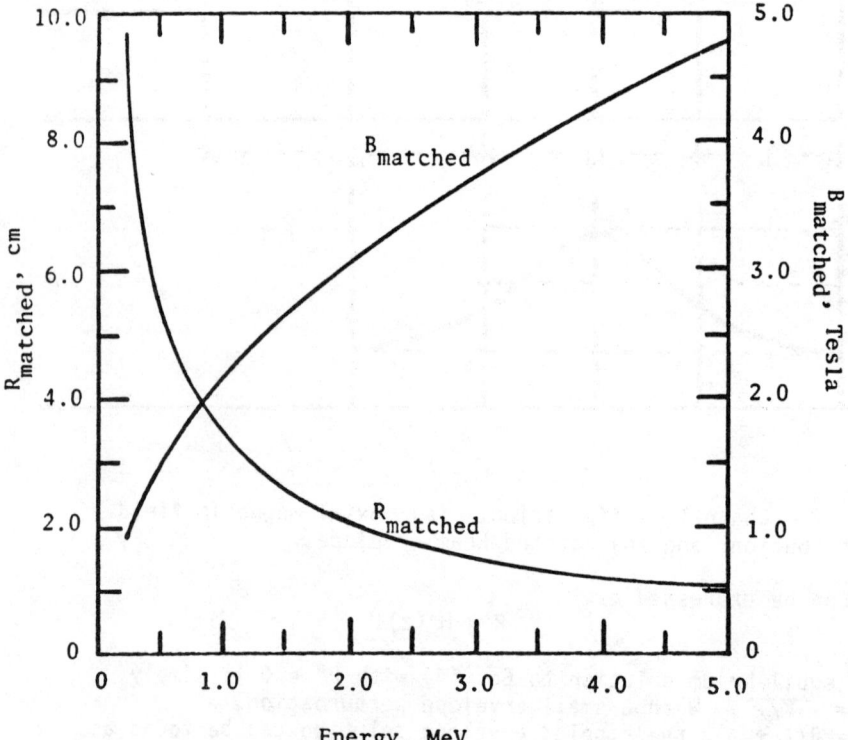

Fig. 3. Matched beam radius and corresponding magnetic field for 100 A proton beam versus beam energy.

In summary, the matched beam with accelerating fields in a periodic focusing channel requires corresponding magnetic fields for each period. It is expensive, and it requires too many controls even if computer control is adopted for a large scale system. We suggest from this study that independently controlled magnets should be employed only for few periods at the low energy end, about 1.0-1.5 MeV in this exercise from Fig. 3. The rest of magnets are then powered with the same magnetic strength with unmatched conditions, since the beam radius is reduced enough for mismatched envelope expansions.[11]

REFERENCES

* This work is supported by the IR Fund at NSWC.

1. J. E. Leiss, IEEE Trans. NS-26, 3870 (1979).
2. D. Keefe, paper presented at this symposium.
3. R. J. Briggs, in Proc. 1984 Linear Acc. Conf. (Darmstadt, W. Germany), GSI-84-11, 507 (1984).
4. W. Namkung and H. S. Uhm, IEEE Trans. NS-32, 3515 (1985).
5. H. S. Uhm, K. T. Nguyen, and W. Namkung, IEEE Trans. NS-32, 3518 (1985).
6. W. L. Gardener, et al., Rev. Sci. Instrum. 53, 424 (1982).
7. See, for example, M. Reiser, paper presented at this symposium.
8. M. Reiser, Part. Acc. 8, 167 (1978).
9. J. R. Pierce, in Theory and Design of Electron Beams, (Van Nostrand, New York, 1954), Chapter 9.
10. J. D. Lawson, in Applied Charged Particle Optics, Part C (Ed. A. Septier, Academic, New York, 1983), p. 22.
11. J. Struckmeier and M. Reiser, Part. Acc. 14, 227 (1984).

PARTICLE TRACKING IN SOLENOID CHANNELS

C. R. Prior
Rutherford Appleton Laboratory, Chilton, Didcot, Oxfordshire, U.K.

ABSTRACT

Computer simulation of the transport of high-intensity beams through periodic focusing systems of magnetic lenses poses problems not found with conventional hard-edged multipole channels. The presence of a variable longitudinal magnetic field means that time-step integration methods have to be modified and third-order terms, which often have a marked effect on the behaviour of the beam, included. In order to help understand observed effects in the experimental channel at the University of Maryland, three different analytical models of solenoid fields have been incorporated in the two-dimensional particle code TRACK2D, developed at the Rutherford Appleton Laboratory. Each model produces such well-known phenomena as sharp rings in the beam cross-section or a concentrated centre with a surrounding halo, which are both direct results of the spherical aberrations. Tracking through channels with perfectly conducting boundary walls of arbitrary geometry is permitted by the use of finite element techniques in the space-charge routines. In particular, simulation runs in which the beam is off-centre, so feeling the effects of image charges in a cylindrical bounding pipe, have been made, and transmission properties over a range of values of $\sigma°$ (zero-current phase advance) investigated in an attempt to explain measured results.

INTRODUCTION

The benefits of flexible and trustworthy computer tracking codes have been widely recognised for a long time not only through their use in predicting what might happen in an experiment which is too costly to set up but also in helping to diagnose problems and indicate solutions when the unexpected happens in existing machines. The Heavy Ion Fusion programme falls mainly into the former category but several small experiments have been set up to look at the problems of transporting beams under the effects of very large space-charge forces and in all aspects simulation codes are finding more and more use. As part of the computational program for HIF at the Rutherford Appleton Laboratory, a package of codes (both two- and three-dimensional) is being developed with the somewhat ambitious aim of simulating the motion of a beam emerging from a linac in its passage through the processes of multiturn injection into storage rings (in both phase planes) and then in a highly non-linear state through to its focus on the target pellet. The codes are however able to handle other problems as well and here we describe the use of the two-dimensional (x-y) version to look at beam transport in periodic solenoid channels using the apparatus at the University of Maryland as a model. Three analytical solenoid fields included in the program are discussed and a brief indication is given of the derivation of the equations of motion and their conversion to a form suitable for tracking when non-zero longitudinal magnetic fields are present. Some of the ideas behind the Poisson solver are included. Finally we detail the results of the simulation runs giving particular attention to those obtained for the behaviour of space-charge dominated electron beams under the effects of spherical aberrations and misaligments in the solenoid lenses.

SOLENOID MODELLING

The multi-particle tracking code TRACK2D represents the development of a computer program that originally dealt only with hard-edged multipole channels and simulated fringing fields in a very primitive way. A simple model of a solenoid field was first introduced to provide computational aid when a small experiment comprising a single magnetic lens was set up at the Rutherford Appleton Laboratory to investigate emittance enhancement using a system of wire grids[1]. Since then the code has been fully adapted to deal with overlapping fields in a channel of lenses and to study effects of spherical aberrations in a rigorous fashion.

Three analytic models of solenoid fields are at present incorporated. Mathematically these are described by

$$B(z) = B_0/(1+z^2/a^2) \qquad (1a)$$
$$B(z) = B_0 \operatorname{sech}(z/b) \qquad (1b)$$
$$B(z) = B_0 \exp(-z^2/b^2)/(1+z^2/a^2) \qquad (1c)$$

where $B(z)$ is the longitudinal magnetic field on the axis of the solenoid, B_0 is the peak field at the centre of the lens, and a and b are parameters. From the point of view of the mathematics, model (1a), the Glaser or bell-shaped model, is the easiest to handle. The single-particle equations of motion through a lens can be solved exactly, the focus found and a formula derived for the spherical aberration coefficient[2]. However, the field does not die away with increasing $|z|$ sufficiently rapidly for it to be a good model for practical purposes. The Grivet model (1b)[3] can usually be arranged to fit the measured data much better or the exponential factor in (1c) can be used to reduce the long tails of the Glaser field. Broadly speaking, all three models exhibit similar effects in the behaviour of a charged-particle beam although model (1a) is probably of academic interest only. The relative profiles are shown in figure 1.

The fields off the axis can be derived by integrating Maxwell's equations $\operatorname{curl}\underline{B} = 0 = \operatorname{div}\underline{B}$ and matching the solution to the particular model to be used. In principle this can be done exactly (and certainly numerically) but in practice, and in order to reduce computer run-time, it is only necessary to obtain an approximation. By systematically integrating the equations to progressively higher orders in the radial distance parameter r, one finds that

$$B_z(r,z) = B - B''r^2/4 + O(r^4)$$
$$B_r(r,z) = -B'r/2 + B'''r^3/16 + O(r^5)$$

where $B' = dB/dz$. The second terms in each of these series are those responsible for the spherical aberration effects in the beam.

TRACKING PROCEDURE

To simulate the motion of a transverse cross-section of the beam through a machine it is first necessary to convert the Lorentz force equations into a form suitable for tracking. We denote by $\underline{u}$ the (constant) velocity of the cross-section through the channel and write $\underline{v} = \underline{u}+\Delta\underline{v}$ for the velocity of an individual particle. By looking at the fields in the rest-frame of the beam and then transforming back to the laboratory frame, one first derives the equation

$$\frac{m_0}{q}\frac{d}{dt}(\gamma_v \underline{v}) = \underline{F} + \underline{E}_{\parallel} + (1-\underline{u}^2/c^2)\underline{E}_{\perp} + \Delta\underline{v} \times (\underline{u} \times \underline{E}_{\perp})/c^2 \qquad (2)$$

where $\gamma_v = \gamma(\underline{v}) = (1-\underline{v}^2/c^2)^{-1/2}$, $\underline{F}$ denotes the external forces and $\parallel$ and $\perp$ refer to components parallel and perpendicular to $\underline{u}$ respectively. We assume in the two dimensional code that there is no longitudinal electric field ($\underline{E}_{\parallel}=0$), which is a valid assumption for a long beam focused over a distance large compared to its radius.

Component equations are obtained by writing $\underline{u}=(0,0,\beta c)$, $\Delta\underline{v}=(\dot{x},\dot{y},\dot{z})$ where the dot denotes differentiation with respect to time, and defining $x'=\dot{x}/(\beta c+\dot{z})$ etc. A little algebra results in

$$\frac{m_0 \gamma_v}{q}(\beta c+\dot{z})x'' = x'y'B_x - (1+x'^2)B_y + y'B_z$$
$$+[1/(\beta c+\dot{z})-\beta(1+x'^2)/c]E_x - x'y'\beta E_y/c \qquad (3a)$$

$$\frac{m_0 \gamma_v}{q}(\beta c+\dot{z})y'' = (1+y'^2)B_x - x'y'B_y - x'B_z$$
$$+[1/(\beta c+\dot{z})-\beta(1+y'^2)/c]E_y - x'y'\beta E_x/c. \qquad (3b)$$

Terms in $\dot{z}$ can be eliminated using the energy equation

$$m_0 \gamma_v c^2 + e\phi = \text{constant}. \qquad (4)$$

For our purposes, this need only be carried out as far as terms in x'^2 and y'^2. The result is two equations of the form

$$x'' = \omega B_z y' + f_x \qquad (5a)$$
$$y'' = -\omega B_z x' + f_y \qquad (5b)$$

where $\omega=q/m_0\gamma\beta c$, $\gamma=\gamma_u$ and $\underline{f}$ incorporates all second and higher order terms and terms independent of x' and y'.

The step from the mathematical to the simulation model is to approximate the mathematical equations (5) by the algebraic equations required for numerical computations. This is usually accomplished by means of a leapfrog finite-difference scheme in which positions and velocities are stored half a timestep out of phase and leapfrog over each other forward in time. Accuracy is maintained to second order in the step length. However this method cannot be used to the same accuracy when the expression for $\underline{x}''$ involves $\underline{x}'$ explicitly. In particular this will be the case for solenoid channels ($B_z \neq 0$), for models of quadrupole fringing fields ($B_z \neq 0$) or for systems in which terms in $\underline{x}'$ higher than the first (aberrations) are to be included. In general, for equations of the form $\underline{x}'' = \underline{F}(s,\underline{x},\underline{x}')$, $\underline{x}' = d\underline{x}/ds$ an implicit approximation such as

$$\underline{x}'_{n+\frac{1}{2}} - \underline{x}'_{n-\frac{1}{2}} = h[\underline{F}(s_n,\underline{x}_n,\underline{x}'_{n+\frac{1}{2}})+\underline{F}(s_n,\underline{x}_n,\underline{x}'_{n-\frac{1}{2}})]/2 \qquad (6a)$$

$$\underline{x}_{n+1} - \underline{x}_n = h\underline{x}'_{n+\frac{1}{2}} \qquad (6b)$$

where s = nh, has to be used. This preserves accuracy to second order in the step-length h, and, although its implicit nature looks unpromising, it turns out to be possible, because of the form of equations (5), to rearrange it to obtain an explicit expression for the new $\underline{x}'$ values (level n+1/2) in terms of the old (level n-1/2). Sufficient accuracy can be maintained to preserve the third-order aberration terms and the resultant equations can be written

$$x'_{n+\frac{1}{2}} = [(1-\Omega^2h^2)x'_{n-\frac{1}{2}} + 2\Omega hy'_{n-\frac{1}{2}} + hf_x + h^2\Omega f_y]/(1+\Omega^2h^2) \qquad (7a)$$

$$y'_{n+\frac{1}{2}} = [(1-\Omega^2h^2)y'_{n-\frac{1}{2}} - 2\Omega hx'_{n-\frac{1}{2}} + hf_y - h^2\Omega f_x]/(1+\Omega^2h^2). \qquad (7b)$$

The quantity $\Omega=\omega B_z/2$ is proportional to the Larmor frequency. These equations can be shown to be stable against round-off errors in the computer and to the small errors generated in the initial step necessary to get from $(\underline{x}_0,\underline{\dot{x}}_0)$ to $(\underline{x}_0,\underline{x}'_{\frac{1}{2}})$ before tracking can begin.

CALCULATION OF THE SPACE-CHARGE FORCES

The space-charge forces within the beam are computed using a Poisson solver based on finite element techniques, which replaces the differential equation ($\nabla^2\phi=-\rho/\epsilon_0$) for the potential ϕ by the equivalent variational problem

$$\delta\Pi(\phi) = 0; \qquad \Pi(\phi) = \int_A |\nabla\phi|^2 dS - \int_A \rho\phi/\epsilon_0 dS - \int_{\partial A} \bar{\phi}_\nu \phi ds. \qquad (8)$$

Values of $\phi(=\bar{\phi})$ and its normal derivatives $(=\bar{\phi}_\nu)$, or combinations of both, can be prescribed on the boundaries. A mesh of elements of a predetermined shape (mainly triangular in TRACK2D) covers the beam and surrounding domain, and different polynomial functions ϕ with unknown coefficients are fitted to each. Equation (8) leads to a set of simultaneous linear equations for the unknowns which is inverted using the method of triple factoring. The approach possesses several advantages over conventional finite difference schemes: boundaries of almost any shape can be included; the two-dimensional x-y coordinate system can be fairly easily adapted for other purposes to a two-dimensional r-z version or a full three-dimensional version (as in the companion program TRACK3D); and, as the bulk of the calculations depends only on the mesh and not the particle coordinates, a simple calculation of the charge density contributions ρ and the final part of the triple factoring inversion is all that is required during tracking until TRACK2D judges that the beam has altered so much that a new mesh is necessary to maintain accuracy. Some idea of the saving can be gained from the observation that a full call to the routine with a mesh of just over 3000 triangular elements and 8000 macro-particles takes from 1-2 seconds (Amdahl Atlas-10 time) and a shortened call using a previously defined mesh takes of the order of a few milliseconds.

APPLICATIONS TO SOLENOID SYSTEMS

The tracking code TRACK2D works in harmony with an analytical program KVBL which integrates the envelope equations for a Kapchinskij-Vladimirskij (K-V) particle distribution. The latter includes an optimising routine which is used to find matched beam parameters and vary current, kinetic energy and field strengths so as to produce the model system for full simulation. It is worth pointing out that, even analytically, solenoid systems based on the models (1) are far harder to treat than quadrupole systems; indeed, many of the methods generally used for, say, matching frequently tend to diverge.

The clearest illustration that TRACK2D works effectively for solenoid systems was found in work carried out in connection with an experiment at the Rutherford Appleton Laboratory into the possible use of wire grids to control beam emittance[1]. The Grivet model was used here and in several subsequent runs designed to investigate the effects of aberrations and non-linearities in space-charge dominated electron beams. In exact accord with experiment, the influence of the third order solenoid terms causing spherical aberrations was found to be characterised by the alternating formation in the beam of a fairly sparse central core with a sharply defined outer ring followed by an inversion (usually at a waist and in a less prominent form at low current levels) to a concentrated centre surrounded by a cloud of particles. Such phenomena can be seen in figures 3a-3f.

Subsequently, the code has been used to look at periodic solenoid and quadrupole channels over a range of zero-current phase advance (σ^0) and with various models of input distribution. The region $\sigma^0 \geq 90°$ is of particular interest as it is recognised to be theoretically unstable for certain ranges of the depressed tune σ.[4] For such values it has been observed that third order effects generally bring about an earlier onset of instability and that, for equivalent systems using the same criterion to define an unstable beam, the instability seems to occur very slightly sooner in a quadrupole channel than in a solenoid system.

More recently, computational backup being provided to help understand the problems caused by misalignments in the experimental solenoid channel at the University of Maryland[5] has utilised TRACK2D to its full capabilities. Model (1c) (the bell-shaped model with the exponential factor reducing the tails) was chosen for the simulation runs with parameters a = 4.4cm, b = 3.24cm best fitting the observed fields. The program KVBL first computed the peak field B_0 for the required σ^0, then found the matched beam size for a given electron current of 220mA and kinetic energy 5keV, and finally optimised the fields of two initial matching lenses whose purpose is to take the beam from the gun (diameter 1.27cm) into the periodic system. With these values and an initial semi-gaussian (thermal) particle distribution, TRACK2D was then used to simulate passage of the beam down the channel under non-linear space-charge forces and aberration effects assuming a perfectly conducting bounding pipe of radius 1.85cm. The first sequence of runs was performed with the input beam placed 1.5mm off-centre, so that image charge forces were called into play. Ignoring lower values of σ^0 for which the matched beam size exceeds the pipe radius, the results for $\sigma^0 = 45°$ show a total loss of about 20% of the beam after 36 lenses (most of it in the first 12 cells). As σ^0 is increased, the loss decreases until in the range $70°\leq\sigma^0\leq 90°$ there is complete or almost complete transmission. Thereafter increasing loss is encountered again as σ^0 increases further until at $\sigma^0=120°$ it has reached 35%. The transmission curve (figure 2) is smooth, in contrast to observational results at Maryland which show sharp dips in transmission at $\sigma^0=83°$, 96° and 100°[5].

In general the tracking runs for a given σ^0 reveal that in the cases showing beam loss the initial amplitude of the oscillation of the beam centre is maintained for several solenoids as the outermost particles are absorbed at the bounding pipe but that the off-centring eventually tends to die away, the loss rate diminishes and there is a suggestion that the remaining beam reaches a state that is fairly well matched to the channel. The larger σ^0 the faster this state is reached - it is certainly predicted by the codes for $\sigma^0=120°$ and looks to be happening for $\sigma^0=100°$ although the 36-cell channel is not really long enough to be sure. Sample phase-space plots for the case $\sigma^0=120°$ are shown in figures 3a-3f.

In all instances an increase in r.m.s. emittance by factors of 3 and upwards is predicted even in the cases of 100% transmission. Such large increases have been observed experimentally at the University of Maryland, although not quite to this magnitude, and indicate that off-centring can be an important consideration in the control of emittance in the beam.

Runs with different initial displacements have also been performed. These support the view that the loss levels are not much affected by an off-centring of up to 2mm, but beyond 2mm transmission is very much reduced. Thus, for example, at $\sigma^0=96°$, the loss for an initial displacement of 1.5mm is about 2% but for a displacement of 4mm it rises to just over 40%.

CONCLUSIONS

It would appear from these results that the instances of sharply reduced transmission at particular values of σ^0 in the Maryland channel are not caused simply by an initial misalignment of the electron gun of less than about 2mm. Insufficient tracking runs have been performed to date for a decision to be made for values between 2mm and 4mm but the evidence so far accumulated shows that the predicted magnitudes of the losses are more closely related to those actually measured although there may still be no sharp dips in the transmission curve. It is clear however that TRACK2D is providing useful computational back-up to the experimental work.

REFERENCES

1. J.D. Lawson et al, IEEE Trans.Nucl.Sci. Vol.NS-$\underline{30}$, 2537, 1983.
2. W. Glaser, Z. Physik, Vol.$\underline{117}$, pp 285-315, 1941.
3. P. Grivet, J. Phys. Radium $\underline{13}$, 1A-9A, 1952.
4. I. Hofmann, L.J. Laslett, L. Smith and I. Haber, Particle Accelerators, $\underline{13}$, 145, 1983.
5. M. Reiser, J. McAdoo, D. Kehne, K. Low, J. Lawson and C. Prior, elsewhere in these proceedings.

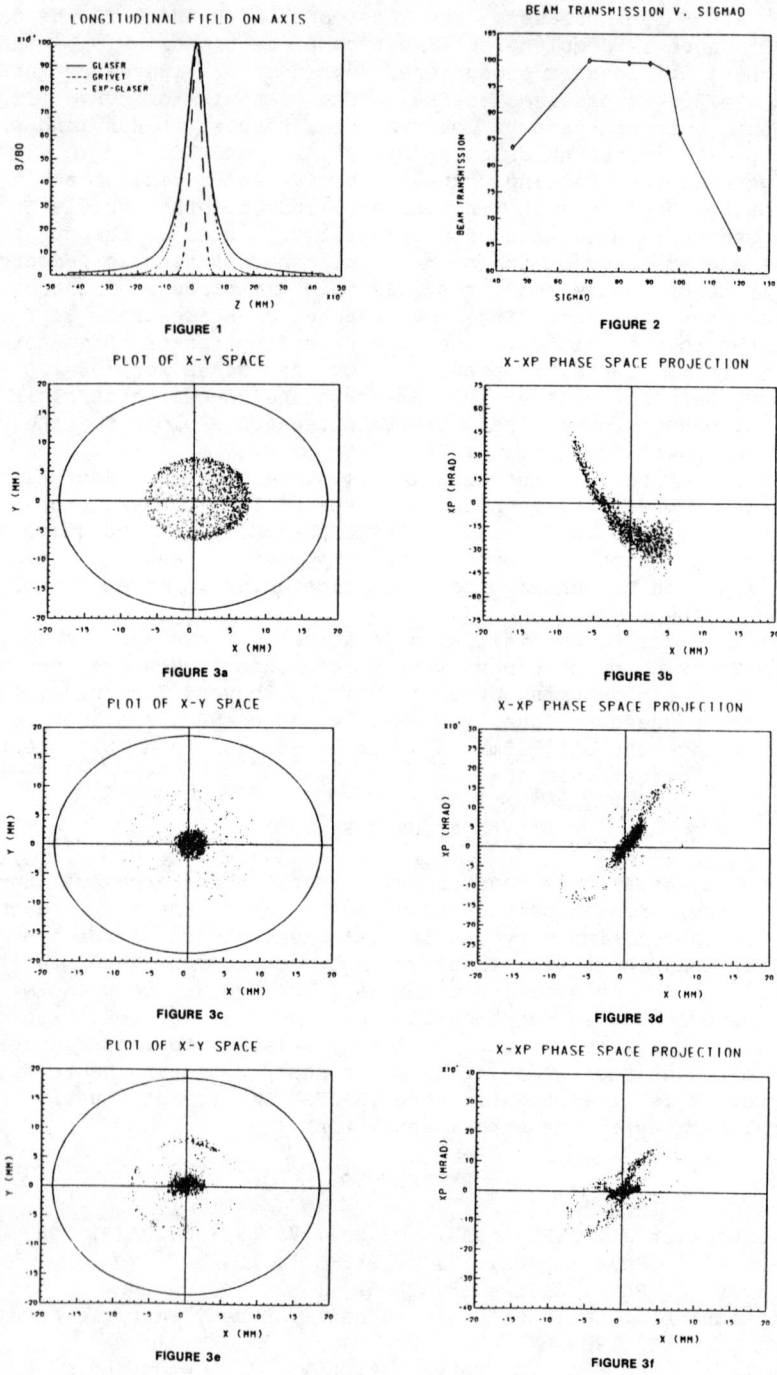

PONDEROMOTIVE CONFINEMENT OF ION BEAMS IN A CYLINDRICAL WAVEGUIDE

Celso Grebogi* and Han S. Uhm
Naval Surface Weapons Center
White Oak, Silver Spring, Maryland 20903-5000

ABSTRACT

Ion beam confinement by a ponderomotive force is investigated. The beam is assumed to be propagating along a cylindrical drift tube and the ponderomotive force results from a fast oscillating TE_{01} waveguide field. Formulas for the r.f. power and energy density in the waveguide needed to confine a solid ion beam are derived. Numerical estimates show that the amounts of r.f. power already available within the present technology are sufficient to confine ion beams with typical kinetic energy and typical density.

INTRODUCTION

It is often desirable to transport a large ion current between the beam source and some other location where the beam can be utilized. However, as the beam propagates, it tends to disperse unless external confining forces are applied. In this work, we investigate the use of the ponderomotive force[1], resulting from fast oscillating applied electromagnetic fields, as such confining force.

We consider an ion beam propagating in a cylindrical drift tube. The proper choice of an electromagnetic wave to be used for beam confinement in this geometry is the TE_{01} waveguide mode. The reason for this choice is the need for a wave with minimum electric field on the cylinder axis and with the gradient of the field pointing towards the center of the cylinder in the region of the beam.

We first obtain an expression for the relativistic ponderomotive force of a TE_{01} mode and then derive the radial force on the particles of the beam due to self-fields and pinching. Finally, we derive formulas for the minimum r.f. power and energy needed to confine a given beam and present a numerical estimate for a typical experimental situation.

PONDEROMOTIVE FORCE

In Reference 2, we derived a general expression for the relativistic ponderomotive Hamiltonian of TE_{0n} waveguide modes in cylindrical geometry (cf. also Reference 3 for the ideas and structure underlying the ponderomotive Hamiltonian theory)[2,3]. We found that the slow time-dependent Hamiltonian can be written as

---

*Permanent Address: Laboratory for Plasma and Fusion Energy Studies
University of Maryland
College Park, Maryland 20742

where
$$K = K_0 + K_2, \tag{1}$$
$$K_0 = e\phi(R,t) + mc^2\Gamma(V) \tag{2}$$
and
$$K_2 = \frac{e^2|\tilde{E}_\phi(R,t)|^2}{\Gamma(V)m\omega^2}. \tag{3}$$

The Hamiltonian (1) is the oscillation center[4] Hamiltonian and contains only nonoscillatory terms. The oscillation center dynamical variables (R,V) are averaged variables. We allow for slow time-dependence in the electric potential (which includes the beam self-fields) $\phi(R,t)$ and in the wave field $\tilde{E}_\phi(R,t)$. The relativistic factor Γ is still defined as

$$\Gamma = (1 + \frac{V^2}{c^2})^{-1/2}. \tag{4}$$

Observe that the ponderomotive Hamiltonian K_2, as given by Eq. (3), is a velocity dependent quantity and, hence, it is a generalization of the ponderomotive potential. For a TE_{01} mode, the waveguide field is[2]

$$\tilde{E}_\phi(R) = \frac{i\omega E}{cp} J_1(pR), \tag{5}$$

where J_1 is the first order Bessel function and p obeys the dispersion relation

$$p^2 R_c^2 = (\frac{\omega^2}{c^2} - k^2) R_c^2 = (3.832)^2, \tag{6}$$

R_c being the waveguide radius, and E is the wave amplitude.

The ponderomotive force, from Hamilton's equations, is the gradient of Eq. (3), or

$$\underline{F}_p(R) = -\frac{2e^2|E|^2}{\Gamma mc^2 p} J_1(PR)J_1'(PR)\hat{R}. \tag{7}$$

FORCE DUE TO SELF-FIELDS AND PINCHING

The radial force due to beam self-fields and pinching is, from Lorentz equation, given by

$$\underline{F}_S(R) = e(E_R^S - \beta B_\phi^S)\hat{R}, \tag{8}$$

where $E_R^S = -\partial\phi^S/\partial R$ and, for a uniform beam, $B_\phi^S = -\partial A_Z^S/\partial R$. We calculate ϕ^S and A_Z^S from Maxwell's equations by choosing the proper boundary conditions. For a beam with uniform density profile, we obtain the following potentials inside the beam

$$\phi^S(R) = 2eN_b \ln(\frac{Rc}{R_b}) + N_b(1 - \frac{R^2}{R_b^2}) \qquad (9)$$

and

$$A_Z^S(R) = \beta \phi^S(R), \qquad (10)$$

where $\beta = V/c$, N_b is the beam charge per unit length, and R_b is the beam radius. The substitution of Eqs. (9) and (10) into Eq. (8) yields

$$\underline{F}_S(R) = \frac{2Nbe^2 R}{\Gamma^2 R_b^2} \hat{R}. \qquad (11)$$

MICROWAVE POWER AND ENERGY NEEDED TO CONFINE THE ION BEAM

For confinement, we require that the sum of the radial forces should vanish at $R = R_b$, or,

$$\underline{F}_S(R_b) + \underline{F}_P(R_b) = 0 \qquad (12)$$

Substituting Eqs. (7) and (11) into (12) gives

$$N_b(R_b) = \frac{\Gamma |E|^2 R_b}{mc^2 p} J_1(pR_b) J_1'(pR_b) . \qquad (13)$$

To obtain the maximum beam charge per unit length which is possible to confine for a given ω, E, and Γ, we require that $(dN_b/dR_b) = 0$ at R_b^{max}, or,

$$J_1(x) J_1'(x) + x [J_1'(x)^2 + J_1(x) J_1''(x)] = 0 \qquad (14)$$

where $x = p R_b$. The numerical solution of Eq. (14) gives $pR_b^{max} \simeq 1.18$, which upon substitution into Eq. (13) yields

$$N_b^{max} \simeq \frac{0.15 \, \Gamma \, |E|^2}{mc^2 p^2}. \qquad (15)$$

To understand this maximization process, in Fig. 1 we plot F_S and $|F_P|$ versus pR_b. Note that the force balance condition $F_S = |F_P|$ can give zero, one, or two solutions, depending how many crossings the F_S and $|F_P|$ curves have. The condition that maximizes N_b corresponds to having one solution, as depicted in Fig. 1.

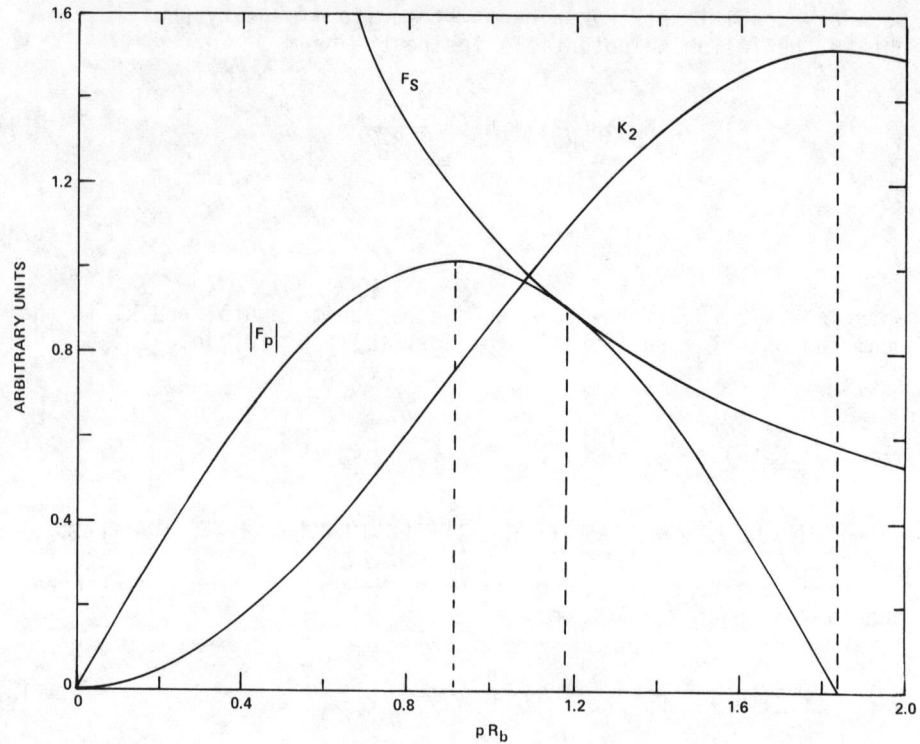

Fig. 1 Plot of F_s, $|F_p|$, and K_2 versus pR_b.

From Eq. (15), we obtain the maximum beam current, $I_{max} = ec\beta N_b^{max}$, or,

$$I_{max} = \frac{0.15 \, e\beta \, \Gamma \, |E|^2}{mcp^2}. \tag{16}$$

The r.f. power for the waveguide is calculated from[5]

$$P_{rf} = \frac{kc^2}{8\pi\omega} \int_0^{R_c} |\tilde{E}_\phi|^2 2\pi R dR. \tag{17}$$

Using Eq. (5) and integrating Eq. (17) gives

$$P_{rf} = \frac{\omega k R_c^2 |E|^2}{8p^2} [J_1'(pR_c)]^2. \tag{18}$$

By comparing Eqs. (16) and (18) and using Eq. (6), we obtain the desired result

$$P_{rf} = \frac{0.13 \, mc\omega k R_c^2}{e\beta\Gamma} I_{max}. \tag{19}$$

For a waveguide of length L, the total energy inside the waveguide is[5]

$$U_{rf} = \frac{\omega L}{kc^2} P_{rf}. \tag{20}$$

Hence, Eqs. (19) and (20) give the r.f. power and energy, respectively, required to confine an ion beam with current I_{max} in a cylindrical waveguide.

NUMERICAL ESTIMATE AND CONCLUSION

Assume a 1 Amp ion beam with $\Gamma = 3$ propagating in a waveguide with $R_c = 1$ cm and $L = 5000$ cm. For a waveguide mode with $k = 3 \times 10^{-3}$ cm^{-1} [and, hence, $\omega \approx 10^{11}$ rad/sec by Eq. (6)], Eqs. (19) and (20) give $P_{rf} \approx 465$ kWatt and $U_{rf} \approx 85$J, respectively.

The above values for P_{rf} and U_{rf} are reasonable and well within the present available technology.

ACKNOWLEDGEMENT

This work was supported by the NSWC Independent Research Fund.

REFERENCES

1. S. Yadavalli, J. Vac. Sci. Technol. **15**, 868 (1978) and references therein.

2. C. Grebogi and H. S. Uhm, accepted in Phys. Rev. A.

3. C. Grebogi and R. G. Littlejohn, Phys. Fluids **27**, 1996 (1984).

4. R. L. Dewar, Phys. Fluids **16**, 1102 (1973).

5. J. D. Jackson, Classical Electrodynamics (Wiley, 1962), Sec. 8.5.

GRID-CONTROLLED METAL ION SOURCES FOR HEAVY ION FUSION ACCELERATORS

L.K. Len, S. Humphries, Jr. and C. Burkhart
Institute for Accelerator and Plasma Beam Technology
University of New Mexico
Albuquerque, New Mexico 87131

ABSTRACT

A variety of metal ions can be generated using vacuum arcs, but due to the nature of these arcs, the flux generated fluctuates in time. We have successfully employed electrostatically biased grids to control the plasma and to provide a well-behaved, space charge limited ion source. The grid prevents the plasma from entering the extraction gap before the main voltage pulse is applied. The extracted ion current is space charge limited, resulting in a constant output current even though the ion flux from the vacuum arc source varies considerably. There are several advantages over conventional sources. For instance, thermionic sources are faced with heating problems for large area configurations, while gas-injection sources cause prefill problems because they take too long to reach equilibrium. We have performed extraction experiments with aluminum and indium arc sources. We have extracted 300 mA of pure Al$^+$ at 30 kV for 10 μs. The normalized beam emittance has been measured to be 3 x 10^{-7} π-m-rad.

INTRODUCTION

The metal vapor vacuum arc is one of the most versatile methods for the generation of metal ions for various research and industrial applications. Its operation is simple, reproducible and reliable. Many different types of ion can be readily generated by simply changing its cathode material [1,2]. Cathode materials that we have used successfully to generate their respective ions thus far include C, Mg, Al, Ti, Ni, Cu, Zn and In. Although it is capable of producing a respectably high ion flux, the vacuum arc by itself is nevertheless not directly applicable to highbrightness beam generation. This is because the flux level produced by such arcs is intrinsically very spiky, ie., significant flux fluctuation in short time scales. The long time overall flux level is very reproducible. Figure 1 shows three typical ion flux signals for aluminum, copper and indium as recorded by biased flux cup. Each frame is a three trace overlay. Even the smoothest of them is far from being an ideal source for accelerator. The optical quality of a beam extracted from a plasma depends very much on the plasma anode surface. To obtain a good quality beam, one needs a plasma anode surface that is concave towards the cathode [3] and extraction condition which perfectly matches the plasma ion source. In the event there is a mismatch, the plasma anode surface will bulge one way or another thereby deteriorating the beam quality.

This paper describes an experimental technique that has been used successfully to eliminate the rapid flux variation in such sources, making them appropraite for use as ion sources for particle accelerators. The technique involves the introduction of appropriately biased grids to separate the ions electrostatically from the electrons in the plasma. A space charge limited, constant current beam can then be extracted from this grid-controlled plasma ion source so long as the extraction gap and voltage are such that the source limit is not exceeded by the space charge limit.

The basic principle of the operation of the grid-controlled ion source is illustrated in Fig. 2. The plasma grid defines the potential of the plasma and the anode grid at the extraction gap is biased to a negative potential with respect to the plasma grid. For this source to work properly, the negative bias must be large enough to turn all the electrons from the plasma. This of course is determined by the plasma electron temperature. To satisfy the above requirement, one needs a bias voltage given by

$$eV_g > (kT_e) \ln(m_i/m_e)$$

where V_g is the magnitude of the anode voltage relative to the plasma grid, T_e is the electron temperature and m_i, m_e are the ion and electron masses. For an electron temperature of 10 eV, V_g should be 100V or more for light through intermediate mass ions. This negative bias causes the electrons in the plasma to turn back, allowing only the ions to enter the extraction gap as depicted in Fig. 2. Before the application of the extraction voltage, the ions that enter the extraction gap are forced to reflex because of the formation of a virtual anode. There is thus no significant buildup of particles in the gap before the arrival of the extraction voltage. The grid-controlled plasma anode thus behaves in manner similar to a thermionic anode. When the extraction voltage is applied, the ion flow is space charge limited and is insensitive to source fluctuations. The beam optics are determined entirely by the shape of the plasma grid.

Experimental results verifying space charge limited ion beam extraction will be presented for Al+ and In+. Pulsed Al+ beams of up to 15 mA/cm² in a 1.8 cm gap at 30 kV have been produced. Beam divergence measurements showed an upper limit on normalized emittance of ϵ_n = 3 x 10⁻⁷ π-m-rad.

APPARATUS

Fig. 3 shows the experimental apparatus and the electrical circuit components. A standard conical pyrex glass cross (15 cm diameter) pumped by a 6 inch ion pump provided a clean vacuum environment for performing the experiments. Typical operating pressure was 5×10^{-6} torr.

The metal vacuum arc was constructed from a stainless steel tube on which a cathode of desired material can be mounted. Inserted inside this stainless tube was a piece of rigid coaxial tubing which was used as a trigger for the vacuum arc source. The anode was also made of stainless steel. It served as a flange for mounting the source to the plasma expansion chamber. The vacuum arc was operated with a gap of 0.8 mm. We have learnt from our experience that the flux output would be reduced with smaller gaps. On the other hand, if the gap was too large, the discharge would be very erratic resulting in higher electron temperature. This in turn could lead to grid shorting at the plasma anode. The vacuum arc was triggered by the plasma generated on the insulator of the rigid coaxial tubing when the Krytron was fired dumping the 0.1 μF capacitor which had been charged to 5 kV. The energy was coupled through a ferrite isolation transformer. Once the main arc was initiated by this trigger plasma, it was sustained by an 8-element pulse forming network (PFN). This 5 Ω PFN provided a square pulse of 100 μs. It could supply a constant current ranging from 100 to 500 amperes. Fig. 4 shows a typical PFN current pulse and the plasma ion flux measured behind the cathode grid. The vacuum arc cathode used in this case was aluminum.

The plasma grid was mounted at the end of the plasma chamber. All the grids used in the experiment were made from 100x100 stainless steel mesh. The mesh had a transparency factor of 0.8. All three grids can be easily replaced in the event they were damaged in the experiment. They survived over 1000 pulses with no visible damage. The anode grid bias was provided by a capacitor with a series resistor to protect the grids in case of a short.

The ion beam extraction voltage was provided by a pulse transformer. The transformer was comprised of three laminated silicon steel cores. These cores had an inner diameter of 30 cm, outer diameter of 66 cm and a thickness of 5 cm, giving a total volt-second product of 0.083 V-s. An ignitron switch was used to dump the capacitor through a 5 Ω resistor into the primary, which consisted of just one single turn. The cores magnetically switched out a flat extraction voltage pulse whose pulselength was dependent on the voltage in accordance with the volt-second product. The cores had to be reset after each pulse. This was done with a different capacitor charged to the opposite polarity as the extraction capacitor.

Since the cathode and its supporting flange were at high voltage, diagnostic probes were also floated up to the same high voltage. To avoid having to float the oscilloscope, the signal cables were wound around the secondary of the silicon steel cores to provide voltage isolation. The noise level was very low. Good signal traces could be recorded on unshielded oscilloscopes.

The extracted beam angular divergence was studied using a channel-electron-multiplier-array (CEMA). Only time integrated measurements have been performed. A beam mask plate with four pinholes of diameter 0.36 mm, each 8 mm from the axis, was positioned at the cathode. The CEMA was located 37.2 cm downstream. The images of the beamlets were amplified and transferred to photographic film outside the vacuum system via fibre-optic coupler.

RESULTS

The top trace of Fig. 4 shows total driving current in the vacuum arc using an 8-element PFN. The bottom trace is the plasma ion flux from an aluminum arc. The plasma flux was measured with a biased ion flux collector located behind the cathode grid. The 20 μs time delay between the onset of the arc current and the ion flux signal was the time taken by the plasma to traverse the 18.5 cm from the plasma source to the probe. Since our source was operated at very low repetition rate, the arc had to undergo a cleaning phase at the beginning of the pulse, during which contaminants due to adsorbed gases may be present [1]. To avoid extracting these contaminants, the ion beam extraction voltage in this experiment was applied somewhere in the middle part of the pulse, ie., 40 μs after the initiation of the vacuum arc.

Space charge limited ion flow in the extraction gap is demonstrated in the sequence of measurements in Fig. 5. All the traces displayed are 2 trace overlays. The anode grid bias voltage was set at -150 V in all cases. Only the extraction voltage was varied. In the first two frames of Fig. 5, where the extraction voltage were relatively low, the space charge limits were well below the source limits and the extracted ion currents were identical. The reproducibility was excellent. As the extraction voltage was elevated, the space charge limit approached the source limit and irregularities began to show up in the bottom traces. These irregularities occurred whenever there was a cross over between the space charge limit and the source limit. Fig. 5e, in particular, shows very clearly the effect of the source fluctuation when the space charge demand could not be met.

Extensive extraction measurements had been performed on the aluminum source. The results were plotted in Fig. 6. The gap size was 1.8 cm and the anode grid bias was

−150 V. Measurements for indium plasma source has also been plotted in the same log-log graph. It can be seen that both sets of data follow the $V^{3/2}$ lines as predicted by Child's Law. It should also be noted that for the same extraction voltage, the aluminum flux is twice as high, also in agreement with space charge flow.

The beam divergence measurement was performed using an aperture plate located at the cathode and a channel-electron-multiplier-array (CEMA). Fig. 7 shows two sets of photographic output from the CEMA detector. Extraction voltage for both cases were 27 keV and the Al^+ ion flux was 15 mA/cm^2. In the case of Fig. 7a, the anode grid was biased at a relatively high voltage of −250 V. The beamlets formed well-defined images with diameter of 0.7 cm. The fact that a well-defined image was obtained for a beam implied that the anode plasma surface was highly stable. The pattern of the images was somewhat distorted when compared to the aperture pattern. This was due to space charge effect and error in the alignment of the extraction anode and cathode. The alignment was made in reference to the pyrex glass cross which is not perfectly symmetric. The normalized emittance inferred from the above measurement has a value of 3×10^{-7} π-m-rad.

The picture in Fig. 7b was taken with an anode grid bias of only −100 V. The images were not as well-defined as for Fig. 7a. The halos found in the images implied that the plasma anode surface was irregularly shaped, a result of the plasma anode surface bulging into the extraction gap thereby deteriorating the optics.

SUMMARY

The use of metal vacuum arc ion sources in conjunction with the grid-controlled technique can be used to provide a whole spectrum of ion sources which are suitable for applications in the heavy ion fusion accelerator programs. In principle, ions of any solid conductors and semiconductors can be generated. We have demonstrated the extraction of space charge limited, constant current ion beams from such sources. An Al^+ beam of 15 mA/cm^2 (300 mA total beam current) had been extracted at 30 kV with normalized beam emittance of 3×10^{-7} π-m-rad. It should be possible to improve on the beam quality by optimizing the shape of the extraction anode.

This work was supported by the U.S. Department of Energy under contract number DE AC0383 ER13138 and by Los Alamos National Laboratory under contract number 9X54-K-66711.

REFERENCES

1. L.K. Len, C. Burkhart, G. Cooper, S. Humphries, Jr., M. Savage and D. Woodall, to be published in IEEE Trans. on Plasma Science, June, 1986.
2. I.G. Brown, J.E. Gavin and R.A. MacGill, to be published in Appl. Phys. Lett.
3. S. Humphries, Jr., C. Burkhart, S. Coffey, G. Cooper, L.K. Len, M. Savage, D. Woodall, H. Rutkowski, H. Oona and R. Shurter, J. Appl. Phys., Vol. 59, 1790 (1986)

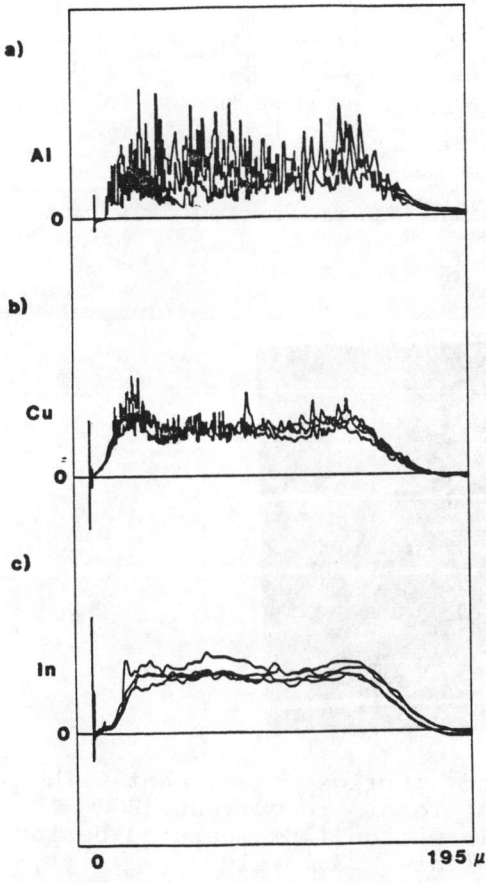

Fig. 1. Typical ion fluxes from metal vacuum arc sources. Note the short time flux variation.

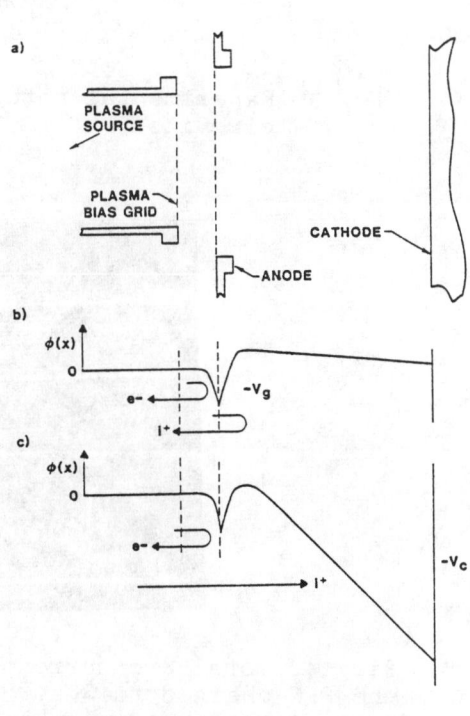

Fig. 2. Schematics of the grid-controlled ion source. a) Grid geometry. b) Potential and particle flow with no extraction voltage. c) With extraction voltage.

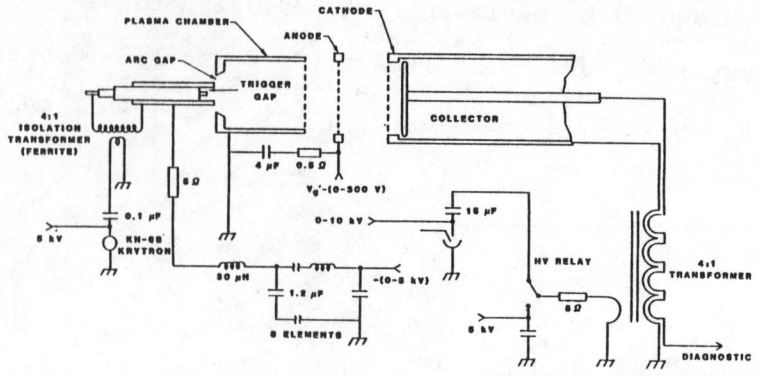

Fig. 3. Experimental setup with its associated driving circuits.

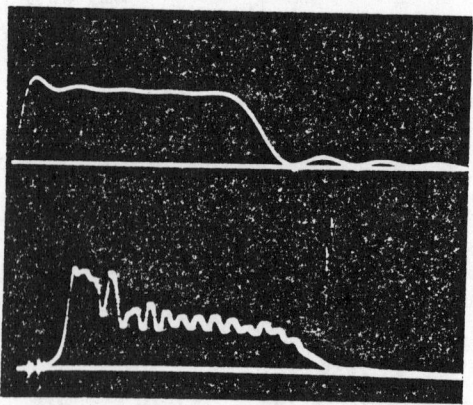

Fig. 4. Total arc current and ion flux measurement with the PFN charged to 3 kV. Top: Total arc current, 200 A/div. Bottom: Available plasma ion flux measured behind the cathode grid, 12.4 mA/cm² /div., 2 μs/div.

321

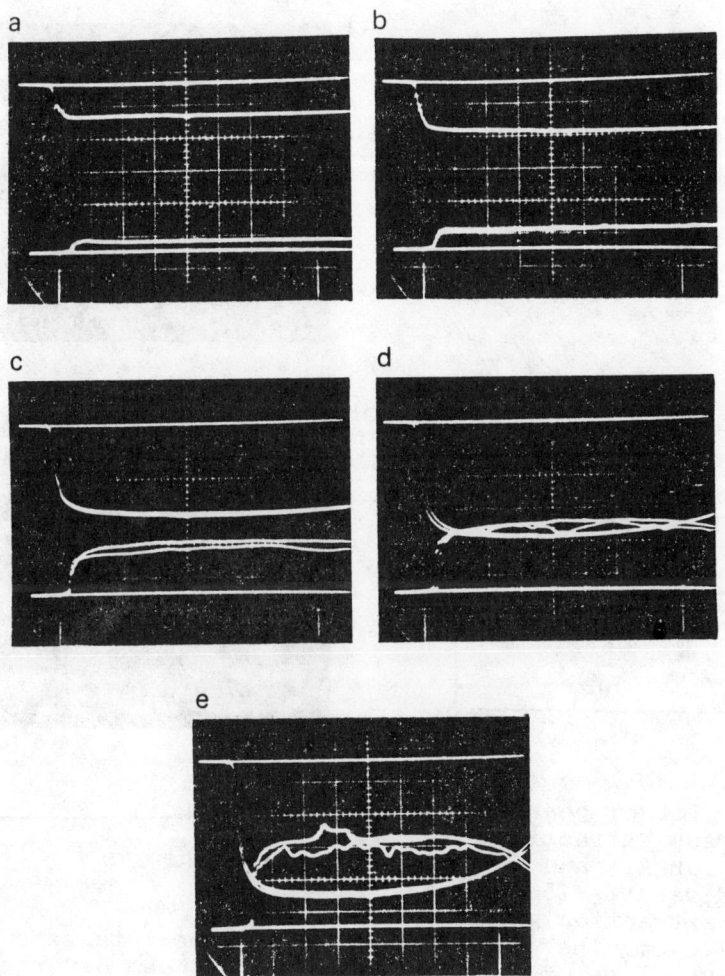

Fig. 5. Two trace overlays of extraction voltage and extracted ion beam density for an aluminum plasma source. The anode grid voltage was -150 V for all cases. Top: Extraction voltage, 4 kV/div. Bottom: Extracted ion current density, 6.2 mA/cm^2/div., 2 μs/div.

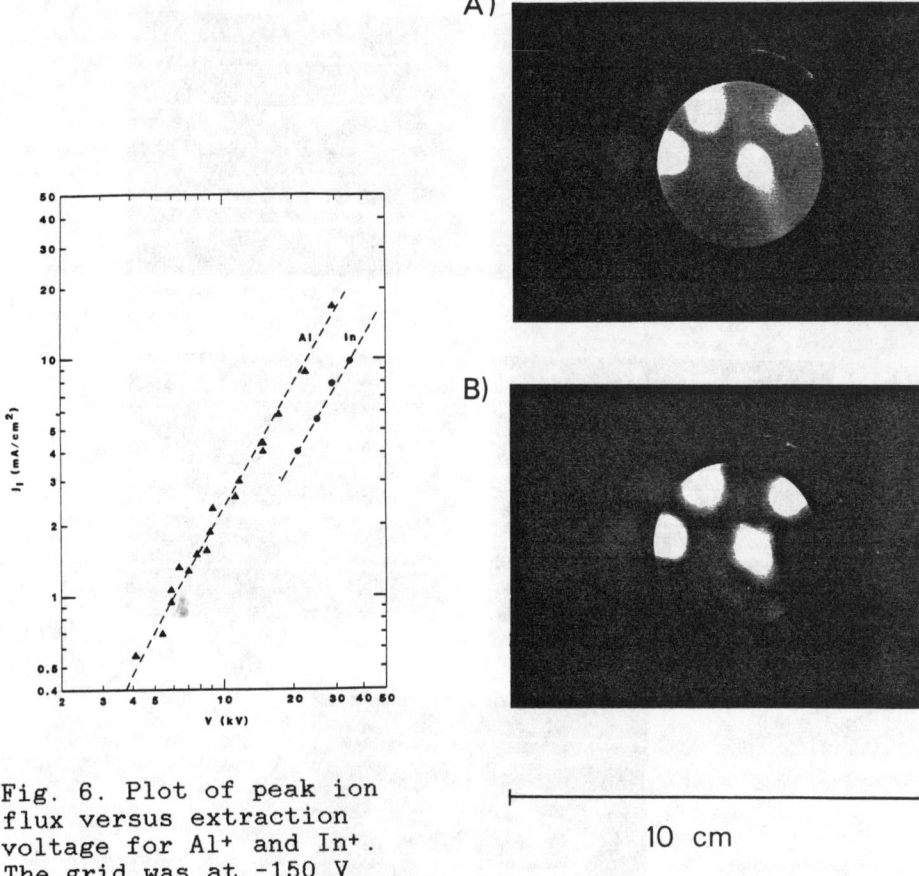

Fig. 6. Plot of peak ion flux versus extraction voltage for Al+ and In+. The grid was at -150 V and the extraction gap set at 1.8 cm. The dashed lines have slope of $V^{3/2}$.

Fig. 7. Time-integrated picture of beam divergence measurement for Al+ ions, extraction voltage was 27 kV, ion current density was 15 mA/cm^2. a) Anode grid voltage at -250 V. b) Anode grid voltage at -100 V.

PROGRESS ON THE LOS ALAMOS HEAVY-ION INJECTOR*

D. C. Wilson, K. B. Riepe, E. O. Ballard, E. A. Meyer
R. P. Shurter, and F. W. Van Haaften
Los Alamos National Laboratory, Los Alamos, NM 87545

S. Humphries, Jr.
University of New Mexico, Albuquerque, NM 87131

ABSTRACT

Heavy-ion fusion using an induction linac requires injection of multiple high-current beams from a pulsed electrostatic accelerator at as high a voltage as practical. Los Alamos National Laboratory is developing a 16-beam, 2-MeV, pulsed electrostatic accelerator for Al^+ ions. The ion source will use a pulsed metal vapor arc plasma. A biased grid will control plasma flux into the ion extraction region. This source has achieved a normalized emittance of $\epsilon_n < 3 \cdot 10^{-7} \pi$-m-rad with Al^+ ions. An 800 kV Marx prototype with a laser fired diverter is being assembled. The ceramic accelerating column sections have been brazed and leak tested. Voltage hold off on a brazed sample was more than doubled by selective removal of the Ticusil braze fillet extending along the ceramic. A scaled test module held 250 kV for 50 μs, giving confidence that the full module can hold 175 kV per section. The pressure vessel should be received in June 1986. High-voltage testing of a 1 MV column will begin by early 1987.

INTRODUCTION

As part of the U.S. Department of Energy's Heavy-Ion Accelerator Research (HIFAR) program Los Alamos National Laboratory is developing and building a multiple beam injector to produce high-voltage, high-current input beams needed for induction linear accelerators being pursued by Lawrence Berkeley Laboratory. In this paper we discuss progress in the development, design, and construction of this injector including source development, pulse power design, column fabrication, and high-voltage testing.

The current design criteria for this accelerator are:

Number of Beams - 16
Particle Energy - 2 MeV
Current per Beam - 150 mA
Ion Mass - 27 (Al^+)
Pulse Length - 6μs
Beam Normalized Emittance - 4×10^{-7} πm-rad
Energy and Current Flatness - 0.1 percent

---

*Work performed under the auspices of the USDOE.

Originally this electrostatic accelerator was designed as a prototype injector to feed a 125 MeV induction linac for a High-Temperature Experiment (HTE). However, the injector will now be used as a test bed for source and injector development and to supply beams for lower energy experimental induction accelerators. As a consequence a number of the specifications have been refined. The original HTE injector could use ion masses of 20 to 40 AMU while we have focused on 27 for this machine. The flatness criteria on the beam energy and current are now treated as developmental goals. Higher currents per beam will be pursued later, as required.

The injector may be needed as early as 1989 for an Induction Linac System Experiment (ILSE). In this case an ion with a charge to mass ratio of about 10 is needed, for example C^+ or Al^{++}, rather than 20 to 40. The current desired would be the maximum transportable in an electrostatic quadrupole focusing system, or about 0.2 μC/m. For a 2 MV accelerator this would be about 1000 ma/beam of either C^+ or Al^{++}. These may require either larger diameter beams, or an alternative to aperture focusing in the higher energy portions of the injector. ILSE would require a shorter pulse, only 1μs. At this stage we are proceeding to test the injector at the design specifications and study the modifications to the source, electrodes, and pulsed power needed to achieve these objectives.

The injector design is shown in Fig. 1. Sixteen separate beams are accelerated electrostatically through apertures in flat electrodes inside a 1-m-long accelerating column of 85 percent alumina ceramic rings brazed to niobium grading rings.

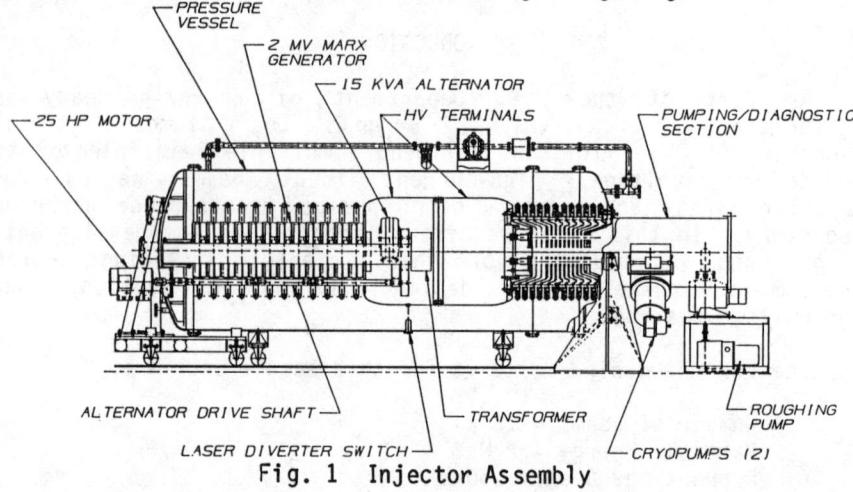

Fig. 1 Injector Assembly

The column voltage is created and maintained by a 2 MV Marx generator with a special circuit designed to control voltage pulse flatness across a water resistor voltage divider until diversion by a laser triggered spark gap. The individual sources are powered by a motor driven alternator contained in the Marx high-voltage dome. The accelerating column is made of aluminum oxide rings brazed to niobium high-voltage feedthrough rings. The Marx generator and the accelerating column are cantilevered from opposite ends of a pressurized containment vessel. The column, Marx, and high-voltage dome are insulated by SF_6 gas at 3 atm pressure contained in a stainless steel pressure vessel.

The injector is being completed in two phases. First a single beam will be accelerated to 1 MeV to test voltage standoff, pulsed power operation, and the vacuum arc source. The second phase, a multiple beam 2-MeV injector, will add more pulsed power, an additional column module, additional ion optics, and improved sources. We expect to begin testing at 1 MV in 1987 and complete Phase I by the end of 1987. We plan to complete Phase II in 1988.

ION SOURCE

The ion source for this injector is being developed by the University of New Mexico.[1] We have concentrated efforts on metal vapor vacuum arcs that have the following advantages: 1) rapid turn-on, 2) simplicity of fabrication and operation, 3) generation of clean plasma with little species contamination, 4) production of a wide variety of ions from metals and semiconductors, 5) potentially high ion flux, and importantly 6) drastically reduces the column heat loading compared to hot sources.

Our experiments have shown that this source meets our objectives. Both aluminum and indium plasmas have been generated. A biased grid electrostatically separates ions from electrons and prevents plasma prefill in the extraction gap. This allows rapid and quiet beam initiation. A 20 cm^2 anode produced peak current densities of 15 mA/cm^2 for Al^+ and In^+. Time of flight measurements indicated that beam ions were predominantly in the +1 ionization state with no observable species contamination. A divergence measurement with a channel-electron-multiplier array implies a normalized emittance of $\epsilon_n < 3 \times 10^{-7} \pi$-m-rad. This exceeds our design requirements and those for both the HTE accelerator and ILSE.

PULSED POWER SUPPLY

Several changes have recently been made in the pulsed power design[1] that improve the predicted performance.

Beam transport considerations require that beam energy flatness, out of the injector, be extremely good--the goal is ±0.05 percent. The largest perturbation on beam voltage is voltage droop of the Marx, due to the charge flow (current x time). The beam current is not large, but when current in the Marx charge resistors and trigger resistors is taken into account, the total charge flow is too large to be compensated by making the Marx capacitance larger because of weight limitations. In our initial design for the Marx, droop was compensated by putting negatively charged stages at the bottom of the Marx. Shunt resistors around the negative stages caused the voltage on these stages to droop at a relatively high rate, so that the total droop on the negative stages was equal in magnitude (but opposite in sign) to the droop on the rest of the Marx, thus compensating for the Marx droop. The disadvantage of this technique was that several negative stages were required to keep the negative exponential fall in a sufficiently linear region, thus requiring more stages to get the 2.0 MV desired. In the new design[2] the first stages are an LC inversion generator[3] with output voltage waveform of the form $(1-\cos(\omega t))$. The linear portion of this waveform can be used to compensate for Marx droop. This circuit gives better compensation than the RC decay circuit, and has the added advantage of adding to the output voltage, rather than subtracting from it.

The Marx triggering has also been changed. The size of the Marx inter-stage trigger coupling resistors is determined by a trade-off between better triggering and Marx voltage droop rate. The minimum value of a trigger coupling resistor that is required to keep the droop small is in the neighborhood of 20 kOhm. Testing of the Marx with this magnitude of trigger coupling gives extremely large delays and jitter. Past experience with this type of Marx generator indicates that trigger coupling resistors must be no larger than 1 kOhm. A new trigger coupling scheme, using a small resistance in series with a small (~1 nF) coupling capacitor in place of the large resistor, has been tried with very satisfactory results in a small model Marx generator. We will be trying this soon in the full injector Marx.

ACCELERATING COLUMN

The accelerating column[4] is made up of a stack of alumina rings brazed to niobium electrodes that feed the voltage from grading rings on the outside of the column to the accelerating electrodes inside. The voltage on the grading rings is determined by a liquid resistor voltage divider.

Recent analysis has shown that, because of the capacitance of the column to the tank walls, the voltage division along the column at the beginning of the pulse does not follow the resistive division with sufficient accuracy. The circuit model of the column used in a computer circuit analysis is shown in Fig. 2. The capacitances were calculated using an electric field analysis program. The resistive divider is designed so that, with 2.0 MV on the column, each of the 11 accelerating gaps has 175 kV except the gap nearest the anode, which has 75 kV for reasons having to do with the ion optics. The transient circuit solution gives 280 kV on the first (75 kV) gap, 230 kV on the next gap, and 100 kV on the gap nearest ground. These voltages approach the equilibrium values determined by the resistive voltage division with time constants on the order of 10-15 µs. These transient voltage errors cause three problems: the ion optics in the column will not be correct, with potential effects on beam emittance and divergence; the beam edge may hit an electrode; and the overvoltage on the first few gaps will increase the probability of electrical flashover.

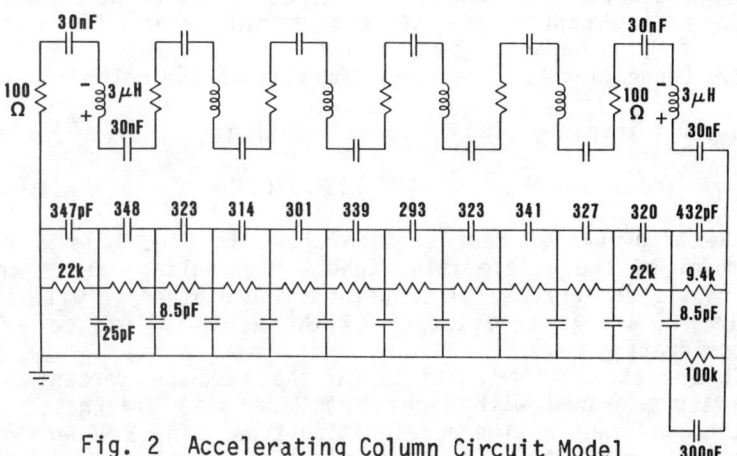

Fig. 2 Accelerating Column Circuit Model

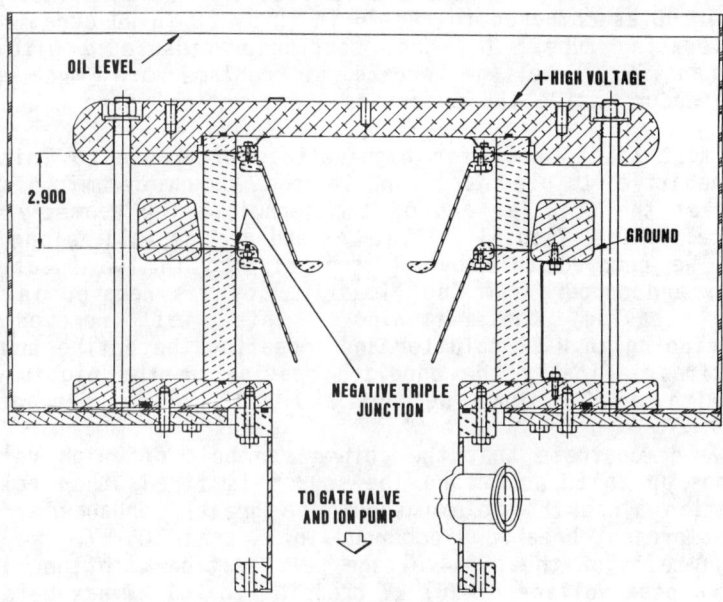

Fig. 3 High-Voltage Test Assembly

Several approaches have been considered to alleviate this problem. They include

1. Reducing the value of grading resistors and injecting ions from the source when the voltages reach equilibrium.
2. Increasing the series resistance in the Marx.
3. Adding capacitance between grading rings to compensate for the capacitance to ground.
4. Connecting Marx sections directly to the column.

We are still investigating the best solution.

HIGH VOLTAGE TEST SECTION

The high-voltage test section shown in Fig. 3 is a reduced size model of the accelerating tube. High-voltage tests on this model should be very useful in determining whether we will be able to hold off 175 kV across each of the 2.900-in-long ceramics in the accelerating tube.

This test section consists of the same 85 percent alumina rings vacuum brazed with Ticusil to the same thickness niobium feed through rings as the accelerating tube. The 2.900-in ceramic length and the electrode geometry for field shaping at the ceramic ID is also the same. However, the test section ceramics are 8-in ID by 10-in OD as compared to the 26-in ID by 28-in OD ceramics on the accelerating tube. The test section is insulated with oil rather than SF_6. Voltage breakdown problems are much more likely to occur inside the evacuated test section where electrons move freely.

The most likely spot for high-voltage breakdown to initiate is the negative triple junction at the vacuum-ceramic-niobium interface at the negative end of the ceramic. The geometry here is critical. The Ticusil fillet extended 2-mm all along the ceramic. We removed this by electrostripping in two solutions. The silver and copper from the Ticusil alloy was removed in NaCN solution, leaving the titanium. This was removed by electrostripping in NaCl solution and repeating the entire process several times. Finally the anodized coating on the niobium was removed with a 50 volume percent HNO_3, 10 volume percent HF solution.

If we demonstrate that the column can hold off high voltage for periods up to 50 µs before the source is fired, then voltage equilibration along the column should be greatly enhanced. With the fillet present breakdown occurred in 1 µs at 103 kV. We have just begun tests on the high-voltage test section with the fillet removed. A peak voltage of 267 kV drooping to 251 kV has held for 50 µs with essentially no voltage conditioning.

These test sections results are quite promising for the full accelerating column with much larger area. In some cases, voltage breakdown scales inversely with the one tenth power of electrode area. If this applies to ceramic columns and 50 µs pulses, then a voltage of only 250 kV is required on this small scale test to confirm voltage standoff on the 2 MV accelerating column, with 11 gaps at 175 kV. We exceeded this value.

REFERENCES

1. L. Len et al, Grid-Controlled Metal Ion Sources for Heavy Ion Fusion Accelerators, these proceedings.
2. Suggested by Ian Smith, Pulse Sciences, Inc.
3. R. A. Fitch, "Marx and Marx-Like High-Voltage Generators," Fourth Symposium on Engineering Problems of Fusion Research, Wash. DC (April 1977) 190-198.
4. E. O. Ballard, E. A. Meyer, H. L. Rutkowski, R. P. Shurter, F. W. Van Haaften, K. B. Riepe, "Design Status of Heavy-Ion Injector Program," 1985 Particle Accelerator Conference Proceedings, IEEE Trans. Nucl. Sci. NS-32, 1788 (1986).

CHARGE AND CURRENT NEUTRALIZATION PHYSICS OF A HEAVY ION BEAM DURING FINAL TRANSPORT

G. R. Magelssen and D. W. Forslund
Los Alamos National Laboratory, Los Alamos, New Mexico 87545

ABSTRACT

Heavy ion fusion requires high power to be focussed onto a small pellet. If the reactor chamber pressure is below 10^{-4}-10^{-5} Torr, beam compression will be limited by space charge unless neutralized by co-moving electrons. If higher chamber pressures are used, the heavy ion beam will create a significant number of background electrons during its propagation and will undergo stripping. The background electrons could provide the neutralization required for high beam intensities. In this paper we will focus on the physics associated with propagation through a fully ionized hydrogen plasma, so background electron generation is not included. One-dimensional electrostatic and two-dimensional fully electromagnetic particle-in-cell simulations are presented.

If a background plasma is present, we find that coinjected electrons whose purpose is to charge and current neutralize the ion beam become two-stream unstable and no longer provide the thermally cool neutralization required. Further, we find that the ion induced background electron temperature is very sensitive to the ion beam to background electron charge density ratio.

INTRODUCTION

Heavy ion fusion requires high beam power to be focussed onto a small pellet - intensities of 100-1000 TW/cm^2 may be required. If the reactor chamber pressure is below 10^{-4} - 10^{-5} Torr, beam compression will be limited by space charge unless neutralized by comoving electrons.[1] If higher chamber pressures are to be used, the heavy ion beam will create a significant number of background electrons during its propagation[2-3] and will undergo stripping.[4] The background electrons could provide the neutralization for high beam intensities.

Previous studies have avoided the background pressure region from about 10^{-4} - 10^{-1} Torr because the ion beam will become two-stream unstable.[5-7] However, recent analytical estimates that included ion beam stripping, ion space charge, beam convergence, and electron generation suggest that the two-stream instability growth is too small - the electric field is too weak - to cause a significant change in the beam's waist size.[8] Charge and current neutralization physics were not included.

In the past the neutralization problem was studied at pressures on the order of or greater than 1 Torr. As a result, the ion beam charge density was always significantly smaller than the background electron density except possibly at the front of the beam.[9-11] At background densities of 10^{-4} - 10^{-2} Torr, the beam charge density can become greater than and possibly significantly greater than the background electron density as the beam approaches the target. In this paper our primary purpose is to present results for this low pressure region of the heating of reactor chamber electrons by a heavy ion beam during its propagation to a target. We will not address the question of the amount of ion beam energy loss or the effect of the beam-generated electric and magnetic fields on the beam's waist. We focus on the electron temperatures generated. If the background electrons are to provide the neutralization required for a small focus at high ion currents, high electron temperatures must be avoided.[1]

Our approach was to first study the propagation problem in the electrostatic case to understand whether electron heating was significant and under what conditions. Our results suggested that heating could be important; so, two-dimensional fully electromagnetic simulations were done to determine whether the added dimension had a significant impact on the one-dimensional results. Two approximations were made to simplify the calculations. Neither the background electron generation nor the ion beam stripping were included in our simulations. The possible effects of these approximations are discussed in the summary.

The orientation of this paper is the following. First we present one-dimensional electrostatic simulation results that suggest significant electron heating can occur where the heating is examined as a function of the ion beam to electron charge density ratio. Results with a background plasma density gradient and with a comoving electron beam are also presented. This is followed by fully relativistic and electromagnetic two-dimensional simulation results that confirm the sensitivity of the background electron heating to the ion beam to background electron charge density ratio.

ONE-DIMENSIONAL SIMULATIONS

We present one-dimensional electrostatic simulations of electron heating by an ion beam under four different conditions. The simulation results were obtained using the Los Alamos code Wave, a two-dimensional, fully relativistic and electromagnetic code that self-consistently solves the Newton-Lorentz equations of motion and Maxwell's equations for ions and electrons with the particle-in-cell technique.[12] The presentation of results will be brief emphasizing qualitative aspects of the simulations. Details have been left for a later paper.

The first simulation is ion propagation through a dense plasma into vacuum. The electron and ion density were an order of magnitude greater than the beam's peak charge density. This high density plasma that is located at the left boundary exists in all of our simulations. Thermal electrons are continuously injected from the left boundary to maintain charge neutrality. The plasma represents a grid placed in the ion beam's path near the entrance port to the reactor chamber. In all simulations the heavy ion beam is injected from the left boundary, propagates in the x direction, and is absorbed at the right boundary. Electrons are absorbed at the right boundary while background ions are reflected at all boundaries. The boundary condition for the electric field was aperiodic, metallic-like. In all simulations the mass-to-charge ratio was 20,000 for the heavy ions and 1836 for the background ions; the injected ion charge density profile was

$$A\{ \exp(-t/t_2) - \exp(-t/t_1) \}^3 \qquad (1)$$

where A, t_1, and t_2 are constants. For all simulations the ion drift velocity was $v_d = c/3$ and the momentum spread 0.003 - a value of 0.001 gave similar results. There were 20 cells per wavelength assumed to be $2\pi v_d/\omega_{pe}$. This is the value associated with the maximum growth rate for the ion-electron two-stream instability. In all simulations, the t_1 and t_2 coefficients were 20 and 300, respectively. Figures 1 and 2 illustrate the electron heating at time 200. The front of the beam was located at the position 60. In all figures time is normalized to ω_{pe}^{-1}, length to c/ω_{pe}, and the relativistic particle momentum P to $m_e c^2$. The ω_{pe}, c, and m_e are the electron plasma frequency, the speed of light, and the electron mass, respectively. Substantial heating of electrons occurs at early times creating a high energy tail in the momentum distribution. The electron distribution (arbitrary units) versus momentum is given in Fig. 1.

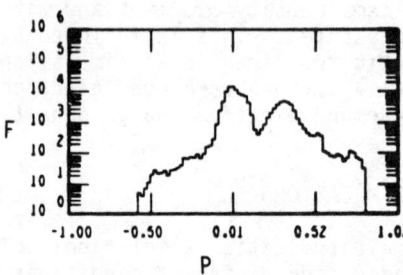

Fig. 1 Electron density F versus momentum P at t = 200 (vacuum).

This shows the electron return current with an average velocity about equal to and a thermal velocity greater than v_d. Figure 2 is a plot of the electron momentum in the x direction as a function of the distance x. A virtual cathode-like structure is evident as was seen and described by S. Humpheries et al.[13] However, the cathode structure exists only near the dense plasma; so, a different description of the downstream electrons is needed. Filling the vacuum region with a plasma with a density equal to 1/10th the peak ion beam charge density had little effect on the cathode structure, or the return current average and thermal velocity. Figure 3 is a plot analogous to Fig. 2 and illustrates the similarity in the cathode structure and heating for these two plasma conditions.

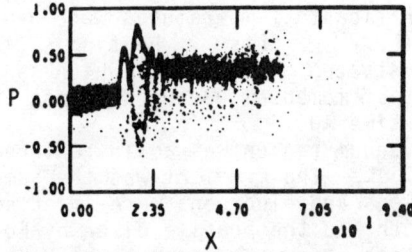 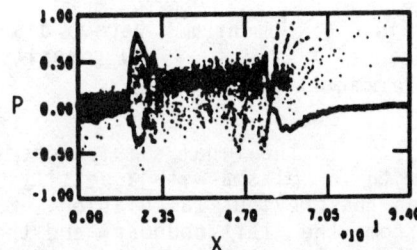

Fig. 2 Momentum P versus distance X at t = 200 (vacuum).

Fig. 3 Momentum P versus distance X at t = 200 (plasma one-tenth beam's peak density).

In the second simulation the vacuum region was replaced with a uniform plasma that was twice the beam's peak density. Figures 4 and 5 illustrate the dramatic change in the electron temperature. These plots are snapshots at 200 the time of Figs. 1 and 2. The electron charge density versus momentum is shown in Fig. 4. The return current and background distributions have merged making it difficult to determine an average momentum associated with the return current. Figure 5 displays the momentum versus distance. The cathode structure is no longer evident and the number of high energy electrons greatly reduced.

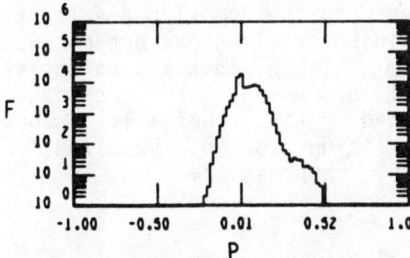 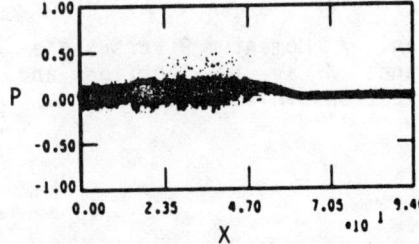

Fig. 4 Electron density F versus momentum P at t = 200 (plasma twice beam's peak density).

Fig. 5 Momentum P versus distance X at t = 200 (plasma twice beam's peak density).

In the third simulation the vacuum region was replaced with a plasma with a linear density gradient. The density was twice the beam's peak charge density at the interface with the dense plasma and a tenth this peak density at the right boundary. The density gradiant was used to gain some understanding of beam convergence in one dimension. Initially the background electron heating was similar to that of the previous simulation - fairly low temperatures prevailed. However, when the beam charge density became greater than the electron density, the electron temperature rose. By the time the beam reached the right boundary, significant temperatures were generated. This heating is illustrated in Fig. 6 which is a plot of momentum versus distance at the time 400.

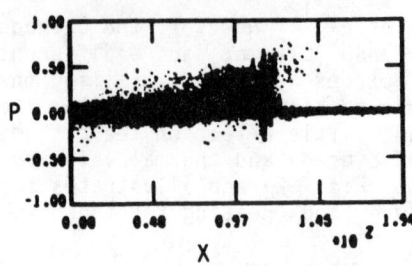

Fig. 6 Momentum P versus distance X at t = 400 (density gradient).

In the final simulation, the vacuum region was again replaced with a plasma with a density gradient. The gradient was the same as the previous calculation. Both ions and electrons were injected from the left boundary and they both had the profile given by Eq. (1). The temperature of the electron beam was chosen to be 2 KeV to minimize the electron-electron two-stream instability. If the injected electrons remain trapped in the ion beam, higher initial temperatures will not provide charge neutralization of a converging ion beam.[1] The electron and ion drift velocities were the same. The gradient was used to provide some understanding of the effects of beam convergence. For example, both the electron-electron and the ion-electron two-stream instability are resonant processes and can be modified by density gradients in either the beam or background plasma. Figure 7 illustrates what happens to the comoving electrons. The injected electrons became two-stream unstable, lost a substantial amount of energy, and no longer provided a thermally cool neutralizing current required for ion beam compression.

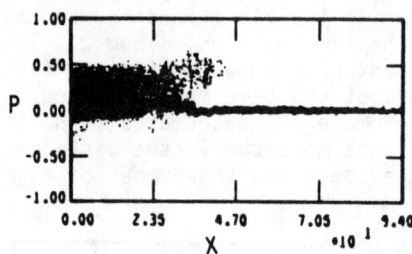

Fig. 7 Momentum P versus distance X at t = 120 (ions and electrons injected).

TWO-DIMENSIONAL SIMULATIONS

The two-dimensional simulations had the same boundary conditions in the x-direction as the one-dimensional calculations. The fields and particles had periodic boundary conditions in the y spatial dimension. The injected ion beam had a gaussian density profile in y and the density profile of Eq. (1) in time. The t_1 and t_2 were 20 and 300, respectively. The beam propagated in the x-direction with a speed of c/3 and a momentum spread of 0.003. For the simulations presented significant electric fields existed in the x and y directions as well as a magnetic field in the z direction. The results from two different plasma conditions are presented. As in the one-dimensional calculations, a dense plasma was placed at the left boundary in the x dimension.

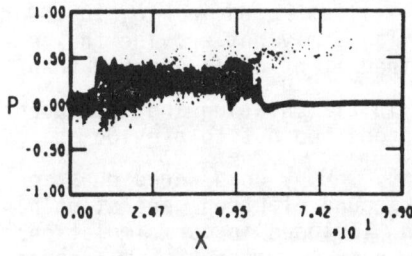

Fig. 8 Momentum P versus distance X at t = 200 for 2-D simulation (plasma one-tenth beam's peak density).

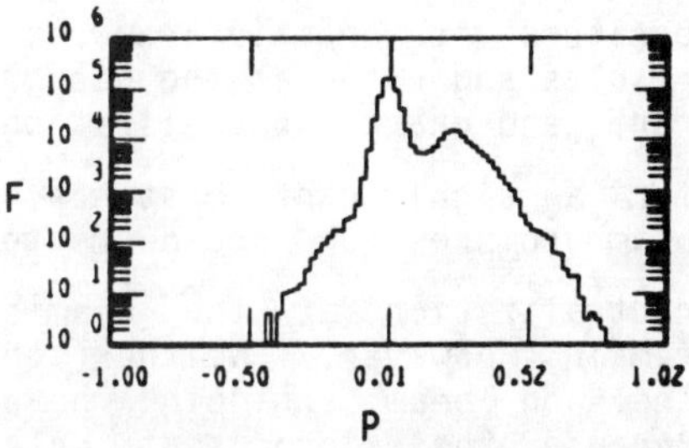

Fig. 9 Electron density F versus momentum P at t = 200 for 2-D simulation (plasma one-tenth beam's peak density).

In the first simulation a uniform plasma one-tenth the beam's peak density was located to the right of the dense plasma. Figure 8 shows the plasma condition analogous to Fig. 3. The cathode structure is again evident as well as significant electron heating although somewhat reduced from the one-dimensional result. The electron density as a function of momentum is given in Fig. 9. The thermal velocity associated with the return current is the order of or greater than v_d. In the second simulation the uniform plasma density was increased to twice the beam's peak density.

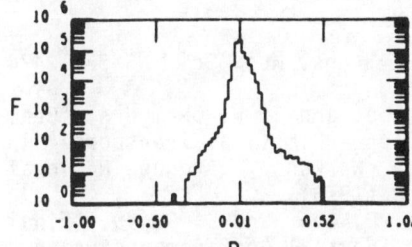

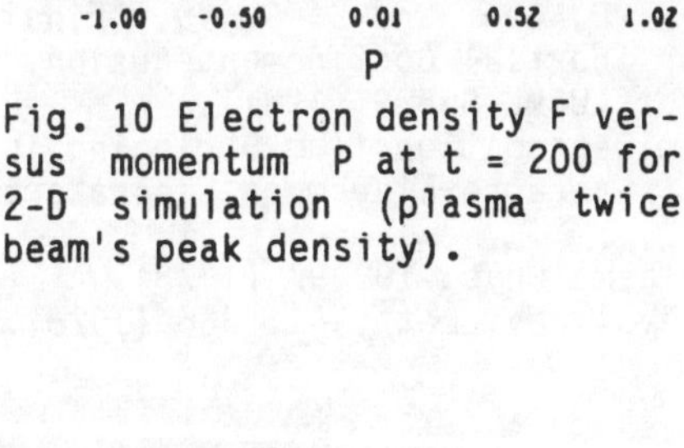

Fig. 10 Electron density F versus momentum P at t = 200 for 2-D simulation (plasma twice beam's peak density).

beam's peak charge density. As in the one-dimensional calculation, a dramatic reduction in the electron temperature occurs. A plot analogous to Fig. 4 is depicted in Fig. 10.

SUMMARY

Our results suggest that if a plasma is present, coinjected electrons whose purpose is to charge and current neutralize the ion beam become two-stream unstable, become hot, and no longer provide the thermally cool electrons required. We also find that the ion beam induced background electron temperature is very sensitive to the ion beam charge density (n_b) to background electron density (n_e) ratio reaching a temperature on the order of or greater than the ion's drift speed for $n_b/n_e = 10$. At a ratio of 1/2, the temperature is greatly reduced as is illustrated in Fig. 10. Our results suggest that the background plasma may not provide the current and charge neutralization required if n_b/n_e is greater than 1 for a significant distance. The initial heating and subsequent beam compression could create electrons too hot to provide charge neutralization at the beam's waist.[1] Many unanswered questions remain, however. Neither the background electron generation nor the ion beam stripping have been included in our simulations. Because these processes determine n_b/n_e, without them accurate predictions of the electron temperatures will not be possible. A time and spatial variation of n_b/n_e could also modify the streaming instability growth rates. Finally, a simulation of a two-dimensional converging beam needs to be done to determine if electrons initially trapped remain trapped as the beam compresses.

REFERENCES

1. D. S. Lemons and M. E. Jones, IEEE Trans. Nuc. Sci. NS-32, 2474 (1985).
2. S. S. Yu, H. L. Buchanan, E. P. Lee, and F. W. Chambers, "Beam Propagation Through a Gaseous Reactor - Classical Transport," in Proceedings of Heavy Ion Fusion Workshop, Argonne National Laboratory report ANL-79-41, p. 403 (1979).
3. W. A. Barletta, W. M. Fawley, E. P. Lee, and S. S. Yu, "Final Transport of Heavy Ion Beams for Inertial Confinement Fusion," Lawrence Livermore Laboratory report UCRL-85689 (1981).
4. W. A. Barletta, "Heavy Ion Beam Degradation from Stripping in Near Vacuum Reactor Chambers," Lawrence Livermore Laboratory report UCRL-85651 (1981).
5. S. Jorna and J. B. Thompson, J. Plasma Phys. 19, 97 (1978).
6. R. F. Hubbard and D. A. Tidman, Phys. Rev. Lett. 41, 866 (1978).

7. Y. K. Kim and G. R. Magelssen (editors), Report on the Workshop on Atomic and Plasma Physics Requirements for Heavy Ion Fusion, Argonne National Laboratory report ANL-80-17 (1980).
8. P. Stroud, Laser and Particle Beams $\underline{4}$, 230 (1986).
9. D. A. Tidman, ERDA Summer Study of Heavy Ions for Inertial Fusion, Lawrence Livermore Laboratory report LBL-5543, p. 52 (1976).
10. D. Mosher, Ref. 9 p. 59.
11. P. F. Ottinger, D. Mosher and Shyke A. Goldstein, Phys. Fluids $\underline{24}$, 164 (1981).
12. R. L. Morse and C. W. Nielson, Phys. Fluids $\underline{14}$, 830 (1971).
13. S. Humphries Jr., T. R. Lockner, J. W. Poukey, and J. P. Quintenz, Phys. Rev. Lett. $\underline{46}$, 995 (1981).

BEAM TRANSPORT EXPERIMENTS USING A RFQ CHANNEL[*]

N. Zoubek, K. Langbein, P. Junior, A. Schempp, H. Klein, G. Riehl
Institut für Angewandte Physik, Universität Frankfurt/M.
Robert-Mayer-Str. 2-4, D-6000 Frankfurt am Main 1, FRG

ABSTRACT

A beam transport experiment has been built up, where a He ion beam is transported in a long RFQ channel with electrodes, which are not modulated. In these experiments parameters as input current, particle energy, injection geometry and electrode voltage (corresponding to different phase advances) can be varied. The maximum beam transmission depends strongly on proper adjustment of the injection system to the quadrupole axis as well as on space charge conditions. Especially at high phase advances, where coupling resonances are generated by space charge, mismatch leads to significant emittance growth. First results will be reported.

INTRODUCTION

The transversal current limit of a RFQ channel was investigated using a four rod resonator[1], where the electrodes were not modulated. Fig. 1 shows the arrangement. The beam transport system consisted of a duoplasmatron ion source with accel-decel extraction, which for beam matching purposes was followed by two einzel lenses. At both their ends the RFQ electrodes were provided with matching fingers, a Faraday cup and an emittance measurement equipment[2] served for beam diagnostics.
The rf-voltage determining the transverse phase advance σ_o could be varied such that σ_o ranged from $0°$ to $180°$, and the beam current turned out maximal at about $90°$. Emittance growth was measured at rather large phase advances and could be based on imperfect matching conditions. Corresponding slopes of beam currents as functions of electrode voltage could be explained by making simple beam dynamical assumptions.

CHOICE OF PARAMETERS

The maximum beam current in a periodic quadrupole channel is given by Reiser's formula[3,4]

[*]work supported by BMFT

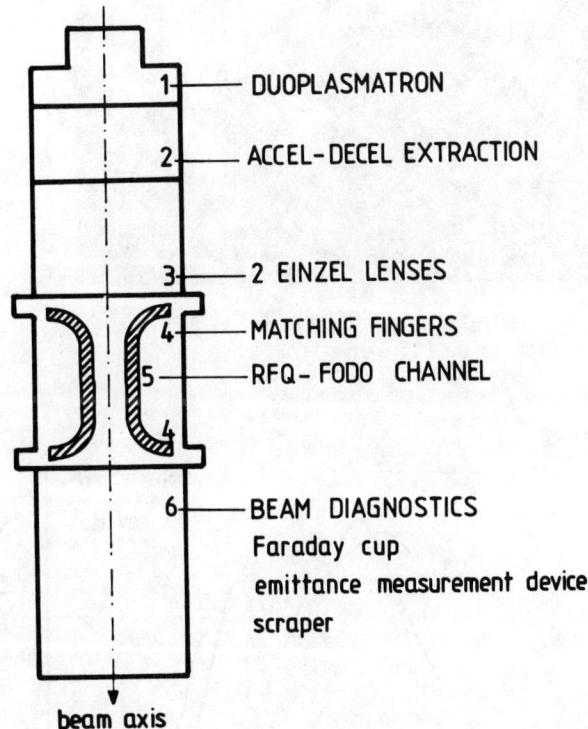

Fig.1 Scheme of the RFQ Transport Experiment

Fig.2 Spiral loaded Transport RFQ with matching fingers

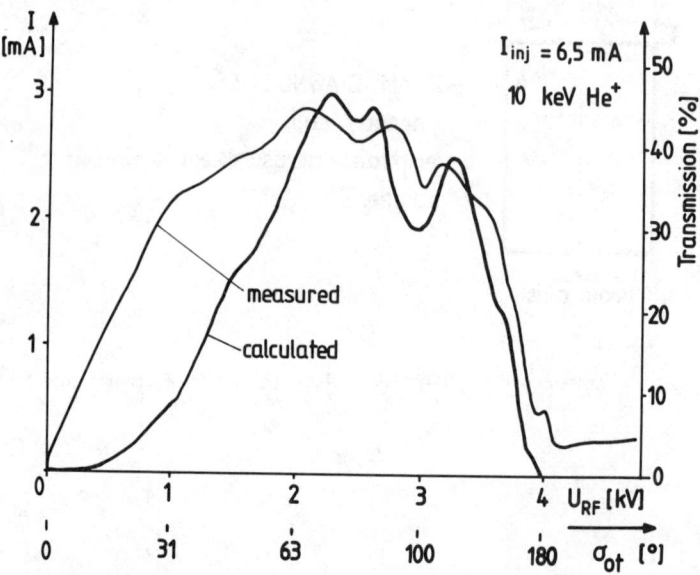

Fig.3 Transmitted current as function of rf-voltage resp. phase advance

$$I = I_0 \frac{\omega^2 \bar{v r}^2 \sigma_0^2}{8c^3} (1 - \frac{\sigma^2}{\sigma_0^2}) \text{ with } I_0 = \frac{4\pi\varepsilon_0 mc^3}{e} \quad (1)$$

where $\bar{r}$ stands for the mean beam radius, σ_0 for the zero current phase advance per rf-cycle, v for particle velocity, $f=\omega/2\pi$ for the frequency, and σ/σ_0 for the tune depression. In order to bring the current limit $I(\sigma=0)$ down to values obtainable with the ion source, our choices were f=16.5MHz, which requires spiral supports for the RFQ electrodes (Fig. 2), a 300 W transmitter corresponding to an electrode voltage of more than 4kV for $\sigma_0 = 180^0$. This low power necessary is due to the high Rp-value (500kΩ) of the resonator.
Further choices were He^+-ions, where the perveance of the ion source was $0.5 \cdot 10^{-8}$ A/V$^{3/2}$. H^+ or H_2^+ ions proved unfavourable, because it was impossible to extract a clean beam with uniform specific charge. The current limit of Eq.(1) then amounts to about 10mA at $\sigma=0$ and $\sigma_0=90^0$. Table 1 lists the experimental parameters.

Table 1: Experimental parameters of the RFQ channel

Resonance frequency (MHz)	16.5
Rp-value (kΩ)	500
Aperture radius (mm)	4.3
Resonator length (cm)	65
Cycles at 10 keV He$^+$ ($\beta\lambda$)	12
matching in/out sections ($\beta\lambda$)	2

MATCHING

In order to match the cylindrical dc beam to the time dependent and extremely astigmatic injection conditions of the RFQ, the field gradient must increase from zero to the final value, thus the distance of the electrodes along the matching section from the optical axis should accord as well as possible to this demand.
We investigated optimum lengths for either emittance or space charge dominated beams at several slopes of the field gradient and at different phase advances of the periodic channel[5]. Here the following assumptions were made:
1. the charge distribution is KV[6]
2. currents and acceptances match to the corresponding currents and emittances of the ion source.
Optimum injection conditions were determined by transformation of the matched beam envelopes of the periodic channel backwards to the input of the matching system. Results were tested with the PARMTEQ code[7].

At small phase advances σ_0 a matcher length of $1\beta\lambda$ was sufficient for a transmission of almost 100 %. At larger σ_0, $1\beta\lambda$ still proved optimal, however, for the space charge dominated beam, there was a decrease in transmission, e.g. 50 % at $\sigma_0 = 90^0$ On the other hand, the emittance dominated beam showed improved transmission with longer matchers, 2 or $3\beta\lambda$, for instance. This behaviour of the space charged dominated beam is due to it's increased diameter and corresponding losses, when outer particles touch the matching fingers. As the orientation of the rotating acceptance ellipses, depending on the rf phase at the input of the periodic system, does not depend on space charge, the consequence is that the mismatch is principally caused by the matcher.

EXPERIMENTS AND CALCULATIONS

In Fig. 3 the measured current is plotted versus the rf voltage resp. σ_0. Evidently maximum transmission turns out with $\sigma_0 = 90^0$. The calculated curve gives the result by solving the KV envelope equations with respect to particle losses along the channel by an input emittance of 180 mm mrad. Especially at higher phase advances both curves agree well in regard to maximum current as well as the slope. The "round" form of the overall transmission curve indicates that the transports were associated with losses. Thus Reiser's formula, Eq.(1), valid for matched beams without loss, predicts larger currents than observed here.

Fig. 4 demonstrates a further experiment, where (before injection) the beam was collimated thin filamentlike by screening with a diaphragma aperture. This beam practically showed no losses up to the boundary $\sigma_0 = 180^0$, corresponding to an electrode voltage of 4kV. Measured and calculated beam emittances at different rf voltages are presented in Fig. 5. Fig. 6 shows distinctions in transmission behaviour of H^+, H_2^+, H_3^+, and He^+ ions, the lighter the ion the larger it's σ_0 and the earlier the loss.

Then a "scraper" experiment was performed, i.e. a metal shield as beam collector was held stepwise into the beam behind the RFQ. The collected scraper current was measured and calculated. Fig. 7 shows characteristic sequences of maxima and minima. Different voltage resp. phase advance results in different numbers of betatron oscillations of the center of the imperfectly matched beam and the envelope ripple. Let us assume the scraper registrates beam particles only far away off the beam axis (position 0.4cm in Fig. 7). So when the beam center is near the axis or the ripple is near a minimum the scraper current is smaller compared to the case, when the beam center is further off the axis or the ripple is near a maximum.

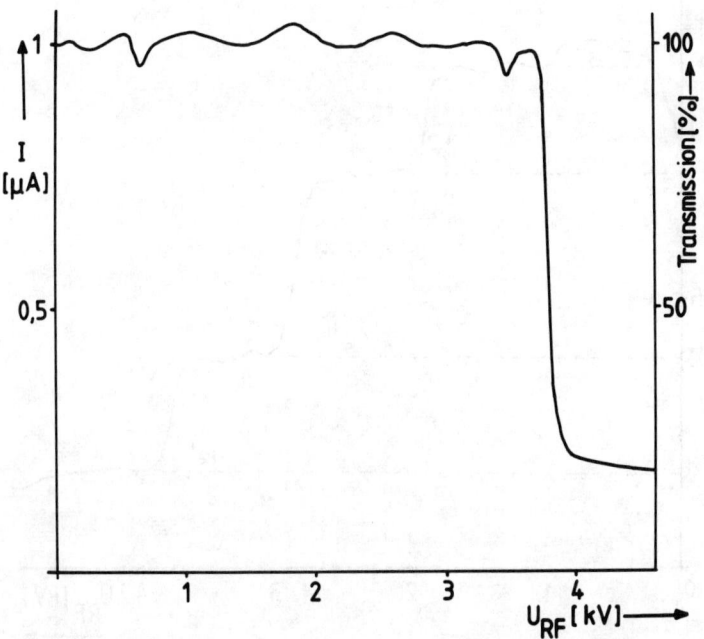

Fig.4 Transmission curve of "pencil" beam

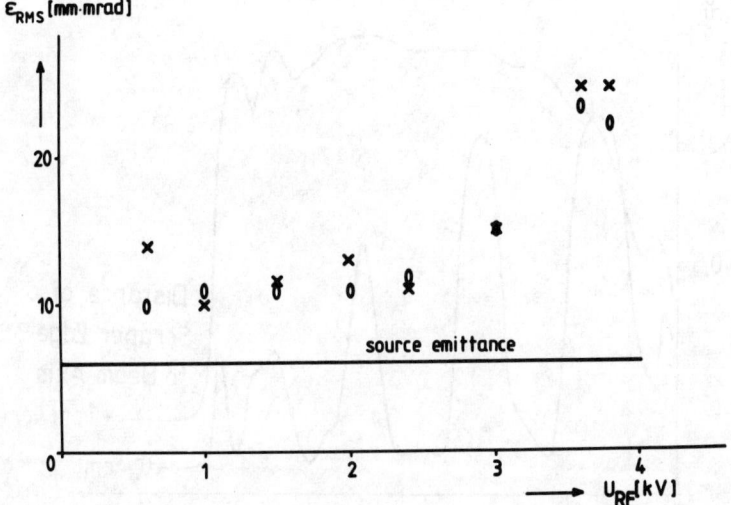

Fig.5 Measured (o) and calculated (x) emittances of a "pencil" beam

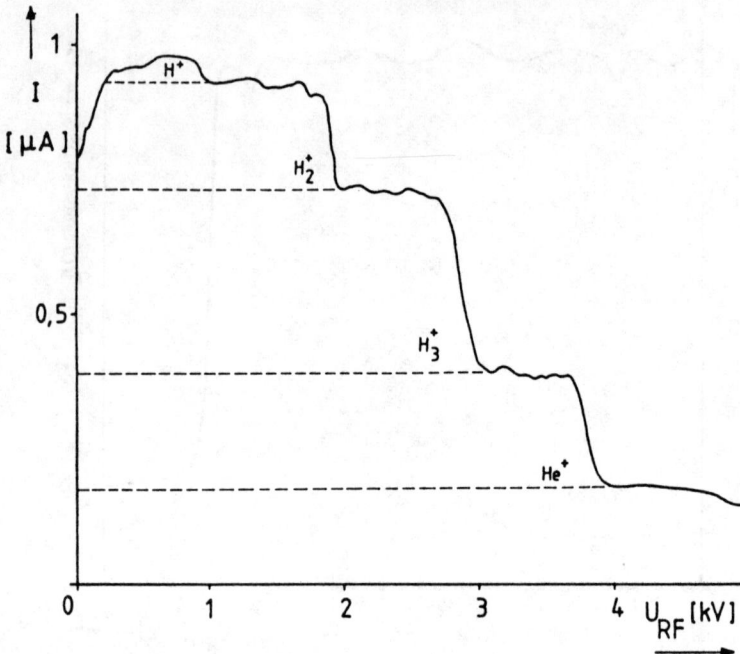

Fig.6 Total current of a mixture of ions

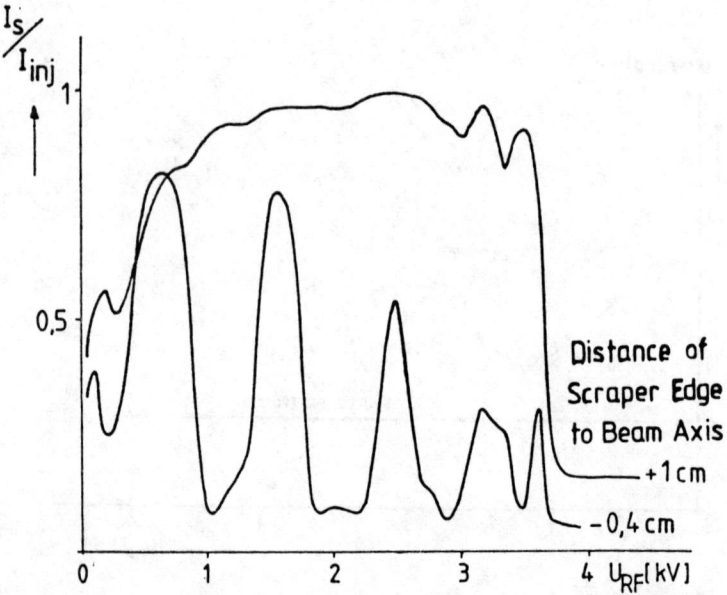

Fig.7 Ratio of scraper current I_S to injected current I_{inj} as function of rf-voltage at two different scraper positions

SUMMARY

In a RFQ channel with 12 cycles the measured emittance growth and low transmitted beam currents, smaller than calculated with Reiser's formula especially at large phase advances, are explained by principal imperfect beam matching. Due to this inconvenience we have not yet arrived at our destination, namely to investigate instabilities experimentally at $\sigma_0 > 90°$ and $\sigma/\sigma_0 < 0.4$.

REFERENCES

1. A. Schempp, H. Klein, H. Deitinghoff, P. Junior, M. Ferch, A. Gerhard, K. Langbein, N. Zoubek, Proc. 1984 Linac Conf. Seeheim GSI-84-11 (1984) p.100.

2. G. Riehl, Diploma Thesis, Univ. Frankfurt 1985.

3. M. Reiser, Particle Accelerators 8, 167 (1978).

4. P. Junior, Particle Accelerators 13, 231 (1983).

5. R. Becker, N. Zoubek, Proc. 1984 Linac Conf. Seeheim GSI-84-11 (1984), p.342.

6. I.M. Kapchinskij, W. Vladimirskij, Proc. Int. Conf. High Energy Accelerators CERN, Geneve 1959, p.274.

7. W. Neumann, Int. Rep. 83-25 Inst. f. Angew. Phys. Frankfurt, 1983.

RFQ DEVELOPMENT AND SPARKING EXPERIMENTS IN FRANKFURT[*]

A. Schempp, H. Klein, M. Ferch, A. Gerhard, M. Kurz
Institut für Angewandte Physik, Universität Frankfurt/M.
Robert-Mayer-Str. 2-4, D-6000 Frankfurt am Main 1, FRG

ABSTRACT

The Four-Rod-RFQ-structure has been developed in Frankfurt as a heavy ion fusion accelerator. The structure as well as new experimental results concerning the sparking limit are reported.

INTRODUCTION

For heavy ion fusion, ions with low e/m, low velocity, and high beam current are accelerated in a tree-like accelerator system.

For the first branches, RFQ-structures will be used because they offer very attractive features: simultaneous bunching, acceleration and focusing of high intensity ion beams at low velocities. So in all new high-current accelerator designs a RFQ-structure is applied.

For protons and light ions the 4-vane RFQ is the structure mostly used; frequencies as low as 80-100 MHz as chosen in FMIT and TALL seem to be the lower limit however. For low charged heavy ions a frequency of 10-30 MHz must be chosen which results in a too large diameter for the 4-vane resonator. At GSI, a Split Coaxial heavy Ion RFQ is being built for U ions[1].

FOUR-ROD RFQ

We developed the Four-Rod structure which is a simple alternative to the 4-vane for protons and light ions and can be applied for low charged heavy ions too[2]. The resonator's basic cell consists of two $\lambda/2$ oscillators excited in the transversal π-mode to give the proper quadrupole field distribution between the electrodes. The accelerating structure consists of a chain of these cells operating in longitudinal O-mode. Fig. 1 shows two cells of a linear version of this structure. Optimizing this structure with respect to shunt impedance results in equidistant arrangement of the stems. The shunt impedance η of the Four-Rod RFQ is astonishingly high and especially favourable for high frequencies. Fig. 2 shows the optimal η as a function of the operating frequency. For comparison the values of existing 4-vane RFQs are added.

[*] Work supported by BMFT

TABLE 1

Parameters of the Four-Rod RFQ Prototype

resonance frequency	202.56 MHz
beam energy	18 - 750 KeV
rod-to-rod voltage	70.5 KV
rf-power	50 KW
Rp-value	100 KΩ
quality-factor	3500
duty-factor	$2.5 \cdot 10^{-5}$
repetition frequency	1 Hz
beam pulse-length	35 μs
beam current (33% of Imax)	20 mA
tank diameter	25 cm
structure length	117.7 cm
number of $\beta\lambda/2$ cells	133
average aperture-radius	5 mm
synchr. phase	$-90°$ ... $-30°$
modulation factor	1 ... 1.88
focusing parameter	6.5

TABLE 2

Parameters of the CRNL Four-Rod RFQ

resonance frequency	108.5 MHz
Rp-value	250 KΩ
quality-factor	6000
tank diameter	50 cm
length	94 cm
apertureradius	8.5 mm
rod-radius	6.4 mm
max. el. field strength (200 KW)	30 MV/m

TABLE 3

Parameters of Kilpatrick voltage (U_{KP}), breakdown voltage (U_B) and gap length (gap) for the different gap spaces and -geometries.

U_{KP} (KV)	U_B (KV)	gap (mm)	electrode geometry (see fig. 8)
50.26	136	4	1a
50.26	145	4	1b
50.26	131	4	1c
60.00	170	5	1a
60.00	201	5	1c
60.00	180	5	1d
71.40	205	6	1a
71.40	316	6	1b
71.40	290	6	1c

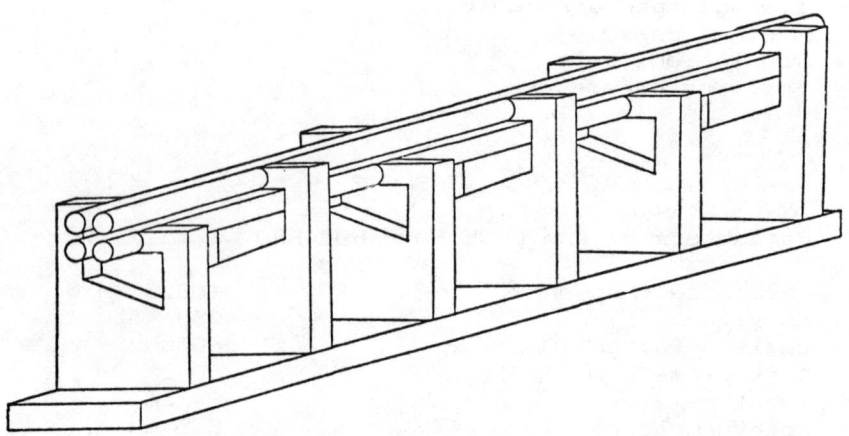

Fig. 1. Four-Rod $\lambda/2$ RFQ-resonator with linear arrangement of the stems.

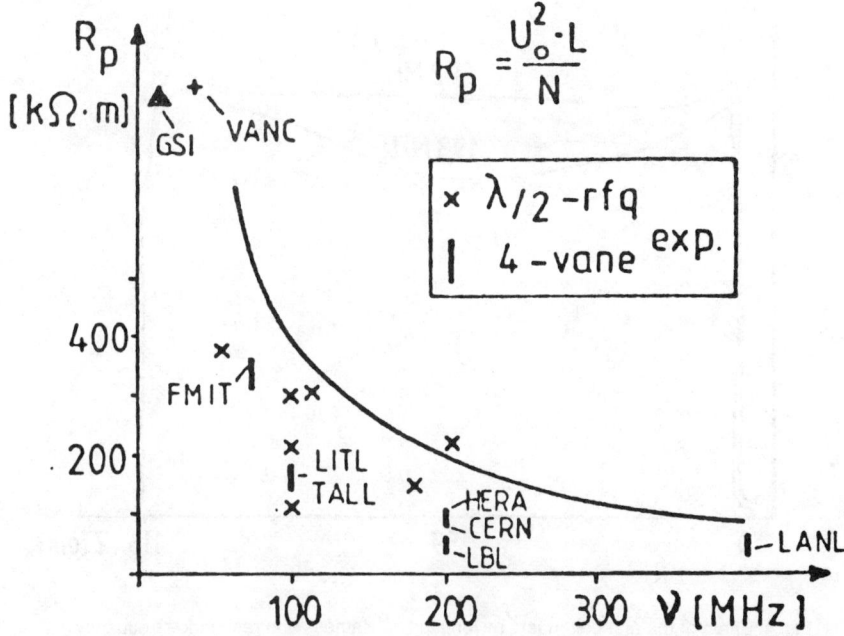

Fig. 2. Optimal shunt impedance of the Four-Rod RFQ as a function of operating frequency.

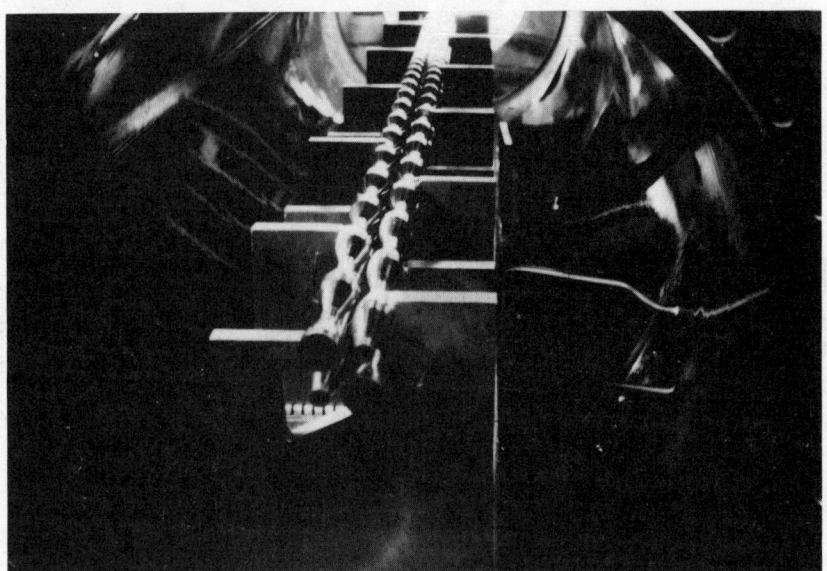

Fig. 3. High power prototype Four-Rod mounted in the vacuum tank.

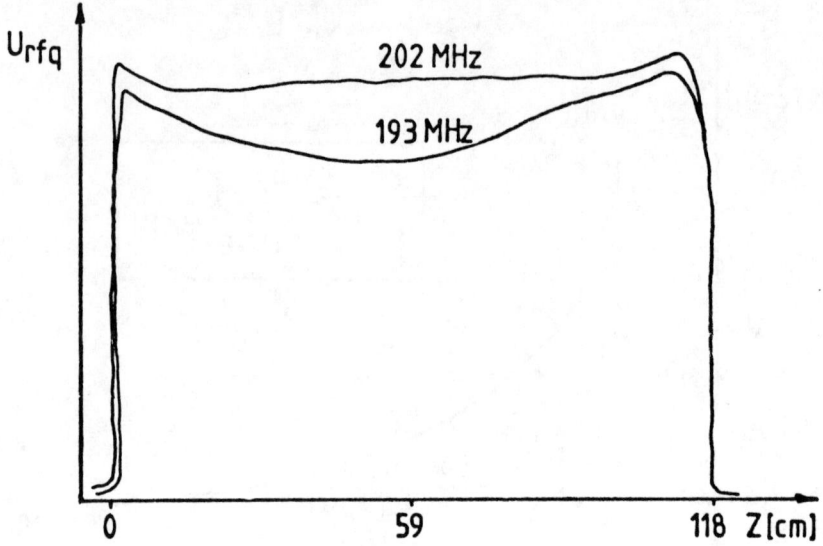

Fig. 4. Effect of the end-cell-tuners on longitudinal flatness and resonance frequency.

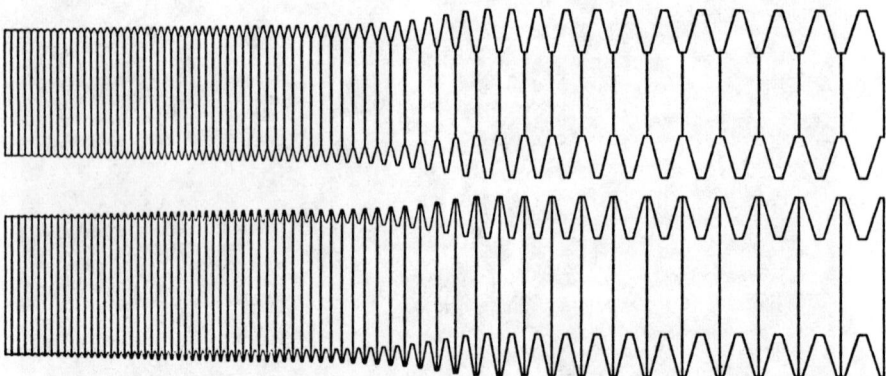

Fig. 5. Trapezoidally modulated electrodes of the Four-Rod light ion prototype RFQ.

A prototype for light ions has already been built. The beam dynamic parameters of the 4-Vane RFQ for HERA[3] had been taken for this Four-Rod structure too, which enables a perfect comparison of the structure's characteristics and the beam behaviour. Table 1 shows the general parameters of this prototype RFQ which has been built at our institute in Frankfurt (see Fig.3). The structure consists of a massive water-cooled ground bar with fourteen specially shaped stems to carry the quadrupole electrodes. The copper plated vacuum tank is provided with a precise steel base plate with again fourteen bores so that the structure can be mounted easily from the outside. Cooling water is supplied through two of the screw bolts at both ends. Frequency and longitudinal flatness-tuning can be achieved in one operation by varying the inductance of the end cells with adjustable tuning bars between the stems. Fig. 4 shows the effect which has been measured with unmodulated electrodes. In Fig. 5 a schematic of the modulated electrodes is shown; the sinusoidal shape was approximated by trapezoidal segments which can be realized on a lathe relatively easy.

In first rf-tests with 1.5 KW cw rf-power no problems occured. After high power tests in pulsed operation up to the design voltage the Four-Rod RFQ will be tested in the HERA-beamline in Hamburg.

A high power test cavity with a Four-Rod structure has been built by the Chalk River National Laboratory, Canada. Because of the direct cooling of the electrodes and stems the Four-Rod structure is a promising candidate for the breeder cw-accelerator programme[4]. The resonator which has been assembled in Canada was sent to Frankfurt recently. Preparations for the forthcoming high power tests with our 108 MHz power-transmitter up to 200 KW have almost finished. (Parameters see Table 2).

In a first step we will examine properties of cw operation. With a shorter version we then might be able to increase the field strength far over two times the Kilpatrick limit to study sparking in this configuration.

A spiral loaded Four-Rod structure with a frequency of 17 MHz is used in our beam transport experiment[5], and it is in discussion now to employ such a structure (27 MHz) as a second stage after the first stripper in the planned high current injector system at GSI .

The Four-Rod is the only RFQ-structure which can accelerate several beams in parallel in one tank by combining e.g. two identical resonators to provide two or four beamlines (Fig.6) and make a direct use of funneling[7].

Since it has certain advantages like good efficiency, compact dimensions, flexible rf-design, relatively simple

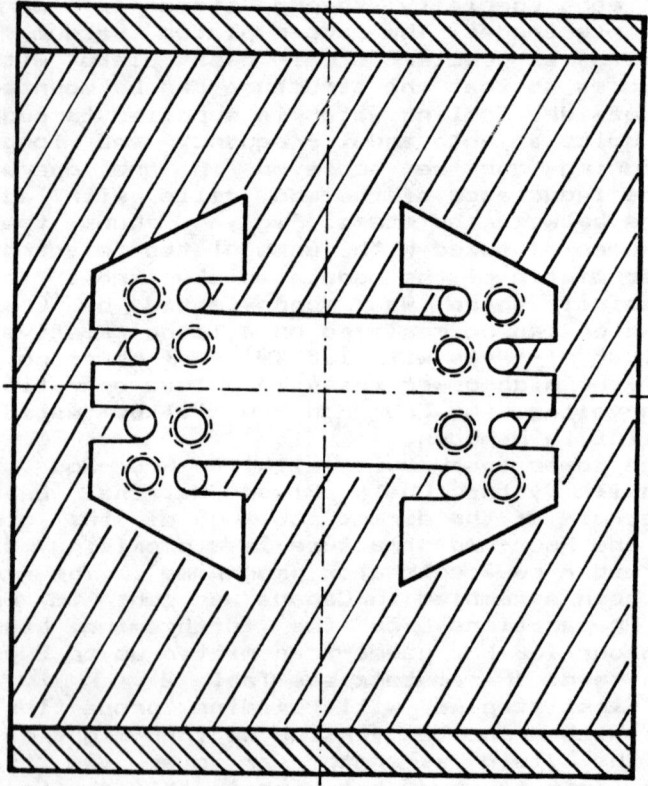

Fig. 6. Possible Arrangement of a multi-beam accelerator using Four-Rod resonators.

manufacturing and easy high duty factor operation this new structure offers a wide range of possible application.

SPARKING EXPERIMENTS

Theoretical investigations show in first approximation that the maximum ion beam current for an RFQ is only dependent on the maximum applicable electrode voltage which is limited by sparking [8]. For most applications high rf fields well above the Kilpatrick criterion[9] are required. Operating at these high fields makes it necessary to investigate a reliable breakdown threshold.

All experiments have been done in a 108.5 MHz quarterwave coaxial resonator[10], where the test electrodes have been brazed to the inner conductor and the tank end plate. The gap width can be adjusted from outside the tank without breaking the vacuum, the resonance frequency is tunable into the transmitter frequency of 108.5 Mhz by two slow tuners. The elcetrodes were mechanically polished and ultrasonically cleaned with methanol. For optical observation of the gap region a TV-camera and a video recorder were used. Furthermore a quadrupol mass spectrometer was attached to determine the concentration of different gas components in dependence of the gap voltage[11], pulse length and duty cycle.

The results showed that the Kilpatrick[9] threshold indeed could be exceeded considerably. The experimental results of the breakdown voltage are depending on the gap width, the electrode geometry, the pulse length and the duty cycle. The experiments with three different electrode geometries have shown an increase of the breakdown voltage from plane to cylindrical electrodes[12]. An increasing pulse length reduces the breakdown voltage. For greater gap width the breakdown voltage comes closer to the Kilpatrick threshold. The results are plottet in a Kilpatrick diagram (see Fig.7).

In all experiments the electrodes were made of standard industrial OFHC copper. For machining a CrCu alloy (components 99.17 % Cu, 0.75 % Cr, 0.08 % Zr) is far better suited than copper, but sparking tests with electrodes made of this material showed a decrease of breakdown voltage of about 10%. Another experiment with the same material, but rectangular, plane electrodes - as they are used in the Meqalac accelerator, gave the same result[13]. To find out the final design of new Meqalac cavities and to study the influence of sharp and rounded edges sparking tests have been done with a set of flat electrodes. The rounded edges (radius 2mm) improved the breakdown threshold by roughly 10%[14], see Table 3 and

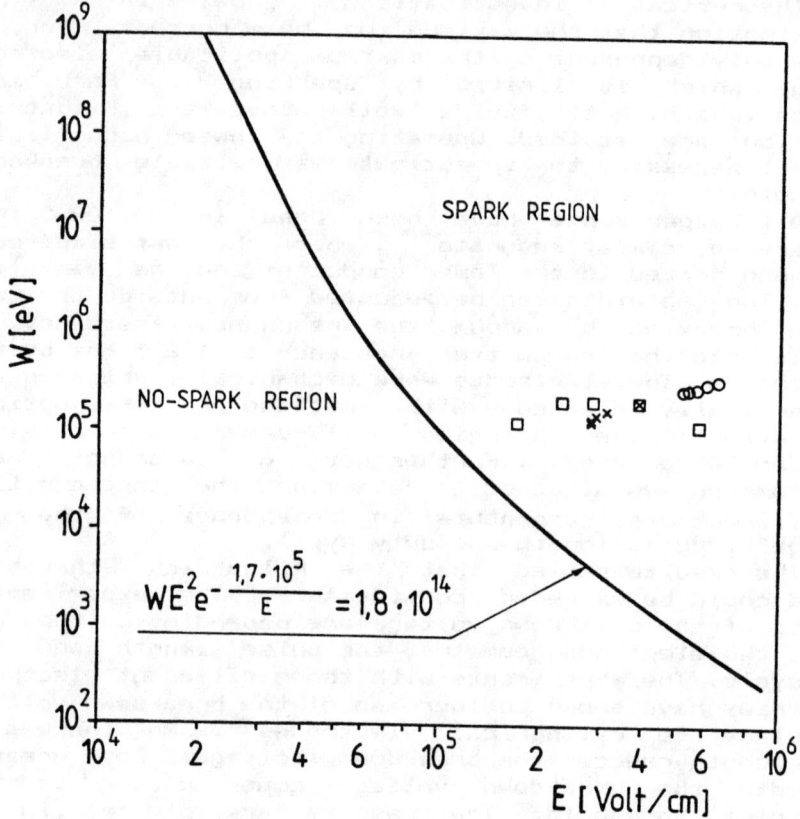

Fig. 7. Comparison of our measurements with Kilpatrick criterion. W is the maximum ion energy at the cathode, including the transit time factor given by the Kilpatrick versus the cathode gradient E. Pulse length τ 1,2,3,4,5 msec, repetition freq. 45 Hz. O = rods, 0 1 cm, gap space 0.5 cm, X = disks, 0 7 cm gap distance 0.5 cm, □ = disks, 0 7 cm, pulse length 1 msec, gap 0.2, 0.5, 0.7, 1.0, 1.8 cm, rep. freq. 45 Hz.

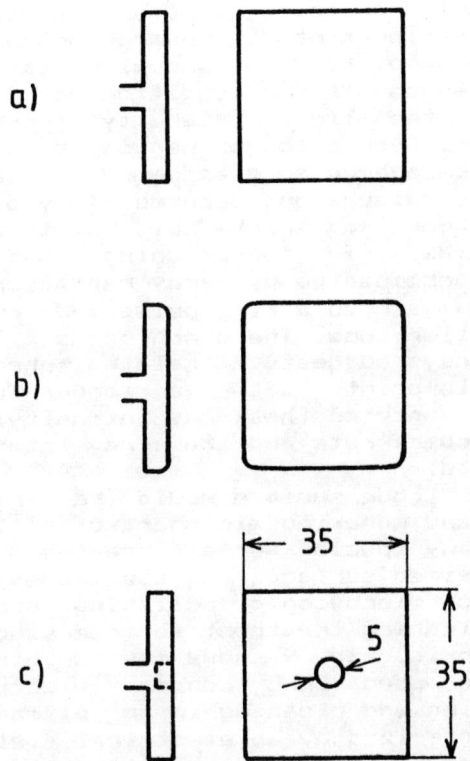

Fig. 8. Set of plane electrodes with an area of 35×35 mm electrode material OFHC copper. a) sharp edges b) round edges, edge radius 2mm c)sharp edges with a beam aperture 0 5 mm.

Fig.8.

The electrodes were carefully conditioned. Conditioning[15,16] started at very low power levels, pulsed and cw operation was increased by small increments. The experiments showed that by this kind of conditioning the sparking limit could be considerably increased. During conditioning small glowpoints[17] appeared on the electrode surfaces. These glowpoints were randomly distributed and their location changed after reconditioning. There was a wide variation in the light intensity increasing with rising power level. Over a longer period of conditioning the glowpoints disappeared more and more. In addition to the glowpoints microdischarges occured. They did not lead to a voltage breakdown (as sparks do), but to an enhanced H pressure. The rf conditioning and breakdown measurements are accompanied by x-ray radiation. Up to now the radiation was measured during pulse and cw operation with a Geiger-Mueller tube. The synchronous appearance of glowpoints and x-rays suggests a related phenomenon but the reducing of glowpoints after a longer conditioning period has no influence on the x-ray intensity. The time relation between spark rate and the x-ray intensity could not yet be measured.

The ideal electrode surface would be free of both microprotrusions and adhering microparticles[18]. Until now we have not done any special surface treatment. In order to achieve the desired surfaces the electrodes should be first subjected to microscopic polishing procedure and then to a final cleaning treatment for removing all traces of superficial debris. Fig. 9 shows the original surface finishing of the used OFHC copper electrode after mechanical polishing and cleaning in an ultrasonic bath. Sparking is characterized by an electrical field breakdown and a flash between the electrode surfaces and produces an erosion with craters (dimension .1 mm), see Fig. 10.

The mass spectra of residual gases in our experiment show a water spectrum with the dissociated parts of water (H_1, H_2, OH^-) and typical lines for hydrocarbons, nitrogen, carbondioxide and pumping oil. The spectra between $1 > m > 70$ showed no significant dependence on the gap voltage. Only the spectra for H concentration as function of the gap voltage show a significant behaviour. Figs.11 and 12 give a comparison of H_2 - spectrum with and without conditioning. After conditioning the peaks at the low voltages at 40KV disappeared and no glowpoints have been observed in this region. Sparking started at a level of about 240 KV whereas glowpoints could be observed already at 210 kV. During the measurements the pressure varied between 10^{-6} and $4 \cdot 10^{-6}$ hPa, Fig.6 shows a H_2-spectrum for 1 and 5 msec pulse length with a duty cycle of 5 % after

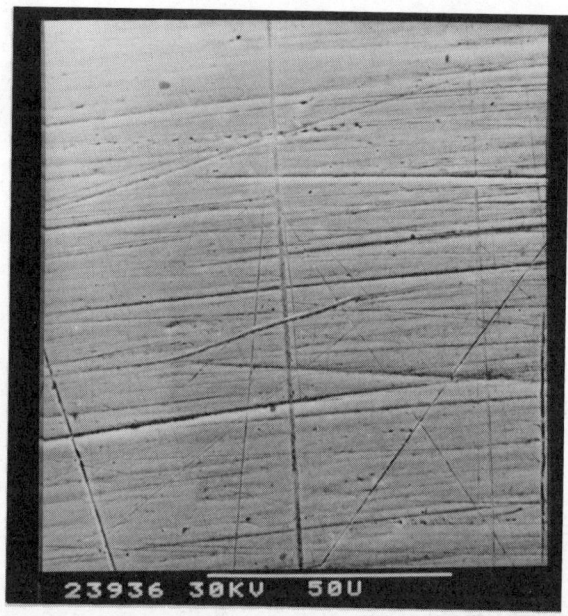

Fig. 9. Electron scanning microscope photograph of original surface finish of a OFHC copper electrode with polishing traces.

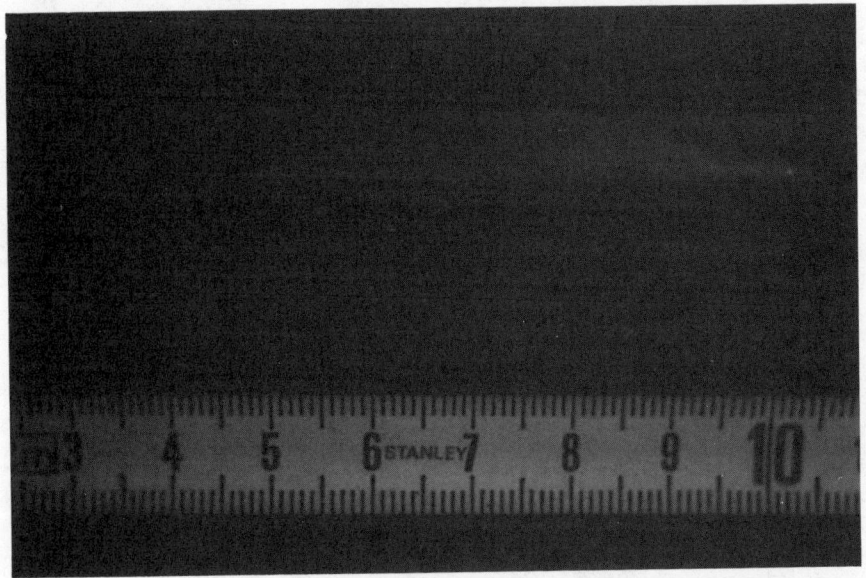

Fig. 10. Craters on the electrode surface after breakdown measurements, crater diameter < 0.5 mm.

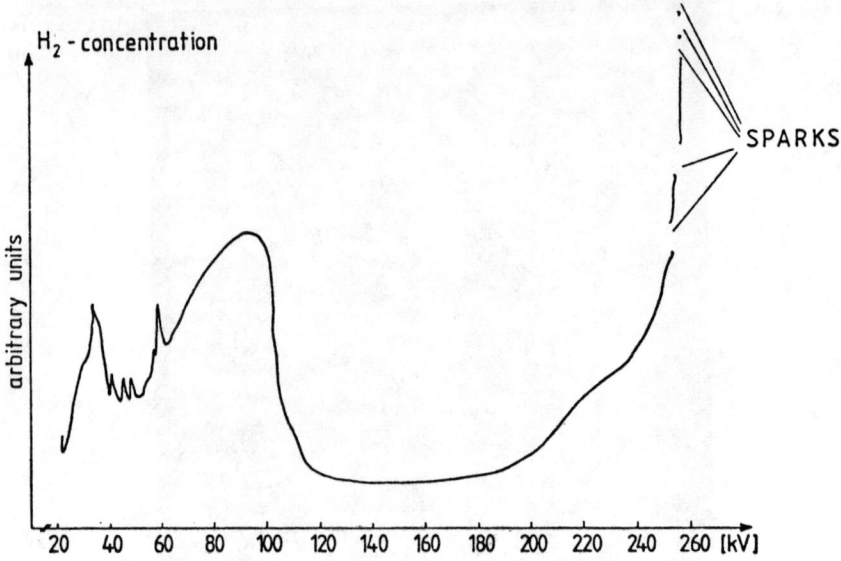

Fig. 11. H_2-concentration versus gap voltage for unconditioned electrodes. $f = 108.5$ MHz, $\tau = 1$ msec, dc = 5% $4 \cdot 10^{-6} < P < 10^{-6}$ hPa.

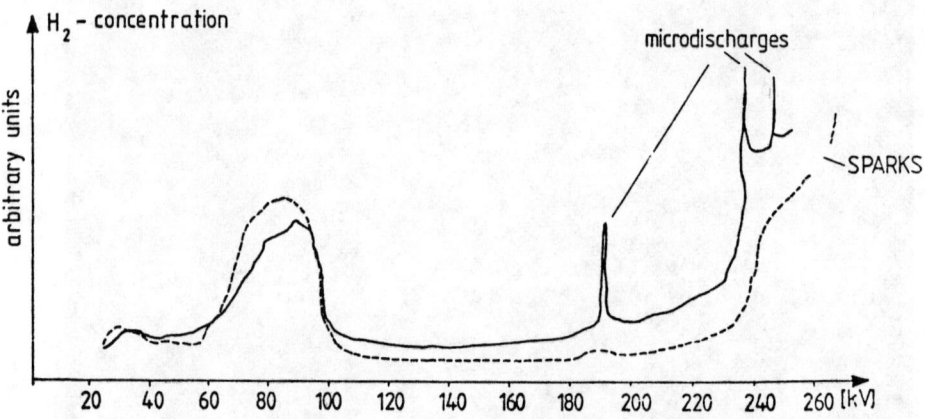

Fig. 12. H_2-concentration versus gap voltage for conditioned electrodes. Parameters as above except: (—) $\tau = 1$ msec, (- - -) $\tau = 5$ msec.

24 hours of conditioning. No spark could be observed for 1 msec pulses up to 240 kV in contrast to the unconditioned electrodes. They do not lead to a voltage breakdown, but to an enhanced H_2 pressure. Sparks appear at about 260 kV for 5 mec pulse length.

REFERENCES

1. R. W. Mueller, Proc. Radioactive Beam Workshop, Vancouver 1985.
2. A. Schempp et al., Nucl. Instr. & Meth. B10/11, 831-834 (1985)
3. A. Schempp et al., IEEE Trans. Nucl. Sci. Vol. NS-32 No. 5 (1985)
4. R. M. Hutcheon, Proc. 1984 Lin. Acc. Conf. Seeheim, GSI-84-11, 94 (1984).
5. N. Zoubek et al., elsewhere this conference
6. GSI int. note # 070586 (1986).
7. A. Schempp, Int.-Rep. 85-6, Inst. f. Angew. Physik., Univ. of Frankfurt, 1985.
8. P. Junior, Part. Acc., Vol. 13, 231 (1983).
9. W. D. Kilpatrick, Univ. of. California, Berkeley, UCRL-2321 (1953).
10. A. Gerhard, IEEE Trans. EL IN EL-20 no.4,709 (1985).
11. A. Gruber, diploma thesis, Univ. of Frankfurt,568 (1985).
12. M. Daehne, diploma thesis, Univ. of Frankfurt,253 (1982).
13. R. Thomae et al. The FOM Meqalac-Project status report,194 (1985).
14. A. Gerhard, M. Kurz, GSI Darmstadt to be published
15. D. Boehne et al. IEEE Trans. Nucl. Sci., NS-18 (1971).
16. J. Halbritter IEEE Trans. Elect. Ins., El-18. (1983).
17. R. Hutcheon et al., Proc. 1984 Lin. Acc. Con. Seeheim
18. R. V. Latham et al. IEEE Trans. El-18, (1983).

CHARGE STATE EVOLUTION AND ENERGY LOSSES IN A BEAM-PLASMA INTERACTION EXPERIMENT

R. Dei-Cas, J.M. Guihaumé, M. Beau, M.A. Beuve, J.F. Glicenstein,
J.P. Laget, C. Moreau, J.P. Mosnier, M. Renaud
CEA, SPTN, B.P. N° 2 - 91680 Bruyères-le-Châtel

R. Barchewitz
Laboratoire de Chimie Physique - Université Paris VI -
11, rue P. et M. Curie 75005 Paris

M. Cukier
Laboratoire de Physique des Gaz et des Plasmas
Université Paris XI 91405 Orsay

ABSTRACT

To study the charge state and energy evolutions of heavy ions travelling through a laser created plasma, a numerical code of the Monte-Carlo type has been written. In this paper, we describe the physical processes included in this code and some results relevant to our experimental facility are presented. Preliminary results on beam and plasma diagnostics are given.

I - INTRODUCTION

The main aims and a description of the experiment were presented elsewhere[1,2], and in this paper we will report briefly on beam foil spectroscopy (section II) and we will mainly describe the numerical simulation code[3] (section III) which has been written to help us in the analysis of the experimental results which we expect to get very soon.

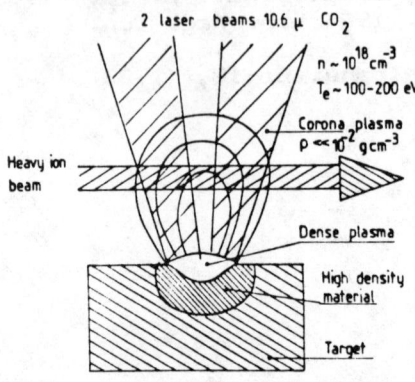

Fig.1. Principle of the beam-plasma interaction experiment.

Figure 1 shows the concept of the beam-plasma interaction experiment. In such a scheme, the charge state distribution at the exit will depend on many physical processes (see section III) which are not precisely known. Therefore it is helpful, for the data analysis, to have a guide describing the experiment as closely as possible. This is the reason why we have used a Monte-Carlo technique to follow the history of many particles (100 to 1000); in this way, the charge state distribution (and not only the mean effective charge state) and the Coulomb scattering process can be described. Also, in this trajectory code, magnetic and electric fields inside or outside the plasma target can be taken into account.

Some technical difficulties have been encountered which have slowed down the progress of the overall experiment and, in section IV, we will report on the present status of the main components : beam, laser and diagnostics.

II - BEAM-FOIL SPECTROSCOPY STUDIES

The X-ray spectrometer and the particle charge analyser have been used for beam-foil spectroscopy studies and figure 2 shows as an example, the measured and calculated X-ray spectra[4] [5] in the 1820 to 1860 eV range. The high resolution spectra obtained in these beam-foil experiments and the use of MCDF (multi-configurations Dirac-Fock) calculations[6] to compute line energies and fluorescence yields have allowed us to identify many 2p → 1s satellite lines (see Ref.4).

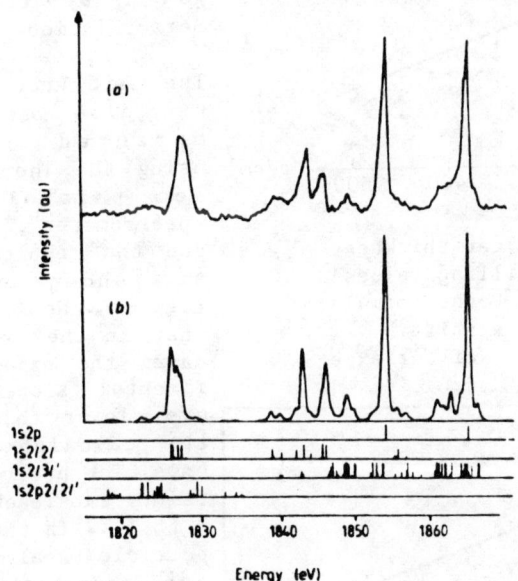

Fig.2 K_α satellite spectrum obtained after excitation of a 44 MeV Si^{7+} beam passing through a 5 µg cm^{-2} carbon foil: (a) measured (b) calculated.

The radiative lifetime of excited states can be very short and, for example, shorter than the projectile time of flight inside the target. Therefore, comparing the X-ray line shifts of a resonance and intercombination lines, it should be possible to measure the dynamical screening effect of the target i.e the modification of the projectile Coulomb field by the target free electron gas. Indeed, the resonance line has a short lifetime and is de-excited inside the target; on the contrary, the intercombination line has a longer life time and should not be sensitive to the dynamical screening effect. In fact, the observed line shifts[5] seem to be, within the error bars, entirely due to the energy loss of the projectile in contradiction with other measurements[7]. It should be emphasized that this is still an open question. From the observed X-ray line shift, it is possible to calculate the energy lost in the target by using the relation between the emitted and observed wavelengths (Doppler effect):

$$\lambda_{em} = \lambda_{obs} \frac{(1-\beta^2)^{1/2}}{1-\beta\cos\theta}$$

where θ is the observation angle. The experimental points are plotted on figure 3 and compared to the exit energy deduced from the Northcliffe and Schilling tables and calculated from the Bethe formula where the mean effective charge is assumed to be given by the Betz relation.

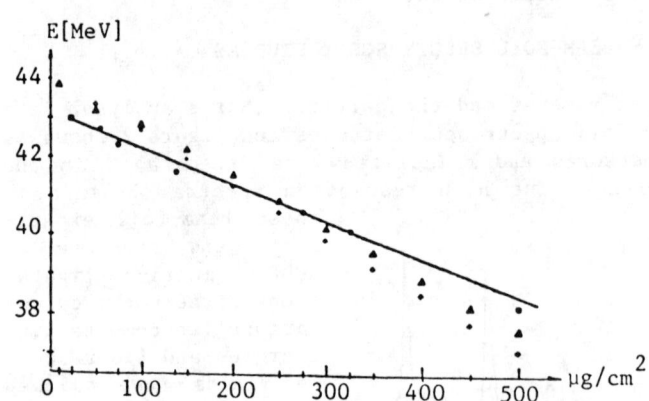

Fig.3 Exit energy versus target thickness
△ Northcliffe and Schilling values
+ Calculated from the Bethe formula
• deduced from the line shift.
The solid line is a fit of our experimental data.

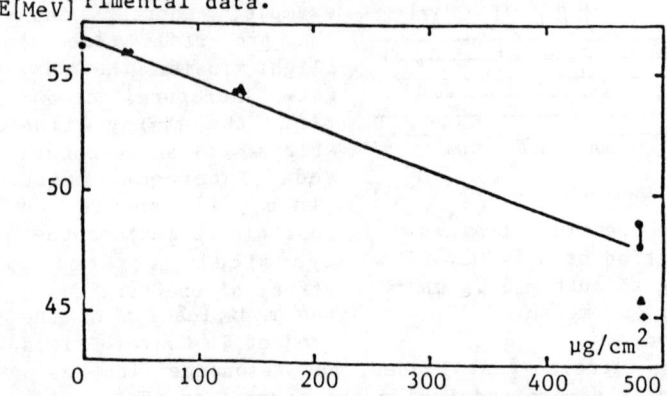

Fig.4 Exit energy measured with the particle analyser, • experimental values

The exit energy has also been measured by using the Thomson parabola spectrometer[1],[2] and the results are shown on figure 4. Notice that in the two cases the experimental slopes are slower than the calculated ones. The uncertainty can reach 3 to 7 %. In the particle analyser case, the energy loss is given by the different Z trace positions as shown on figure 5. These traces on the detection plane are relative to target thicknesses of 5 and 88 µg/cm^2. These ρR values are relevant to our expected plasma parameters. The trace displacement is due to the

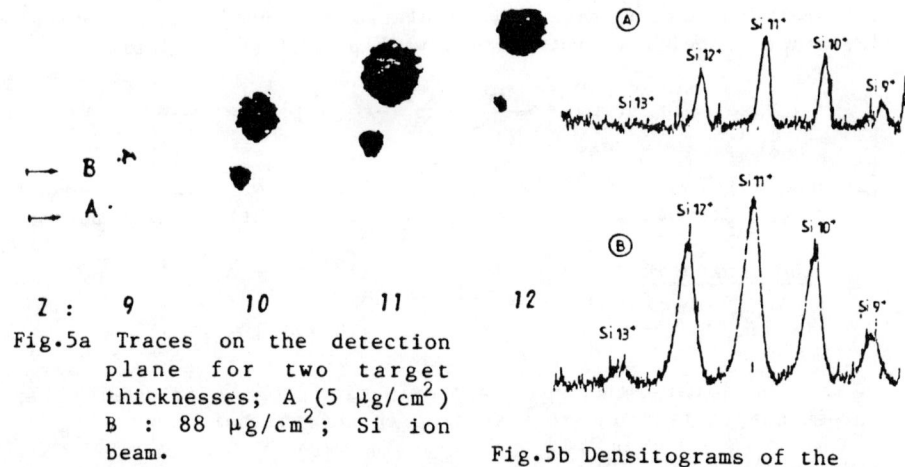

Fig.5a Traces on the detection plane for two target thicknesses; A (5 µg/cm^2) B : 88 µg/cm^2; Si ion beam.

Fig.5b Densitograms of the traces.

energy losses and the Coulomb scattering is responsible for the trace broadening. This is the kind of results that we expect to have with a plasma target except that the charge state distribution should be displaced on the high-Z side.

III - NUMERICAL SIMULATION CODE[3]

3-1 Atomic physical processes

In the heavy ion particle interaction with a plasma, many atomic physical processes play a role, namely, electron and ion ionization, bound-bound recombination (charge exchange), radiative and dielectronic recombinations. In the region of interest, the three body recombination is negligible. In a first step, excitation and de-excitation processes are neglected, mainly due to the fact that we do not have useful cross sections.

- <u>Electron ionization</u> $X^{q+} + e \rightarrow X^{(q+1)+} + 2e$

The Lotz formula is chosen for the ionization cross section where the electron energy is replaced by the relative energy $1/2\ mg^2$; g being the relative velocity between the projectile ($\vec{v}_1$) and the Maxwell velocity electron distribution : $g = |\vec{v}_1 - \vec{w}_2|$

$$\sigma_{ie}(g) = 4.5\ 10^{-14} \sum_{i=1}^{N} q_i\ \frac{\ln(1/2\ mg^2/U_i)}{1/2\ mg^2\ U_i}\ cm^2$$

U_i being the binding energy of the q_i electrons of the ith level among the N considered levels : N = 1 for H - and He - like ions; N = 2 for Li- and Ne - like ions and N = 3 for the others.

The reaction rate is calculated for projectiles having a constant velocity v_1 interacting with a Maxwellian plasma :

$$f_2(w_2) = \frac{\ell^3}{\pi^{3/2}} \exp(-\ell^2 w_2^2) \; ; \; \ell^2 = \frac{m}{2kT}$$

In our case $v_1 \gg v_{the}$ (v_{the}, thermal electron velocity) and the reaction rate has to be averaged over the electron distribution.

- **Ion ionization** $X^{q+} + A_j^{p+} \longrightarrow X^{(q+1)+} + A_j^{(p-1)+}$

$$X^{(q+1)+} + A_j^{p+} + e$$

The binary encounter approximation[8] (BEA) is used to express the ionization cross section in the following way :

$$\sigma_{nj} = q_n Z_{jn}^{*2} \frac{6.5 \; 10^{-14}}{U_n^2} G(V_r)$$

σ_{nj} being the projectile ionization (for the n level) cross section by the j plasma ion species. q_n is the electron number in the level n having a binding energy U_n; $G(V_r)$ is a tabulated function of the ratio v_1/v_{en}, where v_{en} is the mean electron orbital velocity of the nth level. The effective charge Z_{jn}^* is defined from the G.S.Z. (Green, Sellin, Zackor) model[9].

- **Bound-bound recombination** (charge exchange)

$$X^{q+} + A^{p+} \rightarrow X^{(q-1)+} + A^{(p+1)+}$$

The electron capture of a H-like ion of charge p and level k by an ion of mass A, energy E and charge state q on the level n is given, in the O.B.K.[10] (Oppenheimer, Brinkmann, Kramers) approximation, by the relation :

$$\sigma_{nk} = \frac{3.36 \; 10^{-11} (pq)^5}{n^3 k^5 \frac{E}{A} \left[\frac{p^2}{2k^2} + \frac{q^2}{2n^2} + 10^{-5} \frac{E}{A} + \frac{A}{1.610^{-4} E} \left(\frac{p^2}{k^2} - \frac{q^2}{n^2}\right)^2\right]^5}$$

and the total cross section :

$$\sigma_{q,q-1} = \sum_{n=1}^{\infty} \sum_{k=1}^{\infty} N_k \; \sigma_{nk}$$

where N_k is the electron number in the k level.

Radiative recombination $X^{q+} + e \rightarrow X^{(q-1)+} + h\nu$

The radiative recombination cross section is calculated for H-like ions by the following formula, assuming a Gaunt factor of unity and replacing the electron energy by its relative energy $(1/2\ mg^2)$.

$$\sigma_{RR} = 2.11\ 10^{-22}\ \frac{h\nu_1}{(1/2\ mg^2 + \frac{h\nu_1}{n^2})}\ \frac{h\nu_1}{1/2\ mg^2}\ \frac{1}{n^3}.$$

with $h\nu_1 = 13.6\ q^2$ eV.

The reaction rate is obtained by summing up the n levels and integrating over the electron distribution.

- <u>Dielectronic recombinaison</u>

$$X^{q+} + e \rightarrow (X^{(q-1)+})^{**} \rightarrow X^{(q+1)+} + h\nu$$

No general formulation is available for this process and a constant value of $\alpha_{RR} = 10^{-11}$ cm^3 s^{-1} has been assumed.

Table 1 gives typical values for a 64 MeV Cu projectile travelling through an Al-plasma assumed to be in a coronal regime with an electron temperature of 200 eV and Al $^{10+}$ - Al $^{11+}$ ions.

Table I Reaction rates for Cu^{+7} and Cu^{+14} at 64 MeV and T_e=200 eV

	α_{ie}	α_{RR}	$\alpha_{ij}(Al^{10+})$	$\alpha_{ij}(Al^{11+})$	$\alpha_{q,q-1}(Al^{10+})$	$\alpha_{q,q-1}(Al^{11+})$
Cu^{+7}	5.57 10^{-9}	2.86 10^{-13}	1.44 10^{-6}	1.46 10^{-6}	4.81 10^{-8}	8.14 10^{-9}
Cu^{+14}	3.92 10^{-10}	2.10^{-12}	2.22 10^{-7}	2.15 10^{-7}	3.48 10^{-7}	6.19 10^{-8}

It is clear from these values that ion-ion reactions will play an important role.

The charge of the projectile is calculated along the ion path inside the plasma by using a random process (see section 3-3) and figure 6 shows as an example the charge state evolution for one particular run. The charge state distribution, at the plasma exit, for 100 Cu^{+7} ions of 50 MeV travelling through an Al laser plasma, having a 200 eV electron temperature and a line density of the order of 5 10^{18} cm^{-2}, is shown on figure 7. The mean charge state is of the order of 15-16 when the Betz formula should give a mean Z of 11-12.

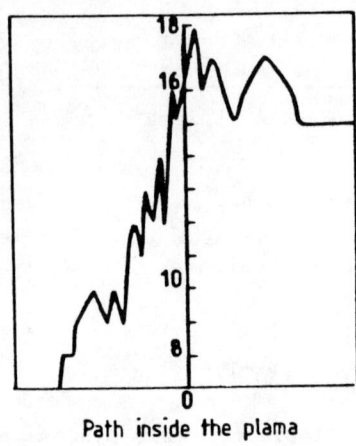 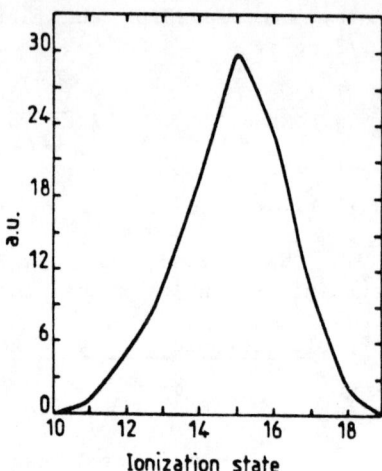

Figure 6 : Charge state evolution inside the plasma along the projectile path

Figure 7 : Charge state distribution at the plasma exit

3-2 Energy losses

Each individual particle, with its local charge state calculated as indicated in section 3-1, is slowed down continuously by computing at each integration step the energy lost on ions and on free and bound electrons.

- Energy loss on bound electrons

The Bethe formula is used at large velocity where the mean energy excitation for ions having more than 4 bound electrons is taken from Garbet's work[11]. At low velocity, the Sudden Impulse Approximation is assumed.

$$\left(-\frac{dE}{dx}\right)_B = 2.37 \; 10^{-10} \left(\frac{E}{A}\right)^{\frac{1}{2}} q^2 \; \Sigma_j \; n_j \; (Z_c - Z_j) \; \ln \left(\frac{2.2 \; 10^{-3} \; E}{A \; I_j}\right)$$

and

$$\left(-\frac{dE}{dx}\right)_{SIA} = 2.22 \; 10^{-15} \left(\frac{E}{A}\right)^{\frac{1}{2}} q^2 \; \Sigma_j \; n_j \; (Z_c - Z_j) \; \frac{1}{E_{jn}} \left<\frac{1}{p}\right>_{jn}$$

E_{jn} is the binding energy of the last occupied shell and $\left<\frac{1}{p}\right>_{jn} = 2I_{jn}(o)$ was tabulated in reference 12.

- Energy loss on free electrons

A general formulation valuable in all the velocity range has been formulated by Maynard and Deutsch[13] by using the Born Random Phase Approximation and we use their relation :

$$\left(-\frac{dE}{dx}\right)_f = 2.37 \; 10^{-10} \; q^2 \; \frac{A}{E} \; n_e \; \Psi(u) \; \frac{u^4 + 0.482}{u^2 + 1.5} \times \ln\left(2.1 \; 10^3 \; \frac{1}{b_{min}} \sqrt{\frac{T_e}{n_e}}\right)$$

with $u = \left(\frac{EA_e}{AT_e}\right)^{\frac{1}{2}}$; $\Psi(u) = \text{erf}(u) - \frac{2u}{\sqrt{\Pi}} \exp(-u^2)$

$$b_{min} = \text{Max}\left(\frac{q \; e^2}{m \langle g \rangle^2}, \frac{h}{m \langle g \rangle}\right)$$

- Energy loss on ions

The loss on plasma ions is very small in our case where the projectile exit energy has to be significant to be measured. We use the following relation[14] :

$$\left(-\frac{dE}{dx}\right)_i = 1.29 \; 10^{-13} \; \frac{q^2 A}{E A_p} \; \Psi(u') \; \sum_j n_j \; z_j^2 \times \ln\left(7.03 \; 10^9 \; \frac{A_p E}{(A+A_p)q z_j} \sqrt{\frac{T_e}{m_e}}\right)$$

with $u' = \left(\frac{EA_p}{AT_e}\right)^{\frac{1}{2}}$

3-3 Simulation code

The plasma target parameters are deduced from a 1D Lagrangian code[1] and in the region of interest (at a few mm in front of the target plane), the electron temperature is assumed to be constant and the density profile of the type :

$$n_e(y_e, r) = 4.25 \; 10^{19} \exp(-4.25 \; y) \exp\left[-\left(\frac{r}{0.25}\right)^2\right]$$

where y is the plasma expansion along the laser line and r the radial position.

The ion species densities are deduced from a coronal model.

The two processes, charge changing and pitch-angle scattering, are taken into account by a random walk method.

The charge changing probability is calculated from the reaction rates given in section 3-1 and the local Z is used in the slowing down process. In the same way, collisions are simulated by changing, after a random number of trajectory integration steps, the velocity vector angle in the following way : the Coulomb scattering time is given by :

$$\tau_x = \frac{<(\Delta v_{1\perp})^2>}{v_1^2}$$

where $<(\Delta v_1\perp)^2>$, the diffusion coefficient, is given in [15]; from this expression one can calculate the variation of the scattering cone angle $d\theta$ after a path $d\ell$ inside the plasma :

$$d\theta = \left[\frac{<(\Delta v_{1\perp})^2> d\ell}{v_1^3} \right]^{\frac{1}{2}}$$

$$d\theta = 3.61 \; 10^{-7} \; \frac{q}{E} \; \ell^{\frac{1}{2}} \left[n_e \ln \Lambda_e (\emptyset(u) - G(u)) + \sum_j n_j Z_j^2 \ln \Lambda_j (\emptyset(u') - G(u')) \right]^{\frac{1}{2}}$$

where $\emptyset(x) = \text{erf}(x)$ and $G(x) = \left[\emptyset(x) - x \emptyset(x) \right]/2 \; x^2$

After a collision, another random choice on this scattering cone is made in order to define new initial conditions for the velocity vector and the trajectory is re-started with these new initial conditions.

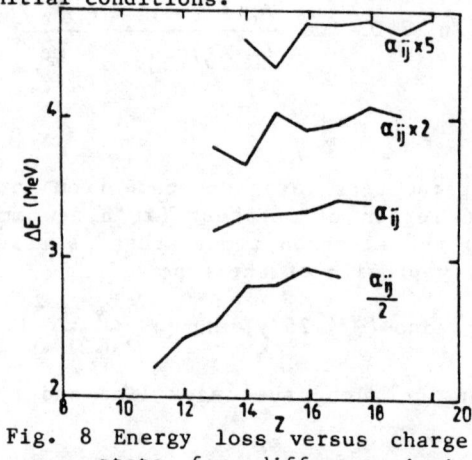

Fig. 8 Energy loss versus charge state for different ionization cross section values

Electric and magnetic fields, inside or ousiside the plasma, can be added. Figure 8 shows an example of mean energy loss as a function of the charge state ; the 4 curves have been calculated for 4 differents values of the ionization cross section to see the sensitivity of the results to cross section uncertainties.

IV - EXPERIMENTAL SET-UP

The plasma experiments have been started recently with a single laser beam. The delivered energy is of the order of 20 J and the beam is focused on a few 100 µm diameter spot on an Al plane target. Figure 9 shows the laser pulse and figure 10 a X-ray-diode signal . These signals were registered by the data acquisition system which is now in operation.

The CO_2 laser pulse is generated via an AsGa electro-optical switch on which a high voltage is applied through a spark-gap fired by the CO_2 laser itself. To avoid the jitter inherent to this kind of gas laser a 1 J Nd - laser has been ordered which will be used to fire the spark-gap in synchronism with the particle beam. At the moment, this timing system is not yet in operation.

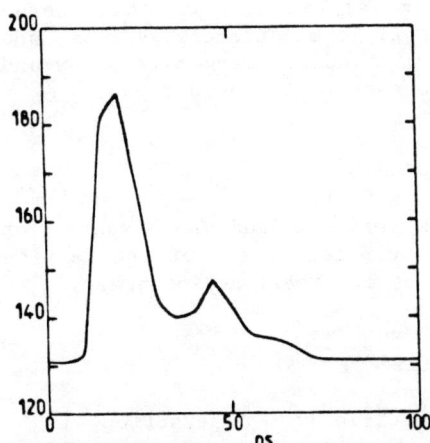

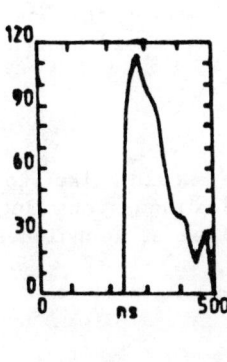

Fig.9 laser pulse   Fig.10 X-Ray diode signal

Different kinds of heavy-ion beams have been transported through the beam line : C, Si, Ni, Cu, and Au. To increase the beam current, a new heavy-ion source has been built and is now under test. The chopping system to select few individual heavy-ion beam pulses is in operation.

The main effort has been concentrated on the development of a special electronic tube able to detect, via an imaging system, the trace positions in a 5 ns pulse from which the Z value and the energy will be obtained. After successful operations with a 30 mm diameter tube, we have built a 70 mm diameter one ; this tube is movable under vacuum in the detection plane. Figure 11 shows that this tube works quite nicely ; one can detect a particle bunch of the mm size containing only 50 - 100 particles.

Fig. 11 Trace obtained with the ionographic tube.

Other diagnostics are still under development : X-ray spectrometer with a diode array instead of a film as detector, Jamin interferometer to measure the density.

V - CONCLUSIONS

All the main components are now ready and although a lot of work has still to be done to increase the laser energy (by activating the second beam), to synchronize the ion beam and the laser and to have all the diagnostics on hand, one should be able to have in the near future preliminary results on beam charge state evolution with a plasma target. To measure energy loss, one needs to work with ions having a mass number larger than 100 and the new source is needed to have enough particles per bunch.

ACKNOWLEDGMENTS

We would like to thank C. Deutsch and G. Maynard for fruitful discussions and all the technical staff of the Service de Physique et Techniques Nucléaires for their assistance.

REFERENCES

1. R. DEI-CAS et al. Heavy-ion particle beam interaction with a hot ionized target - Journal de Physique C8 n° 11, (Nov 1983) p. 179.
2. R. DEI-CAS - Status report on the plasma-particle beam interaction experiment at Bruyères-le-Châtel, 2th Int. Workshop on Atomic Physics for Ion Fusion - RAL - Chilton - Sept. 11-16, (1984)
3. J.M. GUIHAUME. Thesis 24 Sept. 1986.
4. J.P. MOSNIER et al., Beam-foil spectroscopy of 2p - 1s transitions in Si XI to Si XIV ions - J. Phys. B : At. Mol. Phys. 19 25-31,(1986).
5. J.P. MOSNIER. Thesis Paris VI, Spectres d'émission X d'ions silicium par la méthode faisceau-feuille, (1986).
6. J. BRUNEAU. J. Phys. B16, 4135, (1983) and 17,3009 (1984).
7. F. BELL et al. J. Phys. B15,L 443 (1976).
8. J.H. Mc. GUIRE, P. RICHARD. Phys. Rev. A8, 1374 (1973).
9. A.E.S. GREEN et al. Phys. Rev. 184, 1 (1969).
10. R.M. MAY. Phys. Rev. A136, 699 (1964).
11. X.GARBET. Thèse 3ème Cycle Orsay (1984).
12. F. BRIGGS et al. At. Data and Nucl. Data Tables 16, 201 (1975).
13. G. MAYNARD, G. DEUTSCH. J. Phys. 46, 1113 (1985).
14. G. MAYNARD. Thèse 3ème Cycle Orsay (1982).
15. R. DEI-CAS, D. MARTY. Rapport EUR-CEA-FC 726, (1974).

ENERGY DEPOSITION OF HEAVY IONS IN PLASMA

J. Meyer-ter-Vehn, Th. Peter, and R. Arnold
Max-Planck-Institut für Quantenoptik, 8046 Garching, FRG

ABSTRACT

The charge state of heavy ions slowing down in plasma is calculated by solving rate equations. Dielectronic recombination is found to be of particular importance. The heavy ions typically run through a sequence of high non-equilibrium charge states which lead to additional range shortening in plasma and strongly enhanced stopping powers in particular at the end of the range.

NEW DEVELOPMENTS

Stopping of heavy ions in hot, dense matter, the basic beam/target coupling process for inertial confinement fusion (ICF) with heavy ion beams, has not yet been examined experimentally. The very first experiments are now in progress in France and Germany using existing heavy ion beams and dense plasma generated externally by means of a CO_2 laser and z-pinch devices (Orsay, Darmstadt)[1]. With respect to ion energy and plasma parameters, these experiments will be still far from actual ICF situations, but they will test for the first time plasma effects on energy deposition specific to heavy ions.

At GSI/Darmstadt, the world's first opportunity of really generating a plasma with the heavy ion beam itself is now almost at hand with the completion of the high-intensity radio-frequency quadrupole (RFQ) injector. Although only modest plasma parameters are expected with the 10 mA beam of 50 keV/u ions, we are rather excited about the possibility of depositing 1-10 kJ/g over 1 μs into a Xe-gas target, for example. This would produce a $T \cong 1eV$, $n_e \cong 10^{19}/cm^3$ plasma of about 1 mm size which would radiate most of its energy as visible light. In another experiment an attempt will be made to pump an excimer laser with this beam. These will be preparatory steps for heavy ion beam/plasma experiments starting in 1990 with the SIS/ESR facility now under construction at GSI. Heavy ion beam generated plasmas near solid state density and with temperatures in the region of 10-100 eV are expected at this later stage.

The theoretical basis for planning the GSI experiments has been developed over the last few years at MPQ/Garching, studying heavy ion beam/plasma interactions as well as hydrodynamic target response and plasma regimes that can be reached[2,3]. In this contribution, the emphasis is on energy deposition of heavy ions and, in particular, on the effective charge Z_{eff} of the ions when passing through plasma. As new results, we find that Z_{eff} is essentially in non-equilibrium over most of the stopping range and that dielectronic recombination plays a dominant role in determining Z_{eff} in fully ionized matter[3]. Values of Z_{eff} are larger than in cold matter, though smaller than predicted by Nardi and Zinamon[4], and lead to much enhanced range shortening. The z-pinch experiment at GSI is designed predominantly for an investigation of these effects.

THE HEAVY ION STOPPING POWER IN PLASMA

The stopping power of heavy ions for a completely ionized plasma can be written approximately as

$$dE/dx = (eZ_{eff}\omega_{pl}/v_{ion})^2 \cdot G(v_{ion}/v_{th}) \cdot \ln\Lambda \qquad (1)$$

where v_{ion} and Z_{eff} are the velocity and the effective charge of the projectile,

$$\omega_{pl} = (4\pi e^2 n_e/m_e)^{1/2} \qquad (2)$$

is the plasma frequency for an electron density n_e, v_{th} is the thermal electron velocity, the function G is defined as

$$G(x) = \text{erf}(x) - (2x/\sqrt{\pi})\exp(-x^2) \qquad (3)$$

and $\ln\Lambda$ denotes the Coulomb-logarithm with $\Lambda = b_{max}/b_{min}$ being the ratio of the maximum and minimum impact parameters. For a plasma, we use $b_{max} = v_{ion}/\omega_{pl}$ and the appropriate Bohr or Bethe value for b_{min}.

All three factors: Z_{eff}^2, $G(v_{Ion}/v_{th})$, and $\ln\Lambda$ in eq. (1) contribute to the temperature dependence of the stopping power in plasma. Most of the discussion in the literature so far has been concentrated on the effect of $\ln\Lambda$ which is larger in a plasma than in cold target matter due to a larger b_{max}. The effect of the plasma function G becomes important for $v_{ion} < v_{th}$ which occurs either for high temperature plasma or for low ion velocities, e.g. near the end of the ion stopping range. In this region, $G \sim (v_{ion}/v_{th})^3$ and one approaches the friction limit for the stopping power $dE/dx \sim v_{ion}$, whereas for $v_{ion} > v_{th}$ one has $G \cong 1$ and $dE/dx \sim 1/v_{ion}^2$. The maximum of dE/dx is reached for $v_{ion} \cong v_{th}$ provided that Z_{eff} has no strong velocity dependence.

As we will show in the following, we find rather high and constant values of Z_{eff} over most of the stopping range in plasma contrary to the Z_{eff} behaviour in cold targets. In particular, this applies to the GSI Z-pinch experiment where we expect to have a completely ionized hydrogen plasma with a density $n_e \cong 10^{19}/cm^3$ and a temperature T = 10 eV. In Fig. 1, the anticipated dependence of dE/dx on v_{ion} is qualitatively shown for this case and is compared with the dE/dx curve for a corresponding cold gas target with same density. At low v_{ion}, the plasma dE/dx is expected to exceed the cold gas dE/dx by a large factor mainly because of a much larger Z_{eff}. It may thereby provide a sensitive test for the plasma physics involved in low-velocity stopping for the case of highly charged heavy ions in a plasma. The high plasma dE/dx values at low v_{ion} lead to very pronounced Bragg peaks and will be discussed further below.

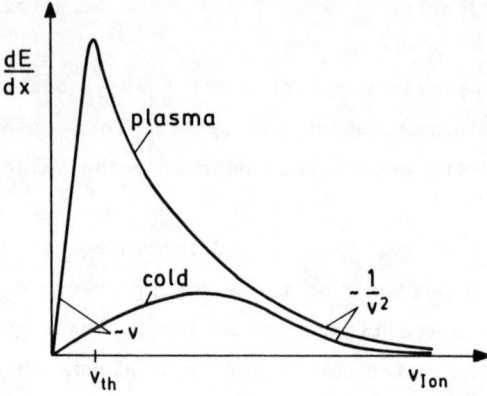

Fig. 1: Comparison of stopping power versus ion velocity for a cold gas target and a plasma target. Schematic drawing, qualitatively valid for the plasma parameters of the GSI Z-pinch experiment.

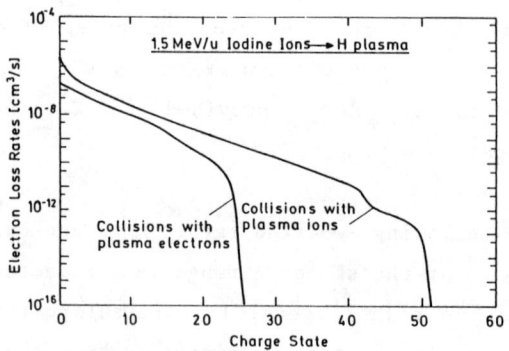

Fig.2a: Electron loss rate coefficients versus charge state for 1.5 MeV/u iodine ions passing through hydrogen plasma of 10 eV temperature and $n_e = 10^{17}/cm^3$ density.

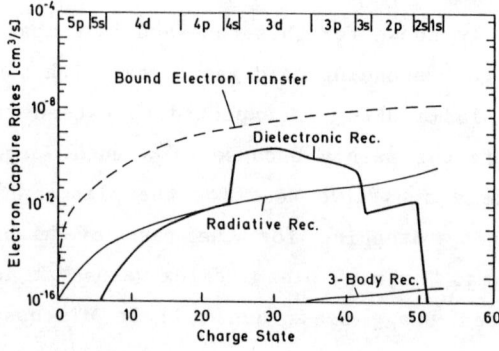

Fig.2b: Electron capture rate coefficients versus charge state. Beam/target system same as for Fig. 2a. For comparison, the rate of bound electron transfer for cold hydrogen gas of same density is also shown.

THE EFFECTIVE CHARGE Z_{eff}

The effective charge of the heavy ion along its stopping path is a key quantity since it enters quadratically $\sim Z_{eff}^2$ into the stopping power formula (1). In most of the published calculations for heavy ion stopping in plasma, semi-empirical formulas for Z_{eff} are used. These were derived for cold matter targets and give $Z_{eff}(v_{ion})$ as a function of the instantaneous ion velocity. Nardi and Zinamon were the first to point out that these formulas do not apply to highly ionized matter and that Z_{eff} may be much larger in plasma than in cold matter[4]. Here, we present new results of Z_{eff} calculations following the heavy ion charge state during slowing down by solving rate equations.

ELECTRON LOSS AND CAPTURE RATES

Calculated rate coefficients of electron loss and capture processes are shown in Figs. 2a and 2b for the reference case of 1.5 Mev/u iodine ions passing through 10 eV hydrogen plasma. The projectile loses electrons predominantly due to collisions with plasma ions and to a smaller extent due to collisions with plasma electrons. In completely ionized plasma, the projectile captures electrons due to radiative recombination and due to dielectronic recombination[3]. The latter was studied recently by the authors showing for the first time that the dielectronic rate exceeds the radiative one and determines the equilibrium charge state of fast ions in plasma. At an electron density $n_e = 10^{17}/cm^3$ chosen for Fig. 2b, 3-body-recombination is negligible; it starts to compete with the other rates only at much higher density and lower ion velocity. For comparison, the rate of bound electron transfer is also shown in Fig. 2b as a dashed line. It represents the most effective recombination process in cold target matter, but cannot contribute in completely ionized matter with no bound electrons left. Details of the calculations are described in ref. 3.

IMPORTANCE OF DIELECTRONIC RECOMBINATION

Balance between loss and capture defines the equilibrium charge Z_{eq}. Results for Z_{eq}^2 versus v_{ion} are given in Fig. 3 for two densities. The step-like form of the curves reflects the strong shell effects inherent to the dielectronic rate α_{DR}. Dielectronic recombination tends to be reduced at high plasma density, and this leads to larger Z_{eq}^2 for $n_e = 10^{21}/cm^3$ in Fig. 3 except at low ion velocity, where 3-body recombination takes over and decreases Z_{eq}^2. The most important high density effect on α_{DR} is collisional ionization of the Rydberg electron occuring during dielectronic recombination and has been included in the present calculation. In Fig. 3, results for the T=10eV, $n_e = 10^{17}/cm^3$ plasma neglecting α_{DR} (upper dash-dotted curve) as well as results for a neutral hydrogen gas of same density (lower dashed curve) have been added for comparison. It is seen that much larger Z_{eq} are obtained for plasma than for cold matter and that dielectronic recombination plays a dominant role.

At this point, it is important to note that the equilibrium charge states, though clearly displaying the relative trends, are not necessarily reached in actual stopping situations. The non-equilibrium charge build-up during the initial stripping phase after incidence of weakly charged heavy ions on the target had been studied by Bailey et al. solving rate equations without accounting for energy loss. In the following, we report on new calculations in which the charge state is obtained together with the energy loss using the rates for Z_{eff} and the stopping power as discussed above.

NON-EQUILIBRIUM CHARGE STATES

The evolution of the effective charge Z_{eff} of initially neutral iodine projectiles penetrating into T = 10eV, $n_e = 10^{17}/cm^3$ hydrogen plasma is shown in Fig. 4 as curves in the Z^2 versus v_{ion} plot. The equilibrium curve (dashed, same as thin curve in Fig. 3)

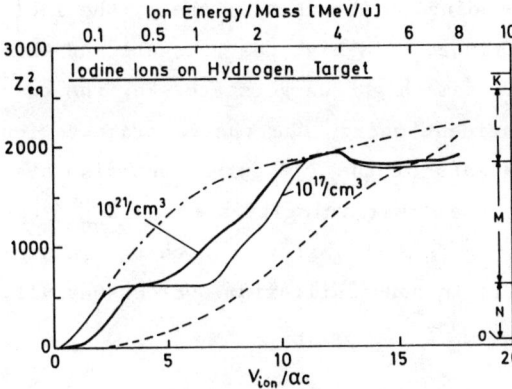

Fig. 3: Equilibrium charge states versus ion velocity for two plasma densities and T=10 eV (solid curves). The dash-dotted curve refers to results without dielectronic recombination, the dashed curve to cold hydrogen with $10^{17}/cm^3$ density in both cases.

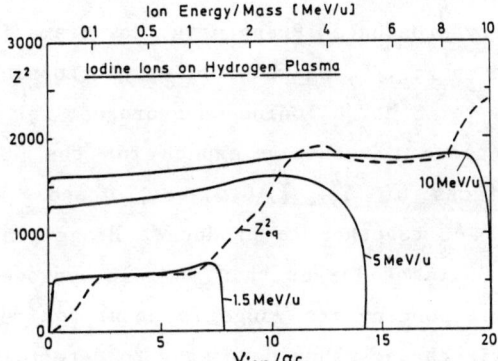

Fig. 4: Non-equilibrium charge states (solid curves) versus ion velocity for T=10 eV, $n_e = 10^{17}/cm^3$ plasma; labels give incident energies. The dashed curve shows the equilibrium values for comrarison.

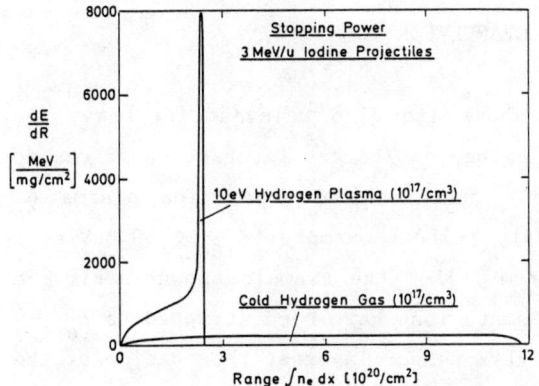

Fig. 5: Stopping power versus range for cold hydrogen gas and 10 eV plasma; same density in both cases.

is given for comparison. Depending on the initial energy the incident projectile is quickly stripped to a high charge state and then slows down keeping almost the same high charge state. For the case of 1.5 MeV/u and 10 MeV/u incident energy the charge trajectories stick for a while to the plateaus of the Z_{eq} curve, but also stay at the high charge level for the lower velocities where Z_{eq} drops to zero. In this region stopping is much faster than charge relaxation. The 5 MeV/u trajectory is in non-equilibrium over essentially all of the stopping range.

The high non-equilibrium charge states tend to further increase the stopping power in plasma, in particular near the end of the range. This leads to a very pronounced Bragg peak. How dramatic these effects may be is shown in Fig. 5, where profiles of stopping power versus range are given for 3 MeV/u iodine on hydrogen. For a T=10eV, $n_e=10^{17}/cm^3$ plasma, close to what we expect for the GSI Z-pinch experiment, the effects of Z^2_{eff}, $G(v_{Ion}/v_{th})$ and $\ln\Lambda$ (compare eq.(1) and Fig. 1) act together to produce a Bragg peak with a stopping power about 30 times larger than in cold hydrogen gas of same density; the corresponding ion range is about 5 times smaller in plasma. Such drastic changes should be easy to detect in the planned experiments.

RELEVANCE FOR ICF

The effects discussed above are also relevant for heavy ion fusion, although in this case the beam/target interaction is characterized by higher projectile energies and deposition plasma of higher temperature and density (HIBALL-example: $E_{ion} \cong 50$ MeV/u, T $\cong 300$ eV, $n_e \cong 10^{21} - 10^{23}/cm^3$). For the example shown in Figs. 6 and 7, it is assumed that bismuth ions have been stripped to $Z_{eff} = 80$ when passing through a high Z tamper layer at the outside of the target and are entering a lithium deposition layer with an energy of 30 MeV/u. The charge state evolution during slow-down in 300 MeV Li-plasma is shown in Fig. 6. Again the effective charge state stays well above the equilibrium values, but tends to follow the Z_{eq}^2 curve more closely in the case of $10^{21}/cm^3$ plasma density than for a density of $10^{17}/cm^3$. Deposition profiles calculated for different temperatures and densities in the lithium layer are plotted in Fig. 7. The plasma cases were obtained with the full model as outlined above. The T = 0 results are taken from the HIBALL calculations. Changing from solid lithium to 300 eV, 0.1 g/cm^3 Li-plasma, a range shortening of more than 30% is obtained which is about half due to a larger lnΛ and half due to enhanced, non-equilibrium charge states. An additional 30% range shortening and the formation of a sharp Bragg peak would be obtained if the density were reduced to 1 µg/cm^3.

CONCLUSIONS

The most prominent result of this contribution is that heavy ions when passing through plasma typically run through a sequence of high non-equilibrium charge states. Semi-empirical formulas derived for cold target material which give Z_{eff} (v_{ion}) as a function of the instantaneous ion velocity are not applicable. Fast heavy ions, incident on a plasma target in a low charge state, are typically stripped on a short time-scale to a high charge state and then tend to keep this high level up to the end of their range. The actual sequence of charge states may strongly deviate from calculated equilibrium values. Only at very high plasma density close to solid state density, the charge state evolution approaches the behaviour known for cold targets. The present findings cause drastic changes in range and stopping power for plasma conditions such as planned for the GSI Z-pinch experiment. They are also expected to lead to additional range shortening in ICF targets, of the same order as the temperature effects originating from the Coulomb logarithm.

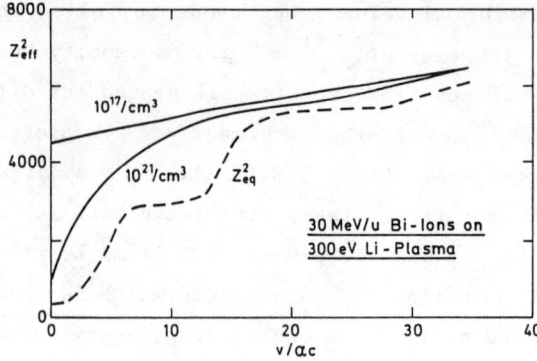

Fig. 6: Non-equilibrium charge states Z^2_{eff} of bismuth ions passing through lithium plasma of different densities (solid curves), compared with equilibrium values Z^2_{eq} (dashed curve).

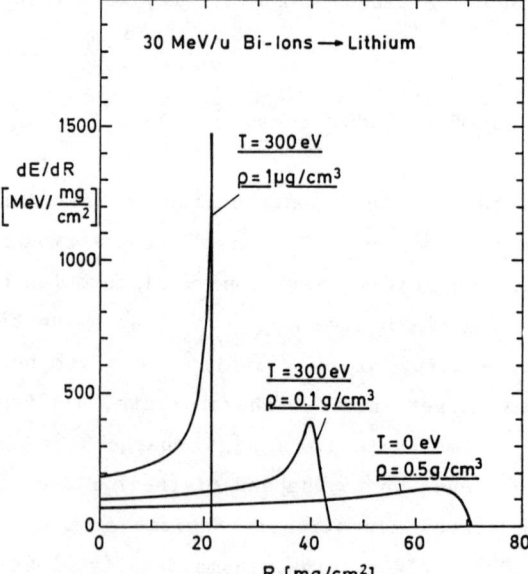

Fig. 7: Stopping power of Bi ions in lithium versus range for three different temperature and density combinations.

REFERENCES

1. See contributions of R. Dei Cas and D. Hoffmann.
2. R.C. Arnold and J.Meyer-ter-Vehn, Rep. Progr. Phys. to be published (1986).
3. Th. Peter, Diploma Thesis, Report MPQ 105 (Nov. 1985); Th. Peter, R.C. Arnold, J. Meyer-ter-Vehn, submitted to Phys. Rev. Letters (1986).
4. E. Nardi and Z. Zinamon, Phys. Rev. Lett. 49, 1241 (1982).
5. D.S. Bailey, Y.T. Lee and R.M. More, Proc. HIF Symposium, Darmstadt, 1982, p. 571.

HEAVY ION BEAM TRANSPORT AND INTERACTION WITH ICF TARGETS

G. Velarde, J.M. Aragonés, J.A. Gago, L. Gámez,
M.C. González, J.J. Honrubia, J.M. Martínez-Val,
E. Mínguez, J.L. Ocaña, R. Otero, J.M. Perlado,
J.M. Santolaya, J.F. Serrano and P.M. Velarde
Instituto de Fusión Nuclear (DENIM), Castellana, 80;
28046 Madrid, Spain

ABSTRACT

Numerical simulation codes provide an essential tool for analyzing the very broad range of concepts and variables considered in ICF targets. In this paper, the relevant processes embodied in the NORCLA code, needed to simulate ICF targets driven by heavy ion beams will be presented.

Atomic physic models developed at DENIM to improve the atomic data needed for ion beam plasma interaction will be explained. Concerning the stopping power, the average ionization potential following a Thomas-Fermi model has been calculated, and results are compared with full quantum calculations.

Finally, a parametric study of multilayered single shell targets driven by heavy ion beams will be shown.

INTRODUCTION

In the analysis of the physics of ICF targets there are several mechanisms that must be taken into account in computer codes. Driver plasma interaction regimes; equation of state for hot and dense plasmas; hydrodynamic equations distinguishing radiation, ion and electron species; multigroup radiation, charged particles and neutron transport coupled to the hydrodynamic equations; atomic data and nuclear reactions, are the most important topics to be included in a simulation code.

These relevant processes have been embodied in our NORCLA code, whose previous versions have been explained in the appropriate bibliography[1,2,3,4,5].

Although our computer code can also be used to analyze targets driven by laser beams, in the present work, capabilities to simulate targets driven by particle beams will be explained.

Besides a short review of these capabilities, this work is going to treat specially stopping power and atomic physics models, topics really important for heavy ion beams.

The stopping power data used in our ion beam target simulation have been recently improved with a better calculation of the average ionization potential in the context of the Thomas-Fermi model. Results are compared with other models; the conclusions will be shown.

Atomic physic models developed at DENIM are also presented, as well as the present state of the DENIM Atomic Table.

Finally as a continuation of previous work on the analysis and optimization of ICF beam and target configurations[4,5], the influence of the energy deposition profile by heavy ions in the different layers of the multilayered targets has been analyzed.

SIMULATION OF ICF TARGETS

The ICF physics and computational developments at DENIM are incorporated in or closely related to the NORCLA code for one-dimensional simulation of plane and spherical ICF targets. The NORCLA code[4] consists of a driver-interfacing routine and two main segments, NORMA and CLARA. Figures 1 and 2 show the simulation models included in both segments.

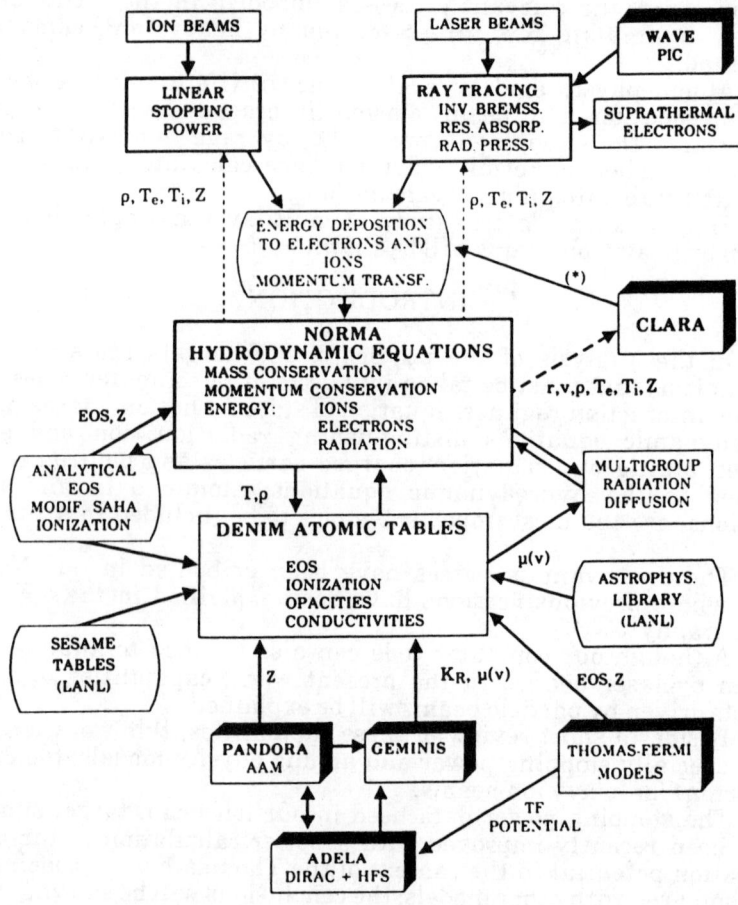

Fig. 1. Simulation models for hydrodynamics in ICF plasmas.

The NORMA segment computes the linear stopping of external ion beams using analytical stopping powers along radial tracks, with space dependent densities and temperatures at selected timestep.

The NORMA hydro solver has been extended to a fully implicit solution of the energy equations in 3 temperatures (electrons, ions and radiation). The implicit solutions in 1 or 2 temperatures (electrons + radiation and ions or material and radiation) are also available. Radiation diffusion can be treated in a grey approximation, with Rosseland mean opacities. The multigroup radiation diffusion set of subroutines have been recently completed in a quasi-implicit coupling with the energy equations, and benchmarks are being completed.

Atomic data (EOS, ionization, conductivities and opacities) are available from the DENIM Atomic Tables, whose structure will be explained in the next section.

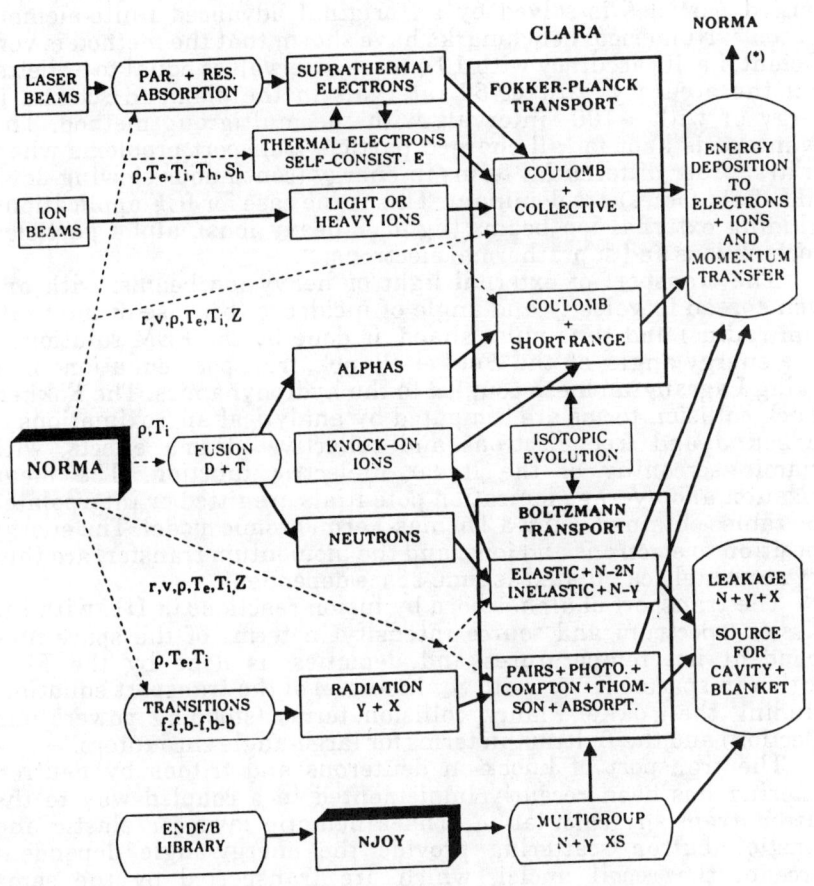

Fig. 2. Simulation models for particle transport in ICF plasmas

A Thomas-Fermi code computes tabulated EOS and ionizations for elements in any T-ρ grid. This model provides also the mean ionization and the average ionization potential for partially ionized atoms in plasmas with any degeneracy, which are used to calculate ion stopping powers in the bound and free electrons of partially ionized plasmas. The model has been extended for calculations of the linear dielectric function in plasmas with any degree of ionization and degeneracy.

The central part of the CLARA segment (figure 2) is the numerical solution of the time-dependent one-dimensional transport equations for charged particles (Fokker-Planck) and neutrons and photons (Boltzmann). Both transport equations are solved on a moving Lagrangian mesh that follows the hydrodynamic mesh motion with detailed space-time dependent plasma properties (different electron and ion densities and temperatures).

The time-dependent Fokker-Planck transport equation for charged particles is solved by an original advanced finite-element method [7]. Numerical benchmarks have shown that the method is very efficient, i.e. its accuracy with 10 energy intervals is equal to or better than the accuracy with 40-50 intervals in the diamond scheme in energy or with ~100 intervals with the multigroup method. This advantage is kept for all charged particle transport problems where the first order differential terms in energy (continuous slowing-down and E-field forces) are dominant. This is the case for ICF applications, including external ion beams (light or heavy ions), alpha particles, knock-on ions and suprathermal electrons.

The transport of external light or heavy ion beams, with any given spread in velocity and angle of incidence (because of the finite beam radius) and time pulse shape, is done by the FEM solution in space-energy-angle of the Fokker-Planck transport equation on a moving Lagrangian mesh coupled to the hydrodynamics. The Fokker-Planck collision terms are computed by analytical approximations of the bound and free electrons and collective plasma effects, with dynamic screening of the linear dielectric function. The mean ionization and average ionization potentials are fitted or interpolated from tables obtained using a Thomas-Fermi atomic model. The energy deposition to electrons and ions, and the momentum transfer, are thus very accurately calculated as time-space dependent.

The transport of alphas born by fusion reactions in DT, with the emission spectrum and source intensity in terms of the space-time dependent ion temperatures and densities, is done by the FEM solution in space-energy (with S_N in angle) of the transport equation, including the Fokker-Planck collision terms (stopping powers and deflection) and the Boltzmann terms for large-angle encounters.

The transport of knock-on deuterons and tritons by neutron scattering has been recently implemented in a coupled way to the neutron transport calculation. The kinematic laws for elastic and inelastic neutron scattering provide the energy-angle dependent source of the recoil nuclei, which are transported by the same procedure as the alpha particles. The CLARA code is being extended for simultaneous transport of neutrons, alphas and knock-ons, calculating not only the energy deposition and momentum transfer but also the transport of mass (thermal ions) and the isotopic evolution by neutron reactions.

The neutron transport capability in the CLARA code[4] was extended to a continuously moving Lagrangian mesh with changing densities and coupled to the hydrodynamics. The mesh motion drag and acceleration terms were introduced later and the multigroup neutron and gamma libraries have also been updated, using the NJOY processing code [8].

The detailed calculation of the isotopic evolution by neutron reactions, for the hydrogen and lithium isotopes, has been recently completed and is being coupled to the knock-on transport procedures for calculating the suprathermal fusions along the ion trajectories.

ATOMIC PHYSIC MODELS

To improve the atomic data used in the first versions of the NORMA code, the scheme of calculation proposed in the figure 3 is used to determine opacities, EOS and conductivities, which are gathered in the DENIM Atomic Table.

This library is being generated for materials such as aluminium europium, gold, lead, deuterium-tritium and lithium. Those data available in the SESAME library[6] are being compared with these models, and those not available are being generated. This library works on line with our hydrocode NORMA.

At present, not all the models have been completed. Those using LTE conditions are working and generating atomic data, such as energy levels, populations and average ionization, needed for ion beam plasma interaction. However, those using NLTE conditions, are being developed, and benchmarks are being completed, comparing results with models used at ILE (Osaka) and NRL. Conclusions are very optimistic, taking into account several simplified hypotheses. When the final coupling between LIRA and PLEYADE codes is finished, the results should be better.

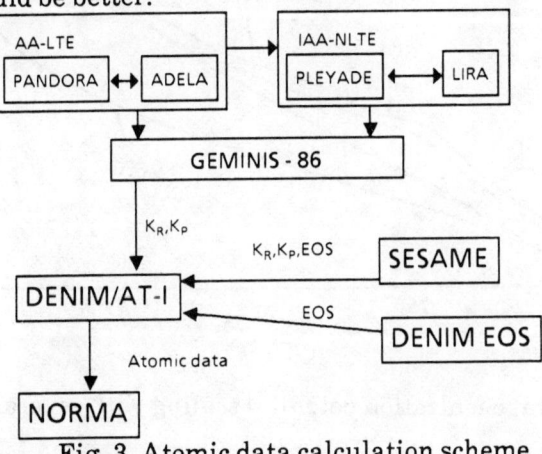

Fig. 3. Atomic data calculation scheme

ION BEAM PLASMA INTERACTION

The first step in the simulation of ICF systems driven by charged particle beams is to obtain the energy deposition profile by zone coming from a power pulse of ions. Two models have been adopted in our NORMA code to determine this energy deposition profile.

The first model (STOP program)[4] calculates that energy depending on the ion-beam driver (light ion or heavy ions). Because the stopping medium is composed of partially ionized high atomic number elements, the energy deposition has two terms corresponding to bound and free electrons. STOP also distinguishes between high ion energy-mass ratios and low energies, using a classical Bohr expression in the first case and the LSS model in the second one.

The starting point of the second model is the linear Fokker-Planck equation written in a Lagrangian frame, and incorporated in our transport theory segment CLARA, described in the former section.

The use of the time dependent Fokker-Planck equation in the ion beam energy deposition has been considered for the proper treatment of ion beams focusing. Numerical simulation shows the important role played by the focusing and the energy spread parameter on ion beam target coupling.

From the physical viewpoint, the major issue concerning the ion beam-plasma interaction is the stopping power. Following the classical theory, a model for this issue has been developed and is being

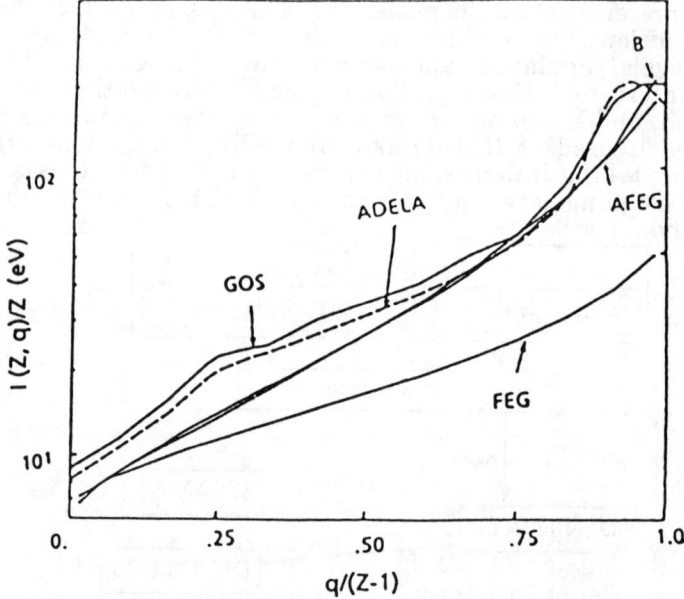

Fig. 4. Average ionization potential scaling with several models

used for both ion beam energy deposition and fusion product transport. The free electron contribution to the stopping power is obtained from the linear dielectric response theory for a plasma with an arbitrary degeneracy. The electron degeneracy, obtained from a TF model, is included for fusion product transport at the first stages of the ignition phase, when the fuel is dense and cold.

The bound electron contribution is obtained from the Bethe-Bloch formula using the TF model for average ionization potential (I). The formulation of I includes the dependence on ion velocity and the plasma effects. The ionization potential corresponding to isolated ground state aluminium ions is shown in figure 4, where the results of several models are compared. The B-labeled curve stands for the potential obtained with our model, and the AFEG[9] for the augmented free electron gas model which forces the hydrogenic scaling of free electron gas FEG potentials. The B, FEG and AFEG curves have been obtained from the TF electron density. The ADELA-labeled curve has been obtained through the AFEG theory using the electron density calculated with the ADELA code. In this case results are better than those from the B curve.

TARGETS DRIVEN BY HEAVY ION BEAMS

As a continuation of previous work[10,11], an attempt to establish physically based criteria other than a simple analysis of the aspect ratios of the different material zones has been made because simple analysis frequently leads to inappropriate or, at least, quite partial design conclusions and trends.

For this study a reference heavy ion pulse and multilayered target configuration have been considered, whose defining characteristics have been compiled in Table I.

Table I. Reference target structure

Material	Mass (mg)	Density (g.cm^{-3})	External Radius (cm)
Pb	190	11.614	0.3
Al	60	2.707	0.2847754
Pb	15	11.614	0.2611145
DT	1	.21	0.2595983
Void	-	-	0.2538489

In order to complete this study the properties of the ion beam driver have been slightly altered in the sense of changing the peak power of 720 TW used in previous simulations, to a peak power of 500 TW (more reasonable in view of the capabilities of the accelerators). Under this initial condition, a first analysis has been performed dealing directly with the abovementioned influence of the ion energy deposition profile on the target dynamics and characterizing the results through rather simplified criteria such as maximum temperature and maximum optical thickness reached in the fusion fuel. As recognized elsewhere[4,11] simple criteria are not to be taken as definitive in the analysis of the performance of ICF targets as they refer only to the end of their compression phase, but they provide some estimate of the capability of the fusion fuel to obtain an appreciable burnup.

The analysis took as variable parameter the ratio of mass of aluminium to mass of internal lead in order to establish the conditions under which a better compression of the fusion fuel could take place in compatibility with a high enough mass of internal lead needed in the burnup phase.

From previous simulations of the dynamics of ICF targets such as those considered[4] and also from a pure theoretical analysis of the conditions under which an efficient compression of the fusion fuel can take place, the bulk energy carried by the driver must be released on the "pusher" zone. In order to get this condition, and if the energy of these ions has to remain constant at the begining of the compression process (medium cold), one has to allow some part of this energy to penetrate to deeper zones, which as a consequence of the conducting properties of the matter, can lead to an undesirable fusion fuel preheating.

For a given set of data concerning the driver energy pulse and the base geometry of the target under compression the above mentioned ratio has been analyzed, and optimal values have been obtained for different mass values of aluminium close to those of the reference case (Table I).

As a complement to this study, starting from the configuration recognized as optimal in it (considering fixed ion energy), the effect of the different design characteristics of the target have been analyzed and some relevant consequences have been obtained concerning the possible features of the heavy ion beams as ICF drivers.

As a first point in the analysis the mass of the internal shielding-tamper can be drastically reduced as its shielding function is practically eliminated by avoiding the initial enhanced penetration of the beam. As a consequence of this, the mentioned mass can be reduced to the minimum compatible with its later function as fuel explosion tamper. This gives, in fact, significant improvements in the hydrodynamic performance of the pellet under compression but the tendency is limited certainly by the requirements imposed by the burn phase.

The second important consequence of the analysis is that, even in the cases not optimized from the just mentioned point of view, the adjustment of the deposition profile to a given pattern improves in an effective way the dynamics of the target under compression, and configurations not able to obtain significant fuel burnup become very suitable ones once the energy deposition profile has been conveniently tailored.

As a consequence of this fact and as previously mentioned, the driver input energy required for the attainment of a significant fuel burnup (or energy gain) can be sensitively lowered, and, under the appropriate tailoring of the energy deposition profile, can be set reasonably low to improve the prospects of the accelerator comunity. See the gain curve presented in figure 5.

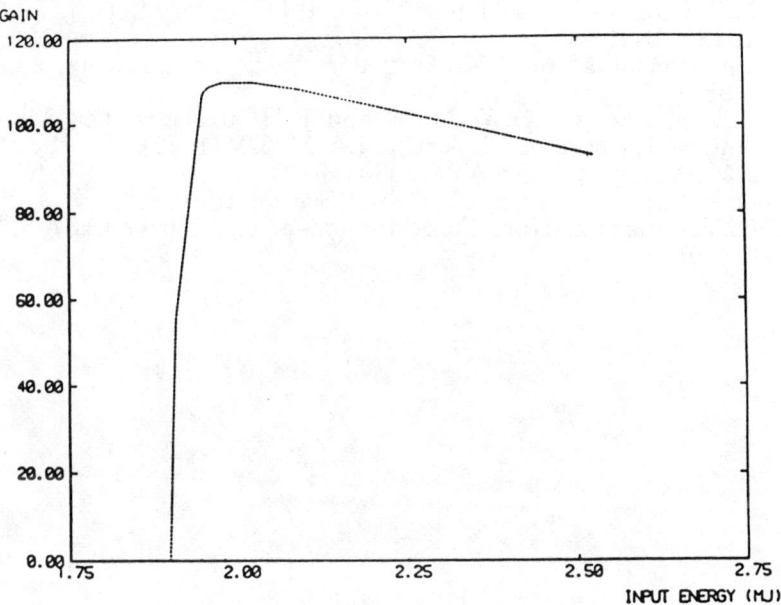

Fig.5. Gain curve versus driver energy

This leads directly to the conclusion that tuning ion beams in voltage (energy) as a function of time is an important tool for the attainment of high gains in the compression of ICF targets.

CONCLUSIONS

In this work all capabilities of NORCLA to study heavy ion beam targets have been explained. References to atomic physic models and ion-beam plasma interaction have special emphasis. The ion beam interaction has been treated with more detail than the ion beam transport, which is a matter of present development focused on the calculation of the effective charge.

Conclusions about parametric studies are really optimistic if an appropriate tailoring of the energy deposition profile can be assumed, showing a reduction of the driver input energy required for the attainment of a significant fuel burnup or energy gain. More optimistic results will be obtained using all the capabilities of the NORCLA code.

REFERENCES

1. G. Velarde et al., Atomkernenergie/Kerntechnik 32, 58 (1978).
2. G. Velarde et al., Atomkernenergie/Kerntechnik 35, 58 (1980).
3. G. Velarde et al., Atomkernenergie/Kerntechnik 36, 213 (1980).
4. G. Velarde et al., Atomkernenergie/Kerntechnik, 44,3, (1984).
5. G. Velarde et al., Laser & Particle Beams, 4,349 (1986).
6. N.G. Cooper ed. Los Alamos National Laboratory Report No. LA-7313-MS (1983).
7. J.J. Honrubia and J.M. Aragonés, Nucl. Sci. and Eng. 93,386 (1986)
8. R.E. McFarlane, D.W. Muir, and R.M. Boicourt, Los Alamos National Laboratory Report No.LA-9303-M (1982).
9. J.M. Peek, Phys. Rev. A 26, 1030 (1982)
10. G. Velarde et al. Trans. Am. Nucl. Soc. 46, 190 (1984)
11. G. Velarde et al. Nucl. Fus. Supplement 10th Fusion IAEA 3, 121 (1985).

BEAM PLASMA INTERACTION EXPERIMENTS AT GSI

D.H.H. Hoffmann, J. Jacoby, H. Wahl, and K. Weyrich,
Max Planck Institut für Quantenoptik, D-8046 Garching and
GSI-Darmstadt, D-6100 Darmstadt, Germany

R. Noll
Fraunhofer Institut für Lasertechnik, D-5100 Aachen, Germany

C.R. Haas and B. Weikl
Lehrstuhl für Lasertechnik
RWTH Aachen, D-5100 Aachen, Germany

ABSTRACT

The present status of experimental facilities at GSI to investigate interaction phenomena of heavy ion particle beams and a hot dense plasma is described. There are currently two experimental scenarios that are persued at GSI. In one experiment the heavy ion beam of the UNILAC is injected into an externally created plasma to investigate energy loss and charge state distribution of heavy ions traversing a plasma target. The other experiment will use a very intense particle beam accelerated by five RFQ sections to heat gas targets up to temperatures of some eV. This experiment will produce a very homogenous and dense plasma.

INTRODUCTION

In the near future the development of very powerful heavy ion accelerators - as planned at GSI with a heavy ion synchrotron (SIS) injecting into a storage ring (ESR) with a beam cooling facility - will open the possibility to create a hot dense plasma by irradiating solid matter with intense heavy ion beams. Detailed experimental and theoretical knowledge of the interaction processes of intense heavy ion beams with matter is therefore one prerequisite to investigations of matter under the condition of very high energy density. Results in this field are of basic interest to plasma- and atomic physics as well, and applications range from inertial confinement fusion to heavy ion pumped short wavelength lasers.

The basic interaction processes of heavy ions with a plasma, leading to increased energy loss of heavy ion projectiles can already be studied experimentally by coupling an external plasma source to the accelerator. Therefore a z-pinch which is carefully designed to meet the requirements of an accelerator based experiment, will be set up at the UNILAC to provide a plasma target. Main emphasis in this experiment will be placed upon measurements of the energy loss and charge state distribution of heavy ions traversing the plasma.

In a second experiment we aim to produce a dense plasma with temperatures in the eV range, irradiating a pulsed gas puff target with a high intensity Kr and Xe beam (45 keV/u) from the five RFQ sections that are already completed at GSI. These sections will later form a part of the high current injector to SIS.

THE Z-PINCH EXPERIMENT

Heavy ion beams that are currently available at the GSI-UNILAC and other accelerator laboratories as well are by some orders of magnitude too low in intensity to create a target plasma. For experimental investigations of heavy ion beam plasma interactions it is therefore necessary to provide a plasma source of sufficient density and temperature.

There are currently several experiments planned to study beam plasma interactions using an external plasma[1]. A laser produced plasma coupled to a tandem accelerator will be used by a group[2] at Bruyeres-le-Chatel (France). At the tandem accelerator at Orsay another beam plasma experiment will be performed[3], using a linear discharge in Hydrogen gas as plasma source.

Light to medium heavy ions with energies up to a few MeV/u are available at Tandem accelerators. At these energies projectiles like Carbon are already or almost fully ionized. Therefore an effect of enhanced ion stopping due to a higher effective charge state Z_{eff} is negligible. At the GSI-UNILAC all heavy ions up to Uranium are available as projectiles and the energy range extends from 0.2 to 20 MeV/u.

1) Ion Plasma Interaction

The main aim of the experiment is to measure energy loss and charge state distributions of heavy ion beams traversing a plasma target. In a fully ionized plasma which consists of free electrons and bare nuclei capture of electrons to the heavy ion projectile is largely reduced because only very few channels for electron capture are open, e.g. radiative electron capture (REC) and dielectronic recombination (DR). The cross sections for these processes are by some orders of magnitude lower than Coulomb capture cross sections.

The energy loss (dE/dx) of a particle traversing matter with the velocity v_{ion} is well described by the Bethe formula:

$$dE/dx = -(eZ_{eff}\omega_p/v_{ion})^2 \, 1/2 \, \ln\Lambda. \tag{1}$$

For bound electrons the Coulomb logarithm Λ reads (nonrlativistically):

$$\Lambda = 2m_e v^2_{ion}/I_{av}.$$

For a target medium consisting of free electrons and bare nuclei the average ionization energy of bound electrons I_{av} is replaced by the plasma frequency ω_p, and thus

$$\Lambda = m_e v^2_{ion}/\hbar\omega_p$$

The free electron density in the z-pinch at the time of maximum compression is of the order of $n_e \cong 10^{19} cm^{-3}$, this yields a plasma frequency of $\omega_p = (4\pi n_e e^2/m_e)^{1/2} \cong 2 \times 10^{12}$ s^{-1}; and therefore $\hbar\omega_p \cong 0.1$ eV. At electron densities of $n_e = 10^{19}$ cm^{-3} and lower the difference between the average ionization energy $I_{av}(\cong 10 Z_T$ eV) and $\hbar\omega_p$ (≤ 0.1 eV) is already considerable, however these factors contribute only logarithmically to the energy loss.

The next important quantity in the Bethe formula is the effective charge state Z_{eff}. For solid material at room temperature the equilibrium charge state is reasonably well described by the Betz formula[4]

$$Z_{eff} = Z_0\{1 - C_1 \exp[-v_{ion}/(\alpha c Z_0^\gamma)]\} \quad (2)$$

or similar semiempirical formulas. The parameters C_1 and γ in eq. 2 are adjusted to fit experimental data. Inside the target the effective charge state Z_{eff} is determined by a dynamic equilibrium of ionization and electron capture processes. Only dielectronic recombination and radiative electron capture are important for the capture of free electrons into the heavy ion projectile. Three body recombination is only important close to solid state density or above. Radiative electron capture cross sections for heavy ions are very well established[5,6] experimentally and theoretically as well. Dielectronic recombination with fast moving heavy ions has been investigated very little until recently[7]. Ionization processes of the projectile occur through Coulomb collisions with nuclei and free electrons, and the cross sections for these processes are known with sufficient precision[8-11]. Taking into account all relevant ionization and capture processes and applying "state of the art" cross sections, average charge states of heavy ions in cold matter and plasmas can be calculated by solving the rate equations[7,12]. Fig. 1 shows the charge state of Iodine ions in a cold Hydrogen gas and a hot Hydrogen plasma, respectively. The calculation was done by T. Peter (ref. 7). At low temperatures the energy dependence of the charge state is a smooth function of the projectile energy (Betz formula), whereas at high temperatures the projectile charge remains at a very high level over a wide range of energy. Accordingly a dramatic effect on the energy deposition profile of the ions is expected. Fig. 2 compares the energy loss as a function of range for Iodine ions at 1.5 MeV/u in Hydrogen gas and plasma. The calculation for the plasma case is again taken from ref. 6, and the cold gas case is according to Northcliff and Schilling (ref. 13). Due to the fact, that in a plasma target the charge state remains at a high value until the very end of the range, a pronounced Bragg peak develops even at a comparatively low energy (1.5 MeV/u). The range shortens accordingly. In a cold gas, where bound electrons are available and electron capture reactions have large cross sections the projectile charge state decreases at lower energy and results in a smooth behavior of the energy loss.

The z-pinch experiment will therefore provide an ideal test for the predicted effects of enhanced stopping and increased average charge states of heavy ions in a target medium that is composed of free ions and bare nuclei. It will lead to a deeper understanding of electron capture processes for free electrons into highly charged heavy ions.

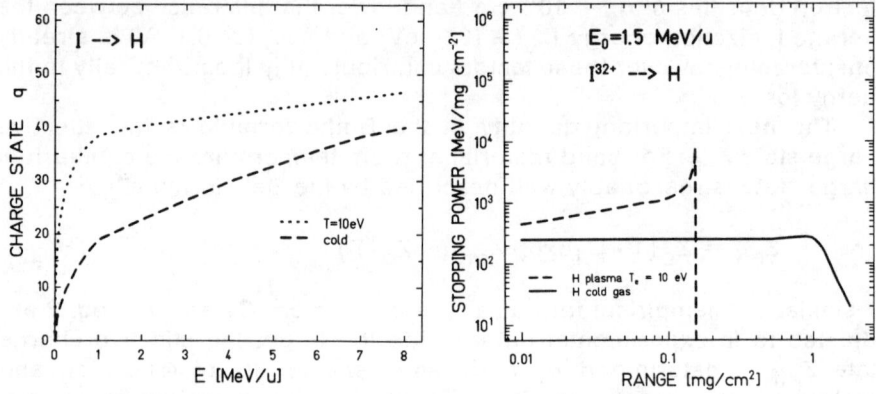

Fig. 1: Charge state of Iodine ions traversing a hot and a cold target, respectively. The calculations are from ref. 4 and 7.

Fig. 2: Stopping power of 1.5 MeV/u Iodine ions incident on a Hydrogen target. Comparison of the hot and cold target case show the importance of range shortening effects in hot matter.

2) Experimental Set-Up

The effective coupling of a plasma generator like the proposed z-pinch to the experimental facility of an accelerator is a very demanding task. Targets that are commonly used for an accelerator experiment usually have long lifetimes. Using a hot dense plasma target with a lifetime in the submicrosecond range and a repetition rate of one plasma shot every few minutes requires that a vast number of events has to be recorded in a very short time. Therefore new experimental techniques and detectors have to be developed.

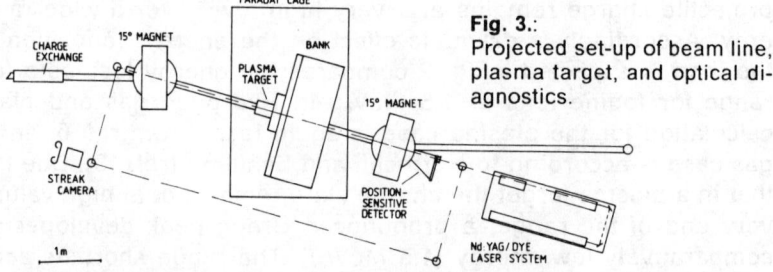

Fig. 3.: Projected set-up of beam line, plasma target, and optical diagnostics.

The Unilac beam has a microstructure of 27 MHz, and thus delivers a microbunch of particles every 37 ns. The duration of the microbunch has a half width of 1-2 ns (without the rebunching option), and filled to the space charge limit may contain up to 10^5 particles if Uranium is selected as projectile. Hence exact triggering and high reproduceability of the plasma ignition phase is a necessary condition for this experiment.

At the "stripping area" of the UNILAC where ion beams with a fixed energy of 1.4 MeV/u are available one beamline has been set aside to accomodate the z-pinch experiment. The beam optics at this beamline were redesigned completely. Fig. 3. shows a schematic outline of the new experimental area, including the optical plasma diagnostics. In this beamline the beam first passes a charge exchange target (either gas or solid), to provide the option to vary the charge state of the projectile over a wide range. Then a magnet (2.2 T) deflects the beam by $15°$ to assure charge state purity of the beam. The plasma target is housed in an rf-tight Faraday cage to shield the experimental area from high intensity rf-pulses during the plasma ignition phase. A second magnet behind the target area bends the beam by another $15°$. This opens up the possibility to inject a laser beam collinear to the ion beam for diagnostic purposes.

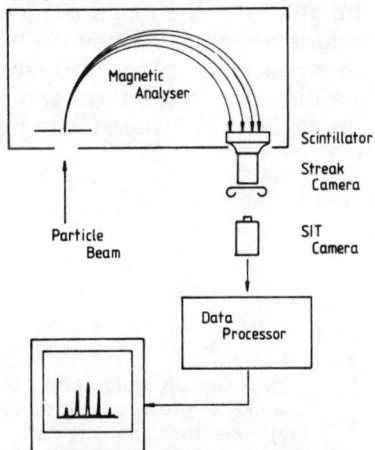

Fig. 4: Particle spectrometer

It is very essential to the experiment, that the z-pinch electrodes leave a windowless entrance and exit for the ion beam. Any window would completely deteriorate the charge state distribution that results from the beam plasma interaction. For this reason differential pumping has to be applied. Time of flight techniques using capacitive phase probes for signal generation will be applied to measure the energy loss of the projectiles. The necessary flight distance of $\cong$ 5m is provided by the beamline extending behind the second magnet. This technique is rather conventional. However, a new detector technique is required to analyze the charge state distribution of a complete beam microbunch in a couple of nanoseconds. The operation principle - a magnetic analyzer combined with a position sensitive detector - is shown in Fig. 4. The second bending magnet (Fig. 3) will serve as analyzer to seperate the charge states. According to their respective charge state the ions hit different positions of a fast scintillator, and the light output is funnelled by a light guide to the slit of a highly light sensitive streak camera with image intensifier. An intensified image of the light intensity distribution is produced on the output phosphor of the streak camera. This output is viewed by another camera and subseqently the image is digitized and reproduced on a monitor. Streak times of 10 µs allow the analysis of charge state spectra of up to 255 consecutive beam bunches. Thus the time history of beam plasma interaction at different stages of the plasma can be followed closely.

3) The Plasma Target

A fully ionized Hydrogen plasma provides a clean experimental test to investigate interaction processes between fast moving heavy ions and a dense ionized target. For a beam plasma experiment the plasma has to meet the following requirements

- **fully ionized**
- **electrons/area $\geq 10^{19}/cm^2$**
- **diameter of plasma column $\geq$ 3 mm**
- **small magnetic field on the axis**

At a density of free electrons of the order of $10^{19} cm^{-3}$ the energy loss of the projectiles is large enough ($\cong$ 20%) to give a measurable effect and the charge state distribution of the projectiles traversing this kind of target is significantly different from a cold target gas where charge exchange cross sections are dominated by transfer reactions of bound electrons. The requirement for the plasma column diameter is due to the beam spot size of approximately 3 mm at the entrance and exit apertures of the plasma tube. In an ideal case the magnetic field vanishes on the axis. Any residual field however should be small enough not to deflect the ion beam off axis. However, if the beam direction is from the anode towards the cathode of the z-pinch the magnetic field associated with the plasma current will have a focussing effect.

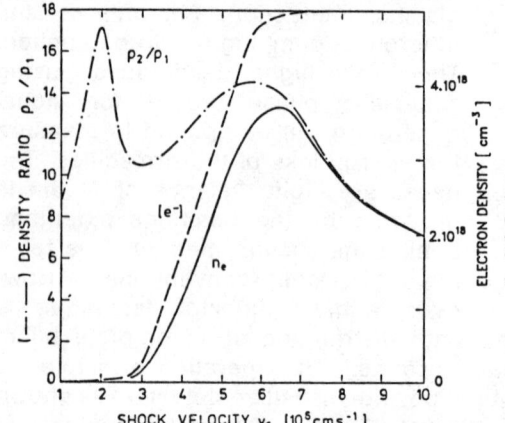

Fig. 5: Density ratio across the shock, degree of ionization and electron density as a function of the shock velocity

In a z-pinch device a plasma is generated using a fast electrical discharge in a cylindrical glass tube with a low pressure gas filling. A capacitor bank of low inductance is switched to the electrodes at the end of the discharge vessel. The plasma is ignited near the inner wall of the glass cylinder. Capacitors feed a fast rising current of typically 10^{12} A/s into the plasma load, and the azimuthal field of the axial current flow inside the plasma compresses the hollow plasma sheath towards the axis. At the end of the radial collapse a plasma cylinder is formed on the axis. The column has a length of 20 cm, with an estimated diameter in the range of 1-2 cm. The plasma lifetime on the axis should be longer than the transit time (12 ns) of 1.4 MeV/u ions for 20 cm. The design criteria of the z-pinch are:

- ignition via sliding discharge in order to create a homogenous plasma sheath.
- pinch time at current zero passage in order to minimize the magnetic field B.
- approximately acceleration free collapse to reduce the growth rate of Rayleigh-Taylor instabilities.
- optimum shock velocity to obtain high electron densities.

During the radial plasma collapse a cylindrical shock wave evolves in front of the magnetic piston. Fig. 5 shows the compression density ratio ρ_2/ρ_1 across the shock, the degree of ionization and the electron density as a function of shock velocity[14]. For shock velocities greater than $v^* = 6 \times 10^6$ cm/s hydrogen gas is fully ionized. The velocity v^* satisfies an energy matching condition. At v^* the kinetic energy of a hydrogen molecule is equal to the sum of dissociation and ionization energy of H_2. Fig. 5 shows that at the shock velocity v^* the electron density behind the compression shock attains a maximum. Hence for an optimum shock velocity v - with respect to high electron densities in a fully ionized plasma - the condition holds $v \geq v^*$.

The discharge dynamics of the z-pinch is simulated with the help of a snowplow model assuming a thin plasma sheath. The model includes the hydrodynamic equation of motion and the circuit equation. Input parameters of the model are the z-pinch geometry and the electrical data of the discharge circuit. The model equations are integrated numerically obtaining the radial position r_p of the plasma sheath and the plasma current I_p as a function of time. Fig. 6 shows a solution, which fits the above specified design criteria.

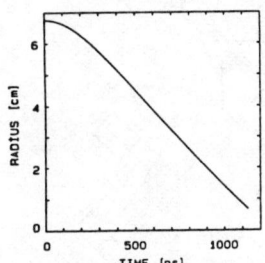
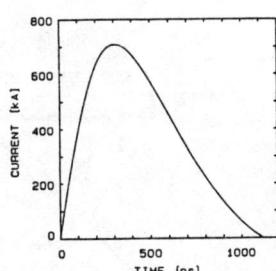

Fig. 6:
Radial position of the plasma sheath and plasma current as function of time

From these calculations we have estimated the parameters of the z-pinch:

BANK ENERGY	10 kJ
BANK VOLTAGE	70 kV
BANK CAPACITANCE	4 µF
BANK INDUCTANCE	15 nH
CURRENT RISE	5×10^{12} A/s
MAX. CURRENT	8×10^5 A
FILLING PRESSURE H_2	5 - 10 mbar
LENGTH	20 cm
RADIUS	$\cong$ 7 cm

The electron density along the path of the ion beam is measured using Mach-Zehnder interferometry, where the fringe displacement is directly related to the number of electrons per area. The density will be measured at the time of the ion bunch interaction with the plasma target.

BEAM HEATED PLASMAS: THE RFQ EXPERIMENT

With the completion of the heavy ion synchrotron (SIS) at GSI a new high current injector consisting of twelve RFQ sections will be available. However, five of these RFQ-sections are already in operation and deliver a high intensity heavy ion beam[15]. By the end of this year a Xe beam (45 keV/u) will be available with an intensity of approximately 30 mA average current in a macropulse. Microbunches with an effective duration of about 10 ns are produced at a repetition frequency of 13.5 MHz ($\cong$ 70 ns intervalls). A precision final focussing system that is currently under construction[16] will focus the beam to a spotsize of 0.6 x 2.5 mm².

The range of 45 keV/u Xe ions in different gases is about 1 mg/cm². A gas target with this areal density requires a pressure of 10 bar and a diameter of 1mm. A windowless gas target with these parameters in a vacuum chamber can be achieved with a pulsed gas puff target. We are constructing such a target, where a fast valve opens a high pressure ($p \cong 100$ bar) reservoir for a short time ($t \leq 1$ ms). The high vacuum in the chamber is maintained by a powerful kryogenic pump and additional turbo pumps.

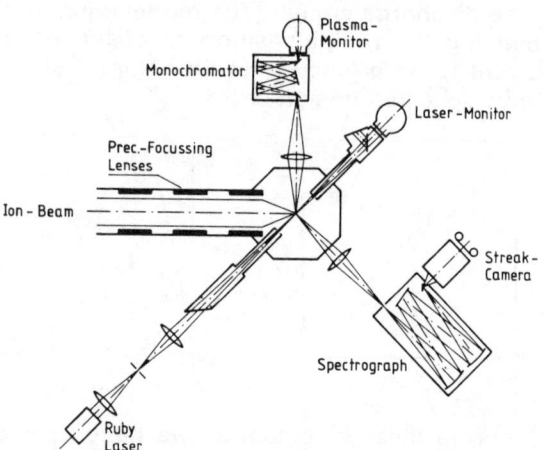

Fig. 7: Set-up for the beam heated plasma experiment with equipment for a Thomson scattering experiment.

The time dependent radiation of the beam heated gas will be monitored with an optical spectrometer coupled to a streak camera. Times relevant for the heating process depend on the velocity of sound of the gas and are in the range of 1 µs. Detailed information about temperature and free electron density will be obtained by a Thomson scattering experiment.

Some characteristic properties distinguish heavy ion beam heated plasmas from other conventional methods of plasma generation. The energy deposition occurs deep inside the target which leads to a uniformly heated volume of target material. The beam currents associated with heavy ion beams are negligible compared to those in magnetically confined plasmas, and there is no acceleration process that leads to instabilities. Hydrodynamic instabilities will finally determine the plasma lifetime which we expect to be of the order of 1 µs.

Thus high intensity heavy ion beams will soon allow the investigation of matter under conditions of high energy density.

REFERENCES

1. C. Deutsch, Ann. Phys.Fr. **11**, 1-111(1986)
2. R. Dei-Cas, J. Bardy, M.A. Beuve, J.P. Laget, A. Menier, and M. Renaud. J. Phys. **C8**, 179(1983)
3. R. Bimbot, S. Della-Negra, D. Gardes, M.F. Rivet, C. Fleurier, B. Dumax, D.H.H. Hoffmann, K. Weyrich, C. Deutsch, G. Maynard, SPQR II: A BEAM-PLASMA INTERACTION EXPERIMENT, elsewhere in these proceedings
4. H.D. Betz, G. Hortig, E. Leischner, Ch. Schmelzer, B. Stadler, and J. Weihrauch. Phys. Lett. **22**, 643(1966)
5. E. Spindler, H.-D. Betz, and F. Bell, Phys. Rev. Lett. **42**, 832 (1979)
6. Anholt, S.A. Andriamonje, E. Morenzoni, Ch. Stoller, J.D. Molitoris, W.E. Meyerhof, H. Bowman, J.-S. Xu, Z.-Z. Xu, J.O. Rasmussen, and D.H.H. Hoffmann, Phys. Rev. Lett. **53**, 234(1984)
7. T. Peter, Thesis, MPQ 105, 1985
8. M. Gryzinski, Phys. Rev. **A138**, 336(1965)
9. W. Lotz, Z. Phys. **206**, 203(1967) and **216**, 241(1968)
10. D.H.H. Hoffmann, C. Brendel, H. Genz, W. Löw, S. Müller, and A. Richter, Z. Phys. **A293** 187(1979) and references therein
11. P. Richard, in Atomic Inner Shell Processes I, edited by B. Crasemann, Academic Press, New York 1975
12. E. Nardi and Z. Zinamon, Phys. Rev. Lett. **49**, 1251(1982)
13. L.C. Northcliffe and R.F. Schilling, Nucl. Data Tables **A7**(1970)
14. R.A. Gross., Rev. of Mod. Phys. **37**, 724(1965)
15. R.W. Müller, U. Kopf, P. Spädtke, and J. Bolle, Status of the Heavy-Ion RFQ Linac, "MAXILAC", elsewhere in these proceedings
16. H. Wollnik, private communication

BEAM LOSSES IN THE STORAGE RING DUE TO CHARGE CHANGING COLLISIONS WITHIN A BEAM OF Xe^+ IONS[*]

F.Melchert, K.Rinn, K.Rink and E.Salzborn
Institut für Kernphysik, Universität Giessen
D-6300 Giessen, West-Germany

ABSTRACT

Employing the crossed-beams technique we have measured absolute cross sections both for electron capture (σ_c) and for ionization (σ_i) in collisions between two Xe^+ ions in the center-of-mass energy range from 8.7 keV to 54.7 keV. The resulting beam loss cross section $\sigma_L = 2(2\sigma_c + \sigma_i)$ increases with collision energy from $1 \cdot 10^{-16} cm^2$ to $7 \cdot 10^{-16} cm^2$. Taking into account the design parameters of the heavy ion fusion scenario HIBALL, the beam loss due to ion-ion collisions in the storage ring can be estimated to be ≈ 6 per cent.

INTRODUCTION

High current heavy ion beams of GeV energy, as projected for igniting a DT pellet, may suffer from severe intensity losses due to ion-ion interactions within the beam pulses in the storage ring. In order to obtain an accurate assessment of the expected loss rates the cross sections both for

electron capture (σ_c): $A^+ + A^+ \rightarrow A^0 + A^{2+}$ (1)

and for

ionization (σ_i): $A^+ + A^+ \rightarrow A^+ + A^{2+} + e$ (2)

have to be known as a function of the relative velocity.

[*]Work supported by BMFT

The total beam loss cross section σ_L is given by

$$\sigma_L = 2(2\sigma_c + \sigma_i) \qquad (3)$$

since 2 particles are lost per interaction in the electron capture reaction (1) and, furthermore, both ions simultaneously act both as projectile and target particles.

In the last decade substantial attention has been directed towards collisions between positive ions. And only since 1977 have reliable experimental data for a number of different collision systems become available[1,2]. However, most of the data involve collisions between light ions. Furthermore, only the total A^{2+} production cross sections $\sigma_{2+} = \sigma_c + \sigma_i$ have in general been measured and not the cross sections σ_c and σ_i separately.

EXPERIMENTAL RESULTS

In view of the small data basis available so far, we have set up a crossed-beams experiment for measuring the cross sections σ_c and σ_i for collisions between two Xe^+ ions at center-of-mass energies ranging from 8.7 keV to 54.7 keV. A detailed description of the apparatus and the measuring procedures has been given in previous papers[3,4]. In short, two momentum-analyzed Xe^+ ion beams of adjustable energies (up to 150 keV and up to 15 keV, respectively) are arranged to intersect at an angle of 45° in an ultra-high vacuum region of about $1 \cdot 10^{-10}$ mbar. The collision products formed in both beams are analyzed with respect to their charge states by electrostatic deflectors downstream of the interaction region. The parent Xe^+ ion beams are recorded by biased Faraday cups, whereas the reaction products (Xe^0 and Xe^{2+}) are counted by single-particle detectors.

Although the experimental approach, in principle, appears to be straightforward, inherent difficulties result from the low ion densities obtainable in the two ion beams. Since the residual gas density, even at pressures of $1 \cdot 10^{-10}$ mbar, exceeds the ion beam densities by orders of magnitude, a low signal rate (~Hz) has to be detected in the presence of a large background rate (~kHz) of reaction products originating from ion collisions with residual gas particles.

A coincidence technique was employed to separate signal from background events in measuring the cross section σ_c for electron capture. The cross section σ_i for ionization is obtained from the difference $\sigma_i = \sigma_{2+} - \sigma_c$ with σ_{2+} being the cross section for the total Xe^{2+} ion production (sum of reactions (1) and (2)). In measuring σ_{2+} a beam pulsing technique was employed to discriminate signal from background events.

The measured cross sections σ_c and σ_{2+} showed an unexpected abrupt increase with decreasing CM energy below about 20 keV. We found that this effect is due to metastable $(Xe^+)^*$ ions in the parent ion beam. By means of a method described by Vujović et al[5] we could considerably decrease the fraction of metastable ions in the slow Xe$^+$ parent beam. To this end a gas cell was inserted inside the analyzing magnet and the apparent cross sections σ_c^a and σ_{2+}^a were measured as a function of the pressure in this cell. Both of these apparent cross sections decreased with increasing pressure and finally attained equilibrium values which were independent of the gas species admitted. This suggests that metastable Xe$^+$ states, which lie approximately 11 - 15 eV above the ground state, are preferentially removed from the ion beam by one-electron loss collisions. The equilibrium values of σ_c and σ_{2+} exhibit the expected energy dependence. They join

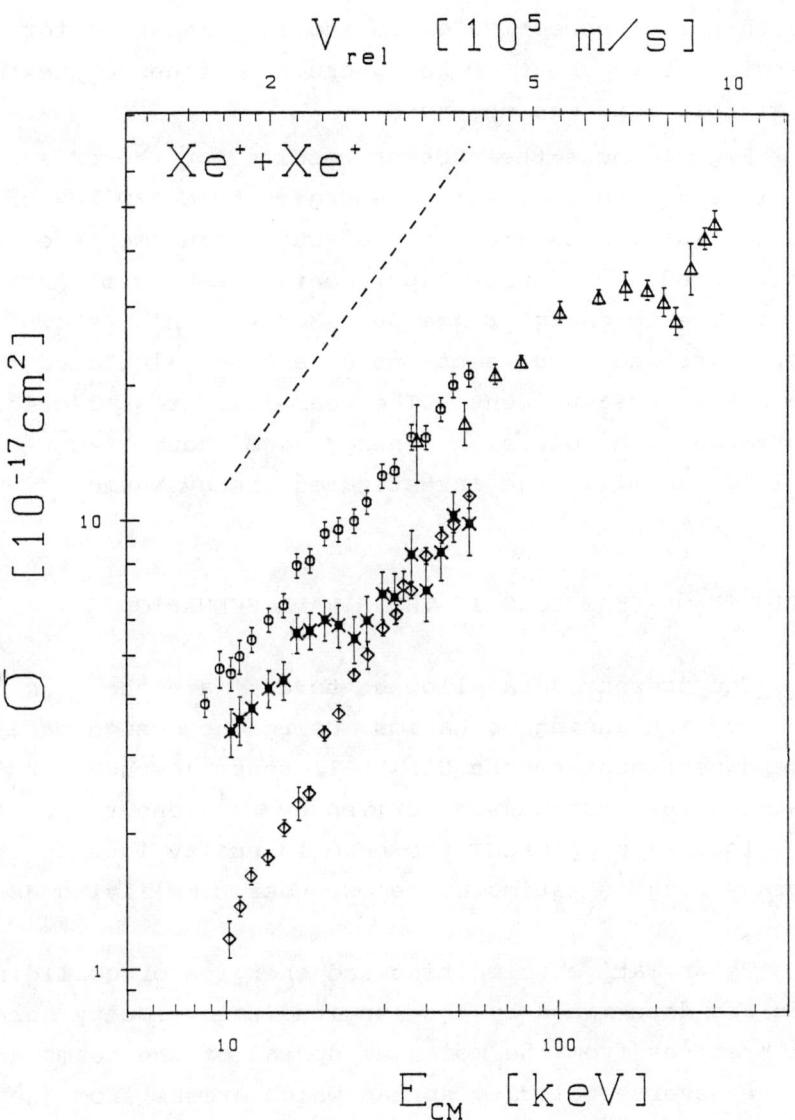

Fig. 1: Cross sections for $Xe^+ + Xe^+$ collisions:

◊   electron capture σ_c
✶   ionization σ_i } this work
o Xe^{2+} production $\sigma_{2+} = \sigma_c + \sigma_i$
----   beam loss $\sigma_L = 2(2\sigma_c + \sigma_i)$
△ Xe^{2+} production $\sigma_{2+} = \sigma_c + \sigma_i$ (Angel et al[6])

smoothly the respective cross sections measured for CM energies above about 20 keV where no influence due to metastable ions was observed.

Fig. 1 shows the present results for the cross sections σ_c and σ_{2+}. For CM energies below 20 keV the equilibrium values are plotted. Our measurements of σ_{2+} are in good accord with experimental results of Angel et al[6] in the CM energy range 38 - 303 keV. Also shown in Fig. 1 are the cross sections σ_i and σ_L calculated from the present measurements. The beam loss cross section σ_L increases with collision energy from about $1 \cdot 10^{-16}$ cm^2 to $7 \cdot 10^{-16}$ cm^2 in the investigated energy range.

ESTIMATE OF BEAM LOSS IN THE HIBALL SCENARIO

The present data allow us to estimate the beam loss in heavy ion fusion scenarios. Using the design parameters pertinent to the HIBALL-II reactor study[7], but assuming that fusion were driven by Xe$^+$ ions rather than Bi$^+$ ions as projected, the beam intensity loss in the storage ring is estimated below under simplifying assumptions.

The relative velocities and energies of colliding ions are determined by the longitudinal velocity spread which arises from the momentum spread of the beam, and the transverse velocity spread which arises from the betatron oscillations about the equilibrium orbit.

According to King and Lawson[8] the relative longitudinal energy is given to good approximation by

$$\Delta E_L \simeq \frac{1}{2} A \cdot U \cdot \beta^2 \cdot (\delta p/p)^2 \qquad (4)$$

where A is the ion mass number (A = 132 for Xe), U = 931.48 MeV is the atomic mass unit, $\beta = v/c$, and

$\delta p/p$ is the full momentum spread. With the HIBALL-II storage ring design parameters (β = 0.30875 and $\delta p/p = \pm 1 \cdot 10^{-4}$) we obtain from eq. (4) for the relative energy of the longitudinal motion $\Delta E_L \simeq$ 240 eV.

The relative transverse energy is given[8] by

$$\Delta E_T = 2 \, A \cdot U \cdot F^2 \cdot (\varepsilon_n/y_n)^2 \cdot \beta \cdot \gamma \qquad (5)$$

where the quantity F may be regarded as a form factor (which usually is in the region of 2), $\gamma = (1-\beta^2)^{-1/2}$, and $\varepsilon_n = \beta \cdot \gamma \cdot \varepsilon$ denotes the normalized emittance. The normalized beam amplitude $y_n = (\beta \cdot \gamma)^{1/2} \cdot y_{max}$ can be expressed[9] by $y_n = (\varepsilon_n \cdot \bar{R}/\nu_H)^{1/2}$ where $\bar{R}$ is the average storage ring radius, and ν_H denotes the number of betatron oscillations per turn. With the HIBALL-II storage ring parameters ($\varepsilon = 30 \cdot 10^{-6}$ m, $\bar{R}$ = 118 m, ν_H = 9.85, and γ = 1.051) we obtain from eq. (5) for the relative energy of the transverse motion $\Delta E_T \simeq$ 260 keV.

The maximum relative energy in $Xe^+ + Xe^+$ collisions is therefore given by $\Delta E_{max} = \Delta E_L + \Delta E_T \simeq \Delta E_T \simeq$ 260 keV which corresponds to a center-of-mass energy $E_{cm} \simeq$ 130 keV. From Fig. 1 we obtain an (extrapolated) beam loss cross section $\sigma_L = 1.5 \cdot 10^{-15}$ cm^2 at E_{cm} = 130 keV.

According to Le Duff and Maidment[10] the beam lifetime τ in the storage ring is given by

$$\tau = \pi \cdot \frac{qe}{I} \cdot \frac{\varepsilon^{1/2} \cdot \beta_{eff}^{3/2} \cdot \gamma}{\sigma_{L,CM}} \qquad (6)$$

where q is the charge state of the ion beam, e is the charge of the electron, I is the ion beam current, and $\sigma_{L,CM}$ denotes the beam loss cross section at CM energies.

The effective value of the envelope function β can be expressed[9] by $\beta_{eff} = \bar{R}/\nu_H$. Equation (6) has been derived under some simplifying assumptions the most important of which are:

a) The relative velocity is that due to betatron motion only and is an average relative velocity.

b) The beam loss cross section is independent of the relative velocity.

c) The particle density is assumed to be uniform.

Using the same values for $\varepsilon, \gamma, \bar{R}$, and ν_H as before, and taking $\sigma_{L,CM} = 1.5 \cdot 10^{-15}$ cm^2 and I = 12.5 A, we obtain from eq. (6) a beam lifetime τ = 64 ms. With the maximum HIBALL II storage time t = 4 ms we estimate the beam loss to be $1 - \exp(-t/\tau) \simeq 6$ %.

This estimate certainly represents an upper limit to what is likely to be achieved. Only a small number of ions will approach the maximum relative energy where the loss cross section σ_L is largest. Most collisions will occur at lower relative energies. To obtain a more accurate estimate of the beam loss, numerical calculations should be performed using realistic density distributions of ions in 6-dimensional phase space.

The authors gratefully acknowledge fruitful discussions with Dr. R.W. Müller, Dr. I. Hofmann and Dr. N.M. King.

REFERENCES

1. H.B.Gilbody, in "Physics of Electronic and Atomic Collisions", ed. S.Datz (North Holland Publ.Comp. 1982), p. 223

2. K.Dolder and B.Peart, Rep.Prog.Phys. 48, 1283 (1985)

3. K.Rinn, F.Melchert and E.Salzborn, J.Phys.B 18, 3783 (1985)

4. K.Rinn, F.Melchert, K.Rink and E. Salzborn, J.Phys.B, in print

5. M.Vujović, M.Matić, B.Cobić and Y.S.Goordeev, J.Phys.B 5, 2085 (1972)

6. G.C.Angel, K.F.Dunn, P.A.Neill and H.B.Gilbody, J.Phys.B 13, L 391 (1980)

7. HIBALL-II: An Improved Conceptual Heavy Ion Driven Fusion Reactor Study, Kernforschungszentrum Karlsruhe, Report KfK 3840, July 1985

8. N.M.King and J.D.Lawson, private communication (HIF/P6 Rev.1)

9. R.W.Müller (GSI), private communication

10. J.R.Le Duff and J.R.Maidment, Proc.HIF workshop, Berkeley 1979, LBL - 10301 (1980), p. 292

SPQR II: A BEAM-PLASMA INTERACTION EXPERIMENT

R. Bimbot, S. Della-Negra, D. Gardès, M.F. Rivet
I.P.N., Bât. 100, University Paris XI, 91405 ORSAY

C. Fleurier, B. Dumax
G.R.E.M.I./C.N.R.S., 45046 ORLEANS-LA-SOURCE

D.H.H. Hoffman and K. Weyrich
G.S.I., Postfach 110541, 6100-DARMSTADT 11

C. Deutsch, G. Maynard
L.P.G.P., Bât. 212, University Paris XI, 91405 ORSAY

ABSTRACT

SPQRII is an interaction experiment designed to probe energy -and charge- exchange of C^{n+} ions at 2 MeV/a.m.u., flowing through a fully ionized plasma column of hydrogen with $n\ell = 10^{19}$ e-cm^{-2} at $T = 5$ eV.
One expects a factor of two enhanced stopping over the cold gas case.

INTRODUCTION

We present a novel benchmark designed for testing the enhanced stopping of point ions projectiles when they interact with the free electrons in a nearly completely ionized hydrogen discharge. Basic goals include in situ measurements of charge exchange and energy loss for incoming ions of potential interest in assessing the validity of particle-driven ICF[1,2]. We are currently using the Tandem facility **L'EMPEREUR** at ORSAY, to produce highly stripped ions with an energy a few MeV/a.m.u.

The corresponding beam is then intercepted by a linear discharge of hydrogen where the stopping is essentially due to encounters with free electrons with a few eV temperature.

The corresponding plasma has already been carefully diagnosed through laser interferometry[3] and Stark broadening[3] of H_α and H_β lines arising from bound-bound transitions in the remaining excited neutral atoms.

We thus expect an accurate and simultaneous determination of the plasma parameters (n_e, T) altogether with the characteristics of ejected particles (energy and charge).

These latter are then deduced from accurate measurements of the energy loss (within 5 %) through a 9 m long time-of-flight device[8], in line with the plasma column.

Again, we expect to capitalize on a large body of accurate[4-6] measurements for stopping powers in cold gases and solids.

A BIT OF NUMEROLOGY

All HIF scenarios claim that current densities up to 10 kA/cm^2 are required to achieve a breakeven[1].

Nevertheless, even in these unusual conditions, the average ion-ion distance in the beam remains much larger than the electron fluid screening lengths, displayed in Table I. These latter are deduced either from Debye-Hückel theory with

$$\lambda_D = 6.90 \left(\frac{T(K)}{n(cm^{-3})}\right) \quad , \quad k_B T \gg \varepsilon_F \quad , \tag{1}$$

and the Fermi energy

$$\varepsilon_F = \frac{1.84}{r^2} \quad , \quad \{r_s = \left(\frac{3}{4\pi n_e}\right)^{1/3} a_0^{-1}\}$$

or from the Thomas-Fermi expression $\sim (0.61 \, r_s)^{1/2}$, at low enough temperature. These data remain always much smaller than the ion interparticle distances in the beams considered in table II.

Table I - Screening lengths (Å) in dense electron fluid

n_e \ T_e	10^4 K	10^5 K	10^6 K	10^7 K	10^8 K
10^{21} cm^{-3}	2.2	6.9	22	69	
10^{22} cm^{-3}	0.8	2.2	6.9	22	69
10^{23} cm^{-3}	0.5	0.7	2.2	6.9	22
10^{24} cm^{-3}	0.35	0.35	0.7	2.2	6.9
10^{25} cm^{-3}	0.24	0.24	0.25	0.7	2.2
10^{26} cm^{-3}	0.16	0.16	0.16	0.2	0.7

Table II - Beam parameters (maximum power 250 TW/cm^2)

	$\beta = V_1/c$	E_1	Interparticle distance
Proton	0.05-0.15	1-10 MeV	60-90 Å
Bi^{2+}	0.3	30 GeV	2 500 Å

These simple considerations allow us to reduce the beam-target interaction to an ion-target one, by neglecting collective aspects in a first approach.

This welcome simplification should nevertheless be taken with a grain of salt for the case of protons.

The corresponding current densities may well range up to MA/cm^2, so that some cautions should be exercised in every practical stiuation dealing with LIB.

With these minor restrictions taken into account, one is entitled to make use of conventional wisdom according to which intense beams are likely to appear dilute in target.

Had we considered intense electron beams, the collective phenomena would not have been so easily eliminated. For instance, potential wells can develop in the plasma produced in the heating of a thin foil, so that incoming projectiles are likely to be trapped, and accelerated backward after several bouncing periods.

ENHANCED STOPPING IN PLASMAS

The celebrated 3B (Bohr-Bethe-Bloch) formula for the stopping of nonrelativistic point charges by isolated atom or cold matter may be straightforwardly extended to hot and partially ionized material. Then, bound and free electrons, and also to a very limited extent, plasma ions, contribute to the projectile stopping. Restricting our attention to the overwhelming electron contribution, one thus writes

$$-\frac{dE}{dx} = \frac{4\pi(Z_1 e^2)^2 n}{m_e V_1^2} \{(Z_T - \overline{Z}) \ln \Lambda_B + \overline{Z} \ln \Lambda_F\} \qquad (2)$$

where 1 refers to projectile effective charge and velocity.
Z_T and Z respectively denote target atomic number and average ionization (number of free electrons/nucleus). For high (with respect to target electron velocities but not yet relativistic) velocities,

$$\Lambda_B = \frac{2m_e V_1^2}{I_{av}} \qquad (3)$$

with I_{av} = mean excitation energy
while

$$\Lambda_F = \frac{m_e V_1^2}{\hbar \omega_p} \qquad (4)$$

where ω_p = target plasma frequency.

For illustration purposes, Λ_B and Λ_F are plotted in column 2 and 3 in Table III for an incoming ion with an energy ~ 0.5 MeV/a.m.u. interacting with 30 cm of fully ionized Hydrogen with $n = 4 \times 10^{17}$ cm^{-3}.

Table III- Energy losses (MeV) in cold and fully ionized Hydrogen with $n\ell = 10^{19}$ e-cm^2 and a beam initial energy = 0.5 MeV/a

Projectile charge Z_1	Gas	Plasma
H$^+$	0.0182	0.050
He^{2+}	0.0652	0.1773
C^{4+}	0.3455	0.9407
Si^{7+}	0.9034	2.443
U^{92+}	3.606	9.661

One thus witnesses a 300% stopping enhancement in plasma as compared to the equivalent one in cold gas.

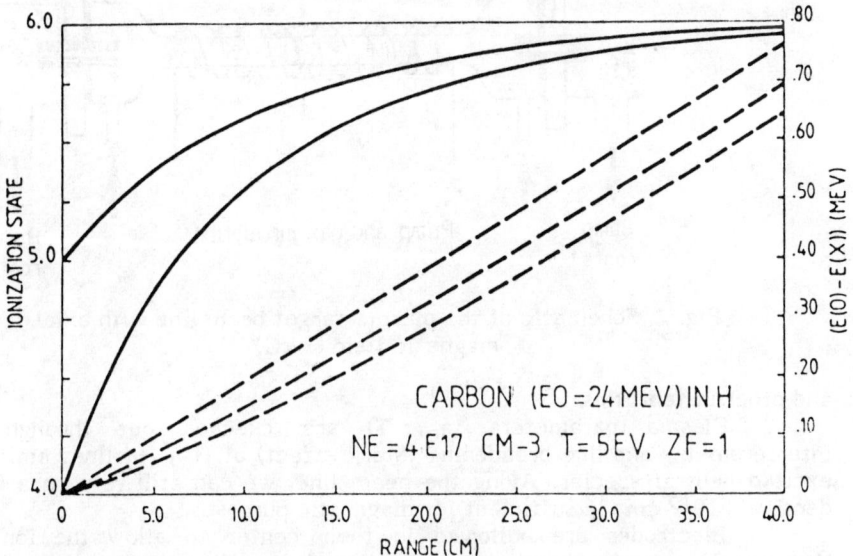

Fig. 1. Effective charges and energy losses of C^{4+} ions with 2 MeV/a.m.u. in fully ionized hydrogen

In Fig. 1, we display results more closely related to the SPQRII interaction experiment[7]. One can see how the effective charge results from a dynamic balance between ionization and recombination. However, the latter is much less likely to occur in a plasma. The important point explains a marked discrepancy between the effective charge in plasmas and the corresponding, in cold matter, where the usual expression (Brown-Moak-Betz)

$$Z_{eff} = Z \{1 - 1.034 \exp [-(v_1/2.19 \; 10^8 \; cm/s) \; Z^{-0.688}]\} \quad (5)$$

applies. Z is the projectile atomic number.

PLASMA PRODUCTION

Our basic goal is to simulate an ablation corona resulting from an implosion conducted through intense ion beams. This is achieved with the discharge tube pictured on Fig. 2. Low pressure hydrogen (10 Torr) is introduced in a quartz capsule (neck diameter = 25 cm, length = 40 cm). High power electric discharge (20 kA) ignites a plasma. A free electron density $\sim 4-5 \times 10^{17}$ e-cm^{-3} is then established for 25 µsec. An axial and steady 2 kG magnetic field prevents the onset of Rayleigh-Taylor

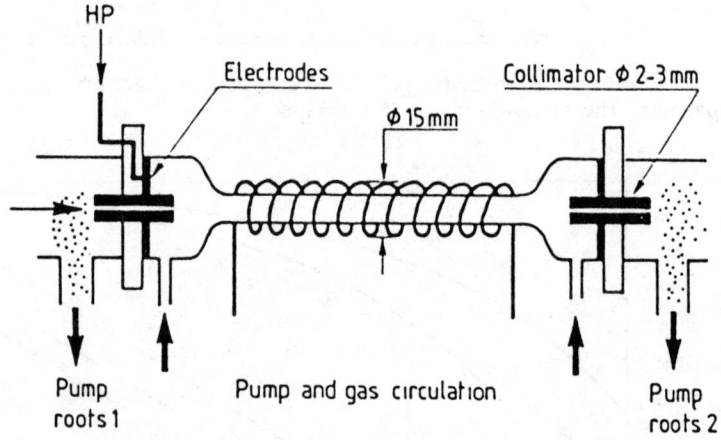

Fig. 2. Schematic of the plasma-target beam line with axial magnetic field (2 kG)

and other instabilities.

Plasma parameters (n_e, T) are ckecked out through laser interferometry and line broadening (Stark effect) of H_α, H_β lines emitted by excited neutral species. Along the beam line, we can still rely on a neutral density $\sim 10^{15}$ cm^{-3}, sufficient for diagnostic purposes.

Electrodes are hollowed in their center to allow the ion beam through. Moreover, there are no solid plasma boundaries. An efficient differential pumping (Fig. 3) secures a swift circulation of helium or hydrogen.

The primary pump works at 17 ℓ/sec. The roots at 140 ℓ/sec, while the booster operates at 2000 ℓ/sec in air or 6000 ℓ/sec in H and He.

The whole set-up is located, together with the capacitor bank in a Faraday cage.

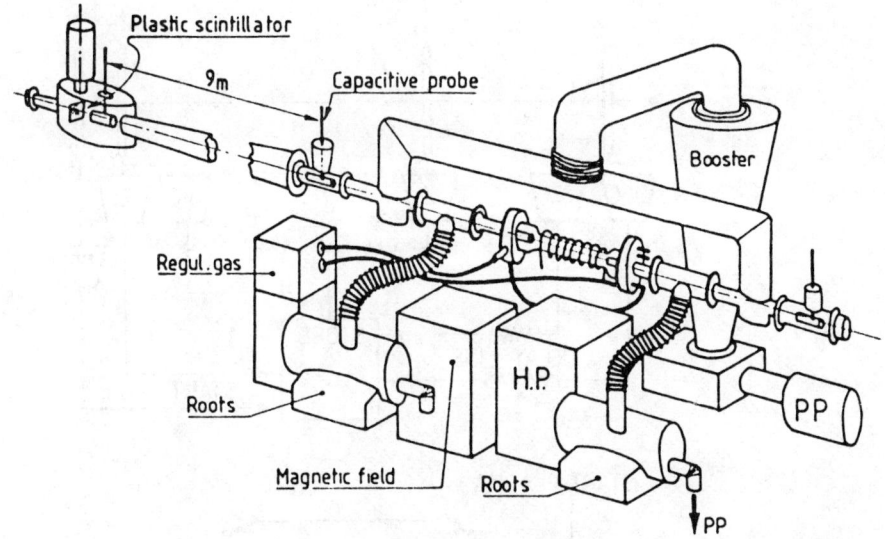

Fig. 3. General set-up including discharge, differential pumping and time-of-flight

TIME - OF - FLIGHT

Four time pick-ups (Fig. 4) through inductive probes (phasensonds) are displayed along the beam line[8]. Times of flight are then measured with and without gas, and also before and after the plasma column. One thus estimates straggling with the aid of a 9 m. T.O.F. (i.e. 397 nsec for ^{12}C at 40 Mev). The expected energy loss $\sim$ a few hundred keV, demands a rigorous T.O.F. calibration.

Preliminary trials are required to optimize the beam pulse shape. The whole line has also to be shielded electromagnetically against plasma noise. According to Fig. 4, the time sequence works as follows:
.t_0: zero cross-over signal locates the beam pulse centroid before interaction. One thus determines the width of the incident pulse before interaction, and the T.O.F. analysis gets triggered out.
.t_1: zero cross-over time pick off yields the centroid of the pulse after interaction.
.t_2: zero cross-over yields pulse centroid after a 9 m flight.
.t_3: plastic scintillator checks out stop signal.

The overall discharge duration is $\sim$200 µS. It can be repeated every 2 minutes.

The T.O.F. should allow a 5% accuracy for the parameters of ejected particles. This is less than the uncertainty $\sim$20% on the plasma parameters (n_e,T) themselves.

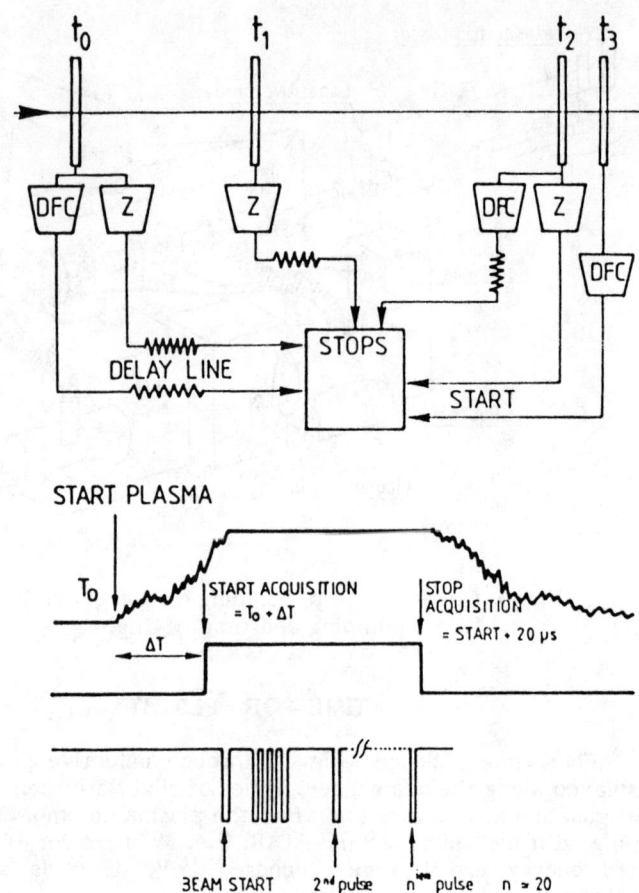

Fig. 4. Expected time sequence: 4 pick-up times with beam-target plasma synchronization

REFERENCES

1. C. Deutsch, Ann. Phys. Fr. **11**, 1 (1986).
2. C. Deutsch, G. Maynard and H. Minoo, in <u>Laser Interaction and Related Plasma Phenomena</u>, (Plenum, N.Y., 1984), p. 1013.
3. C. Fleurier and P. Legall, J. Phys. **B17**, 3311 (1984).
. F. Hubert, A. Fleury, R. Bimbot and D. Gardès, Ann. Phys. Fr. **5**, 3 (1980).

5. P.H. Mokler, D.H.H. Hoffmann, W.A. Schönfeldt, D. Maor, W.E. Meyerhof and Z. Stachura, Nucl. Instr. Meth., **B4**, 34 (1984).
6. R. Bimbot, H. Gauvin, I. Orliange, R. Anne, G. Bastin and F. Hubert, IPN O-DRE-86-06, Nucl. Inst. Methods (to appear).
7. G. Maynard, Private Communication, April 1985, Orsay.
8. P. Sigmund and K.B. Winterbon, Nucl. Instr. Meth., **119**, 541 (1974).

RADIATIVE TRANSFER IN STATISTICALLY HETEROGENEOUS MIXTURES

C. Deutsch and D. Vanderhaegen*
Laboratoire de Physique des Gaz et des Plasmas
Université Paris XI, 91405 ORSAY, France.

ABSTRACT

With a simple statistical model for two-component mixtures recently introduced by Pomraning, we solve exactly three problems of radiative transfer in heterogeneous media.

INTRODUCTION

The importance of radiation transport in inertial confinement fusion (ICF) target calculations has been discussed many times[6]. It has been pointed out that radiation plays a much more important role in ion-beam-driven targets than in the laser-fusion targets. This is because of the fact that in laser-fusion targets, suprathermal electron transport in general is more important than radiation transport and electron thermal conduction. In ion-beam fusion targets, on the other hand, radiation transport and the electron thermal conduction are the only means of energy transfer. The ions deposit energy in high-density target material which makes electron thermal conduction very ineffective. In the absorption region of ion-beam targets the temperature increases to a few hundred eV. At these temperatures the radiation transport dominates the electron thermal conduction.

The presence of very different materials in the structure of fusion pellets produces Rayleigh-Taylor instabilities near the interfaces[5,6]. These instabilities create mixing zones which are spatially localized. They can greatly influence radiative transfer and modify hydrodynamic compression[3,4].

This type of heterogeneous mixture can be statistically described to derive the evolution of the mean radiative intensity and an effective opacity. Pomraning has recently proposed a simple statistical model for a two-components A-B medium. We present the exact solution for three radiation transfer problems: pure absorption, absorption-emission with a diffusion approximation and two-temperatures medium. The model is finally extended to include a linear composition gradient.

MODEL

Levermore et al.[1] describes the statistical mixture of A and B as an isotropic and spatially homogeneous Markov process. There are two states A and B, with mean lengths λ_A and λ_B (Fig. 1). Then, we can construct[2] the matrix of conditional transition probabilities P(x), by means of second-kind Kolmogorov equations :

*ALso: C.E.N.-Limeil, Limeil-Brevannes, France.

$$dP/dx = W\,P \qquad (1)$$

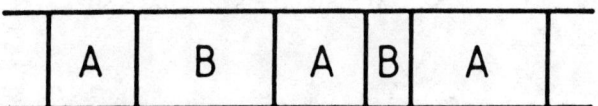

Fig. 1. Randomly distributed binary mixture

$P_{ij}(x)$ is the probability for the medium to have a i subscript at position x, knowing that its subscript is j at position x = 0.
$\overline{\overline{W}}$ is the transition matrix

$$\overline{\overline{W}} = \begin{pmatrix} -\dfrac{1}{\lambda_A} & \dfrac{1}{\lambda_B} \\ \dfrac{1}{\lambda_A} & -\dfrac{1}{\lambda_B} \end{pmatrix} \qquad (2)$$

with
$$P_{ij}(0) = \delta_{ij} \qquad i,j = A,B \qquad (3)$$

So:

$$P_{ii}(x) = \left(\dfrac{\lambda_i}{\lambda_A + \lambda_B} - 1\right)(1 - e^{-x/\lambda_P}) + 1$$

$$P_{ij}(x) = \left(1 - \dfrac{\lambda_i}{\lambda_A + \lambda_B}\right)(1 - e^{-x/\lambda_P}) \quad \text{if } i \neq j \qquad (4)$$

with:

$$\dfrac{1}{\lambda_P} = \dfrac{1}{\lambda_A} + \dfrac{1}{\lambda_B} \qquad (5)$$

From P(x), we obtain the mean value of any g by :

$$\langle g \rangle = P_A\, g_A + P_B\, g_B \qquad (6)$$

with g_A (resp. g_B) the value of g in the A (resp. B) medium and

$$P_A = \dfrac{\lambda_A}{\lambda_A + \lambda_B} \qquad (7)$$

$$P_B = \dfrac{\lambda_B}{\lambda_A + \lambda_B}$$

We can also calculate the correlation function between g and h

$$C_{gh}(x) = \langle (g(x) - \langle g \rangle)(h(x) - \langle h \rangle) \rangle \qquad (8)$$

so that

$$C_{gh}(x) = P_A P_B (g_A - g_B)(h_A - h_B) e^{-x/\lambda_p} \qquad (9)$$

and λ_p can be considered as a correlation length.

EXACT SOLUTION

We are going to calculate the evolution of the mean value of the intensity $\langle I \rangle(x)$, which is the statistical mean value of $I(x)$, $I(x)$ is the solution of a stationary, frequency independent, transport equation[7]

$$\frac{dI}{dx} = F(I, \sigma(x), x) \qquad (10)$$

where $I(x)$ is the specific radiative intensity
$F(I, \sigma(x), x)$ is the source-absorption term
$\sigma(x)$ is the local opacity of the material

given by

$$\sigma(x) = \frac{\sigma_A + \sigma_B}{2} + \frac{\sigma_A - \sigma_B}{2} \xi(x) \qquad (11)$$

where σ_A (resp. σ_B) is the opacity of the A (resp. B) medium and $\xi(x)$ is a non-symmetric dichotomic process i-e it takes the values +1 or -1 according to the characteristic lengths λ_A and λ_B.

$\sigma(x)$ is a Markovian process so the joint process (I, σ) is Markovian too. We can define its density probability $P(I, \sigma, x)$ which is the probability of $I(x)$ and $\sigma(x)$ to be equal to I and σ at abscissa x.

The knowledge of P allows us to calculate $\langle I \rangle$ by averaging over the values of I and σ.

$$\langle I \rangle = \sum_{i=A,B} \int P_i(I,x) \, I \, dI \qquad (12)$$

where

$$P_i(I,x) = P(I, \sigma_i, x) \qquad i = A, B$$

Now we must establish the evolution equation of P.

If we choose a peculiar realization Σ of σ, we can write the Liouville Eq. (10) for $P(I, \Sigma, x)$

$$\frac{\partial \tilde{P}}{\partial x} = -\frac{\partial}{\partial I}(F(I,\Sigma,x)\tilde{P}) \qquad (13)$$

To deal with the whole process σ, we must add the Markovian contribution WP.

So $P_i(I,x)$ satisfies

$$\frac{\partial P_i}{\partial x} = -\frac{\partial}{\partial I}(F(I,\sigma_i,x)P_i) + \overline{(WP)}_i \qquad (14)$$

We can use this method to solve radiation transfer problems.

PURE ABSORPTION

We consider a semi-infinite medium with an incident radiation intensity I_0 at $x = 0$ and no source inside the medium. Then we have:

$$F(I,\sigma,x) = -\sigma(x)I \qquad (15)$$

Introducing the moments

$$m_i = \int I\, P_i\, dI \qquad i = A,B \qquad (16)$$

and integrating by parts in Eq. (14) we get

$$\frac{\partial \overline{m}}{\partial x} = \overline{\overline{T}}\, \overline{m} \qquad (17)$$

with

$$\overline{\overline{T}} = \begin{pmatrix} -\sigma_A - \frac{1}{\lambda_A} & \frac{1}{\lambda_B} \\ \frac{1}{\lambda_A} & -\sigma_B - \frac{1}{\lambda_B} \end{pmatrix} \qquad (18)$$

$$\overline{m} = \begin{pmatrix} m_A \\ m_B \end{pmatrix}$$

So $\langle I \rangle(x) = m_A(x) + m_B(x)$ (19)

A straightforward algebraic calculation yields

$$\langle I \rangle(x) = I_0\, S\, e^{\lambda_+ x} + (1-S)\, e^{\lambda_- x} \qquad , \qquad (20)$$

with

$$\lambda_\pm = -[\frac{\sigma_A + \sigma_B}{2} + \frac{1}{2}(\frac{1}{\lambda_A} + \frac{1}{\lambda_B})] \pm \sqrt{\Delta'} \qquad (21)$$

$$\Delta' = \frac{1}{4}(\frac{1}{\lambda_A} + \frac{1}{\lambda_B})^2 + (\frac{\sigma_A - \sigma_B}{2})^2 - (\frac{\sigma_A - \sigma_B}{2})(\frac{1}{\lambda_B} - \frac{1}{\lambda_A})$$

$$S = \frac{1}{2\sqrt{\Delta'}}\left[\frac{1}{2}(\frac{1}{\lambda_A} + \frac{1}{\lambda_B}) + \sqrt{\Delta'} - (\frac{\sigma_A - \sigma_B}{2})\lambda_p(\frac{1}{\lambda_B} - \frac{1}{\lambda_A})\right]$$

we can check that:
- $\lambda_\pm$ are negative
- when $\sigma_A = \sigma_B = \sigma$, the solution is $\langle I \rangle(x) = I_0 e^{-\sigma x}$
- when there is only one medium left, $p_A = 0$

$$\langle I \rangle = I_0 e^{-\sigma_B x}$$

Asymptotically when $x \to \infty$, ones obtains

$$\langle I \rangle = I_0 S e^{\lambda_+ x} \qquad (22)$$

So we cannot recover the proper evolution of $\langle I \rangle(x)$, with a single constant effective opacity, we also have to renormalize the incident radiation intensity from I_0 to $S\,I_0$. Otherwise, we can define a mean absorption length ℓ_m by

$$\ell_m = \frac{I}{I_0}\int_0^\infty x\, \frac{dI}{dx}\, dx \qquad (23)$$

So

$$\ell_m^{-1} = \langle \sigma \rangle - \frac{\text{Var}\,\sigma}{\frac{1}{\lambda_p} + \sigma_A + \sigma_B - \langle\sigma\rangle} \qquad (24)$$

with

$$\langle \sigma \rangle = p_A \sigma_A + p_B \sigma_B$$
$$\text{Var}\,\sigma = p_A p_B (\sigma_A - \sigma_B)^2. \qquad (25)$$

ABSORPTION-EMISSION

The stationary radiation transfer equation reads

$$\vec{\Omega}\cdot\vec{\nabla} I = -\sigma(\vec{r})\,[I - B(\vec{r})] \qquad (26)$$

where $I(\vec{\Omega},\vec{r})$ is the specific radiative intensity at position $\vec{r}$ and in the direction $\vec{\Omega}$

$\sigma(\vec{r})\,B(\vec{r})$ is the source function according to Kirchoff law.

In this case $B(\vec{r})$ is a deterministic function which is supposed to be known.

Without scattering, we can single out a given direction. So, if we define s, as the coordinate along direction $\vec{\Omega}$ Eq. (26) becomes

$$\frac{\partial I}{\partial s} = -\sigma(s)\,[I - B(s)] \qquad (27)$$

Where is always given by Eq. (11).

Taking $I = B$ at $s = 0$, one obtains after solving the analogue of Eq. (14)

$$<I>(s) = B(s) - \int_0^s ds'\,(\alpha e^{\lambda_+ s'} + \beta e^{\lambda_- s'})\,\frac{\partial B}{\partial s}(s-s') \qquad (28)$$

with

$$\alpha = \frac{1}{2} + \frac{(p_A - p_B)}{4\sqrt{\Delta'}}(\sigma_B - \sigma_A) + \frac{\frac{1}{\lambda}p}{4\sqrt{\Delta'}} \qquad (29)$$

$$\beta = \frac{1}{2} - \frac{(p_A - p_B)}{4\sqrt{\Delta'}}(\sigma_B - \sigma_A) - \frac{\frac{1}{\lambda}p}{4\sqrt{\Delta'}}$$

If we introduce a diffusion approximation

$$\frac{\partial B}{\partial s}(s - s') = \frac{\partial B}{\partial s}(s) \quad,$$

we find the effective opacity

$$<I>(s) = B(s) - \frac{1}{\sigma_{eff}}\,\frac{\partial B}{\partial s} \qquad (30)$$

with

$$\sigma_{eff} = \langle\sigma\rangle - \frac{\text{Var }\sigma}{\frac{1}{\lambda_p} + \sigma_A + \sigma_B - \langle\sigma\rangle} \quad (31)$$

Notice that σ_{eff} is equal to ℓ_m^{-1} defined in Eq. (23).

TWO TEMPERATURES

Here, we examine the case of a mixture of an A-medium at temperature T_A emitting $\sigma_A B_A$ and a B-medium at temperature T_B emitting $\sigma_B B_B$.

The radiative transfer equation reads

$$\frac{\partial I}{\partial s} = -\sigma(s)(I - B(s)) \quad (32)$$

with

$$\sigma(s)B(s) = \frac{\sigma_A B_A + \sigma_B B_B}{2} + \frac{\sigma_A B_A - \sigma_B B_B}{2}\xi(s) \quad (33)$$

if there is no incident radiation then

$$\langle I \rangle = \langle B \rangle \frac{1 + \frac{\langle \sigma B \rangle}{\lambda_p \sigma_A \sigma_B \langle B \rangle}}{1 + \frac{\langle \sigma \rangle}{\lambda_p \lambda_A \lambda_B}} \quad (34)$$

with

$$\langle B \rangle = p_A B_A + p_B B_B$$
$$\langle \sigma B \rangle = p_A \sigma_A B_A + p_B \sigma_B B_B \quad (35)$$

So even for $\lambda_p \to 0$ we obtain

$$\langle I \rangle = \frac{\langle \sigma B \rangle}{\langle \sigma \rangle} \quad (36)$$

which can be distinct from $\langle B \rangle$. In this non equilibrium case the stationary hypothesis is likely to be invalid.

LINEAR GRADIENT

Now, we extend the above formalism, and include a linear composition gradient[8],

$$P_A(x) = P_A(o) + ax \ . \tag{37}$$

For pure absorption, the mean intensity $\langle F \rangle(x)$ satisfies

$$\langle\langle\sigma\rangle\rangle = \lambda_p \left(\frac{\sigma_B}{\lambda_A} + \frac{\sigma_A}{\lambda_B}\right)$$

$$\frac{d^2\langle I\rangle}{dx^2} + \left(\sigma_A + \sigma_B + \frac{1}{\lambda_p}\right)\frac{d\langle I\rangle}{dx} + \left(\sigma_A \sigma_B + \frac{\langle\langle\sigma(x)\rangle\rangle}{\lambda_p}\right)\langle I\rangle = 0 \tag{38}$$

with boundary conditions

$$\langle I\rangle(o) = I_o$$

$$\frac{d\langle I\rangle}{dx}(o) = -\langle\sigma\rangle(o) I_o \ . \tag{39}$$

Let us introduce

$$I' = \langle I\rangle \exp\left(\left(\sigma_A + \sigma_B + \frac{1}{\lambda_p}\right)\frac{x}{2}\right) \tag{40}$$

Then Eq. (38) becomes the Airy equation

$$\frac{d^2 I'}{dx^2} - \omega^2(x) I' = 0 \tag{41}$$

where

$$\omega^2(x) = \left(\frac{\sigma_A - \sigma_B}{2}\right)^2 + \left(\frac{1}{2\lambda_p}\right)^2 + \frac{\sigma_A - \sigma_B}{2\lambda_p}(\bar{P}_B - \bar{P}_A) \times$$

$$\bar{P}_A(x) = 1 - \bar{P}_B(x) = \frac{\lambda_A(x)}{\lambda_A(x) + \lambda_B(x)}$$

and

$$I'(o) = I_o \tag{42}$$

$$\frac{dI'}{dx}(o) = ((\frac{\sigma_A + \sigma_B + \frac{1}{\lambda_p}}{2}) - <\sigma>(o)) \qquad (43)$$

Solutions for Eq. (38) are plotted in Fig. 2. d is the mixing zone length. The two inhomogeneous profiles 1 and 2 converge toward the same absorption at large distance. It should be appreciated that for strong gradient (Fig. 2), $<I>(x)$ is asymmetric with respect to the homogeneous result 3.

This is due to the obvious fact that the interface resulting from a turning over is not statistically equivalent to the initial one.

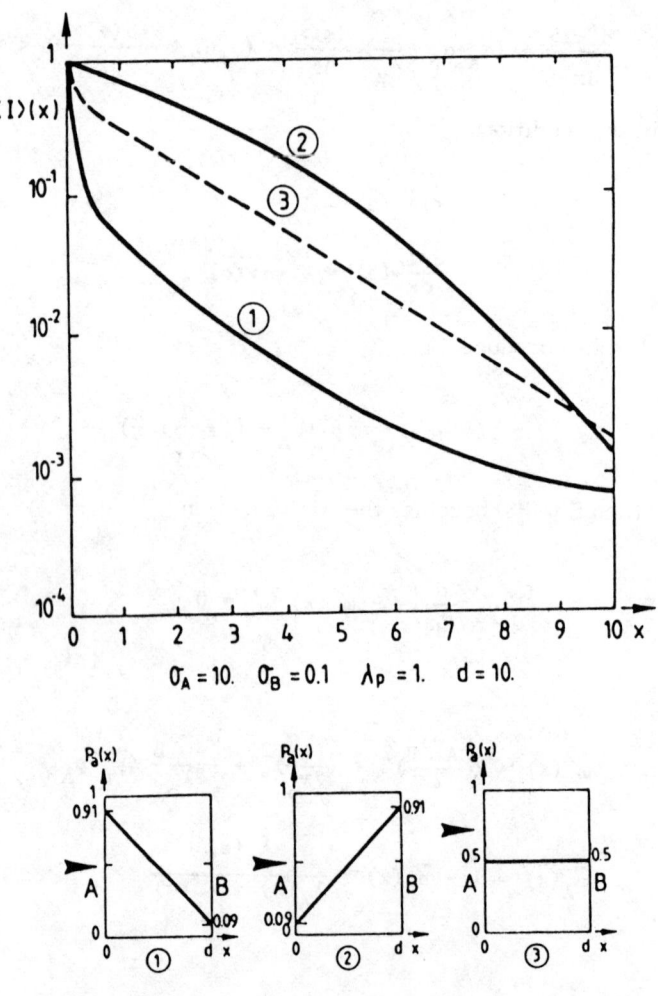

Fig. 2. Solution of Eq.(38) for three gradient profiles.

On the other hand, if the gradient slope decreases, curves 1 and 2 tend to become more symmetrical around 3.

It is also worthwhile to compare the effective local opacity

$$(<\sigma> = p_A(x)\sigma_A + p_B(x)\sigma_B)$$

$$\sigma_{eff}(x) = \frac{\sigma_A \sigma_B + \frac{\sigma_A}{\lambda_B(x)} + \frac{\sigma_B}{\lambda_A(x)}}{\sigma_A + \sigma_B - <\sigma> + \frac{1}{\lambda_A(x)} + \frac{1}{\lambda_B(x)}} \qquad (44)$$

valid in the diffusion approximation, with the 'true' local value (see Fig. 3)

$$\sigma_v(x) = \frac{-1}{<I>(x)} \frac{d<I>(x)}{dx} \quad , \qquad (45)$$

The results are markedly distinct.

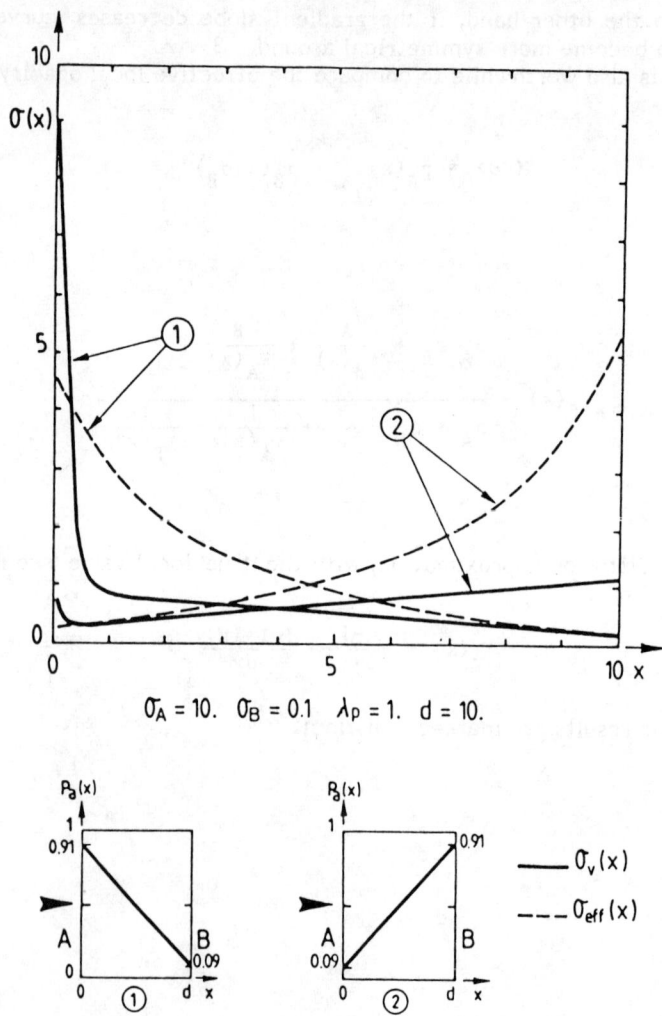

Fig. 3. Effective opacity σ_{eff} (44) and true opacity σ_V (45) given for two composition gradients

REFERENCES

1. C.D. Levermore, G.C. Pomraning, C.L. Sunzo and J. Wong, Submitted to Phys. Rev. A.
2. D. Vanderhaegen, J. Quant. Spectr. Radiat. Transfer to appear.
3. I.B. Kosarev, I.V. Nemchinov, and V.N. Rodinov, Sov. Phys. Dokl., **17**, 886 (1973).
4. For a pedestrian introduction to Rayleigh-Taylor instability see: D.H. Sharp, Physica **12D**, 3 (1984).

5. H.J. Küll, Phys. Rev. **A33**, 1957 (1986).
6. N.A. Tahir and K.A. Long, Phys. Fluids, **29**, 1282 (1986).
7. N.G. Van Kampen, Stochastic Processes in Physics and Chemistry, N.H.P.C., Amsterdam, 1983.
8. D. Vanderhaegen, to be published.

GAIN SCALING LAWS FOR HIF

G. R. Magelssen
Los Alamos National Laboratory, Los Alamos, New Mexico 87545

ABSTRACT

The relationship among target gain (thermonuclear energy-released/ion energy incident) and such factors as the total ion energy, the ion beam radius, the ion range, the ion power, the beam geometry, and the released target debris are critical in assessing the feasibility of a particular heavy ion inertial confinement fusion reactor concept.
Scaling relationships that allow target gain to be calculated from the target hydrodynamic coupling efficiency (η), the target radius (R), and the ion energy incident on the target (E_i) are presented. These relations include scaling laws for the required peak power and ion energy, the fuel fractional burnup, and the fraction of energy released in charged particles and x-rays as a function of R, η, and E_i. Relations have been developed for two single-shell target concepts, one requiring two-sided and the other symmetric ion illumination.

INTRODUCTION

There are many factors that impact the design of a heavy ion fusion reactor facility. The length of the linear accelerator and its cost are functions of the ion charge state, the ion current, and the ion kinetic energy. The number of beam lines and their geometrical orientation around the reactor chamber depend on the target illumination and power requirement. The total ion energy required to ignite the target depends on the beam waist size and the ion kinetic energy. The size and type of reactor chamber depend on the thermonuclear energy released and the partitioning of this energy in x-rays, charged particles, and neutrons. The purpose of this paper is to present scaling relations among target, driver, and reactor parameters.

Curves that give target gain as a function of ion beam energy and the parameter $r_i^{3/2} x$ (r_i is the beam radius and x the range) have been published by Bangerter et al.[1] and Lindl and Mark.[2] Scaling relations have also been published.[3-6] These relations have helped improve our understanding of the gain curves by illustrating how parameters such as the cold fuel isentrope (ξ) (the temperature of the cold fuel at ignition) and the hydrodynamic coupling efficiency (η) (energy in the fuel at ignition/ion energy absorbed) impact gain. However, scaling relations for a particular target concept have not been published; so, the scaling for such parameters as illumination geometry, fuel mass, and target thermonuclear debris have remained unknown.

In this paper scaling relationships that allow target gain to be calculated from the target hydrodynamic coupling efficiency (η), the target radius (R), and the ion energy incident on the target (E_i) are presented. These relations include scaling laws for the required peak power and ion energy, the fuel fractional burnup, and the fraction of energy released in charged particles and x-rays as a function of R, η, and E. Relations have been developed for two single-shell target concepts, one requiring two-sided and the other symmetric ion illumination.

ION ENERGY AND POWER RELATIONS

The power requirement will depend on the amount of mass to be accelerated. The mass is proportional to the product of R^2 and the ion range x (g/cm^2); so, the ion energy absorbed E_i is

$$\eta E_i \propto R^2 \Delta t \tag{1}$$

where Δt is the time of the main power pulse and η is the efficiency of coupling ion energy to the capsule. Notice that η contains both the efficiency of coupling ion energy to the pusher and fuel kinetic energy, and the efficiency of coupling this kinetic energy to the fuel at ignition. Also we have assumed that the energy incident and the energy absorbed are the same. As yet no experimental evidence suggests otherwise.

We have studied the optimal gain as a function of the capsule radius and the cold fuel isentrope ξ that denotes deviation from complete degeneracy, and have found that the maximum velocity v associated with the optimum gain scales as:

$$v \propto (\xi/R)^{1/4}. \tag{2}$$

The Δt will depend on the target collapse time so,

$$\Delta t \propto R/v. \tag{3}$$

Combining Eqs. (1), (2), and (3) gives the scaling relation we wanted. We have

$$E_i \propto R^{13/4}/(\xi^{1/4}\eta). \tag{4}$$

The power P can then be expressed as:

$$P \propto R^2/\eta. \qquad (5)$$

CAPSULE ENERGETICS

Our model of capsule energetics uses many of the ideas discussed by Meyer-ter-Vehn[3] and Rosen[4], and has features first presented by Kidder[5] and Bodner.[6] Consider the ignition conditions for a capsule with radius R -- the distance from capsule center to the outer surface of the ablator, and a convergence ratio γ (R/R_f) -- R_f is the fuel radius at ignition. We divide the fuel into a hot ignition region that can be described by an ideal gas and a highly compressed, low entropy region described as a degenerate electron plasma with the pressure:

$$p_c = 2.3 \times 10^{12} \xi \rho_c^{5/3} \qquad (6)$$

in cgs units. As before the parameter ξ denotes the deviation from complete degeneracy and labels different isentropes.

The thermonuclear energy is

$$E = q_{DT} M_f \Theta \qquad (7)$$

where the specific DT fusion energy $q_{DT} = 3.34 \times 10^{11}$ J/g, the total fuel mass $M_f = M_s + M_c$, and the fraction of burned fuel

$$\Theta = H_f/(H_b + H_f) \qquad (8)$$

where $H_f = \rho_s R_s + \rho_c R_c$. The R_s and ρ_s and the R_c and ρ_c are the hot and cold fuel radius and density, respectively. The total fuel radius is R_f. The H_b is a constant that depends on the amount of fuel tamping and the fuel reaction rate. For our concepts H_b is approximately 3.

To find scaling relations, we use the fact that for high-gain targets the fuel ρr is greater than 1.[3-4] This allows us to write the approximate expressions:

$$E_f \propto p_c R_f^3, \qquad (9)$$

$$\rho_c R_c \propto M_c/R_f^2, \text{ and} \qquad (10)$$

$$E = q_{DT} M_c \Theta \qquad (11)$$

where $\Theta = \rho_c R_c/(3.2 + \rho_c R_c)$ and $E_f = \eta E_i$. We also have

$$M_c \propto (R_f^2 E_f/\xi)^{3/5}. \qquad (12)$$

These ideas were used to write a simple code to study gain as a function of γ. We found that the value of γ that gave maximum gain was proportional to $R^{-1/2}$, a relation also found by Rosen.[4] With this relation for γ, Eq. (12) becomes

$$M_c \propto (R^3 E_f/\xi)^{3/5}, \text{ and} \qquad (13)$$

$$\rho_c R_c \propto (E_f/\xi)^{3/5}/R^{6/5}. \qquad (14)$$

SCALING LAWS

The previous sections give most of the information needed to determine gain as a function of the relevant target and ion beam parameters. For example, combining Eqs. (9), (10), (11), (13), and (14) gives the yield as a function of R, E_i, η, and ξ. Thus, the gain can be written as

$$G = C_1 \{R^3 \eta_i/\xi\}^{3/5} \Theta/E_i \qquad (15)$$

where $\Theta = \rho r/(3.2 + \rho r)$ and

$$\rho r = C_2 \{\eta E_i/\xi\}^{3/5}/R^{6/5} \quad (g/cm^2). \qquad (16)$$

From Eqs. (4) and (5) we also have

$$E_i = C_3 R^{13/4}/(\xi^{1/4} \eta) \quad (MJ), \text{ and} \qquad (17)$$

$$P = C_4 R^2/\eta \quad (TW). \qquad (18)$$

We have found the constant coefficients and η values for two target concepts. Concept A requires two-sided ion irradiation and concept B symmetric illumination.

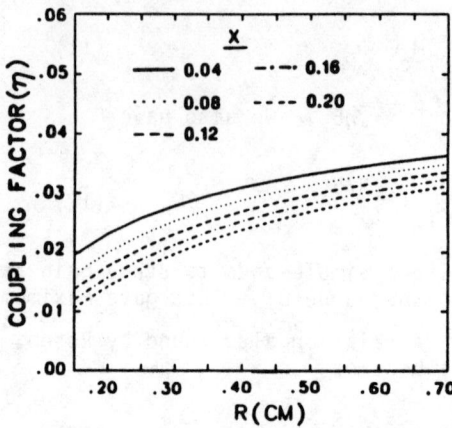

Fig. 1. Hydrodynamic coupling efficiency η as a function of R (cm) and x (g/cm^2) for concept A.

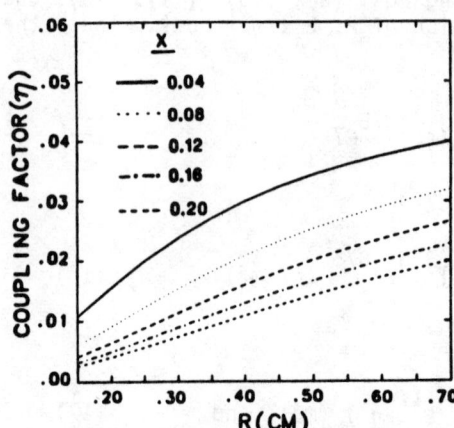

Fig. 2. Hydrodynamic coupling efficiency η as a function of R (cm) and x (g/cm^2) for concept B.

The η values and their associated constant coefficients were determined from target calculations and analytical models of the target coupling physics and include symmetry constraints. For a discussion of symmetry issues for direct drive targets see Ref. 7. The values of η are given in graphical form, that is curves of η versus target radius for different values of the ion range. The values of η as a function of R (cm) and x (g/cm^2) for concepts A and B are shown in Figs. 1 and 2, respectively. Target calculations suggest that the fuel isentrope ξ is a function of R. A very approximate expression for ξ is

$$\xi \approx 1.0 + (C_5/R)^2. \qquad (19)$$

This is a very approximate relation because our calculations have shown the functional form to depend on the preheat allowed, the stability criteria, and the pusher and ablator materials. The constant coefficients C_1 - C_5 for concept A are 5.9×10^4, 3.9, 11.6, 205, and 0.28. The corresponding coefficients for concept B are 3.2×10^4, 6.0, 3.44, 73.4, and 0.54. These scaling relations are valid for incident ion energies from 1 to 20 MJ and for ion ranges from 0.035 to 0.2 g/cm^2. They were determined in such a way as to be consistent with the best estimate gain curves published earlier.[1-2]

A COMPARISON WITH PREVIOUS WORK AND A RELATION FOR THE DEBRIS NEUTRON FRACTION

Because most of the capsule energetics was based on previous scaling studies of gain and is most directly related to that of Meyer-ter-Vehn,[3] we make a comparison with his results. We find that our ρr scaling can be written

$$\rho r \propto (\eta E_i)^y / \xi^z \qquad (20)$$

where $y = 3/13$ and $z = 0.6$. The Meyer-ter-Vehn result gave $y = 0.2$ and $z = 0.6$. Equation (20) was found by combining Eqs. (16) and (17) and ignoring the ξ dependence in Eq. (17). If we ignore the velocity dependence on capsule radius, the y values become equivalent and equal to 0.2.

Comparisons with the gain scaling are slightly more difficult. If we approximate the burn fraction Θ with the expression[3-4]

$$\Theta = (\rho r / 12.8)^{1/2}, \qquad (21)$$

we can write Θ as

$$\Theta \propto (\eta E_i)^{3/26} / \xi^{3/10}. \qquad (22)$$

Then, the gain is approximately

$$G \propto (\eta E_i)^w / \xi^p \qquad (23)$$

where $w = 7/26$ and $p = 0.9$. The Meyer-ter-Vehn result gave $w = 0.3$ and $z = 0.9$. Again, if we ignore the velocity dependence on radius, the w's become equivalent and equal to 0.3.

Besides the gain, yield, power and energy, the fuel mass and neutron debris fraction can be estimated. From the target yield (the product of E_{ion} and G) and the fuel burnup fraction, the fuel mass is determined. The neutron fraction of the yield N can be estimated from the fuel ρr value. The fraction is[8]

$$N = 0.8 - 0.04 \rho r. \qquad (24)$$

The author would like to acknowledge useful discussions with R. O. Bangerter and J. W.-K. Mark.

REFERENCES

1. R. O. Bangerter, J. W-K. Mark, and A. R. Thiessen, Phys. Lett. A88, 225 (1982).
2. J. D. Lindl and J. W-K. Mark, Laser and Particle Beams 3, 37 (1985).
3. J. Meyer-Ter-Vehn, Nucl. Fusion Lett. 22, 561 (1982).
4. M. D. Rosen, J. D. Lindl, and A. R. Thiessen, "Simple Models of High-Gain Targets - Comparisons and Generalizations," in Laser Program Annual Report 83, Lawrence Livermore National Laboratory report UCRL-50021-83, p. 3-5 (1983).
5. S. Bodner, "Critical Elements of High Gain Laser Fusion," NRL Memorandum report 4453 (January 21, 1981).
6. R. E. Kidder, Nucl. Fusion 16, 405 (1976).
7. R. A. Sachs et al., Nucl. Fusion 22, 1421 (1982).
8. R. O. Bangerter et al., "Simple Target Models for Ion Beam Fusion Systems Studies," Lawrence Livermore Laboratory report UCID-20578 (1985).

SUBSTANTIAL REDUCTIONS OF INPUT ENERGY AND PEAK POWER REQUIREMENTS IN TARGETS FOR HEAVY ION FUSION*

James W-K. Mark and Yu-Li Pan
Lawrence Livermore National Laboratory, Livermore, California 94550

ABSTRACT

We describe two ways of reducing the requirements of the heavy ion driver for ICF target implosion. Compared to our estimates of target gain not using these methods, the target input energy and peak power may be reduced by about a factor of two with the use of the hybrid-implosion concept. Another factor of two reduction in input energy may be obtained with the use of spin-polarized DT fuel in the ICF target.

INTRODUCTION

We have examined two ways of achieving substantial reductions in the input energy and peak power requirements for heavy ion targets compared to our estimates without these effects. These are: (1) hybrid implosion concept and (2) polarized DT fuel. We outline below the target physics consequences as well as some of the other necessary requirements. In particular, good ion beam illumination symmetry is needed by the hybrid implosion concept. We have used in our simulations the new strategy[1,2] for achieving illumination symmetry.

THE HYBRID-DRIVE IMPLOSION CONCEPT

Our hybrid-implosion concept is designed to use the unique ability of ion beams to deliver energy efficiently and directly into

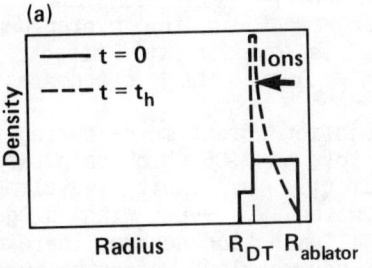 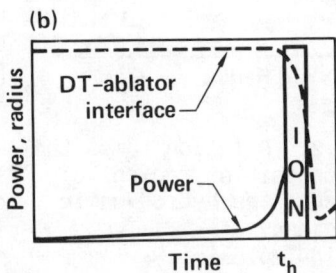

Fig. 1. Hybrid-implosion concept divides the implosion of a high gain ICF target into two phases (see text): (a) Ions deposit their energies in the compressed ablator; (b) Temporal profile of the driver power and the radius-time plot of the DT-ablator interface.

---

*Work performed under the auspices of the U. S. Department of Energy by the Lawrence Livermore National Laboratory under contract number W-7405-ENG-48.

a precompressed ablator. We divide the implosion of a high-gain ICF target into two phases (see Fig. 1): (1) compress the ablator to higher density using short wavelength laser or ion beams; (2) ion driver delivers energy efficiently into the precompressed dense ablator to provide the required ignition velocity with high hydro-efficiency at low convergence ratios.

The physical requirements during the first phase are similar to those demanded by all high gain ICF targets. Careful temporal pulse shaping of the driver power (see Fig. 1b) is necessary to set the DT fuel onto a low adiabat. Concurrently, the ablator is symmetrically compressed to higher density (see Fig. 1a). A high hydrodynamic efficiency is not crucial during this first phase because only about 20% of the total input energy is involved.

The second phase requires that ions be generated and transported efficiently to the compressed ablator. Little or no temporal power pulse shaping of the ion beam driver output is required (Fig. 1b). With the proper choice of ion species and kinetic energy, we can obtain very good coupling of the ion energy to the ablator and thereby maximize the hydrodynamic efficiency. Since about 80% of the input energy is supplied during this second phase, an increase in the hydrodynamic efficiency leads directly to reductions in the input energy and peak power requirements. We can make additional improvements in the hydrodynamic efficiency by tamping the implosion. This can be achieved by the generation of an ablation front on the outside of the ion deposition region (Fig. 2). Moreover, the ablation front can be made to move inward as a function of time to provide dynamic tamping for the ion deposition region. We note, however, that the tamping effect can only provide a modest additional improvement in the hydrodynamic efficiency because the pressure at the ablation front is relatively low. The major improvement in the hydrodynamic efficiency is due to the direct ion deposition into the precompressed ablator.

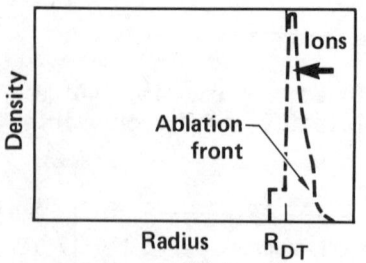

Fig. 2. Ablation tamps the ion deposition region to increase the hydrodynamic efficiency of hybrid-implosion.

The ablation front moves during the implosion of an ICF high gain target driven directly by short wavelength lasers. Normally, even with a good multibeam illumination scheme, there are "residual asymmetries" affected by the focusing geometry and average intensity profile of the laser beams. These beam parameters usually remain fixed during the target implosion. Thus, it would be difficult to have the optimal illumination symmetry as well as high coupling efficiency throughout the implosion with a limited number of beams. In contrast, the thin, dense, compressed ablator shell does not move appreciably during the deposition of the short heavy ion beam pulse in the second phase of the hybrid-implosion concept. For example, in one of our simulations the radius of peak ion deposition, R_{dep}, shifts from its average value by only 13% during the ion-pulse. This is sufficiently small to maintain illumination symmetry. We have also used the theory developed by Mark[2] to determine the choice of beamlet incident angles, beamlet number, currents and spatial shapes to remove most of the asymmetries in the target illumination. The selected axially symmetric illumination scheme is convenient for some reactor designs, particularly those with vertically flowing liquid metals. With this illumination scheme and 32 beamlets, we have attained <2% deposition asymmetries for a convergence ratio of 24. Details are similar to those of the Mark-Lindl, Ref. 1, and will not be discussed here.

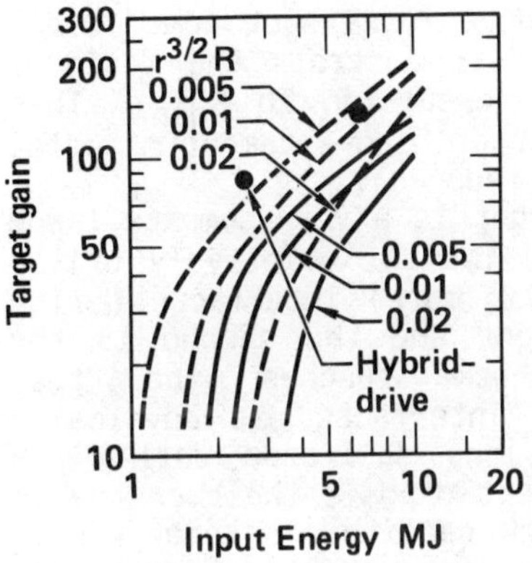

Fig. 3. Gain curves of the single-shell targets for several $r^{3/2}R$ values. The two hybrid-drive points are to be compared to the solid gain curves - unpolarized DT fuel. Dashed gain curves are for targets using 100% polarized DT fuel. Input energy is equal to the total energy deposited in the target.

Figure 3 shows the results of detailed calculations of single shell targets using the hybrid-implosion concept as compared to some of our other targets. Using unpolarized DT fuel, a target gain of about 90 is obtained at 2.5 MJ input energy. We used in the calculations 3 GeV Cs (or 5 GeV Pb) ions with a peak power of 150 TW focused to 3.4 mm spot diameter, D, ($r^{3/2}R \simeq 0.004$). At higher input energy, we obtained a target gain of 140 at 6.6 MJ. This hybrid drive target used D = 5.1 mm with 4 GeV Cs (or 7 GeV Pb ions) at a peak power of 250 TW ($r^{3/2}R \simeq 0.01$, unpolarized DT). The peak power and input energy at constant target gain and $r^{3/2}R$ are about a factor of two lower than those used in the unpolarized DT targets (Fig. 3).

Many proposed targets use direct drive ions (for early examples see Refs. 3-5) to generate efficient target drive. The difference lies in important details such as energy deposition in precompressed ablators and in the treatment of residual direct-drive asymmetries and convergence ratios. Other differences are target materials, in-flight aspect ratios, illumination schemes, etc.

Compared to the target described in Ref. 1 at fixed input energy, the hybrid-implosion target uses a smaller beam spot size and has smaller energy gain. This smaller spot size would place a tighter phase-space constraint on the ion beams. However, this tighter constraint is in part a design trade-off. When the same ablator material is used in the two targets, the hybrid-implosion target has about a factor of two smaller convergence ratio for the ion pulse. (This smaller convergence ratio is also true for hybrid drive versus the unpolarized DT targets of Fig. 3.) The reduction in the convergence ratio is the direct consequence of precompressing the ablator material to higher density and smaller radius before depositing the direct drive ion energy into the ablator material. Convergence ratio is defined here as the radius of peak ion energy deposition, R_{dep}, at the start of the ion pulse divided by the radius of the hot spot at ignition. The two targets also deal with the effects of the residual ion beam asymmetries differently. The same ion beamlet placement scheme is used in the two cases. The hybrid-implosion target tolerates the effects of the residual asymmetries by: (1) reducing the radial excursion in R_{dep} by choosing the proper duration, Δt, for the direct drive ion pulse (too short a Δt would sacrifice the advantage of lower peak power); (2) reducing the effective convergence ratio (further reductions in convergence ratio would reduce focal spot some more). In contrast, the target described in Ref. 1 controls the residual asymmetries by selecting the appropriate excursion in R_{dep}. This is done through choosing the proper amount of high-Z tamper material (convergence ratio has typically not been reduced).

The stability of an ICF high gain target is a very complex issue and is target design dependent. For the typical ablatively driven target, perturbations can grow at two (or more) interfaces during the implosion. One is the ablation front and the second is the fuel-ablator interface when the shock reaches the fuel (Richtmyer-Meshkov). These two unstable interfaces are physically separated by the ablator region. The ions delivered during the second phase of the hybrid-implosion concept deposit their energy in this region at or near the region of peak density and temperature. Since the ablation front is outside of the ion deposition region and this region is exploding during the dominant phase of the target implosion, we expect the deleterious effects due to the perturbations generated at the ablation front to be reduced by the ion deposition.

The hybrid-implosion concept is a new way to reduce the peak power and energy requirements of the heavy ion driver. A number of issues generic to direct drive symmetry and stability requirements need to be explored more fully. Ultimately, the utility of this concept will be determined by factors such as cost and other system issues.

TARGETS USING SPIN-POLARIZED DT FUEL

Spin-polarized DT fuel can be used in single-shell targets driven by ion beams to reduce the input energy requirements. This reduction in the input energy requirements of an ICF target using spin-polarized DT fuel is the direct result of the higher nuclear reaction rate of such fuel (Refs. 6-9). Consequently, for the same areal density of DT fuel, a higher fraction of fuel undergoes thermonuclear burn. Thus the target gain with polarized fuel, at near fixed energy input is increased. Alternatively, for fixed target gain, the energy input is reduced when polarized fuel is used.

For constant target gain and isotropic burn product emission, simple analytical models predict an input energy reduction of between $\delta^{-2.5}$ and $\delta^{-3.0}$, where δ is the $\langle\sigma v\rangle$ multiplier and takes a value of 1.5 or 1.0, respectively, for 100% polarized or unpolarized DT fuel (Refs. 7,10). We define

$$\delta = 1.0 + 0.5(F_D F_T)$$

where F_D and F_T are, respectively, the polarization fractions of deuterons and tritons. We see that δ is only 1.125 for 50% polarization. Thus, significant reductions in the input energy occur only when the fractional polarization is very high.

The results of the simple analytical models suggest that, for completely polarized DT fuel, the input energy requirement can be reduced by a factor of 2.8 - 3.4. However, it is important to recognize that these analytic models assume the fuel has been assembled in the desired manner and ignition has occurred. How, or if, something approximating these idealized conditions can be achieved in an ICF target is not addressed by the models. We have addressed these issues by means of computer simulations. Our results are discussed below.

The target used in our study is similar to the direct drive target proposed by Pan and Hatchett (Ref. 9) for 0.265 μm laser driver. They used a single shell composed of DT wetted foam ablator and liquid/solid DT fuel. Figure 3 shows the gain curves of single-shell targets with and without polarized DT fuel for several $r^{3/2}R$ values. The factor, $r^{3/2}R$, where r is the target radius in cm and R is the beam ion range in g/cm^2, has been used previously to parameterize ion beam target gain (Refs. 11 and 12). A more accurate characterization of target gain requires more parameters than input energy E and $r^{3/2}R$. But for a comparison with additional physics where relative effects are of interest, these parameters should be sufficient (n.b. we assumed $0.1 \leq r/E^{1/3} \leq 0.2$, E in MJ). We note that the convergence ratio remains approximately constant with varying input energy. However, the peak implosion velocity increases with decreasing input energy. This reflects the more stringent ignition requirements of smaller targets. Ignition difficulty is the primary reason for the cutoff in these gain curves near threshold. Consequently, the exact locations of these thresholds are sensitive to the models used to simulate thermonuclear burn. But our conclusions are based only on the relative positions and not the exact locations of the polarized and unpolarized gain curves. Therefore, our results are insensitive to the model used in the computer simulation. At a constant target gain of about 100, we find the input energy requirement is reduced by about a factor of 1.8 for single shell targets using polarized DT fuel and $r^{3/2}R$ of 0.005.

As noted above, significant reductions in the input energy occur only when the fractional polarization is very high. Experimentally, a very high degree of polarization (99%) has been obtained with H_2 molecules at ultra low temperatures. Whether or not one can obtain a high degree of polarization with deuterium and tritium is not known. Experiments are in progress to polarize DT (Ref. 13).

REFERENCES

1. J. W-K. Mark and J. D. Lindl, "Symmetry Issues in a Class of Ion Beam Targets Using Sufficiently Short Direct Drive Pulses," elsewhere in this proceedsings.
2. J. W-K. Mark, Physics Letters, 114A, 458 (1986).
3. M. J. Clauser, Phys. Rev. Lett., 35, 848 (1975).
4. J. D. Lindl and R. O. Bangerter, Lawrence Livermore National Laboratory Report, UCRL-77042 (1975).
5. M. A. Sweeney and M. J. Clauser, App. Phys. Lett., 11, 1 (1975).
6. R. M. Kulsrud, H. P. Furth, E. J. Valeo, M. Goldhaber, Phys. Rev. Lett., 49, 1248 (1982).
7. M. D. Rosen, J. D. Lindl, and A. R. Thiessen, LLNL Laser Program Annual Report - 1983, pp. 3-5 to 3-9 (1984).
8. R. M. More, Phys. Rev. Lett., 51, 396 (1983).
9. Y. L. Pan and S. Hatchett, "Spin Polarized Fuel in ICF Reactor Targets," Lawrence Livermore National Laboratory Report, UCRL-94235, February (1986).
10. K. A. Brueckner and S. Jorna, Rev. Mod. Phys., 46, 325 (1974).
11. R. O. Bangerter, J. W-K. Mark and A. R. Thiessen, Phys. Lett., 88A, 225 (1982).
12. J. D. Lindl and J. W-K. Mark, Laser and Particle Beams, 3, 37 (1985).
13. P. C. Souers, E. M. Fearon, E. R. Mapoles, J. R. Gaines, J. D. Sater and P. A. Fedders, to appear in Jour. Vac. Sci. Tech., May/June (1986).

SYMMETRY ISSUES IN A CLASS OF ION BEAM TARGETS USING SUFFICIENTLY SHORT DIRECT DRIVE PULSES[*]

James W-K. Mark and John D. Lindl
Lawrence Livermore National Laboratory, Livermore, CA 94550

ABSTRACT

Controlling asymmetries in direct drive ion beam targets depends upon the ability to control the effects of residual target asymmetries after an appropriate illumination scheme has already been utilized. We address a class of modified ion beam targets where residual asymmetries are ameliorated. The illumination scheme used is an axially symmetric one[1] convenient for reactor designs. Residual asymmetries are controlled by limiting the radial motion of the radius R_{dep} of peak ion energy deposition. Limiting the motion of R_{dep} is achieved by lengthening the time scale t_s where changes in R_{dep} adversely affect asymmetries. In our example, t_s becomes longer than the duration Δt_D of the entire direct drive pulse train ($t_s > \Delta t_D$).

INTRODUCTION

A number of authors have studied directly driven ion beam targets in inertial confinement fusion energy research (early instances include Refs. 2-4). For direct drive, a judicious placement of beamlets is of course essential. Beyond that, there are normally residual target drive asymmetries left over from beam overlap and beam profile shaping. These residual asymmetries must all be maintained at a tolerable level which is set by target parameters such as convergence ratio.

Careful beam placement and appropriate transverse beam-shape could reduce asymmetries in direct drive targets. But typically, there is a radius near which the illumination symmetry is optimized. In most such targets, this optimal radius eventually becomes displaced relative to the radius R_{dep} where the beam energy deposition reaches a peak. If left unchecked, a consequence is that asymmetries increase after some characteristic time t_s to an extent where target implosion is adversely affected. Viewed in this manner, the above condition on R_{dep} becomes a time constraint in that the direct drive pulse of duration Δt_D must satisfy $\Delta t_D < t_s$. This value t_s could frequently be smaller than the other target timescales such as the one for the total drive pulse.

---

[*]Work performed under the auspices of the U. S. Department of Energy by the Lawrence Livermore National Laboratory under contract number W-7405-ENG-48.

Specifically, the residual asymmetry of one illumination scheme[1] depends upon the deviation of R_{dep} from the beam focal spot radius R_{beam} (considered fixed in this paper). For this illumination scheme, R_{beam} is near the radius where symmetry is optimized. In our detailed example, it turns out that with 32 beamlets and imposing $| R_{dep} - R_{beam} | < 0.08 R_{dep}$, we could already reduce to < 2% the residual deposition asymmetries in a spherical target. In this paper a careful choice of the ratio of inertias in the outer tamper versus that in ablator allows us to lengthen t_s sufficiently. Thus, t_s is prolonged enough so that the entire $\Delta t_D \simeq 62$ ns direct drive pulse train of a specific target satisfies $\Delta t_D < t_s$. (In targets where t_s has not been particularly changed by choices of target structure, we would expect instead that Δt_D be chosen, perhaps shortened, to become less than the existing essentially fixed t_s.)

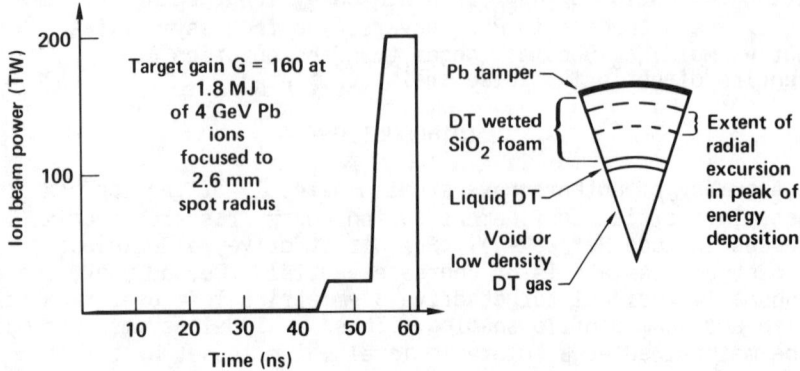

Fig. 1 The direct-drive ion beam target used in the symmetry studies.

THE ION BEAM TARGET

The target in question has the structure illustrated in Fig. 1. Before considering multi-dimensional effects of asymmetries, we first simulated the full hydrodynamics and DT burn of the target using the spherically averaged ion energy deposition. In this initial 1D simulation the beamlets have a realistic distribution of incident angles relative to the target surface but the ion energy deposited at each target radius r is averaged over the sphere with that radius. This can be viewed as simulating the radial distribution of the illumination by an extremely large number of beamlets. We will study later some effects of the residual angular asymmetries of 32 beamlets.

This target gives a 1D gain of G = 160 at 1.8 MJ of input (beam) energy with 2.6 mm spot radius. The drive pulse is illustrated in Fig. 1. We would like to point out that this type of drive pulse can be generated in the final parts of the induction accelerator for heavy ion fusion as part of the process of compressing the pulse in length to reach peak power. In Ref. 5 for example, we show that such a pulse shape could be obtained by using a longitudinal velocity tilt consisting of three continuous linear pieces with different slopes. (Here the "tilt" is in the longitudinal momentum space of the beam pulse and refers to the fact that the beam tail is made to move more rapidly and thus catch up with the beam head so as to achieve peak power.)

BEAM ILLUMINATION SCHEME

The residual asymmetries we consider here arise from the use of a beam placement scheme[1] with axial symmetry about some direction z determined for example for the convenience of fusion chamber design (see Fig. 2).

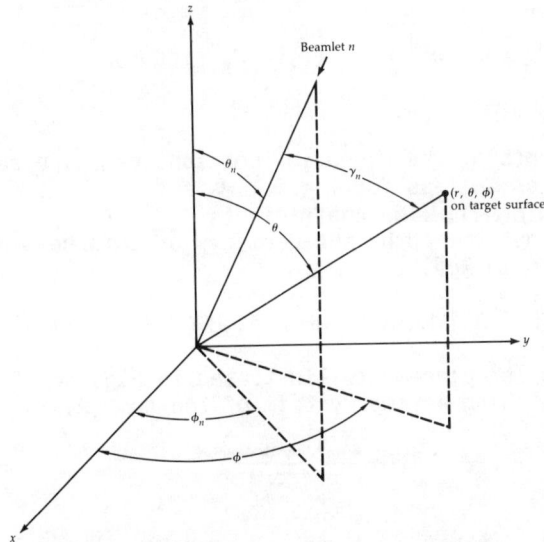

Fig. 2 Geometry used in deriving the results given in Eq. (2). The angle γ_n is measured between the axis of beamlet n and an arbitrary point on the target (r, θ, ϕ).

The beamlet incident directions (θ_j, ϕ_{jk}) all lie in several cones about the z-axis with cone angles θ_j determined by the Legendre-Gaussian quadrature points $x_j = \cos\theta_j$, $j = 1,2,\ldots,L_0$; and the azimuthal angles $\phi_{jk} = 2\pi k/M_j$, $k = 1,2,\ldots,M_j$. Here the constant L_0 determines the number of θ_j angles, and M_j are number of ϕ_{jk} angles for each j. The total number of beamlets is $N_b = \Sigma M_j$. If the n-th beamlet is at (θ_j, ϕ_{jk}), we assume that its deposition profile in a spherical target is

$$E_n(r,\theta,\phi,t) = \frac{w_j}{M_j} \sum_{\ell=0}^{\infty} E_\ell(r,t) P_\ell(\cos\gamma_n) \qquad (1)$$

where the angle γ_n is as illustrated in Fig. 2, $P_\ell(x)$ is a Legendre polynomial and w_j is the Legendre-Gauss quadrature weight corresponding to the quadrature point x_j. The result of summing E_n over all beamlets (over j,k) is [1]

$$E(r,\theta,\phi,t) = A_{0,0} E_0(r,t) + \sum_{\ell=2L_0}^{\infty} A_{\ell,0} E_\ell(r,t) P_\ell(\cos\theta)$$

$$+ 2 \sum_{m=M_0}^{\infty} \sum_{\ell=m}^{\infty} R_e(A_{\ell m} e^{im\phi}) E_\ell(r,t) P_\ell^m(\cos\theta) \qquad (2)$$

where the coefficients of the Legendre-functions cancel exactly for $m = 0$, $1 \le \ell < 2L_0$ as well as for $1 \le m < M_0$ and $\ell \ge m$, where $M_0 = \min(M_j)$. The coefficients $A_{\ell,m}$ of this final sum are determined by the geometry of this beam placement scheme and are given in Ref. 1.

RESIDUAL TARGET ASYMMETRIES OF A 32 BEAMLET ILLUMINATION

For the 32 beamlet placement illustrated in Fig. 4c of Ref. 1, the largest non vanishing asymmetry term of the sum (2) is

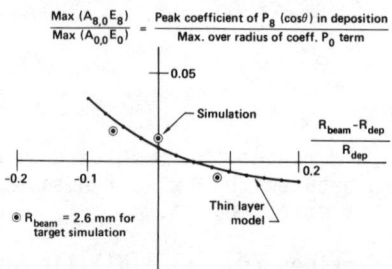

Fig. 3 Residual drive asymmetry is shown as a function of R_{dep}, the radius of peak ion energy deposition in the target. The effect of shift in beam deposition radius versus focal spot radius is noted. Results using a thin deposition layer model (and judicious beam placement) are compared against realistic target simulations.

($A_{8,0}$ E_8) while ($A_{0,0}$ E_0) is the spherically symmetric average deposition. In this paper we outline results of our study on the behavior of this coefficient ($A_{8,0}$ E_8), particularly for the target of Fig. 1.

Firstly, the simple single layer model of Ref. 1 suggests that ($A_{8,0}$ E_8) varies with ($R_{beam} - R_{dep}$) as given by the solid curve of Fig. 3. Numerical results based on our simulations with the target of Fig. 1 give the ($A_{8,0}$ E_8) coefficients as plotted by the circles in Fig. 3. In the single layer model, all information on this coefficient is limited to a point in r-space. But in the target simulations, this coefficient oscillates in radius as indicated in Fig. 4 for a particular time (there are also variations in time). Only the peak value of Fig. 4 is used in Fig. 3. We may point out that the spherically symmetric ($A_{0,0}$ E_0) peaks at zone 45 with gradual decay further out but sharp decay inwards. The asymmetry coefficient ($A_{8,0}$ E_8) peaks rather sharply inside of this radius at zone 42. Presumably this is related to the well known fact that in the deposition of a single beamlet, the particles at the edge of that beamlet penetrates less deeply (larger target radius r) than the particles near the center which penetrate to smaller r values. When summed over many beamlets, it is not surprising that this effect could translate into larger asymmetries inside of the radius of peak energy deposition.

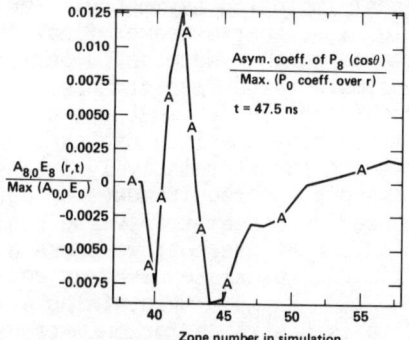

Fig. 4 The asymmetry coefficient $A_{8,0}E_8(r,t)$ actually varies in space and time with changes in signs. It is the peak values at each time which is compared with the thin layer model. The average effect could well be smaller.

In order to discuss the residual asymmetries in a direct drive illumination with 32 beamlets, a first impression is that we would have to introduce a 3D ray-trace even to study the 2D asymmetry term [$A_{8,0}$ E_8 (r,t) P_ℓ $(\cos\theta)$]. However, from our summary Eqs. (1) - (2) of the result of Ref. 1, we see firstly that we only need to obtain E_8 (r,t) from the effective deposition of one beamlet. This we performed by converting the spherically symmetric 1D problem of Fig. 1 into a 2D problem with zones in the (r,θ) directions. The energy of one single beamlet is then deposited in these zones with the beamlet axis along the axis of symmetry. Particularly for heavy ion beams, this operation is quite accurately approximated by a 2D ray trace because the rays of a beamlet approach the target approximately as a cylinder due to the beam approaching a neck at the focal spot. Also the rays inside this cylinder are near parallel because of the small divergence angles involved.

We used 100 angular zones in θ per full circle in the target or roughly 10-15 zones per full cycle in $P_8(\cos\theta)$. We also used 1 to 2×10^4 ion rays per beamlet so that there are almost 10^3 rays per angular zone. This number of rays is necessitated by the fact that we are firstly looking for an effect that is a factor $1-2 \times 10^{-2}$ the size of the spherically averaged main term. Secondly, the major asymmetries are near the end of the ion range and there are a number of radial zones (per angular zone) where these rays finally stop.

Although we have not as yet completed full 2D hydrodynamics, we have taken some pains to anticipate part of its effects in target performance. In particular, since the deposition asymmetries are small, the initial shocks are likely to propagate through the target at roughly the timing of the 1D calculations. For the peak power pulse, the spherically averaged effect of this shock is likely to be weakened and travel slower than indicated by 1D calculations because of the multi-dimensional implosion asymmetry. Thus we have timed the final pulse of Fig. 1 so that a weaker final shock gives even better target performance in 1D. We plan to perform 2D simulations to obtain our best estimate target performance.

In addition, we have not yet evaluated the full effects of random beam displacements from the intended axes. However, steps could be taken to make the target relatively insensitive to beam displacements. For example the requirements on beam-positioning accuracy might be reduced by a beam transverse profile which is rather flat in the middle. At present, we chose a parabolic transverse beam profile due to existence of an analytical model of asymmetries versus ($R_{beam} - R_{dep}$). Maintaining a flatter transverse beam profile is desirable for beam transport in the driver as well as in fusion chambers. We at LLNL have also participated in showing [6] that the transverse profile can be modified as part of the final focus process.

SUMMARY

In summary, we have shown that the beam deposition asymmetries in a spherical target can be kept to rather small levels by (i) judicious choice of beamlet placement and/or current; (ii) choice of the ratio of tamper inertia to ablator inertia (to slow down the rate of change of R_{dep}, the radius of peak ion energy deposition in target).

We appreciate comments by Drs. R. O. Bangerter, D. Munro and Y. L. Pan.

REFERENCES

1. J. W-K. Mark, Phys. Lett., <u>114A</u>, 458 (1986).
2. M. J. Clauser, Phys. Rev. Lett. <u>35</u>, 848 (1975).
3. J. D. Lindl and R. O. Bangerter, Lawrence Livermore National Lab. report UCRL-77042 (1975).
4. M. A. Sweeney and M. J. Clauser, App. Phys. Lett. <u>11</u>, 1 (1975).
5. J. W-K. Mark, D. D-M. Ho, S. T. Brandon, C. L. Chang, A. T. Drobot, A. Faltens, E. P. Lee, and G. A. Krafft, Lawrence Livermore National Lab. report UCRL 94348, Rev. 1 (1986); and elswhere in this proc.
6. W. A. Barletta, W. M. Fawley, D. L. Judd, J. W-K. Mark, and S. S. Yu, "Influence of Target Physics on Ion Accelerators and Reactor Beam Transport in Heavy Ion Fusion" Lawrence Livermore National Lab., Laser Program Annual Report-82, UCRL-50021-82, p. 3-20 (1982).

ON RADIATION ENERGY TRANSPORT IN ION DRIVEN TARGETS

J. Meyer-ter-Vehn, R. Schmalz
Max-Planck-Institut für Quantenoptik
D8046 Garching, FRG

and

R. Ramis
ETSI Aeronauticcos, Universidad Politécnica
E-28006 Madrid / Spain

ABSTRACT

It is proposed to operate the deposition layer of ion beam driven ICF-targets in the radiative regime, defined by $\rho c_T^3 \ll \sigma T^4$. It is characterized by supersonic radiative transport which tends to smooth asymetries occuring in beam energy deposition.

SYMMETRIC IMPLOSION: A KEY PROBLEM FOR ICF

Spherically symmetric target implosion is a crucial requirement for inertial confinement fusion (ICF). With ion beams, spherically symmetric illumination of targets is even more difficult to achieve than with lasers. One therefore has to search for some way of symmetrization inside the target.

In this contribution, it is pointed out that there are two distinct regimes in which a layer for ion beam deposition can be operated: (1) the radiative regime and (2) the hydrodynamic regime. In the radiative regime, radiative transport is faster than sound, deposited energy spreads supersonically to initially non-heated regions, and asymmetries in energy deposition tend to be smoothed. On the other hand, the hydrodynamic regime is characterized by subsonic radiation waves driving shock waves in front of them and deposition irregularities tend to be enhanced by hydrodynamic flow. In the following, these features are explained in more detail. See also ref. 1. The conclusion will be that one should operate deposition layers in the radiative regime. This requires low density material for beam absorption.

DEPOSITION LAYER FOR ION BEAM TARGETS

It is assumed that a typical ion beam target has a layer structure as shown in Fig. 1b. It consists of an absorber sandwiched in between a high-Z tamper facing the ion beam and a pusher which serves here as a payload to be accelerated and might be the fuel itself in a real target. The profile of beam energy deposition is shown in Fig. 1a. Initially the ion beam of some 100 TW/cm^2 penetrates into cold material; the absorber then heats up to some 100 eV, and the deposition front retreats due to range shortening. The energy now spreads from the region of primary deposition through the whole absorber volume, and it is at this stage where

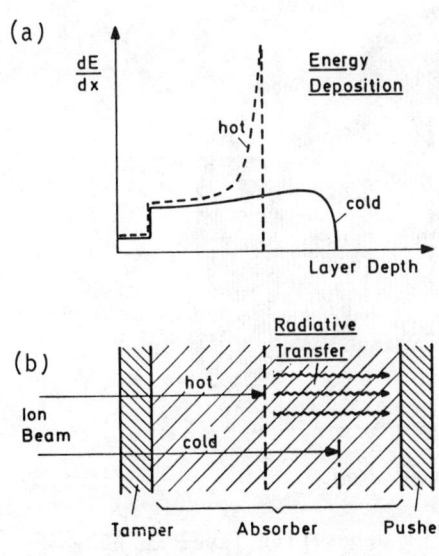

Fig. 1: Typical ion beam deposition layer. Upper part: energy deposition profile for cold and hot layer.

the difference between radiative and hydrodynamic (via shocks) transport matters. The ideal would be that the deposited energy is delocalized instantaneously by very fast transport without perturbing the density and leading to a uniform pressure distribution in the absorber, no matter how irregular the beam deposition is. After a short heating-up phase, the absorber should have an optical depth of order unity, but radiation cannot escape due to high opacity tamper and pusher layers. These layers will be accelerated by radiatively driven ablation as well as the uniform pressure of the absorber plasma. The question is to which extent radiation transport can be used to approach this ideal.

HYDRODYNAMIC AND RADIATIVE REGIMES

A measure for hydrodynamic energy transport in hot matter is $H = \rho c_T^3$ where ρ is the density and c_T the sound velocity; the scale for radiative transport is set by $S = \sigma T^4$ where σ is the Stefan-Boltzmann constant and T the temperature. The dimensionless parameter $H/S = \rho c_T^3/\sigma T^4$ determines which kind of energy transport dominates. For ICF driver beams of $\phi = 10^{14}-10^{15}$ W/cm^2, the temperature of the deposition layer is limited to $T \cong (\phi/\sigma)^{1/4}$ $\cong$ 200-300 eV, and the ratio H/S can be influenced only through the density. The radiative regime, given by $H \ll S$, requires $\rho \lesssim 0.1$ g/cm^3. Of course, decreasing the absorber density means increasing its thickness $d \geq R/\rho$, because the ion range R is fixed, and since the absorber volume cannot be increased indefinitely in a spherical target, because ultimately the overall target efficiency will go down, one has to look for a best compromise.

A SIMULATION

The radiation-hydrodynamics of the deposition layer shown in Fig. 1 has been simulated using the 1D-Code MULTI[2]. Temperature and density evolution are presented for a high density and a low density layer in Figs. 2 and 3, respectively. It is assumed that the plane model layer consists of 30 mg/cm^2 plastic covered on both sides with 10 mg/cm^2 gold. A 100 TW/cm^2 ion beam is incident from

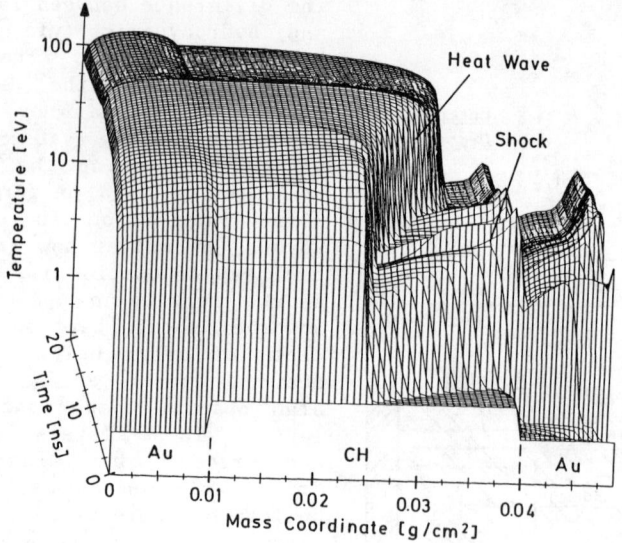

Fig. 2a: Temperature evolution of deposition layer in hydrodynamic regime.

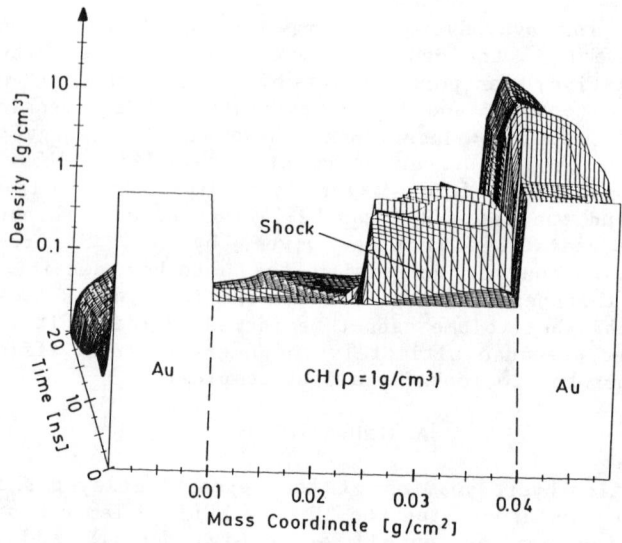

Fig. 2b: Density evolution corresponding to the case of Fig. 2a.

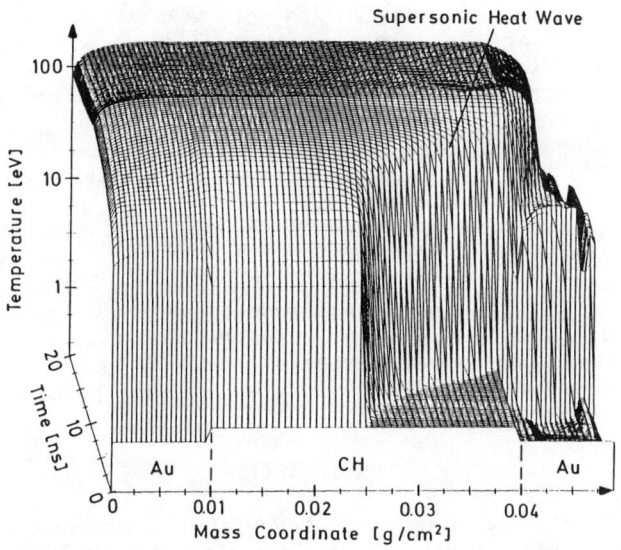

Fig. 3a: Temperature evolution of the deposition layer in the radiative regime.

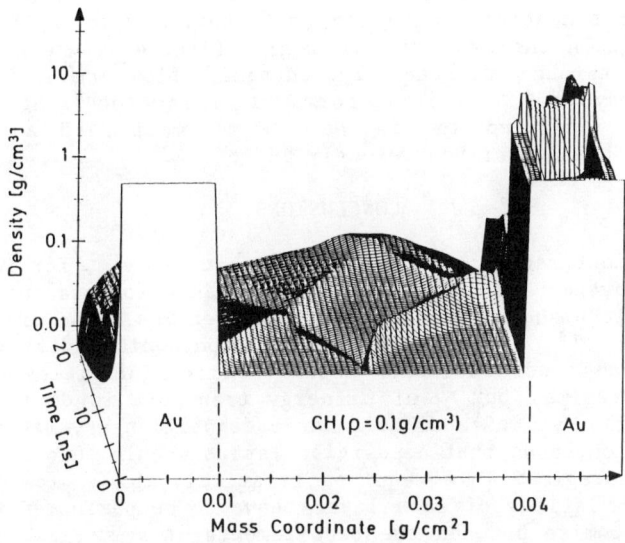

Fig. 3b: Density evolution corresponding to the case of Fig. 3a.

the left with a range of 25 mg/cm² so that beam energy is deposited in the left half of the layer; uniform dE/dx is assumed. In the temperature plots one observes the beam heating on the left and the propagation of heat waves to the right into initially non-heated material.

The case shown in Figs. 2a,b corresponds to the hydrodynamic regime with H/S ≅ 1; the density of the 0.3 mm plastic layer is $\rho = 1$ g/cm³. A strong shock wave is observed to propagate in front of the heat wave; it is best seen as the region of high shock compression in the density plot of Fig. 2b. On the other hand, the radiative regime is demonstrated in Figs. 3a,b. Compared to the first case, the plastic density was reduced to $\rho = 0.1$ g/cm³ corresponding to d = 3 mm and H/S ≅ 0.1. In the temperature plot now a single supersonic heat wave is seen without any shock in front of it, and the density plot shows only minor density perturbations in the absorber regions.

SYMMETRIZATION

The 1D-simulation discussed above cannot give information on smoothing or growth of deposition irregularities. A 2D-simulation would be nessessary for this purpose. In this contribution, we simply describe the difference between the radiative and the hydrodynamic regime qualitatively on the basis of general physics arguments. It is illustrated in Fig. 4. In the radiative regime (Fig. 4a), the mechanism of energy flow beyond the distorted ion beam deposition front is photon diffusion. As in the case of diffusive electronic heat conduction in laser produced plasma, it tends to smooth irregularities; approximately, perturbation amplitudes decay exponentially with the perturbation wavelength as the decay length[3]. In the hydrodynamic regime (Fig. 4b), energy transport occurs through directed hydrodynamic flow which involves inertia. It may lead to violent forms of energy focussing and jet formation as indicated in Fig. 4a. Jet formation in a similar situation has been described recently in ref. 4.

CONCLUSIONS

Two distinct regimes of energy transport relevant for ion beam deposition layers are identified. The dimensionless parameter $\rho c_T^3 / \sigma T^4$ distinguishes between the two regimes. The radiative regime, $\rho c_T^3 \ll \sigma T^4$, is characterized by supersonic radiative heat transport which tends to smooth deposition irregularities. In the hydrodynamic regime, $\rho c_T^3 \gg \sigma T^4$, energy transport is dominated by hydrodynamic flow which tends to enhance deposition asymmetries. It is therefore concluded that deposition layers should be operated in the radiative regime. This requires relatively thick, low density layers. 2D-simulations of such layers have to be performed to find the best compromise between the requirements of symmetrization and driving efficiency.

(a) Radiative Regime ($\rho c_T^3 \ll \sigma T^4$)

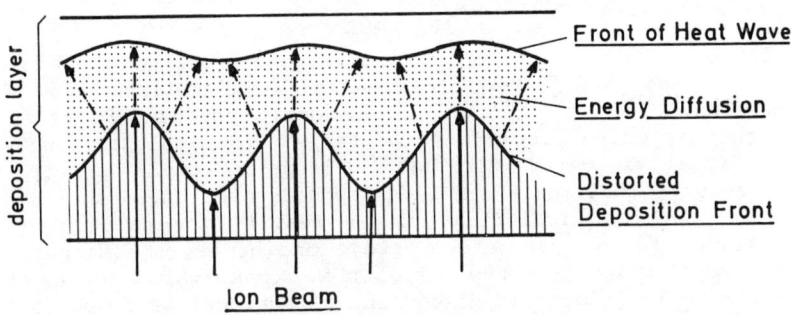

(b) Hydrodynamic Regime ($\rho c_T^3 \gg \sigma T^4$)

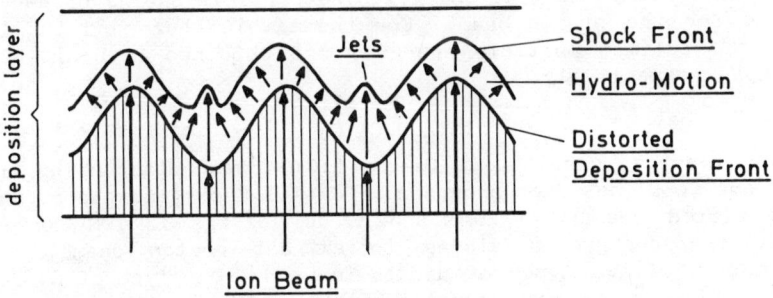

Fig. 4: Schematic drawing of energy transport beyond the distorted deposition region in (a) in the radiative regime and (b) in the hydrodynamic regime.

REFERENCES

1. J. Meyer-ter-Vehn, K. Unterseer, Laser and Particle Beams, 2, 27 (1984).
2. R. Ramis, J. Meyer-ter-Vehn, R. Schmalz, Report MPQ 110 (Feb. 1986).
3. W.M. Manheimer, D.G. Colombant, J.H. Gardner, Phys. Fluids 25, 1644 (1982).
4. S. Atzeni, A. Caruso, V. Pais, to be published in Phys. Lett. A (1986).

HYDRODYNAMIC SIMULATION CODE FOR TARGET IMPLOSION INCLUDING SHOCK WAVES AND CALCULATED TARGET GAIN IRRADIATED BY ION BEAMS

K. Niu
Tokyo Institute of Technology, Midori-ku, Yokohama 227, Japan

ABSTRACT

In the process of target implosion, usually shock waves appear in the fuel layer and have a large influence on fuel compression. So far the one-dimensional hydrodynamic code, by which the implosion process of the target was simulated, included an artificial viscosity which was so large in comparison with the real viscosity that the shock layer spreads over one mesh width of the space in the simulation. As a result, the calculated ion temperature, fusion parameters, etc., and hence, the target gain include some errors. A new calculation scheme which treats shock waves in the fuel layer as discontinuous surfaces is proposed here instead of introducing the artificial viscosity.

The code is applied to targets irradiated by ion beams. The target is a cryogenic hollow shell which consists of three layer of lead, aluminum and deuterium-tritium fuel. For a proton beam, a search is made for the optimum values of the fuel mass, the rate of energy deposition in the Al layer, the particle energy, the target radius and the beam energy. The particle energy of 4MeV is optimum for the proton beam. For the lithium beam and the lead beam, the optimum particle energies are obtained to be 20-30MeV and 3GeV, respectively.

INTRODUCTION

An inertial confinement fusion (ICF) power plant using light ion beams (LIB) has been proposed[1]. Twelve Marx generators, whose total stored energy is 26MJ and diode voltage is 10MV or 5MV, supply the energy to diodes to extract proton beams. The combination of two types of diodes is used. One type of diode is insulated by a radial magnetic field and extracts a rotating ring beam. The other type of diode is the ordinary magnetically insulated one, from which the proton beam is extracted and fills the inner hollow part of the rotating beam. The argon gas filling the reactor cavity neutralizes the change of the proton beams, but does not neutralize the current of the beams. The proton beam is pinched to a small radius by the azimuthal magnetic field, and its propagation is stabilized by the axial magnetic field[2]. The cryogenic hollow shell target of 6mm radius consists of three layers of Pb, Al, DT fuel. The target is biased to a voltage of -1MV, in order to focus the proton beams on the target surface. The ion temperature and ρR of the fuel after the target implosion reach 4.2keV and 7.0g/cm^2, respectively[3]. An output energy of 2.5GJ is released from a target. The reactor is an ADLIB type[4], which consists of an inner rotating cylinder and an outer fixed cylinder. Inside the inner rotating cylinder, the flibe (molten salt of mixture of lithium fluoride and beryllium fluoride) flow, playing the role of coolant and T-breeder. The net plant efficiency is expected to be 33% and the net electric power of 80MW can be supplied from one reactor with an operation frequency of 1Hz.

One of the most important problems in designing an ICF power plant is to estimate the fusion output energy from the target. In order to simulate the implosion motion of the fuel, one-dimensional hydrodynamic codes with artificial viscositites have been employed so far. Instead of the artificial viscosity which leads to a shock wave in the imploding fuel, a shock adoption method is proposed, which applies the Rankin-Hugoniot relations for the shock wave in the fuel.

Lastly the optimum target parameters, such as the fuel mass, energy deposition rate in the pusher layer and optimized particle energy and obtained for the three types of beams, H, Li, and Pb.

IMPROVEMENT OF THE HYDRODYNAMIC SIMULATION CODE

The one-dimensional hydrodynamic simulation code for target implosion includes the driver-beam interaction, diffusion of the fusion products (neutrons and α-particles), radiation energy transfer (by using the local equilibrium type emissivity and opacity, and the averaged radiative energy transport over the ray-paths of multi-groups of wave number), and mass, momentum and energy conservation including sink terms. For the equation of state, the table from the SESAME Library is used[5]. The momentum equation and the energy equations for electrons and ions are

$$\rho \frac{Du}{Dt} = - \frac{\partial(p_i + p_e)}{\partial r} - \frac{\partial V_s}{\partial r} + S_M, \qquad (1)$$

$$\frac{DE_e}{Dt} = - \frac{(E_e + E_i)}{V} \frac{DV}{Dt} - \frac{3n_e k(T_e - T_i)}{2} \nu_{ei} - \frac{1}{r^\alpha} \frac{\partial r^\alpha Q_e}{\partial r} + S_{Ee}, \qquad (2)$$

$$\frac{DE_i}{Dt} = - \frac{(E_i + p_i + q)}{V} \frac{DV}{Dt} + \frac{3n_i k(T_e - T_i)}{2} \nu_{ie} - \frac{1}{r^\alpha} \frac{\partial r^\alpha Q_i}{\partial r} + S_{Ei}, \qquad (3)$$

In these equations, ρ is the density, is the velocity, t is the time, p is the pressure, S_M is the source term for the momentum, r is the radius, E is the internal energy, V is the specific volume, n is the number density, k is the Boltzmann constant, T is the temperature, ν is the collision frequency, Q is the viscous dissipation, S_{Ee} is the source term for the electron energy and S_{Ei} is the source term for the ion energy and D/Dt is the Lagrangian operator. The suffices e and i refer to the electron and the ion, respectively. In eqs.(1)-(3), V_s and q are respectively the viscosity terms given by

$$V_s = \begin{cases} -\frac{\partial q}{\partial r}, & (\frac{\partial u}{\partial r} < 0) \\ 0, & (\frac{\partial u}{\partial r} \geq 0) \end{cases} \qquad (4)$$

$$q = \rho a^2 (\Delta r)^2 \left(\frac{\partial u}{\partial r}\right)^2 + \rho a^2 (\Delta r)^2 u_c \frac{\partial^2 u}{\partial r^2}, \tag{5}$$

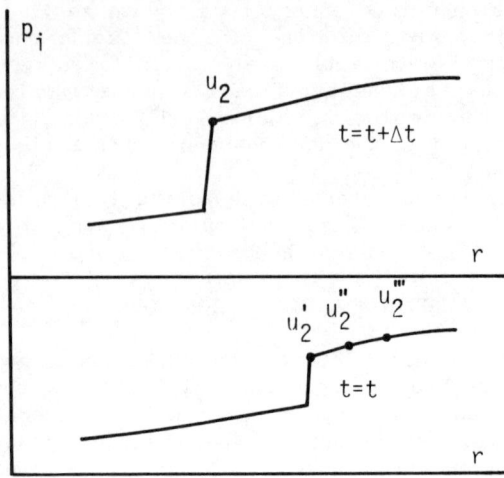

Fig. 1. Shock wave in the fuel layer.

The constant α is chosen as α = 2 for the spherical geometry. In eq.(5), r is the mesh width in space and a is a constant which is chosen as a=0.2.

In the above equations, the artificial viscosity is proportional to $(\Delta r)^2$.

The shock layer extends over one mesh width which is usually much longer than the mean free path on which the real shock wave is formed. The difference between the dissipations by the real and the artificial viscosities in the shock wave leads to the differences in T_i and ρR of the fuel after the implosion.

It is necessary to calculate the energy dissipation correctly at the shock wave in the fuel for the implosion process. The real shock thickness is much smaller than the computational mesh width to which the shock thicknesses in simulations are nearly equal. Here we proposed a method of calculation by which a shock wave is expressed by two spacial points whose distance is roughly equal to the real shock thickness. The Rankine-Hugoniot relations

$$\rho_1(C_s - u_1) = \rho_2(C_s - u_2), \tag{6}$$

$$\rho_1(C_s - u_1)^2 + p_1 = \rho_2(C_s - u_2)^2 + p_2, \tag{7}$$

$$E_1 + p_1/\rho_1 + \tfrac{1}{2}(C_s - u_1)^2 = E_2 + p_2/\rho_2 + \tfrac{1}{2}(C_s - u_2)^2, \tag{8}$$

are applied to the values just before and after the shock waves. In Fig.1, the values ρ_1, u_1, p_1 and E_1 just in front of the shock wave can be obtained by the finite difference equations. We have four equations (three Rankine-Hugoniot relations and one equation of state), while we have five unknown variables ρ_2, u_2, p_2, E_2 and the shock speed C_s. Therefore, if we obtain, for example, u_2 at $t=t+\Delta t$ by using u_2', u_2'' and u_2''' at $t=t$ through the finite difference equations, then we can obtain ρ_2, p_2, E_2 and C_s at

$t=t+\Delta t$. Similarly, if we obtain ρ_2 or E_2 through the finite difference equations at $t=t$, then we obtain u_2, p_2, E_2 and C_s or ρ_2, u_2, p_2 and C_s at $t=t+\Delta t$. If these values of groups are different from other, we modify u_2 or ρ_2 or E_2 and repeat the processes until they converge to definite values of the group. For one shock wave, we must introduce two new spacial mesh points.
 According to the motion of the shock wave, we must interchange mesh points. Thus the simulation code becomes a little tedious and calculations require more time.

ESTIMATION OF FUSION OUTPUT ENERGY FROM TARGET

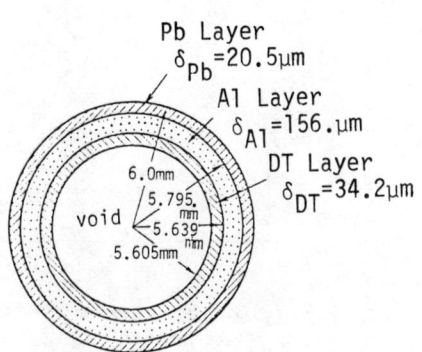

Fig. 2. Target structure of "type 1".

The structure of the "type 1" target is shown in Fig. 2. The target has a cryogenic hollow shell which consists of the three layers of lead, aluminum and DT fuel. The calculated fusion output energy from the target is shown in Figs. 3-8. Fig. 3 shows the fusion output energy E_f versus the mass M_{DT} of the fuel. When $M_{DT}=23.7$mg, E_f becomes maximum. In Fig. 4, E_f is drawn versus C_{Al}, which is the rate of the deposition energy in the Al layer. When $C_{Al}=0.85$, E_f becomes maximum. Fig. 5

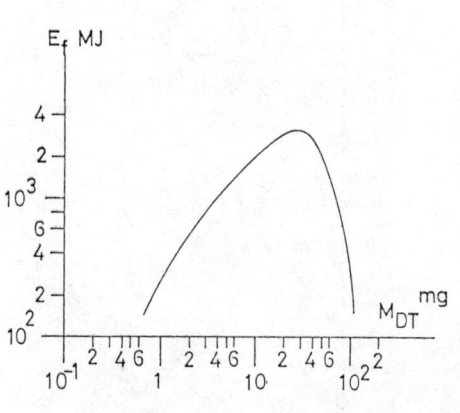

Fig. 3. E_f versus M_{DT} for proton beams.

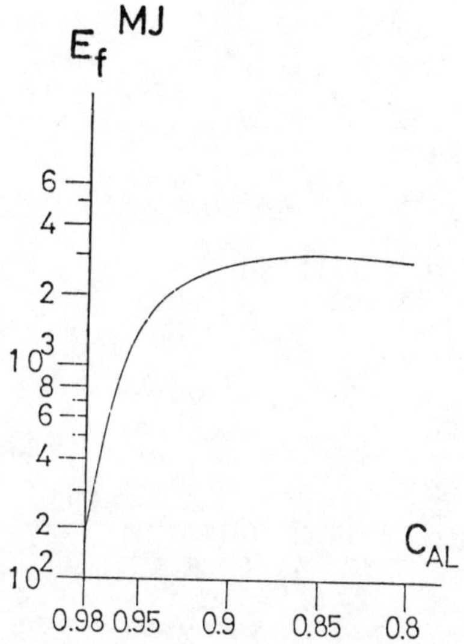

Fig. 4. E_f versus C_{Al} for proton beams.

shows E_f versus the particle energy e_p for a proton beam. When e_p =3.5MeV, E_f becomes maximum. But the dependence of E_f on e_p is not strong. For Figs.3-4, we choose that the beam energy of the

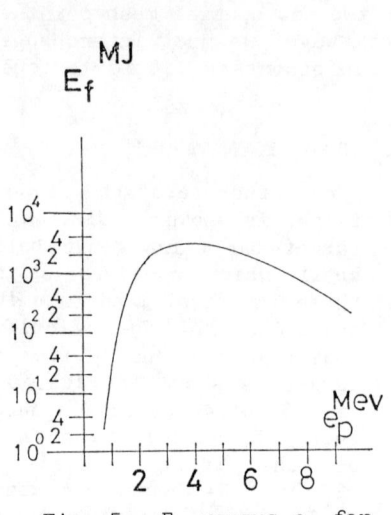

Fig. 5. E_f versus e_p for proton beams.

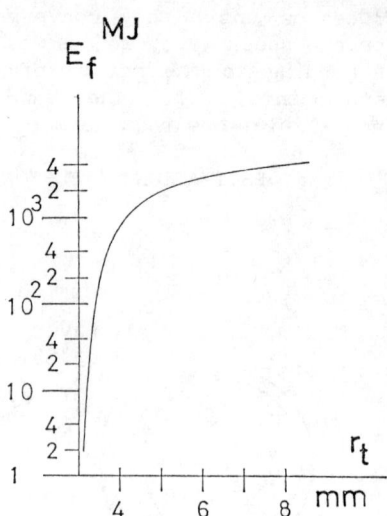

Fig. 6. E_f versus r_t for proton beams.

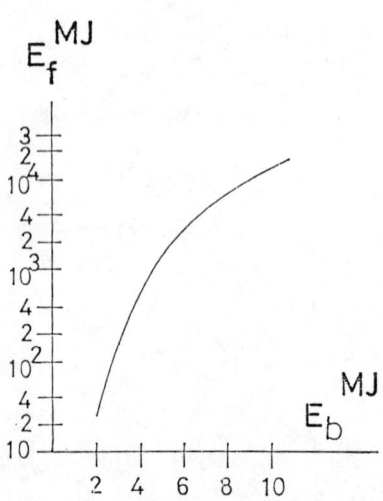

Fig. 7. E_f versus E_b for proton beams.

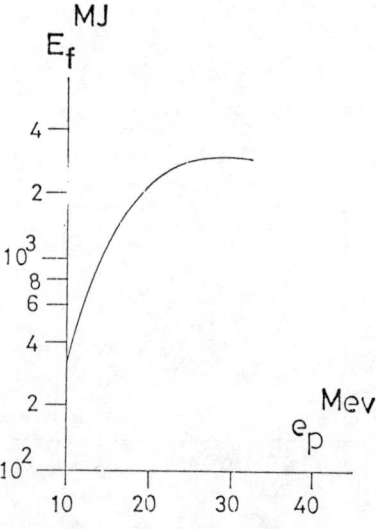

Fig. 8. E_f versus e_p for Lithium beams.

proton is E_b=6MJ, the particle energy is e_p=5.1MeV, the target radius is r_t=6mm and the fuel temperature just before the collision at the center is T_0=1000K. In Figs.4 and 5, we choose that M_{DT}=23.7mg and C_{Al}=0.85. Figure 6 shows that E_f increases with r_t, where E_b=6MJ, M_{DT}=23.7mg, C_{Al}=0.85, e_p=3.5MeV and T_0=1000K are chosen. Figure 7 gives us E_f as a function of E_b for the targets with the optimized parameters.

From Figs.3-7, we conclude that proton beams having E_b=6MJ and e_p=5.1MeV impinging on a target whose radius is r_t=6mm can produce an output energy of E_f=2.4GJ giving an energy gain of G=410. For this optimized target the thickness of the lead layer is δ_{Pb}=20.5μm (M_{Pb}=105mg), the thickness of the aluminum layer is δ_{Al}=156μm (M_{Al}=191mg) and the thickness of the fuel layer is δ_{DT}=27.6μm (M_{DT}=23.7) and hence C_{Al}=0.85.

Figure 8 shows E_f versus the particle energy e_p of the lithium beam when E_b=6MJ, r_t=6mm, T_0=1000K, C_{Al}=0.85 and δ_{DT}=250μm (M_{DT}=21.5mg). When e_p=30.0MeV, E_f becomes maximum and reaches the values of E_f=3.02GJ and G=503.

Figure 9 shows E_f versus e_p for the lead beam, when E_b=4MJ, r_t=5mm and T_0=1000K. The optimum parameters are δ_{DT}=54.6 m (M_{DT}=3.26mg), δ_{Pb}=3.87μm (M_{Pb}=13.7mg), δ_{Al}=49.7μm (M_{Al}=42.2mg) and C_{Al}=0.87. When e_p=1.75GeV, E_f becomes maximum, and the values arrive at E_f=404MJ and G=101.

The curve 1 in Fig.10 shows E_f when the lead beam of E_b=6MJ impinges on the target of "type 1", which has r_t=5mm, and T_0=1000K. For the reactor HIBLIC, the lithium curtain is used as

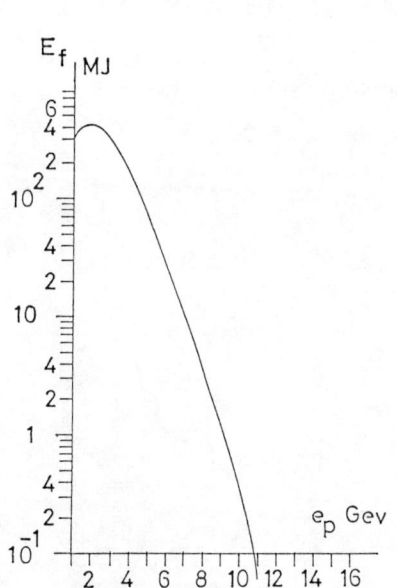

Fig. 9. E_f versus e_p for lead beams.

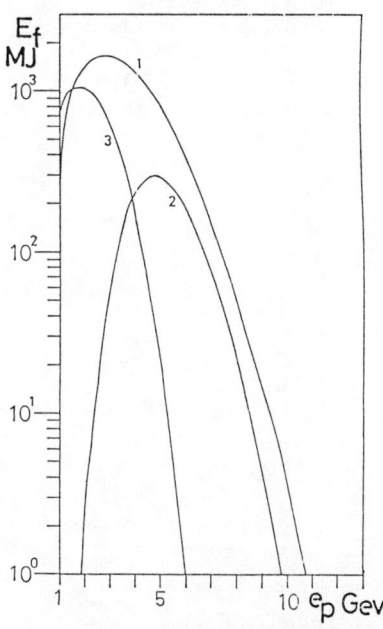

Fig. 10. E_f from three types of targets versus e_p for lead beams.

the coolant and T-breerder inside the solid wall. The lead used as the material of the tamper in the target is not compatiable with the liquid lithium. Therefore, it is hoped that some other material will be substituted for the lead. The target of "type 2" omits the lead layer. Thus the target consists of two layers of Al and DT fuel. The target of "type 3" has the three layers of aluminum, lithium and DT fuel. For the targets of these three types, Fig.10 shows E_f versus e_p. We choose E_b=6MJ, r_t=5mm, T_0=1000K, δ_{Al}=283μm (M_{Al}=240mg) and δ_{DT}=234μm (M_{DT}=14.0mg). When e_p=4.8GeV, we obtain the maximum values of E_f=297MJ and G=50 for the target of "type 2". The curve 3 in Fig.10 shows E_b from the target of "type 3" for E_b=6MJ, r_t=5mm, T_0=1000K, δ_{Al}=22.5 m (M_{Al}=19.1mg), δ_{Li}=133μm (M_{Li}=22.1mg), δ_{DT}=107μm (M_{DT}=6.37mg) and C_{Li}=0.79. When e_p=1.6GeV, we have the optimum values of E_f=1.06GJ and G=160.

Heavy ions allow the use of high particle energy (10-15GeV) and low beam current; however, at high particle energy the material of the beam itself should have a higher atomic number than lead, for exmaple uranium.

References

1. K. Niu and S. Kawata, Fusion Tech. (printing).
2. H. Murakami, T. Aoki, S. Kawata and K. Niu, Laser and Particle Beams **2**, 1 (1984).
3. M. Tamba, N. Nagata, S. Kawata and K. Niu, Laser and Particle Beams **1**, 121 (1983).
4. H. Madarame, S. Iwai, A. Suzuki, Y. Ogata, Y. Oka and S. Tanaka, UTNL-R-0144 (1982).
5. K. S. Holian, La-10160-MS (1984).
6. T. Yamaki, Proc. of 1984 INS Intern'l Symp. 141 (1984).

ACCELERATOR AND FINAL FOCUS MODEL FOR AN INDUCTION LINAC BASED HIF SYSTEM STUDY*

Edward P. Lee
Lawrence Berkeley Laboratory
University of California, Berkeley, CA 94720

ABSTRACT

An overview of the assumptions and models incorporated in the ongoing Induction-Linac-based, HIF System Assessment is presented. Final transport, compression and final focus pose constraints which form a critical link between the accelerator and target requirements. A recent analysis has shown that system costs may be considerably reduced by the use of multiply charged ions. The assumptions underlying this direction are described.

1. INTRODUCTION

Inertial confinement fusion (ICF) requires very high power irradiance and energy deposited on the fusion target which are nearly independent of the driver type. In addition, the depth of deposition must be small (typically $\sim .1$ gm/cm^2 in a stopper material) to produce the high fusion yields required for an economically attractive power plant. The range condition can be met in principle by any ion species, accelerated sufficiently to match the range-energy relation, as well as by short wave-length ($\lesssim 300$ nm) photons. For the heavy-ion driver approach to ICF two conventional but very high current accelerator technologies are being explored. These are the rf linac/storage ring system now studied in W. Germany and Japan, and the induction linac approach of the USA. For both accelerator types the combined considerations of space charge limits and range in dense matter lead to the use of heavy ions of high kinetic energy.

A typical set of final beam parameters suitable for a power reactor, which was adopted in the HIBALL II study,[1] applies equally well to either of the two heavy ion driver types (see Table I). It must be emphasized that cost tradeoffs among the many components of a complete power plant allow a broad range of system parameters (such as repetition rate) to be considered with minor effect on final cost of electricity (COE). Some cost data for HIBALL II are also included in Table I. It should be noted that, although its 47.9 mill/kWh COE is about double that available from existing, on-line coal or fission plants, it is comparable with the estimates from other fusion system studies (e.g. 59.1 mill/kWh for the STARFIRE tokamak in 1984 dollars using current costing methods.) The primary concern at present is not so much the COE, but the magnitude of generating capacity and capital investment of the plant. A 500-1000 MWe fusion plant with this COE is of considerable interest, but it is difficult to achieve, primarily because of the economy of scale associated with all nuclear electric plants. Both the rf linac/storage ring and the induction

---

*This work was supported by the Office of Energy Research, Office of Program Analysis, U.S. Department of Energy under Contract No. DE-AC03-76SF00098.

Table I - Selected HIBALL II Parameters

Pulse Energy	5.0 MJ
Particle Energy	10 GeV
Particle Type	B_i^+ (A = 209)
Pulse Power	250 TW
Pulse Length	20 ns
Rep. Rate per Reactor	5 Hz
Number of Beams per Reactor	20
Net Pulse Charge	500 μC
Relativistic Factor ($\beta\gamma$)	.325
Final Emittance (unnormalized)	3×10^{-5} m-r
Momentum Width at Final Lens	± 1%
Spot Radius	4 mm
Range in Pellet (Pb+Li layers)	.19 gm/cm^2
Convergence Half Angle	≃ 10 mr
Standoff to Final Magnet	8.5 m
Target Gain	87
Net Electric Power (4 Reactors)	3.784 GW
1984 Cost of Electricity (4 Reactors)	47.9 mil/kWh
Direct Cost of Entire Plant	5100 M$
Direct Cost of Driver	2432 M$

linac drivers provide a substantial fraction of total direct capital costs of a plant and scale poorly for lower net electric power. The HIBALL II driver cost is "only" about 48% of total direct capital cost largely because the high rep. rate capability of the accelerator has been exploited in a large (multi-GW) plant with four reactors. Magnetic fusion systems are also very large for reasons of economy of scale as well as physical constraints imposed by the use of low density plasma.

A goal of the ongoing Heavy Ion Fusion Systems Assessment[2] is to find ways of reducing the cost of the driver and other plant components to the point where 500-1000 MWe plants are attractive. Another, related, goal of Heavy Ion Fusion, not covered by the Assessment, is to find a path of development which will lead from the current research to a fusion plant with a minimum of risk and expense. Cost reductions can be achieved in several directions:

a. Reduced cost of materials

b. Innovative use and manufacture of components

c. Optimized match of major system components (e.g. high rep. rate reactor to high rep. rate accelerator)

d. Changes in design limitations resulting from improved understanding of physical constraints.

The present paper addresses only the last of these directions, where there have been significant new developments in the understanding of limits of high current transport in the accelerator system.

2. INDUCTION LINAC SYSTEM

An induction Linac driver is now envisioned as a <u>multiple beamlet</u> transport lattice consisting of (N) closely packed parallel FODO channels. Surrounding the lattice are massive induction cores of ferromagnetic material and associated pulser circuitry which apply a succession of long duration, high voltage pulses to the N parallel beamlets. Longitudinal focussing is also achieved through the detailed timing and shape of the accelerating waveforms (with feedback correction of errors). A multiple beam source of heavy ions operates at 2-3 MV, producing the net charge per pulse required to achieve the desired pellet gain. Initial current (and therefore initial pulse length) are determined by transport limits in the lattice at low energy. The use of a large number of electrostatic quadrupole channels (N ~ 16 - 64) appears to be the least expensive focusing option at low energies (below ~ 50 MV). This is followed by a lower number of superconducting magnetic channels (N ~ 4-16) for the rest of the accelerator. Merging of beams may therefore be required at this transition. Furthermore, some splitting of beams may be required after acceleration to stay within current limits in the final focus system.

The rationale for the use of multiple beams is that it increases the net charge which can be accelerated by a given cross section of core at a fixed accelerating gradient. Alternatively a given amount of charge can be accelerated more rapidly with multiple beams since the pulse length is shortened and a core cross section of specified volt-seconds per meter flux swing can supply an increased gradient. However, an increase in the number of beamlets increases the cost and dimensions of the transport lattice and also increases the cost of the core for given volt-sec product since a larger core volume is required. For a given core cross section ($\propto$ volt-seconds/m), the volume of ferromagnetic material increases as its inside diameter is increased. Hence there is a tradeoff between transport and acceleration costs with an optimum at some finite number of beamlets. The determination of this optimum is a complex problem depending on projected costs of magnets, core, insulators, energy storage, pulsers and fabrication. The induction linac design code LIACEP[3] is used for this purpose.

The choice of superconducting magnets for the bulk of the linac is mandated by the requirement of system efficiency; this must be at least ~ 10% in an ICF driver and ideally > 20% to avoid large circulating power fractions (which result in a high COE). Induction cores are most likely to be constructed from thin laminations of amorphous iron, which is the preferred material due to its excellent electrical characteristics and flux swing. At a projected cost of ~ 4 $/lb (insulated and wound) this is a major cost item for the first 2-4 GV of a typical linac. At higher voltage the cost of pulsers and fabrication of the high gradient column with insulators dominates.

Results from about 400 LIACEP runs, spanning a selected range of beam parameters suitable for a driver, have been fit with a polynomial expression and incorporated into the HIFSA system code. The fitted data are accelerator cost, efficiency and length.

3. FINAL TRANSPORT AND FOCUS

Between the accelerator and the fusion reactor the beamlets are separated and also, if necessary, split. The drift lines leading to the final focus area are 200-600 m in length and used for ballistic compression as well as to match to the final focus configuration of the reactor. The transport lattice is composed of cold bore superconducting quadrupoles, bends, and possibly higher order elements needed to control dispersion. As the beamlets compress, the transport of the high current becomes increasingly demanding, with the large apertures and the close packing of elements especially pronounced immediately before the final focus train.

The final focus system itself has parameters determined largely by the requirements of spot size on target, reactor size, and the neutron, x-ray, and gas fluxes from the reactor. The final focus triplets described by R. Martin[4] are well suited as the basic beam line components. HIBALL II uses magnet trains consisting of a pair of triplets separated by a pair of weak bends used to remove line-of-sight neutrons from the beam transport line. A detailed discussion of shielding requirements for this system is also presented in the HIBALL II Study.

Transport within the reactor vessel has, in most studies, been assumed to take place in near vacuum ($P \lesssim 10^{-4}$ torr Li) to avoid disruption by the two-stream instability, or in a high pressure window ($P \sim 10^{-1}$-10 torr), where the beam is also thought to be stable[5]. HIBALL specifies $P < 10^{-5}$ torr Pb vapor to avoid stripping of beam ions, which would lead to reduced irradiance due to the beam's electric field. Unfortunately, several attractive reactor concepts (CASCADE,[6] HYLIFE[7]) have residual gas pressures in the range 10^{-2}-10^{-3} torr Li at reasonable rep. rates; this pressure must be taken into account both for transport in the reactor and in maintaining vacuum in the final focus lines. Recent calculations[8] show that the two stream mode is benign at these pressures due to the detuning effects of beam convergence. The control of gas flux into the beamlines, and the process of stripping and neutralization in the reactor have not yet been examined in necessary detail, and are dealt with by plausible assumptions in the Heavy Ion Fusion System Assessment.

To produce a small radius (r) on the target, the beamlet emittance (ϵ) must, satisfy

$$\epsilon < r\theta$$

where θ is the beamlet convergence cone half-angle. For HIBALL (r = 4 mm, θ = 10 mr) this condition is $\epsilon < 4 \times 10^{-5}$ m-r, which is 33% larger than the design value of 3×10^{-5} m-r. Allowance must also be made for the effect on spot size of momentum dispersion, various forms of jitter, and space charge induced blow up. A final focus system comprised of quadrupoles and weak bends has dispersion at the target which leads in a practical design based on a pair of triplets to increased spot radius:

$$\Delta r \approx 8 \, L\theta \, \frac{\Delta P}{P} \, ,$$

where L is the distance from pellet to the center of the final quadrupole. Without compensation by higher order elements it is desirable to keep $\Delta P/P \lesssim 10^{-3}$. This is a severe requirement to be met by the accelerator system.

In summary, the requirement of small spot size on target is met by small specified emittance and a set of other focal and reactor constraints which are not currently well understood. The cone half angle θ is set at a value which is determined by a trade-off between factors which drive it towards a low value and those which drive it to a high value. In the first category are dispersion, aberrations, magnet costs, reactor constraints, shielding and beamline vacuum. In the second category are the emittance limit, space charge effects, and jitter control. The range θ = 10-20 mr is the result of compromises among these factors. Aside from the spot size condition, it is economically desirable to make the emittance large, since the limit on transport of high current is found to vary as $\epsilon^{2/3}$.

4. SCALING LAWS FOR TRANSPORTABLE CURRENT

A simple set of formulas relating transportable current to lattice properties were originally derived by A. Maschke.[9] Since these serve as a good framework for the discussion of recent developments in high current transport they are rederived here. The continuous limit approximation for alternate gradient focussing is adopted. Let (σ_0) be the phase advance of a single ion per lattice period of length (2L), η is the lens occupancy factor, (B') or (E') is the pole field gradient, [Bρ] the particle rigidity, and v is its velocity. Then we have approximately[10]

$$\sigma_0 = \frac{\eta(B', E'/v)L^2}{[B\rho]} \qquad (1)$$

The depressed phase advance (σ), which is the actual phase advance per period in the high current beam [mean edge radius (a) and electric current (I)], is given by

$$\sigma^2 = \sigma_0^2 - \frac{2I}{(\beta\gamma)^2 [B\rho]} \left(4\pi\epsilon_0 c^2\right)^{-1} \left(\frac{2L}{a}\right)^2 \qquad (2)$$

Normalized emittance ($\epsilon_n = \beta\gamma\epsilon$) is related to depressed phase advance and radius by

$$\epsilon_n = \beta\gamma \frac{\sigma}{2L} a^2 . \qquad (3)$$

In addition to these three relations we have the definitions

$$[B\rho] = \frac{\beta\gamma M c}{qe} , \qquad (4)$$

$$\beta\gamma = \left[\left(\frac{E}{Mc^2}\right)^2 + 2\left(\frac{E}{Mc^2}\right)\right]^{1/2} ,$$

where $M = M_0 A$ is ion mass, M_0 is the atomic mass unit, q is charge state and E is kinetic energy. Values of β are generally less than about .33 for conceptual drivers, so with $\lesssim 1\%$ error we set

$$\beta\gamma = \sqrt{\frac{2 qeV}{M_0 A c^2}} , \qquad (5)$$

where V is the cumulative accelerating voltage experience by an ion. The magnetic field or electric potential of the quadrupoles, evaluated at the beam radius, are respectively

$$B = B'a \lesssim 3.0 \text{ T} \qquad (6)$$

$$\phi = E'a^2/2 \lesssim 50 \text{ kV} . \qquad (7)$$

In the original treatments of high current transport equations (1)-(3) were solved for I, a, and L as a function of the other parameters; this gave for magnetic lenses:

$$I = (2.89 \text{ MA}) \left(1 - \frac{\sigma^2}{\sigma_0^2}\right) \frac{\sigma_0^{4/3}}{\sigma^{2/3}} (\beta\gamma)^{5/3} \left(\frac{A}{q}\right)^{1/3} \left(nB\epsilon_n\right)^{2/3} , \qquad (8)$$

$$a = (2.32 \text{ m}) \left(\frac{\sigma_0 \epsilon_n^2 A}{nB\sigma^2 q\beta\gamma}\right)^{1/3} , \qquad (9)$$

$$L = (2.68 \text{ m}) \left(\frac{\sigma_0^2 \epsilon_n A^2 \beta\gamma}{n^2 B^2 \sigma q^2}\right)^{1/3} , \qquad (10)$$

with B given in Tesla, ϵ_n in meter-radians, and the phase advance in radians.

In the earlier application of these formulas there was little sound basis for fixing the values of σ_0 and σ. The assumptions, $\sigma_0 \le 90°$ and $\sigma/\sigma_0 \ge 1/\sqrt{2}$ were made somewhat arbitrarily, although the factor $1/\sqrt{2}$ does correspond to an equal split between emittance and charge related forces in an envelope equation. For the HIBALL II parameters ($\beta\gamma = .325$, $A = 209$, $\varepsilon = \varepsilon_n/\beta\gamma = 3\times10^{-5}$ m-r), and setting $\eta B = 1.0$ T, we get $I = 1.29$ kA, which is close to the value reached by a beamlet in final focus (≈ 1.25 kA). In fact the current limit [Eq. (8)] does not actually apply in final focus, where the beamlets are expanded to large radii for the last 180° of phase advance before the reactor chamber is reached. However, it does apply, along with Eqs. (9) and (10) in the transport lines prior to final focus and the induction linac transport lattice. Several developments relating to the use of these transport limits since 1976 are described here:

a. A near exact treatment of the matched envelope was obtained numerically and more recently, analytically.[11] This is incorporated into the optimization code LIACEP.

b. The stability of transverse beam modes, which depends on the values of σ_0 and σ, and the consequences of mode growth were examined with analytical calculations and PIC simulation.

c. Experiments with the Single Beam Transport Experiment (SBTE) at LBL explored the consequences of a broad range of values of σ and σ_0.

d. Scaling laws for the continuous limit formulas point to possible large cost reductions for the driver linac.

5. SCALING LAWS FOR INCREASED CHARGE STATE

Heavy ion driver studies have for several years concentrated on the use of charge state $q = 1$ and the highest available mass ($A \gtrsim 200$), however it has been noted that increased charge state may be desirable in order to lower linac cost and length.[12] It is clear that increased q or decreased A decreases the final cumulative acceleration potential required to reach a final given ion velocity, but it is less clear that, given the constraints of transportable current and range in the target, this is a useful path to take. Examination of Eqs. (5)-(10) shows that increased q and decreased A are equivalent as regards transport for given V. The differences are in the availability of good sources and range in the target at fixed final velocity.

For ion range in the target the situation is clear: at $\beta \approx .3$ an ion of mass number $A = 100$ has about twice the range as an ion of mass number 200. Other things being equal the doubled range would halve the specific energy deposition, and to achieve equal target gain the spot radius would need to be decreased approximately by $\sqrt{2}$.

The only heavy ion sources available at present which can be readily adapted to driver requirements are the contact ionization of Cs and the Mercury vapor arc. However, the metal vapor vacuum arc (MEVVA),[13] which produces copious ions of high brightness in a range of charge states for all metals, is undergoing an impressive development and may be considered as a possible future driver source. The main problem in adaptation appears to be the removal of unwanted charge states from the beam pulse before introduction into the induction linac.

We assume here that the highest mass ions will be used due to their short range, and that charge state can be increased arbitrarily until some transport or focal limit is reached. If σ/σ_0 is small, so that the factor $(1-\sigma^2/\sigma_0^2)$ in Eq. (8) can be replaced by unity, then the following scale relations are found for $q \to q'$:

At each value of voltage V,

$$q \to q' = \alpha q ,\qquad (11)$$

$$V \to V, \quad a \to a, \quad \eta B \to \eta B, \quad \epsilon_n \to \epsilon_n, \quad \sigma_0 \to \sigma_0 ,$$

$$I \to \alpha I, \quad \tau_p \to \tau_p ,$$

$$E \to \alpha E, \quad \beta\gamma \to \alpha^{+1/2} \beta\gamma ,$$

$$\sigma \to \alpha^{-3/4} \sigma, \quad L \to \alpha^{-1/4} L ,$$

$$\text{volt-sec/m} \to \text{volt-sec/m} .$$

The significance of this transformation is that the transported power is increased by the factor α at given V with very little change in the transport lattice. Only the half period has been decreased by the small factor $\alpha^{-1/4}$. The big change is that the depressed phase advance σ is decreased by the factor $\alpha^{-3/4}$. A discussion of phase advance limits is given in Section 6.

There are many possible linac configurations for a given value of q; the low cost optimum is found by LIACEP. One attractive possibility (not optimal) is found by simply applying the transformation [Eq. (11)] to a known configuration with $q = 1$, raising its charge to $q = \alpha$ and eliminating the high voltage portion of the linac so that the final kinetic energy is unchanged. This procedure is expected to yield incremental cost savings for the main portion of the linac of ~ 28% for each doubling of q, and in fact LIACEP verifies this approximate cost scale. A five Megajoule induction linac driver with output similar to HIBALL II is predicted to cost ~ 1500 M$ accelerating charge state +1 and ~ 800 M$ accelerating charge state +4 (1986 dollars). This price does not include the first 50 MV or the final transport and focus lines. The cost of these components is not expected to improve with higher charge state.

At low voltage (V < 50 MV) the current that can be conveniently transported with superconducting quadrupoles is low, and the use of electrostatic quadrupoles is preferred. Unfortunately, the scaling law for increased charge state is not attractive for this form of transport. It is found that the electric line charge density per meter is limited by the value[11]

$$\lambda \leq \left(.0943 \frac{\mu C}{m}\right)\left(\frac{a_m}{.5b}\right)^2 \frac{\phi(b)}{50 \text{ kV}}, \quad (12)$$

where a_m is the maximum radius of the matched beam envelope and b is the pole tip radius. The derivation of Eq. (12) assumes $\sigma = 0$, $\sigma_0 = 60°$, $\eta = 1/2$, and $a_m/a = 1.25$. Adopting the optimistic values $\sigma_0 = 90°$, $\phi(b) = 100$ kV, $a_m/b = .75$, and $a_m/a = 1.44$ we obtain a rough upper limit on transportable line charge density $\lambda \lesssim .5$ μC/m. Hence electric current increases only as $q^{1/2}$, and we are led to consider using a large number of beamlets of small radius, which are merged for the magnetic transport lattice.

The high energy transport and final focus lines may have reduced costs per line for increased charge state since rigidity varies inversely with q. However, this savings is likely to be offset by an increase in the number of lines required to avoid exceeding space charge limits in the final focus and reactor. A rough measure of the minimum number of beamlines is obtained from an examination of the beam envelope equations in vacuum, neglecting all effects except that due to space charge:

$$\frac{d^2 a}{ds^2} = \frac{K}{a}, \quad (13)$$

where K is the dimensionless perveance:

$$K = \frac{2I(4\pi\epsilon_0 c^2)^{-1}}{(\beta\gamma)^2 [B\rho]}. \quad (14)$$

The minimum radius resulting from Eq. (13) is

$$r = a_{lens} \exp(-\theta^2/2K). \quad (15)$$

where $a_{lens} = L\theta$. For a power reactor we expect $L \sim 5\text{-}10$ m, $\theta \sim 10\text{-}20$ mr, and 2-4 mm spot size. To make the influence of space charge negligible we therefore require, in the absence of neutralization,

$$K \lesssim (.1) \theta^2, \quad (16)$$

This condition leads to unacceptably large numbers of beamlets when the charge state exceeds $q = 2$ or 3, so some degree of neutralization must be invoked in general. The figure adopted in the HIF System Assessment is 90% neutrality, either from ionization of residual gas or co-injected electrons. We then have the condition

$$K \lesssim (2.25 \times 10^{-4})(\theta/15 \text{ mr})^2. \quad (17)$$

HIBALL II, with $K = 1.1 \times 10^{-5}$ and $\theta = 10$ mr easily meets this condition without invoking neutralization. In general for fixed total power P (summing over N beamlets), perveance per beamlet varies as

$$K \propto \frac{Pq^2}{N(\beta\gamma)^5 A^2}, \qquad (18)$$

so the condition (17) becomes very restrictive for $q \gtrsim 4\text{-}5$.

6. LIMITS ON PHASE ADVANCE

As mentioned, in the early work on transport limits Maschke adopted the values $\sigma_0 = 90°$ and $\sigma/\sigma_0 = 1/\sqrt{2}$. In fact, it is not immediately apparent from Eqs. (8)-(10) that a higher allowed value of σ_0 and lower allowed σ will result in lowered accelerator costs since the beam radius is also increased as the current increases. If the half period is eliminated between Eqs. (1) and (2) we get the suggestive result

$$I = \frac{(\beta\gamma)^2}{8} 4\pi\epsilon_0 c^2 a \sigma_0 n B, \qquad (19)$$

which shows current to be independent of σ and proportional to σ_0 for fixed a, n, and B. Hence it is good to raise σ_0 as high as possible. The corresponding value of depressed phase advance is given by

$$\sigma = \frac{2\epsilon_n}{a^{3/2}} \left(\frac{Mc\sigma_0}{\beta\gamma q e n B}\right)^{1/2}, \qquad (20)$$

and a lower allowed value for σ permits either a lower normalized emittance or increased q/M.

Since the work of Maschke there have been several developments in the understanding of phase advance limits, which now stand at the values $\sigma_0 \leq 80°$, $\sigma/\sigma_0 > .1$; a brief summary of part of this work is given here:

a. Analytical calculations[14] showed that the Kapchinsky-Vladmirsky (K-V) distribution of phase space variables is unstable in stop bands depending on σ and σ_0. Perturbations of order n in the radial coordinate are potentially unstable for $\sigma_0 > 180°/n$. Simulation studies[14,15] supported this point by demonstrating the onset of the third order and second order (envelope) modes with characteristic phase space distortions. To stabilize these modes the conditions $\sigma_0 \leq 60°$, $\sigma \geq 24°$ were adopted for driver studies during the period 1981-84.

b. Simulation studies performed with realistic (non K-V) distributions [by I. Haber and C. Celata] have shown little evidence of unstable mode growth for $\sigma/\sigma_0 > .1$ and $\sigma_0 < 80°$. The principal diagnostic is the growth of emittance. This empirical result may be the consequence

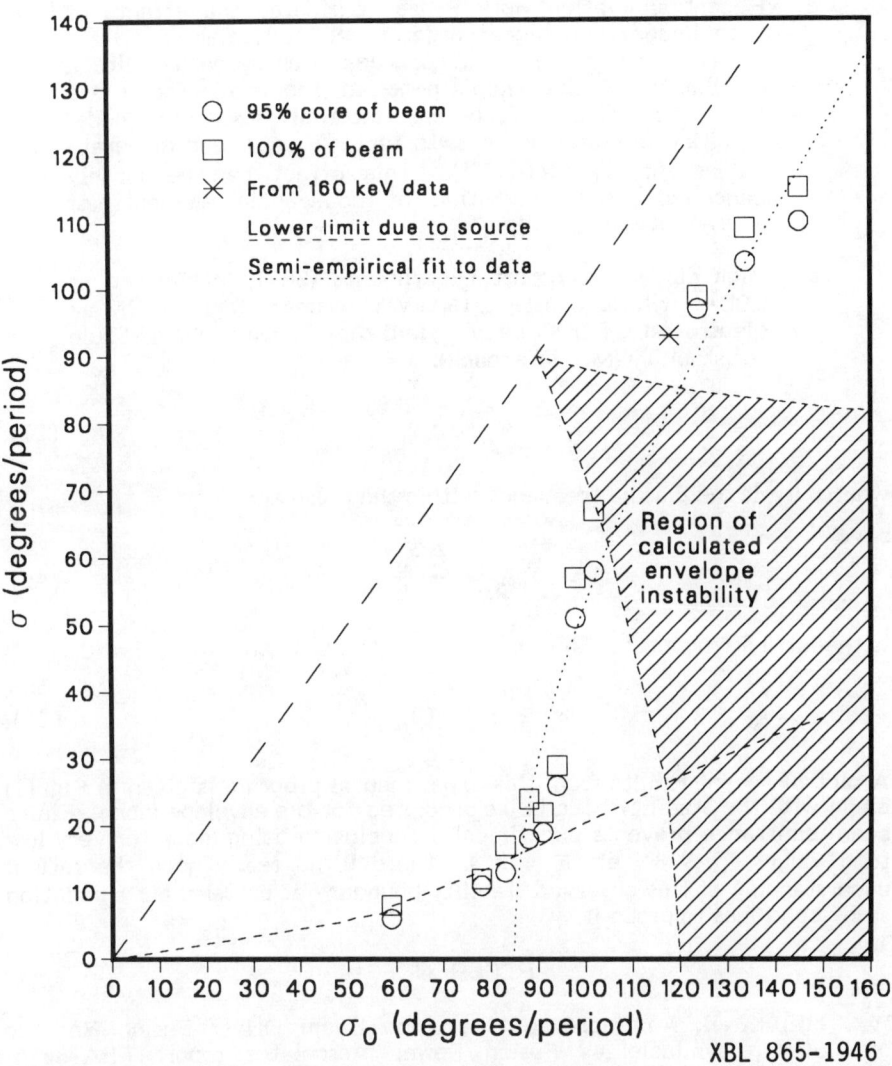

Fig. 1. Zones of predicted and observed instability are depicted in the (σ, σ_0) plane). The cross hatched area corresponds to the unstable envelope mode predicted for the KV distribution. Data points (except for those on the lower broken line) indicate the onset of emittance growth or disruption as σ_0 is increased, with the phenomenological fit $\omega_p = \omega_L/3$ given by the dotted line. The zone below the lower broken line is inaccessible due to the nonzero emittance of the SBTE pulse. (Courtesy of M. Tiefenback)

of the detuning effect of the slightly rounded charge profile of the non-KV distributions, which could damp modes higher than n = 2 (the envelope equations and modes are nearly independent of distribution details).

c. Recent simulation work[16] has considered the effects of both images and higher order focal multipoles which are always present to some degree. For large amplitude oscillations of the beam's centroid, the image forces are found to drive a coherent internal sextupole mode, resulting in emittance growth for $\sigma/\sigma_0 \lesssim .1$ and moderate values of σ_0 (60°-72°). This effect can be largely cancelled by the addition of dodecapole elements of appropriate magnitude.

d. High current transport experiments (SBTE) performed at LBL[17] with a coasting 160 kV Cs$^+$ beam focussed by an electrostatic FODO lattice yield the phenomenological rule for stability (M. Tiefenback):

$$\omega_p \leq \frac{1}{3} \omega_L , \qquad (21)$$

where ω_p is the plasma frequency within the pulse and

$$\omega_L = \frac{2\pi\sigma_0 v}{2L} \qquad (22)$$

is the lattice frequency. This condition may be written

$$\sigma_0^2 - \sigma^2 < (85°)^2 . \qquad (23)$$

A plot of recent results from this experimental program is given in Fig. (1) along with the stability boundaries predicted for the envelope mode. Finite beam emittance prevents experimental conclusion being made for very low tune values ($\sigma \lesssim 8°$ at $\sigma_0 = 60°$). There is no really good theoretical understanding of this observed stability boundary at present, but simulation study continues to probe it.

REFERENCES

1. HIBALL-II, An Improved Heavy Ion Beam Driver Fusion Reactor Study, available as Fusion Power Associates report FPA-84-4, Kernforschungszentrum-Karlsruhe report KfK-3840 and University of Wisconsin report UWFDM-625, (1984).
2. L.M. Wagner et al., Proceedings of the 11th Symposium on Fusion Engineering, Austin, TX, November 1985.
   Also: Various papers in this Proceedings by D. Driemeyer, D. Dudziak, W. Herrmannsfeldt, L. Waganer, and D. Zuckerman.

3. A. Faltens, E. Hoyer, D. Keefe, and L.J. Laslett, Design/Cost Study of an Induction Linac for Heavy Ions for Pellet-Fusion, IEEE Trans. Nuc. Sci., NS-26, No. 3, p. 3106, (1979).
4. Ronald L. Martin, Nuclear Instruments and Methods 187, 271-280, (1981).
5. Proceedings of the Heavy Ion Fusion Workshop, W.B. Herrmannsfeldt (ed.) held October 1979 in Berkeley, CA, issued as LBL-10301, SLAC-PUB-2575, p. 403.
6. J.H. Pitts and I. Maya, Lawrence Livermore Laboratory report UCRL-92558, (1985).
7. J. Blink et al., Lawrence Livermore Laboratory report UCRL-53559, (1985).
8. P. Stroud, LANL report LAUR 85-2809, (1985).
9. E.D. Courant, ERDA Summer Study of Heavy Ions for Inertial Fusion, held in Berkeley, July 1976, issued as LBL-5543, p. 72 [a summary of Maschke's formulation is given].
10. M. Reiser, JAP 52, 555 (1981).
11. E. Lee, T. Fessenden, and L. Laslett, IEEE Trans. Nuc. Sci. 32, no. 5, p. 2489, (1985).
12. A. Faltens, E. Hoyer, and D. Keefe, Proc. of 4th Int. Top. Conf. on High Power Electron and Ion-Beam Research and Technology, Palaiseau, France (1981), also LBL-12409.
13. I. Brown, J. Galvin, and R. Macgill, Ap. Phys. Lett., Vol. 47, 358 (1985).
14. I. Hofmann, L. Laslett, L. Smith, and I. Haber, Particle Accelerators 13, 145-178, (1983).
15. I. Hofmann, Nuclear Instruments and Methods 187 281-287 (1981).
16. C. Celata, I. Haber, L. Laslett, L. Smith, and M. Tiefenback, IEEE Trans. Nuc. Sci. NS-32, no. 5, 2480-2482, (1985).
17. M. Tiefenback and D. Keefe, IEEE Trans. Nuc. Sci. NS-32, no. 5, 2483-2485, (1985).

THE COST OF INDUCTION LINAC DRIVERS FOR INERTIAL FUSION FOR VARIOUS TARGET YIELDS*

Jack Hovingh
Lawrence Livermore National Laboratory, Box 808,
Livermore, CA 94550

V. O. Brady, A. Faltens, D. Keefe, and E. P. Lee
Lawrence Berkeley Laboratory, University of California,
Berkeley, CA 94720

ABSTRACT

The cost of induction linac accelerators for inertial fusion using mass 200 ions at a charge state of +3 for target yields of 300, 600, and 1200 MJ is presented. The ions are injected into the accelerator at 3 MV, and accelerated to the required voltage appropriate to the desired target yield. A cost comparison of the low voltage portion of the accelerator (3-50 MV) is made between a system with 64 and one with 16 superconducting quadrupoles. The design of the low voltage portion which yields the minimum-cost accelerator designs for several target yields and a fusion power of 3000 MW is presented.

INTRODUCTION

An induction linear accelerator that produces an energetic (5 to 20 GeV) beam of heavy (130 to 238 amu) ions is a prime candidate as a driver for inertial fusion. The required accelerator output parameters for an ion species can be determined from the target requirements for a given fusion energy yield, and the cost and efficiency of various accelerator configurations to produce the required output can be determined. In this study we use mass 200 ions.

DETERMINATION OF THE ACCELERATOR OUTPUT PARAMETERS

The required accelerator output parameters for a given target yield can be determined for a single shell target design using the Lindl-Mark gain curves.[1] These include the total energy and, for a given ion species, the emittance and ion kinetic energy. For a given target yield, the output energy, W, is determined based on the upper bound of the Lindl-Mark "best estimate" gain curve. Also determined is the $r^{3/2}R$ parameter where R is the range of the ions in g/cm^2 in the target material and r is the target spot radius which must satisfy

$$0.1 \, W^{1/3} < r < 0.2 \, W^{1/3} \quad (W, \, MJ; \, r, \, cm) \, . \tag{1}$$

---

*This work was supported by the Office of Energy Research, Office of Program Analysis, U.S. Department of Energy under Contract No. DE-AC03-76SF00098.

From the $r^{3/2}R$ parameter and the target spot radius, the desired range can be determined. From this range, the required ion kinetic energy can be specified. From the ion kinetic energy and spot radius, for a given angle of convergence, the maximum normalized emittance of the accelerator beamlets can be determined assuming that it dominates the spot radius. This completes the description of the required accelerator output. Associated with the target gain and beam energy is a peak power requirement which can be independently modulated by the final transport drift lines.

EARLY COST AND PERFORMANCE RESULTS

The cost and performance of the accelerator were determined using the modified cost optimization LIACEP[2] code. We investigated single shell target yields of 300, 600, and 1200 MJ and fusion powers between 1500 and 6000 MW for the singly-charged mass 200 ions. Accelerator configurations accommodating 4, 8, and 16 simultaneous beamlets were studied, with undepressed and depressed tunes (refer to "D. Keefe, these proceedings" for definition) of the transport lattice of $\sigma_0 = 75°$ and $\sigma = 24°$, respectively. The results are given in Table I for an angle of convergence in the chamber of 0.015 radians and a spot radius due only to emittance. The efficiency and costs (in 1979 dollars) of the accelerators for a fusion power of 3000 MW and an initial voltage of 50 MV are also given for accelerator configurations of 4, 8, and 16 beamlets. The efficiencies are greater than 20%, resulting in a ratio of fusion power to accelerator input power greater than 28. The minimum cost designs are near the maximum efficiency designs.

Table I. Accelerator Ouput Characteristics, Efficiencies and 1979$ Costs for 300, 600, and 1200 MJ Target Yields and 3000 MW Fusion Power using 200 amu, q = +1 Ions.
$\phi = 0.5$ MV/m; $\sigma_0 = 75°$, $\sigma = 24°$
Initial Voltage = 50 MV; Spot Radius = $0.1\ W^{1/3}$ cm
Range = R g/cm^2

	300	600	1200
Yield, MJ	300	600	1200
Pulse Rep. Rate, hertz	10	5	2.5
Energy, (W) MJ	2.91	4.25	6.57
Gain (G)	103	141	183
$r^{3/2}R$, 10^3 cm$^{-1/2}$g	7.2	10.4	15.9
Normalized Emittance (ϵ_n), μm-r	7.15	8.65	10.8
Ion Kinetic Energy, (E_i), GeV	10.12	11.46	13.24
Cost, G$			
Beamlets: 4	1.149	1.275	1.483
8	1.107	1.227	1.427
16	1.152	1.276	1.473
Efficiency, (η)%			
Beamlets: 4	21.2	21.5	21.6
8	22.7	24.6	26.2
16	20.7	23.0	25.3

COST REDUCTION STRATEGY

The costs can be reduced by increasing the charge state, increasing the undepressed tune, and decreasing the depressed tune limits. For example, the cost of the 4.25 MJ, 8 beamlet accelerator above 50 MV that produces 11.46 GeV ions can be reduced from 1.227 G$ to 0.6393 G$ (1979$) by increasing the ion charge state to +3, increasing the undepressed tune to 85°, and decreasing the depressed tune to 10.5° while increasing the number of beamlets to 16. From perveance considerations, this accelerator system will require at least 16 beams focussed on target. The cost can be decreased further to 0.5136 G$ by increasing the allowable vacuum surface flashover voltage gradient (ϕ) from 0.5 MV/m used in the Austin Study[2] to 1.0 MV/m used in the Palaiseau Study[3]. The effect of these cost reduction techniques is to reduce the length of the accelerator above 50 MV from 10.7 to 2.23 km, and increase the efficiency from 24.6 to 34.5%. The somewhat longer front end (<50 MV) of the higher charge state option is more than offset by this large length reduction.

The cost of this accelerator can be further reduced from 0.5136 to 0.4826 G$ by double pulsing a 2.125 MJ accelerator. However, the efficiency decreases from 34.5% to 20.8% using current technology. Complete reactor plant system studies[4,5] have shown that the increased balance of plant costs due to the lower efficiency of double pulsing offsets the capital cost advantage of double pulsing[6].

The increase in the charge state (q) of the ions may be made possible by the development of the metal vapor vacuum arc (MEVVA) source which produces large quantities of ions in a range of charge states for most metals.[7] The higher charge state savings are due to the shortening of the accelerator, with savings in the quantity of cores and quadrupoles. Some of the cost savings may be used up by the increased number of beamlets which scales as q^2 in the final focus to meet perveance constraints. These are discussed by Lee.[8] For the case selected for this paper, the number of beamlets from perveance considerations in the final focus does not exceed the number of beamlets in the accelerator.

The increase of the undepressed tune to 85° is speculative. However, there is some experimental evidence that this value of undepressed tune may be achieved.[8]

The use of a vacuum surface flashover voltage gradient of 1 MV/m results in the high acceleration gradients of about 2 MV/m in the final regions of the driver. These high acceleration gradients are adventurous, and caused by the model used to estimate the enhancement of the flashover gradient at short pulse durations.

The use of double pulsing to reduce the cost of the accelerator is most effective for ions with low kinetic energy. Cost savings of 30% can be realized with low kinetic energy (≈ 5 GeV) ions.[9] A possible strategy for a low cost accelerator using low kinetic energy ions may be to use double pulsing coupled with a charge state of +2. This may ease the perveance conditions in the final focus and reduce the number of beamlets in the final focus elements to the target. Advances in tube technology may reduce the power consumption of the pulsers, which will increase the efficiency of the double pulsed accelerator.

ACCELERATOR COST AND PERFORMANCE

Three accelerators were analyzed using LIACEP to give target yields of 300, 600, and 1200 MJ using the minimum spot radius and the upper bound of the best estimate gain curve. The fusion power, which is the product of fusion yield and pulse repetition frequency, was fixed at 3000 MW. The charge state +3, 200 amu ions are injected into the accelerator with a kinetic energy of 9 MeV. This low voltage section of the accelerator consists of 64 beamlets, using superconducting quadrupoles and amorphous iron cores. The transition ion kinetic energy for which it becomes cost effective to combine the 64 beamlets into 16 beamlets is the energy at which the total unit costs for the 64 beamlet system is equal to that of the 16 beamlet system. This transition ion energy is typically between 400 and 600 MeV for the cases considered in this paper. The 64 beamlets are then combined into 16 beamlets, and accelerated to the desired final kinetic energy. The accelerator output characteristics are as shown in Table I, and repeated in Table II.

The undepressed tune σ_0 of 85° and the allowable vacuum surface flashover voltage gradient 1 MV/m are used for these accelerators. The depressed tune for each of the accelerators is given in Table II.

The costs and performance of the accelerators to produce target yields of 300, 600, and 1200 MJ are given in Table II for a fusion power of 3000 MW. The cost of the accelerator increases with the target yield, but the performance, measured as ηG (accelerator efficiency times target gain), also increases, resulting in a lower recirculating power fraction to the accelerator. The costs of the low voltage (<50 MV) section are about 20% of the accelerator costs.

Table II. Accelerator Output Characteristics, Efficiencies and 1979$ Costs for 300, 600, and 1200 MJ Target Yields and 3000 MW Fusion Power using 200 amu, q = +3 Ions.
$\phi = 1.0$ MV/m; $\sigma_0 = 85°$
Initial Voltage = 3 MV; Spot Radius = $0.1 \times W^{1/3}$ cm
Range = R g-cm/cc; N = 16 beamlets, $V > V_c$

Yield, MJ	300	600	1200
Energy, (W) MJ	2.91	4.25	6.57
Gain (G)	103	141	183
$r^{3/2}R$, 10^3 cm$^{-1/2}$g	7.2	10.4	15.9
Emittance (ϵ_n), μm-r	7.15	8.65	10.8
Ion Kinetic Energy, (E_i), GeV	10.12	11.46	13.24
Pulse Repetition Frequency, hertz	10	5	2.5
64 Beamlet Cost to 50 MV, M$	108	124	162
64 to 16 beamlet transition voltage (V_c), MV	133	160	180
ϵ_n/σ, μm-r/degree, $V < V_c$	1.1	0.82	1.1
Depressed Tune (σ), $V < V_c$, degrees	7.5	10.5	10
Total Cost, M$	551.5	633.1	748.7
Total Length, km	1.97	2.22	2.57
Total Efficiency (η)%	26.9	28.7	29.0
ηG	27.7	40.6	52.9

The unit costs (1979$) per volt for a driver which will produce a target yield of 300 MJ are shown in Fig. 1 as a function of the ion energy. At low ion energies, the core costs dominate the total cost. At high ion energies, the structure (including insulators) and pulsers are the more costly units. Integrating the costs over the ion kinetic energy gives the total costs for the complete accelerator. The core costs are about 33% of the total cost of the accelerator. The superconducting magnet costs represent about 23% of the total costs of the accelerator. The structure (including insulators) and the pulsers represent about 17 and 15%, respectively, of the total costs. These cost percentages will change when the costs are in 1985$, as discussed later in this paper.

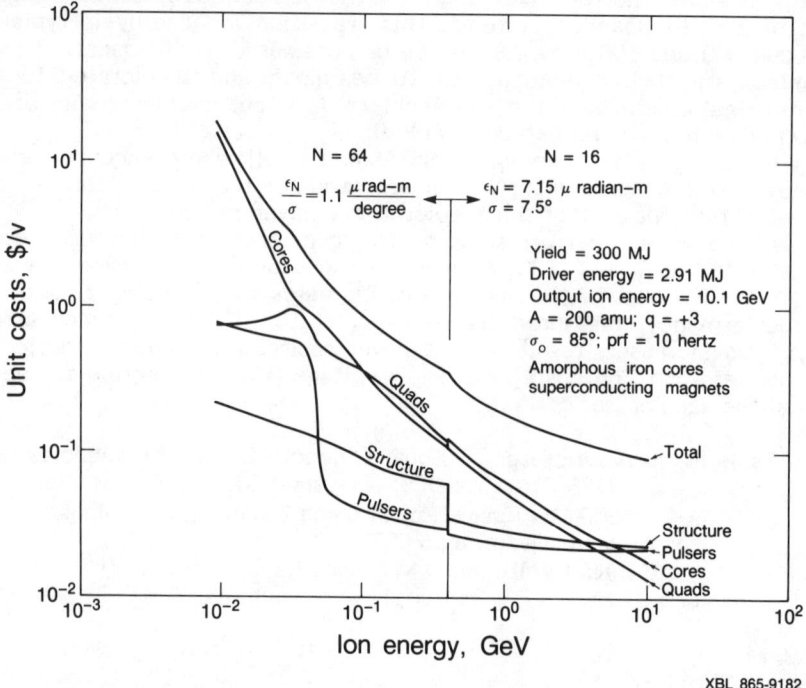

Fig. 1. Distribution of the accelerator costs (1979 dollars per volt) as a function of ion kinetic energy for a 300 MJ target yield producing a fusion power of 3000 MW. The transition ion energy for 64 beamlets to 16 beamlets is 400 MeV (133 MV). The depressed tune is 6.5° below and 7.5° above the transition ion energy.

The results for the low voltage section (<50 MV), as computed by LIACEP and shown in Fig. 1, are not very satisfactory. The cost differential between the 64 beamlet system and the 16 beamlet system is actually larger than currently calculated by LIACEP. This is due in part to not having a maximum velocity tilt ($\delta\beta/\beta$) limit in the code.[10] This limit on the tilt will increase the costs of the low voltage region of the accelerator where the beam length is long by forcing a lower acceleration rate and increasing the cost of the quadrupoles. The effect of the tilt limit will be more severe with the smaller number of beamlets than with the larger number of beamlets. The costs of the pulsers shown in Fig. 1 can be reduced by driving several modules with a single pulser in the region where the ion kinetic energy is less than 60 MeV. This could reduce the pulser cost per volt by perhaps an order of magnitude in the low voltage (<20 MV) region. The LIACEP results show very low superconducting quadrupole fields in the low voltage section of the accelerator due to the constraint that their length to bore ratio must be greater than a minimum specified number. This constraint results in large beamlet diameters, with concurrent large quad and core costs. By relaxing this constraint, the depressed tune could be increased which will increase the quadrupole field and reduce the beamlet diameter, resulting in a reduction in the quad and core costs.[10,11] Also, the use of electrostatic quadrupoles in the low voltage region may decrease the costs. LIACEP does not yet contain a good electrostatic focus system subroutine.

The combining of 64 beamlets into 16 beamlets in space and time may result in a cost savings. This combination of beamlets will result in an increased emittance in the region with the smaller number of beamlets (or conversely, require a reduced emittance in the region with the larger number of beamlets). Thus, there is a maximum number of beamlet combinations that can be allowed that will give the required spot size on target with a given source brightness. In addition, the depressed tune should be held proportional to the emittance. The output emittance is determined from target considerations, and the depressed tune in the high voltage portion of the accelerator is selected to minimize the cost of this portion of the accelerator. The decrease in emittance in the low voltage section due to the combining of beamlets will require a reduction in the depressed tune to minimize the cost in this section. There may be a lower limit to the depressed tune before instabilities occur that may offset some of the cost advantages of combining beamlets.

Additional cost savings can be made by changing the depressed tune along the length of the accelerator.[12] For the case of the 4.25 MJ driver given in Table II with a vacuum surface flashover voltage gradient of 0.5 MV/m, 16 beamlets and an initial ion energy of 150 MeV, the cost savings, by reducing the depressed tune from 10.5 to 8° for ion energies between 200 and 1500 MeV, was greater than 7 M$.

COST ESCALATION

The costs of the accelerators are given for a mature technology in 1979 dollars. The results of the application of cost escalation factors for the various accelerator components from 1979$ to the 1985$ for the three accelerators described in Table II are given in Table III. The pulsers become the most expensive unit in the accelerator, narrowly exceeding the core costs. Placing the new cost factors into LIACEP may result in different accelerator designs to achieve a minimum cost configuration.

Table III. Accelerator Cost Estimates Escalated from 1979 Dollars to 1985 Dollars.
$P = 3000$ MW$_f$; $A = 200$ amu; $q = +3$
$\phi = 1.0$ MV/m; $\sigma_0 = 85°$; $V_0 = 3$ MV

Target Yield, MJ	300	600	1200
Accelerator Energy, MJ	2.91	4.24	6.57
1979 Costs, M$	550	630	750
1985 Costs, M$	710	790	910
Escalation, %	30	24	22
$/J (1985$)	250	190	140

CONCLUSIONS

Induction linac drivers with output energies of 2.91, 4.25, and 6.57 MJ are necessary to obtain yields of 300, 600, and 1200 MJ, respectively, from single shell targets. The accelerators, using 200 amu, charge state +3 ions injected with a kinetic energy of 9 MeV, feature superconducting quadrupoles for beam transport and amorphous iron cores for acceleration.

The costs of these drivers producing 3000 MW of fusion power roughly scaled from 1979 dollars to 1985 dollars decrease from 250 to 140 $/J of driver energy with an increase in target yield from 300 to 1200 MJ. The ratio of fusion power to power into the driver increases from 27.7 to 52.9 with the increase in target yield. These drivers are less than half the cost of the more conservative, charge state +1 drivers.

Further modifications must be made to the cost minimization code LIACEP to better assess the low voltage section of the driver. These modifications include placing a limit on the velocity tilt and reexamining the other constraints that prevent the use of the superconducting quadrupoles at their maximum fields. Finally, the cost algorithms should be modified to reflect the 1985 costs of material and labor.

REFERENCES

1. J. D. Lindl and J. W-K Mark, Revised Gain Curves for Single Shell Ion-Beam Targets, 1982 Laser Program Annual Report, (C.D. Hendricks and G. R. Grow, ed.), Lawrence Livermore Laboratory Report UCRL-50021-82, Livermore, CA, pp. 3-19 (1983).
2. J. Hovingh, V. O. Brady, A. Faltens, E. Hoyer, and E. P. Lee, Cost/Performance Analysis of an Induction Linac Drive System for Inertial Fusion, Proc. 11th Symp. on Fusion Engr., Austin, TX (Nov. 1985); LBL-20613.

3. A. Faltens, E. Hoyer, and D. Keefe, A 3 Megajoule Heavy Ion Fusion Driver, Proc. 4th Int. Top. Conf. on High-Power Electron and Ion-Beam Research and Technology, Palaiseau, France (1981); LBL-12409.
4. L. M. Waganer, D. E. Driemeyer and D. S. Zuckerman, Survey of System Options for Heavy Ion Fusion, these proceedings.
5. D. S. Zuckerman, D. E. Driemeyer, L. M. Waganer, J. Hovingh, E. P. Lee, and K. W. Billman, Performance and Cost Modeling of a Linac-Driver HIF Power Plant, these proceedings.
6. D. E. Driemeyer, L. M. Waganer, and D. S. Zuckerman, Influence of System Optimization Considerations on LIA Drivers, these proceedings.
7. I. Brown, An Intense Metal Ion Beam Source, these proceedings.
8. E. P. Lee, Accelerator and Final Focus Model for an Induction Linac Based HIF System Study, these proceedings.
9. A. Faltens, Multiple Pulsing with Existing Technology, Lawrence Berkeley Laboratory Internal Note, HIFAR-NOTE-39, Berkeley, CA (1985).
10. A. Faltens and L. J. Laslett, Lawrence Berkeley Laboratory, private communication (May 14, 1986).
11. L. J. Laslett and A. Faltens, Revised Tune Prescriptions, Lawrence Berkeley Laboratory Internal Note, HIFAR-NOTE-72, Berkeley, CA (1986).
12. J. Hovingh, Economic Aspects of Changing Depressed Tune with Energy in 4.24 MJ HIF Driver at Voltages above 50 MV, Lawrence Berkeley Laboratory Internal Note, HIFAR-NOTE-58, Berkeley, CA (1986).

SURVEY OF SYSTEM OPTIONS FOR HEAVY ION FUSION*

Lester M. Waganer, Daniel E. Driemeyer and David S. Zuckerman
McDonnell Douglas Astronautics Co., PO Box 516, St. Louis, MO 63166
and Kenneth W. Billman
Titan Systems, Inc, 2685 Marine Way, Mountain View, CA 94043

ABSTRACT

Several potentially attractive system options have been identified for use in a linear induction-driven heavy-ion fusion reactor power plant. These include higher charge-state ions, double-pulsed accelerators, a range of ion species, different target types, innovative cavity approaches and various illumination schemes. Data from the U.S. DOE Heavy-Ion Fusion (HIF) Systems Assessment (HIFSA) Project were integrated into a unified, comprehensive system performance and cost code for a commercial HIF power plant. This code surveyed the above hardware options over the design parameter spaces of multiple accelerator beams, target gain curve parameter ($r^{3/2}R$ or r/R), gain, repetition rate, ion voltage, beam energy and net electric power output. The results indicate that a 1000MWe, linac-driven, HIF power plant can produce electricity on a competitive basis. An innovative triple-charged heavy ion accelerator design is used that greatly reduces the cost (and length) of the accelerator while increasing the efficiency. The ability of the linac accelerator to operate efficiently at repetition rates of 5 to 10 Hertz indicates promise for reactor cavity protection concepts such as granular and wetted wall.

INTRODUCTION

A broad assessment of the merits of a HIF power plant featuring an induction linac driver is underway. The HIFSA Project involves several national laboratories, universities and private industry. The main effort is sponsored by DOE and the systems integration and analysis task is being jointly sponsored by the Electric Power Research Institute and DOE. McDonnell Douglas Astronautics Company and their subcontractor, Titan Systems, Inc. are responsible for integrating the technical efforts of the HIFSA project (see Figure 1). The objectives of the HIFSA project are to identify technical innovations, evaluate their potential for contributing to an economically attractive inertial fusion power plant, and highlight technical areas deserving of R&D support and emphasis. These objectives were thought best evaluated in the context of a commercial power plant, using the cost of electricity as the primary figure of merit. Figure 1 illustrates the four major technical areas to be investigated. The Systems and Integration effort integrated these into a complete and comprehensive power plant model in order to assess the attractiveness of proposed system options. Most of the discussions contained in this paper involve this effort.

---

*Work supported by Electric Power Research Institute and DOE.

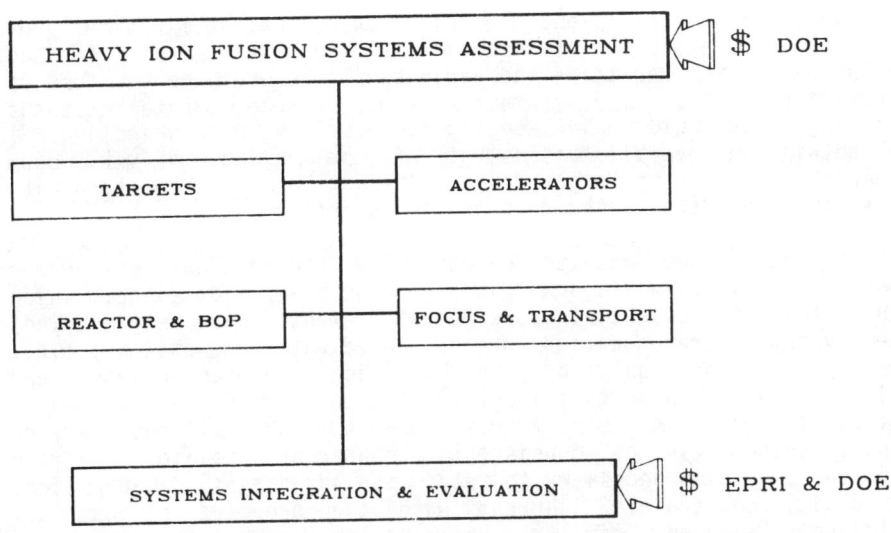

Fig. 1. Heavy Ion Fusion Systems Assessment Organization

PROJECT SCOPE, GUIDELINES, AND METHODOLOGY

The goal of the HIFSA integration effort is to integrate the technical efforts of the participating organizations to enable a comprehensive and consistent assessment of promising and innovative system options. Table 1 is a listing of the more important integration goals. The first goal was to define the expected power plant performance goals against which the plant will be evaluated. An overall system specification was defined which contained the type

Table 1. Goals of the HIFSA Integration Effort

DEFINE POWER PLANT PERFORMANCE GOALS and SPECIFICATIONS
  —*Type of Plant* —*Fuel* —*Net Power* —*Availablity*
  —*Capital Cost* —*Cost of Electricity*
CHARACTERIZE CANDIDATE SYSTEM OPTIONS
DEFINE SYSTEM INTERFACE REQUIREMENTS
MODEL POWER PLANT SYSTEMS
ASSESS PERFORMANCE AND RELATIVE MERITS of VARIOUS OPTION COMBINATIONS

of plant, range of net plant electrical output, plant lifetime, fuel type and assumed plant availability. The main factor for the evaluation of the merits of the various options would be the Cost of Electricity (COE). The parameters of performance, capital cost and operating cost would be influencing factors. The main objective was to obtain an overall assessment of the various driver, beam transport, cavity and target design alternatives <u>in conjunction with one another</u> to identify the more promising options.

The next step was to characterize the candidate system and subsystem options. Figure 2 is a generic power flow diagram which illustrates the essential power plant elements which were modeled. Some systems were modified or not used with some options, e.g., direct convertor only on the Magnetically-Protected Wall and tailored heat transport systems for the particular reactor blanket coolant chosen. An effort was made to model all ancillary or support systems in as consistent a manner as possible, modifying only those systems necessary to retain the advantages, disadvantages and design constraints inherent with the proposed options. All buildings assumed similar construction techniques whenever possible. Similar bases of safety systems and tritium processing were used. Common heat transport mediums and heat exchanges were used for commonality. All systems and options were modeled so they could be incorporated into a systems performance and cost system code. The models were constructed to allow a parametric search over the desired range of investigation. The interface requirements of each system were satisfied to allow the overall code modeling to

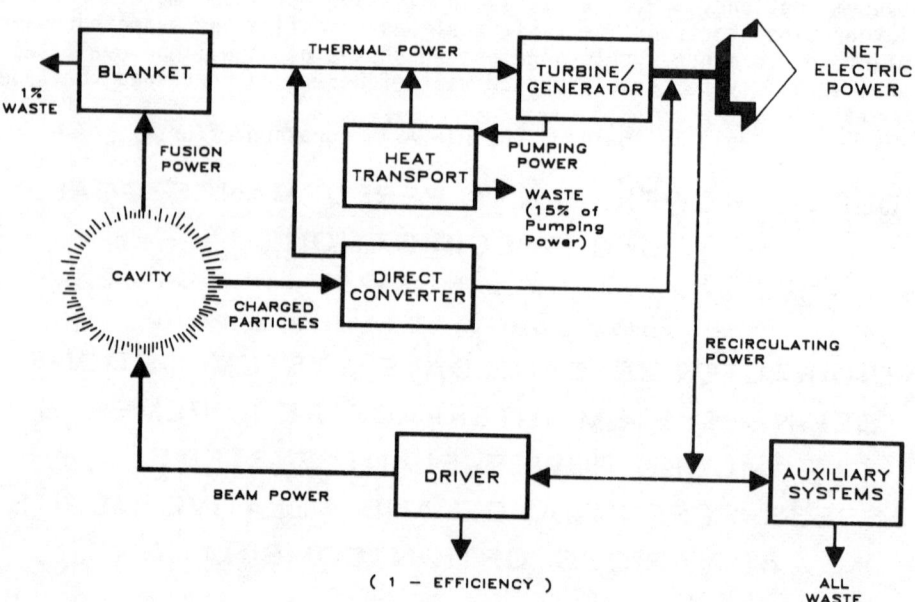

Fig. 2. HIFSA Power Flow Diagram

correctly portray the interactive aspects of the HIF power plant. Much care was taken to assure that there was consistency and that all interface requirements were satisfied for all systems. Systems must interact with each other to correctly model a technically consistent design.

The major system options which were evaluated are shown in Table 2. Various target concepts were considered in addition to the more well-known single shell and double shell targets published in recent papers by Lindl and Mark[1]. These included a symmetric target[2], a range-multiplied target, and an advanced target design - the last two of which assume enhanced target performance. Although DOE provided technical information only on double-sided and symmetric illumination of targets, it was felt by the project team that it was necessary also to investigate the potential cost advantages of a single-sided illumination scheme. The initial project direction on accelerators was on the single pulse, singly-charged ion beam, but the potential advantages of double pulsing and higher charged ions demanded these options be added.

The cavity designs generically represent four of the current inertial confinement fusion design concepts and wall protection schemes. The Liquid Wall concept is representative of the HYLIFE design[3] and the Granular Wall reflects the CASCADE design[4] approach. The remaining cavity approaches have had some substantive engineering analysis but have not been the central focus of a sizeable conceptual design study. We believe that these four concepts represent the broad spectrum of cavity design approaches. Multiple cavities were also considered to see whether multiple cavity power plants would be preferable to a single-cavity approach. The project scope did not allow investigation of all promising options or even a full and complete modeling of all variations of the chosen options, but a majority of the ICF design approaches were considered and evaluated.

Table 2. Elements of the HIF Systems Code (Discrete Options)

TARGET DESIGN	IRRADIATION SCHEME	ACCELERATOR CONFIGURATION	CAVITY DESIGN
SINGLE-SHELL	SINGLE-SIDED	+1 IONS, SINGLE-PULSED	MAG PROT'D WALL
RANGE-MULTIPLIED	DOUBLE-SIDED	+3 IONS, SINGLE-PULSED	LIQUID WALL
DOUBLE-SHELL	PLANAR SYMMETRIC	+3 IONS, DOUBLE-PULSED	GRANULAR WALL
SYMMETRIC	FULLY SYMMETRIC		WETTED WALL
ADVANCED DESIGNS			MULTIPLE CAVITIES

As shown in Table 3, these options were assessed over a range of net electric power, driver repetition rate, beam energy, ion voltage, ion species and gamma (function of target spot size and ion range). Only compatible options and operating conditions were evaluated, e.g., symmetric targets require symmetric irradiation, and the granular and liquid waterfall cavities cannot accommodate more than two beam penetrations (two-sided irradiation).

Table 3. Elements of the HIF Systems Code (Continuous Parameters)

PARAMETER	MINIMUM	NOMINAL	MAXIMUM
NET POWER, MWe	500	1000	1500
DRIVER REP RATE, Hz	1	CAVITY DEPENDENT	20
BEAM ENERGY, MJ	1	3–5	15
ION ENERGY, GeV	5	8–10	20
ION SPECIES, AMU	130	INSENSITIVE	210
GAMMA, $r \sim 1.5/R$	0.01	0.03	0.04

SYSTEMS ASSESSMENT RESULTS

The HIF systems analysis code developed by MDAC evaluated a large database of possible combinations of system options. One data base set which was recently constructed for a rather coarse grid of operational parameters contained 25,000 cases. These were assessed holding many of the independent parameters constant to observe significant trends in the results. The entire database could be searched for the local optima regarding the selected independent options and parameters.

Two of the design options which attracted great interest were the cavity protection scheme and the target design type. A summary chart, Figure 3, shows the relative effects of these two design features. It should be stressed that these data are not necessarily the combination of parameters and options which result in the most optimum Cost of Electricity. But it is a consistent set of parameters which allow a convenient comparison while holding a reasonable number of parameters and options constant and still results in COE values near the respective optima. The data in Figure 3 indicate that the choice of target type for any cavity type generally has a minor effect on the COE. The Magnetically-Protected and the Liquid Wall are slightly more expensive than the two other concepts. In the case of the Magnetically-Protected Wall, this is due to the fact that it has a much larger cavity and surrounding structure than the two more economic concepts. This cavity radius cannot decrease much below 10 meters because of the necessity of limiting sputtering off the carbon walls. The Liquid Wall is limited to repetition rates of two Hertz or less per cavity in order to reestablish the liquid wall protection.

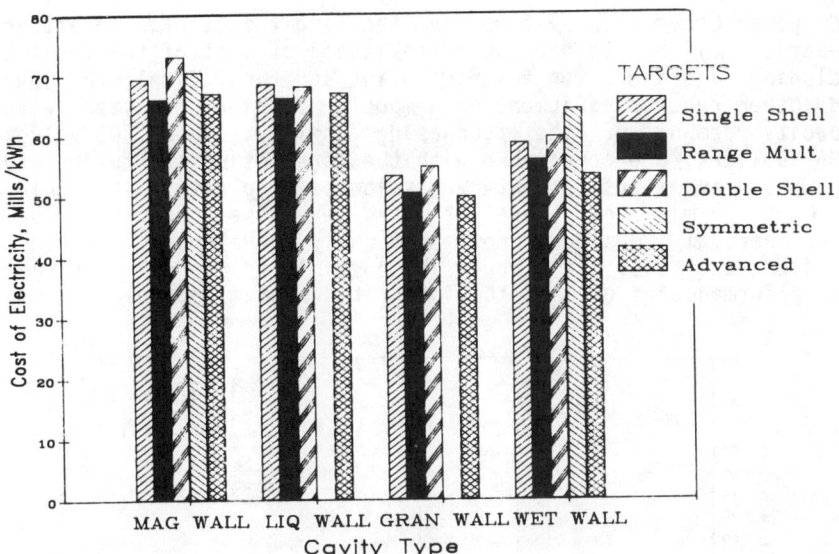

Figure 3. Summary of Near-Optimum HIF Cases
(Variable Cavity Types, Target Types, Ion Masses, Gamma Values
and Repetition Rates For 1000 MWe, 1 Cavity, Two Sided Illumin,
16 Beams and +3 Ions).

The economics of the Granular-protected Wall and Wetted Wall concepts are considered to be very similar. In the Figure 3 and most of the other results, the Granular concept generally is 5 Mills/kWh cheaper than the Wetted Wall. That result stems from the high degree of optimism of the Granular Wall design approach. Also, that approach has assumed a higher efficiency thermal conversion system and that no remote maintenance equipment are required. If these more optimistic assumptions are not realized, the COE advantage of the Granular Wall over the Wetted Wall would be lost. Conversely, if more optimistic performance and operational procedures could be realized on the other three reactor cavity approaches, the COE advantage would be reduced or eliminated between the concepts. The optimistic assumption for the Granular Wall was retained in the analysis to assess the relative advantage of this degree of optimism.

The relative costs of the cavity types are also reflected in Figure 4. In addition, the COE of the Wetted Wall is shown for a range of net electrical power from 500 MWe to 1500 MWe. There is a severe economic disadvantage to building a 500 MWe HIF power plant (96 Mills/kWh to 59 Mills/kWh). By increasing the size of the power plant to 1500 MWe, the COE decreases to 45 Mills/kWh. The capital cost of the driver plant equipment and the power plant buildings do not change much with the net power (small scaling exponent). Items which scale strongly with the power level include the heat transport and power conversion systems, and the target factory. An attractive scenario may be to propose installation of most of the facilities, including a driver, for a larger plant and then install the cavities and other reactor equipment in a modular fashion consistent with the capacity demand of the purchasing utility. These COE values, in 1986 dollars, are comparable with the more optimistic fusion studies and most estimates for future fission power plants. It should also be kept in mind that these estimates were based upon conceptual and pre-conceptual designs, not yet fully developed. Given more detailed analysis, it should be possible to improve both the cost and performance of each of the integrated system options.

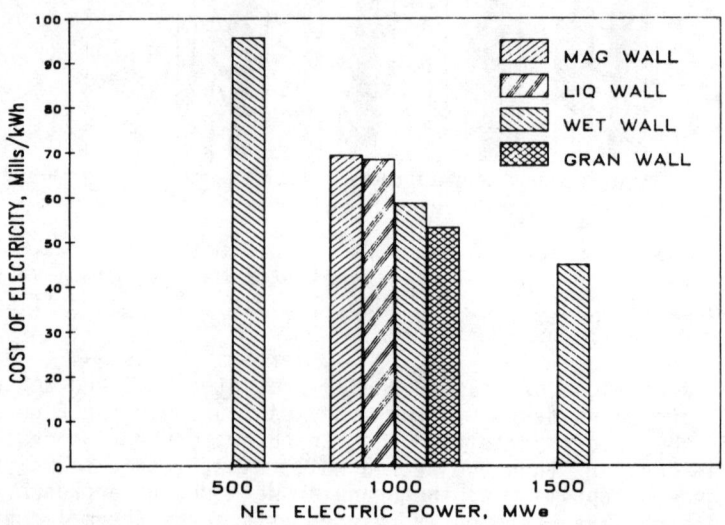

Fig. 4. Net Electric Power and Cavity Type Influence Heavy Ion Fusion Cost of Electricity (Single-Shell Target)

REFERENCES

1. J. D. Lindl and J. W-K. Mark, "Recent Livermore Estimates on the Energy Gain of Cryogenic Single-Shell Ion Beam Targets", International Symposium on Heavy Ion Accelerators and Their Application to Inertial Fusion, Tokyo, Japan, (1984).
2. G. Magelssen, "Gain Scaling Laws for HIF", These Proceedings.
3. W. R. Meier, N. J. Hoffman, M. W. McDowell, "Liquid-Metal Aspects of HYLIFE," Lawrence Livermore Laboratory Report UCRL-84107 (1979).
4. I. Maya, K. R. Schultz, and Project Staff, "Final Report: Inertial Confinement Fusion Reaction Chamber and Power Conversion System Study", GA Technologies Report GA-A17482 (1985).

PERFORMANCE AND COST MODELING OF A LINAC-DRIVEN HIF POWER PLANT*

David S. Zuckerman, Daniel E. Driemeyer, and Lester M. Waganer
McDonnell Douglas Astronautics Co, St. Louis, MO 63166
Jack Hovingh
Lawrence Livermore National Laboratory, Livermore, CA 94550
Edward P. Lee
Lawrence Berkeley Laboratory, Berkeley, CA 94720
Kenneth W. Billman
Titan Systems, Inc, 2685 Marine Way, Mountain View, CA 94043

ABSTRACT

A versatile and powerful systems analysis code has been written to assess the attractiveness of proposed induction linac-driven heavy-ion fusion (LIA-HIF) system options and aid in the assessment of R&D needs. This code was created as a part of the DOE-sponsored Heavy-Ion Fusion Systems Assessment Project (HIFSA)[1]. The code is structured to enable investigation of a large and continuously variable design and operating parameter space. This was possible by modeling the systems with continuous algorithms which define all the necessary interface and system parameters. The code calculates and displays descriptive design, performance, and operational parameters along with the capital costs, annual costs, and cost of electricity. An overview of the code architecture is provided along with a discussion of its application and results.

INTRODUCTION

From the beginning of the HIFSA project, the project team realized that it would be critical to examine the interactions between the various major reactor systems (e.g. driver, target and cavity) in order to determine the best combination of system design parameters. There was an early realization that each system design group could not independently optimize its own system. Rather, the best overall reactor design would consist of a set of compromises between the various systems. In essence, the individual strengths of the different systems would have to be traded off in order to create the best overall power plant design.

As a direct consequence of this realization, McDonnell Douglas Astronautics Company (MDAC) and its subcontractor, Titan Systems Inc., were instructed to perform the systems integration role in the project. Our charter was, in part, to devise an integrated systems computer model which would allow the project to examine the behavior of different systems and determine the sensitivity of the overall power plant to different assumptions, subsystem options and scaling relationships. The result of this effort is the Inertial Confinement Systems and Costing Model (ICCOMO).

---
* Work sponsored by the U.S. Department of Energy and the Electric Power Research Institute

OVERALL SYSTEMS MODEL

The ICCOMO code has seven main modules, as shown in Figure 1. These include the main driver routine plus six subsidiary modules: target design; cavity design; beam transport/final focus; linac design; plant power balance; and plant costing. Each of these modules is self-contained and therefore can be easily removed or replaced. In addition, many of the modules contain several design options, such as different target or cavity types, which can be selected by the user. The structure of the code is essentially once-through, although there are a number of internal convergence and optimization routines which are performed automatically. For instance, the code converges on a desired net electric power by successively adjusting an initial guess for the fusion power. The main input parameters for the ICCOMO code are shown in Table I.

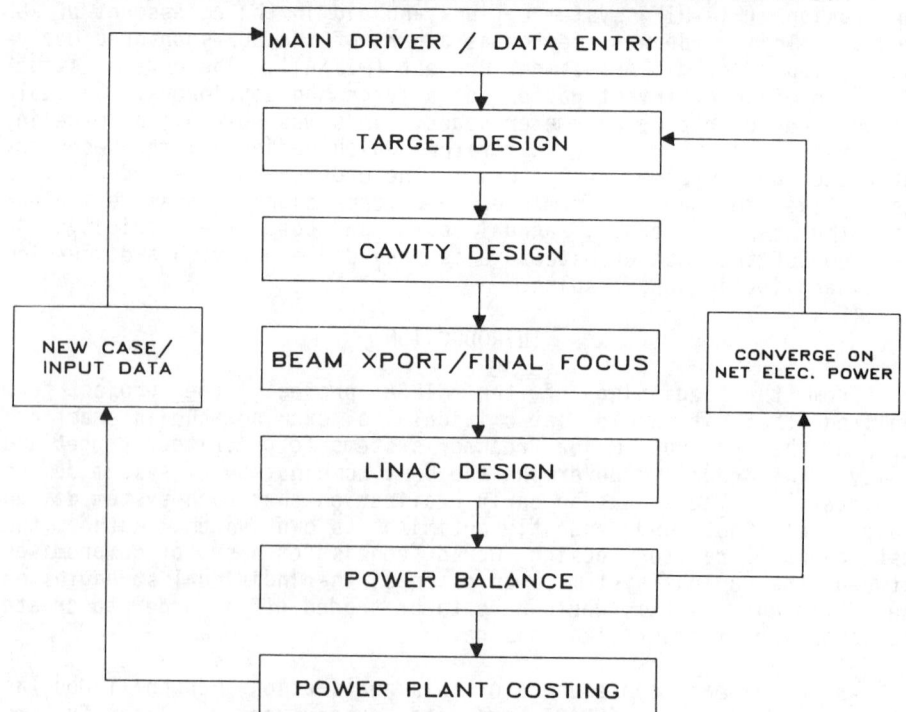

Figure 1. Schematic Flow Diagram of ICCOMO Systems Code.

ICCOMO SYSTEM MODULES

Each of the six system modules is completely separate and self-contained. The target design module contains five target designs, as listed in Table II. The Single-shell and Double-shell designs are based on the gain curves published by Lindl and Mark[2]. The Range-multiplied target is based on the Single-shell target, but the range (R) used to find the correct gamma value ($r^{3/2}R$) is divided by 2.

Table I Main Input Parameters for ICCOMO

SPECIFY:
One of: (1) Gain,   (2) Repetition Rate
           (3) Beam Energy   (4) Gain AND Energy on Target
                                        (no target gain curves)

Either: (1) Gross Thermal   (2) Net Electric Power

Target Type: (1) Single Shell   (2) Range Multiplied
              (3) Double Shell   (4) Symmetric

Cavity Type: (1) Magnetically-   (2) Liquid Waterfall
                  Protected Dry Wall
              (3) Wetted Wall   (4) Granular Wall

Pellet Irradiation Scheme:
              (1) Single Sided   (2) Double Sided
              (3) Multi-Sided Planar (4) Multi-Sided Fully Symmetric

Accelerator Pulse Scheme:
              (1) Single   (2) Double Pulse

Number of Beams in the Accelerator (4,8,16)
Charge State of Heavy Ion (+1 or +3)
Blanket Energy Multiplication Factor
Gain Curve Parameter: $r^{3/2}R$, (R/r for Symmetric target)
    where r = beam spot size, R = ion range in target
Number of Reactor Cavities
Ion Mass (amu)

This has the effect of increasing the gain for a given energy on target (i.e. the gain curves are shifted such that lower energy on target is required for a given gain and gamma). These target gain curves are based on two-sided irradiation. The MDAC/Titan team decided to also allow 1-sided irradiation in order to assess its impact on power plant costs. The Symmetric target design is based on the curves published by Magelssen[3]. This target, as its name implies, must be irradiated in a symmetric manner. The beams are arrayed in either a planar symmetric (16 beams all in plane) or a completely symmetric (16 beams arranged symmetrically both in and out of plane) orientation. Finally, we have incorporated an "advanced" target. This is a gain-multiplied version of the Single-shell target in which the gain curves are multiplied by 2.54. Such a design might be representative of targets incorporating polarized fuel. Each set of curves (Single shell, Double shell and Symmetric) were digitized and curve-fit as a function of energy on target, gain and gamma. The Single-shell target is used as the baseline target in this study.

The cavity design module contains four cavity designs (See Table II), ranging from extremely conservative to very optimistic. The magnetically-protected wall[4] is the most conservative design, using

Table II ICCOMO Systems Options

TARGETS	IRRADIATION SCHEMES	CAVITY/WALL DESIGNS
Single Shell (SS)	1-Sided	Magnetically-Protected
SS Range Multiplied (RM)	2-Sided	Liquid Waterfall
Double Shell (DS)	Planar Symmetric	Wetted Wall
Symmetric (Sym)	Fully Symmetric	Granular Wall
Advanced (Adv)		

ACCELERATOR CONFIGURATIONS	ACCELERATOR ION CHARGE STATES	CAVITY/ACCELERATOR COMBINATIONS
Single Pulse	1+	1 Cavity/1 Accelerator
Double Pulse	3+	Multiple Cavities/ 1 Accelerator

a graphite-lined wall which is protected by a magnetic field which sweeps the charged particles into a thermal dump or direct convertor. The cavity radius is scaled with yield, resulting in relatively high repetition-rate designs (up to 20 Hz) and short wall lifetimes ($\lesssim 5$ years). The liquid waterfall concept is patterned after the HYLIFE design[5]. It is an inherently low repetition-rate design which uses a curtain made up of falling Li jets to attenuate the neutrons and pellet debris. The design is restricted to 2 Hz due to cavity clearing requirements, but the cavity is life-of-plant (LOP). The wetted wall[6] represents a somewhat more optimistic approach, in which a thin Li film ($\lesssim 1$ cm) absorbs charged particles and x-rays before they penetrate the wall. This is a relatively simple design which has a wall lifetime $\lesssim 5$ years and allowable repetition rates up to 10 Hz. The granular wall concept represents the most optimistic of the designs. It is patterned after the CASCADE cavity design[7], and employs a highly optimistic design philosophy. The cavity is LOP and has a repetition rate limit of 10 Hz (based on cavity pumping requirements). The wetted wall design is used as the baseline cavity concept in this study.

The beam-transport/final focus module is the link between the target/cavity designs and the accelerator design. In this module, the required number of beamlets is calculated in order to satisfy perveance and aberration conditions. The required beamline lengths are calculated for compression, beam splitting and final focus, and the final focus magnets are sized. All of the equations used in this module are based on information supplied by Lawrence Berkeley Laboratory[8]. Special beamline sizing equations were incorporated into this module to account for a variety of target irradiation scenarios, cavity designs and accelerator configurations. One, two or multiple beamlines can be simulated, depending on the cavity and target designs selected. If multiple cavities are used, the output of a single accelerator is sequentially shifted from one set of cavity beamlines to another, using appropriate beamline switching. If the accelerator is to be double-pulsed (i.e. two sequential accelerator pulses are, via switching and delay, focused simultaneously onto the same target), appropriate switching and delay lines are incorporated.

The induction linac design used in the linac module is based upon results from the LIACEP linac design code[9]. More than 500 individual LIACEP runs were made in order to examine the parameter space shown in Table III. Some of the parameters, such as charge state and emittance, were examined only at their limiting values. Other parameters were examined at up to five discrete values within the ranges shown. Various normalization coefficients were created and the entire parameter space was curve-fit using multi-term polynomials. The output parameters calculated via the curve-fits are accelerator efficiency; length; and cost. The curve-fits were then checked by comparing the LIACEP output at all points with the curve-fit results. The polynomials were seldom more than 10% in error and the functions were all well-behaved. The accelerator parameters are renormalized if a double-pulse scheme is to be used as opposed to the baseline single-pulsing.

Table III Parameter Space Examined for Induction Linac Model

PARAMETER	MINIMUM VALUE	MAXIMUM VALUE
Repetition Rate (Hz)	1	20
Ion Mass (amu)	130	210
Ion Voltage (GeV)	5	20
Beam Energy (MJ)	1	10
Emittance (microrad-m)	15	30
Number of Beams	4	16
Ion Charge State	1+	3+

The power balance module computes the recirculating power requirements for the driver, reactor cavities, heat transport and auxiliaries; the fusion, thermal and gross electric power; and finally the net electric power. If a particular net electric power output has been selected, this module governs the convergence by varying the fusion power. Convergence is usually achieved within a few iterations.

Finally, the plant costing module calculates the cost of all components and systems, ultimately computing a net bussbar cost of electricity (COE). Costs are taken from a variety of sources, including current nuclear power plant costs and both inertial and magnetic confinement fusion power plant conceptual design studies. Contained in this module is a complex iteration routine which minimizes the cost of the power plant by adjusting the driver parameters. Because of the way in which the beam transport algorithms are formulated, the ion voltage is not limited to a particular value, but rather to a particular range within which all stability conditions are satisfied. At the same time, the ion voltage directly affects the ion range (R) in the target, and hence has an inverse effect on the beam spot size (r) at the target for a fixed value of gamma ($r^{3/2}R$). This in turn affects the convergence angle of the beams and hence the maximum current allowed per beamlet. As current per beamlet goes down, the required number of beamlets goes up, and the final transport and focus costs increase.

Meanwhile, the accelerator cost is varying in a different, complex manner with ion voltage, creating a bucket-shaped cost curve. Thus the cost routine iterates on ion voltage in order to minimize the overall driver cost.

CODE OPERATION AND RESULTS

The ICCOMO code was designed to allow scans of large areas of parameter space. Thus the user can instruct the code to run multiple cases while varying up to nine different parameters: net electric power; cavity type; number of reactor cavities; target type; irradiation scheme; gamma; ion mass; number of beams in the accelerator; and repetition rate, energy on target, or gain. The code can also produce varying degrees of detail in its output. On a case-by-case basis, details of all reactor system parameters and costs are printed out in an annotated, 3-page format. This output consists of 125 different system and cost parameters and thus can become quite cumbersome if more than a few cases are to be examined. For multiple-case studies, a one-line summary of each case can be created which identifies the primary system parameters (e.g. major inputs, powers and system costs). Finally, for large parameter scans, the full output can be sent to a database file in condensed format. This condensed database file can then be used with dBASE III to examine large numbers of cases in a simple and efficient manner.

Each case run requires approximately 5 seconds on an IBM PC/XT with math coprocessor. The database file requires approximately 600 bytes per case. For our studies, we created a 25,000-case database ($\leq$15 MB of stored data). The time required to create the data with ICCOMO and load it into dBASE III was approximately 50 hours. Since ICCOMO is completely automated (including all error-handling processes), the entire database could be set up over a weekend. The following discussion illustrates the types of results which can be obtained with the ICCOMO code.

Figure 2 shows the variation of COE with repetition rate for various targets, using a 1000 MW$_e$ Wetted Wall reactor cavity. All five targets have the same general behavior, resulting from the trade between driver and target manufacturing costs. As repetition rate increases, required energy on target decreases, causing both the accelerator cost and the number of beamlets (and therefore beam transport cost) to decrease. This is most prominent at low repetition rates, since a change from 1 to 3 Hz results in a drop in energy on target by over 40%. Meanwhile, target manufacturing costs increase with the 0.65 power of repetition rate, driving the COE back up at higher repetition rates. The result is a bucket-shaped curve with a steep left side and a more shallow right side. The differences in the relative placements of the curves is due to the relative slopes and placements of the various gain curves.

Figure 3 illustrates the dependence of COE on net electric power for a Wetted Wall reactor cavity. The COE scales as (net electric power)$^{-0.7}$, which is much steeper than other power generation schemes in the 1000 MW$_e$ range. Comparing the baseline Single-shell target to the Advanced target shows that the cost advantage of the Advanced target is more significant at low net electric power than at high. This is a direct result of the fact that at low net electric power, the system is forced to use low gain targets, an area where the gain curves are very steep. At higher net electric power the gains are larger and the curves are much flatter.

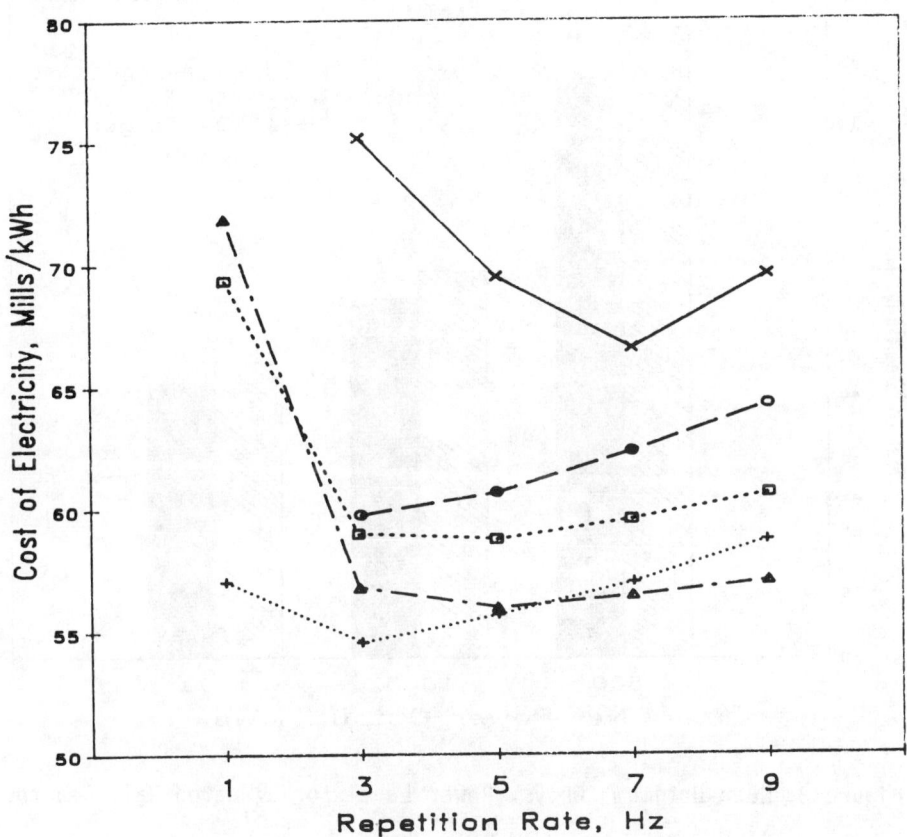

Figure 2. Comparison of Different Target Types for a Wetted Wall, 1000 MW$_e$ Power Plant (x Symmetric; o Double Shell; Single Shell Range Multiplied; + Advanced)

Table IV shows the range of repetition rates for the various targets and the Wetted Wall Cavity which produce low COE's. These ranges represent cases in the database where COE's were found to be within 5% of the local minimum COE for that target ("near-optimum"). In other words, one or more near optimum-COE configurations could be created with any repetition rate in the range listed. As one might expect based on the results shown in Figure 2, the Symmetric target has a more restricted range of allowable repetition rates than do the other targets, while the Advanced target has the broadest available range. Regardless of the target chosen, however, the available range of repetition rates is quite large, implying that the optimum operating region for LIA-HIF power plants may be quite large. This implication is supported by the examination of other cavities and other operating parameters (e.g. gain, beam energy and ion voltage). All critical operating parameters and system options appear to have broad regions in which near-optimum COE's occur.

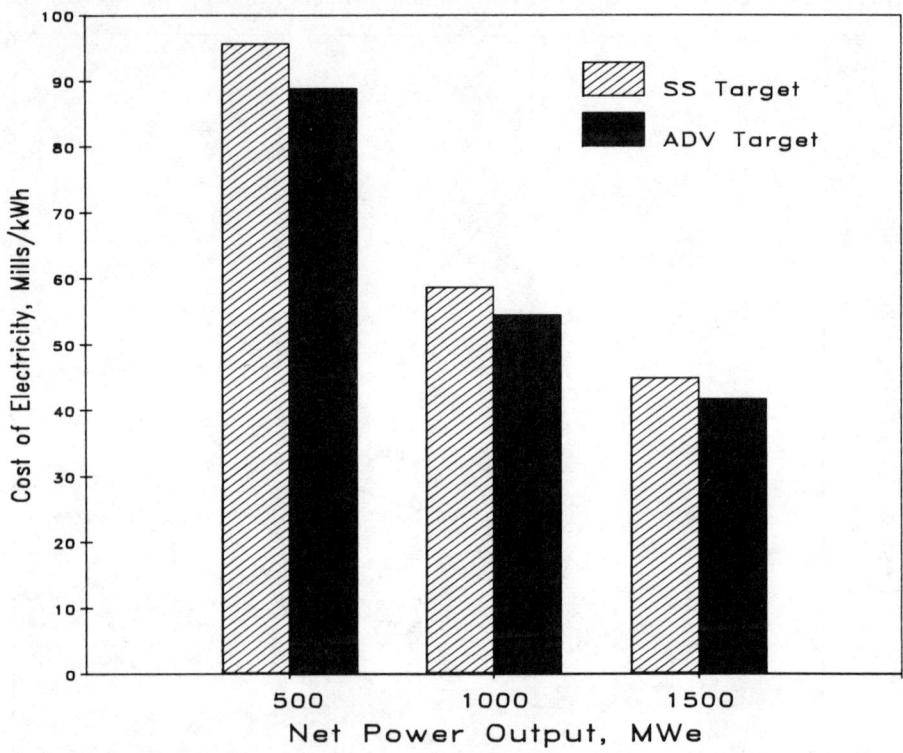

Figure 3. Near-Optimum COE vs. Power Level for a Wetted Wall Reactor.

Table IV HIF Near-Optimum Repetition Rate Ranges
For Wetted Wall Cavity, 1000 MW_e

TARGET TYPE	REPETITION RATE (Hz)	
	MINIMUM	MAXIMUM
Single Shell	4	10
Range Multiplied	4	10
Double Shell	4	8
Symmetric	5	10
Advanced	1	10

CONCLUSIONS

The ICCOMO code is proving to be an extremely valuable tool in the evaluation of LIA-HIF systems and options. We have already used it to identify the relative merits of various target and cavity types, as well as different accelerator designs and configurations. We have been able to narrow the focus of the project to specific areas of parameter space where operation appears most economical. In the future we hope to compare the economic impacts of various physics and technology assumptions, showing the impact on COE of using more optimistic or pessimistic scaling relationships. This in turn should help us to identify the most beneficial and productive avenues for research and development.

REFERENCES

1. D. Dudziak, "Heavy Ion Fusion Systems Assessment," these proceedings.
2. J. D. Lindl and J. W-K. Mark, "Recent Livermore Estimates on the Energy Gain of Cryogenic Single-Shell Ion Beam Targets," International Symposium on Heavy Ion Accelerators and Their Application to Inertial Fusion, Tokyo, Japan, (1984).
3. G. Magelssen, "Gain Scaling Laws for HIF," these proceedings.
4. J. B. Cornwell and J. H. Pendergrass, "Inertial Fusion Reactors and Magnetic Fields," Fusion Tech., Vol 8, No 1, p. 1861 (1986).
5. W. R. Meier, N. J. Hoffman, M. W. McDowell, "Liquid-Metal Aspects of HYLIFE," Lawrence Livermore Laboratory Report UCRL-84107 (1979).
6. J. H. Pendergrass, T. G. Frank and I. O. Bohachevsky, "A Modified Wetted-Wall Inertial Fusion Reactor Concept," 4th Topical Meeting on the Technology of Controlled Nuclear Fusion, King-of-Prussia, PA (1980).
7. I. Maya, K. R. Schultz, and Project Staff, "Final Report: Inertial Confinement Fusion Reaction Chamber and Power Conversion System Study," GA Technologies Report GA-A17482 (1985).
8. E. Lee, "Induction Accelerator Model for HIF Systems," these proceedings.
9. J. Hovingh, A. Faltens, E. Hoyer, and E. Lee, "Cost/Performance Analysis of an Induction Linac Driver System for Inertial Fusion," IEEE 11th Symposium on Fusion Engineering, Austin, Texas (1985).

INFLUENCE OF SYSTEM OPTIMIZATION CONSIDERATIONS ON LIA DRIVERS[*]

Daniel E. Driemeyer, Lester M. Waganer, and David S. Zuckerman
McDonnell Douglas Astronautics Co., St. Louis, MO 63166

ABSTRACT

The value of a systems and cost model lies in its capability to quickly evaluate plant cost and performance for various system/subsystem options and thus to assess the effect of potential design innovations and tradeoffs on the overall cost of electricity (COE). We have developed a Linear-Induction Accelerator driven Heavy-Ion Fusion (LIA-HIF) Inertial Confinement Systems and COsting MOdel (ICCOMO) as part of the Heavy-Ion Fusion Systems Assessment Project (HIFSA). This code has been used to evaluate various combinations of driver, cavity, and target design alternatives in conjunction with one another to determine the cost-sensitivity of different system options and to identify the most promising designs. The most significant result of these studies, to date, is the overall broadness of the optimum region of parameter space. Changes in the performance of one system can generally be compensated for by another system, resulting in a minimal change in the final COE. This paper discusses the tradeoffs leading to this conclusion and their implications relative to LIA design.

INTRODUCTION

Interest in heavy-ion fusion developed primarily as a result of a number of advantages which heavy-ion drivers have relative to competing ICF driver technologies.

- o Higher efficiency (20-35%) which relaxes target gain requirements and allows for more conservative target designs with gains of less than 100.

- o Demonstrated high repetition rates ($>$ 10 Hz) which opens up new regions of parameter space and allows for the possibility of multiple reactor cavities supported by a single driver.

- o Essentially classical beam transport to the target and classical deposition of beam energy in the target ablator shell which improves confidence in the beam energy coupling.

- o Improved confidence in driver reliability (70-80%) based on experience with working accelerators.

- o Operate at large standoff distances ($\sim$10 m) from the target.

---

[*] Work Sponsored by the U.S. Department of Energy and the Electric Power Research Institute.

These advantages are tempered by major concerns about the size and cost of heavy-ion drivers and consequently the overall plant size required for competitively-priced HIF power. The HIBALL studies[1,2] reinforced these concerns by selecting a 5.1 B$ (1984), 3784 MWe, rf-linac driven design point. This has been a major stumbling block for HIF development because it suggests large price tags ($>$1 B$) for experimental test facilities. The HIFSA study[3] was undertaken to identify and quantify innovations which reduce the cost of heavy-ion drivers, leading to more attractive development costs, and to see how competitive HIF could be at other power levels.

OVERVIEW OF SYSTEMS ANALYSIS

In order to assess the cost impact of design innovations identified by the HIFSA team and evaluate the cost sensitivity of various system options (cf., Table 1) in a quantitative manner, the McDonnell Douglas Astronautics Company and its subcontractor, Titan Systems Inc., were assigned the task of developing an integrated systems model. This was accomplished through the development of the ICCOMO code which is discussed in more detail in the paper by D. Zuckerman, et. al.[4] This code has proven to be an extremely valuable tool. It is fast running, requiring only five seconds/case on an IBM PC/XT with a math coprocessor, yet it generates a detailed summary of 125 different system and cost parameters. For extensive scans of parameter space, the full output can be sent to a database file in condensed format. This file can then be read by dBASE III to accommodate analysis of the resulting large amounts of data in a simple and efficient manner.

The information presented here was extracted from a 25,000 case database generated by scanning at several net electric powers, driver repetition rates, driver ion types, and gain curve parameters ($r^{3/2}R$ or R/r for symmetric targets) over the range of discrete system options summarized in Table 1. The reader is again directed to Reference 4 for a description of each of these options and a discussion of the techniques used for modeling them in ICCOMO.

Table 1. Discrete ICCOMO System Options

TARGET DESIGN	IRRADIATION SCHEME	CAVITY/WALL DESIGN
Single-Shell	1-Sided	Magnetically-Protected
Range Multiplied	2-Sided	Liquid Waterfall
Double Shell	Planar Symmetric	Wetted Wall
Symmetric	Completely Symmetric	Granular Wall
Advanced		

ACCELERATOR CONFIGURATION	CAVITY/ACCELERATOR COMBINATION
+1 Ions, Single-Pulsed	Single Cavity, Single Accelerator
+3 Ions, Single-Pulsed	Multiple Cavities, Single Accelerator
+3 Ions, Double-Pulsed	

DISCUSSION OF RESULTS

As was indicated, the primary motivation for the HIFSA study was the identification of design innovations which could reduce the large capital cost associated with LIA's. The principal cost saving idea which was identified during the study was the use of multiply-charged ions (q = +3). A detailed description of the analysis supporting this decision is given in References 5 and 6. However, to summarize, it now appears that it will be possible to develop a good source of multiply-charged ions. Experiments indicate that the MEVVA ion source might well provide adequate current and emittance characteristics for HIF-LIA's. Furthermore, analysis by E. Lee[5] (which has been incorporated in our systems code) indicates that final transport and focusing of +3 ions is feasible. We have thus selected a +3 charge-state accelerator as the baseline for the results presented here.

Figure 1 depicts the savings resulting from the use of +3 ions. This figure compares the major capital account cost breakdown for a power plant based on a +1 charge state accelerator to that for one based on +3. Both cases are for a 1000 MWe power plant with a Wetted Wall cavity, so costs other than the driver are roughly equivalent. However, it should be noted that the +3 charge state LIA is about 50% more efficient than its +1 counterpart which leads to additional savings in BOP costs. We therefore see that the +3 LIA reduces the driver capital cost from 55% to 43% of the total, resulting in a 22% reduction in COE. This savings corresponds to a 710 M$ reduction in direct capital cost. Coupling these savings with those generated by considering advanced targets and the more optimistic Granular Wall cavity option leads to a 1000 MWe HIF power plant that has COE less than 50 Mills/kWh. This is well within the 45-55 Mills/kWh range typical of advanced energy concepts, and it represents a significant improvement in the size and cost previously associated with HIF power plants.

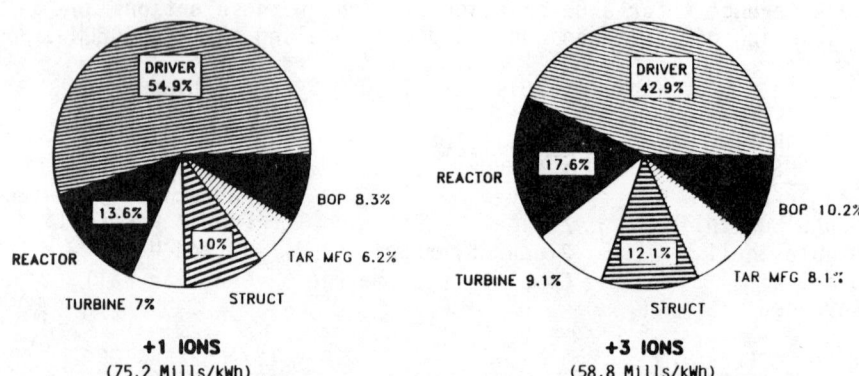

Figure 1. Comparison of +1 to +3 Charge-State Driver for a Single 1000 MWe Wetted Wall Cavity Using a Single-Shell Target With 2-Sided Irradiation.

Figure 2 shows the results of an internal optimization which ICCOMO performs in arriving at a typical design point. This figure highlights another implication of the present study relative to LIA design. The optimization involves the cost of the two major components of the driver, the accelerator and beam delivery systems, as a function of ion energy. Figure 2 shows that the accelerator cost has a very shallow minimum at 10 GeV which turns out to be the optimum energy for the present case. However, the significant feature of this figure is the rapid increase in the beam delivery cost below 8 GeV. This is due to the increase in the number of beamlines required for final transport (NBFT) at low ion energies. This number increases from 34 at 8 GeV to 80 at 5 GeV which causes the beam delivery cost to jump from 71 to 171 M$. As a result, the driver cost does not optimize at low ion energies under most conditions. In fact for the Wetted Wall cavity, over 75% of the near-optimum cases (i.e., cases within 5% of the minimum COE for a particular set of options) have energies in the range of 8-12 GeV.

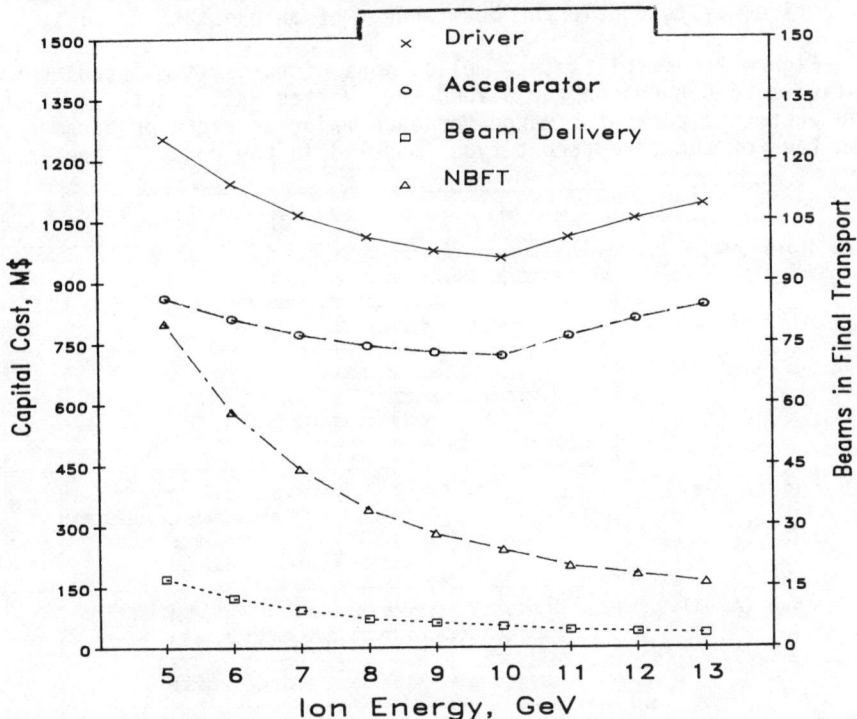

Figure 2. Driver Cost Optimization vs. Ion Energy for a 1000 MWe, Wetted Wall Cavity.

We also investigated the economic benefits of double-pulsing the LIA. Double-pulsing (DP) implies the operation of the linac in a mode where the final required beam current (hence energy) is built up by firing the linac twice in rapid succession with half of the total current each time. This requires that the linac only be capable of handling half the total charge at one time, which generates a significant cost savings. However, it lowers the linac efficiency and increases the cost of the final transport system, because of the need for fast reset of the induction cores and storage of the initial pulse. Lawrence Berkeley Laboratory (LBL) provided cost and efficiency normalization constants for double-pulse operation which were used to model this option in ICCOMO. Studies indicate that DP does reduce driver cost, but this is offset by increased balance-of-plant costs resulting from the lower driver efficiency. We thus find no preference for DP in a reactor application. Subsequent analysis by LBL indicates that DP could be advantageous at lower ion energies (5-7 GeV), but the increasing cost of the beam delivery system at these energies generally precludes operation in this regime. It should be emphasized, however, that DP is an attractive option for an experimental test facility where driver efficiency is not as important. In this application, DP represents a very cost-effective way to double the beam energy of an existing facility.

Figure 3 summarizes the implications of our systems studies relative to LIA design for a 1000 MWe, Wetted Wall reactor. It shows the preferred parameter range for each major accelerator parameter and four of the five target types modeled in the code. A preferred

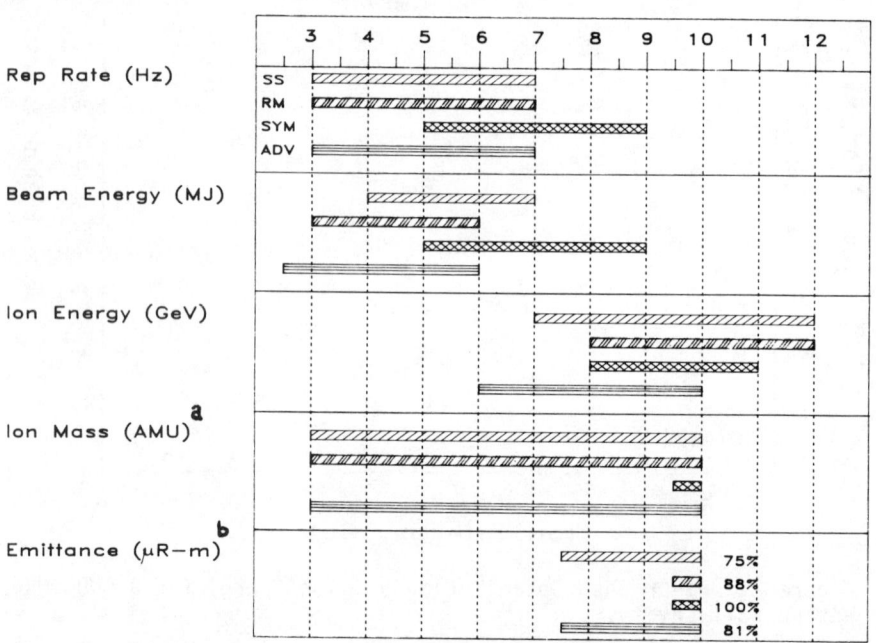

Figure 3. Summary of Near-Optimum Parameter Ranges for a 1000 MWe Wetted Wall Cavity. (a) For Ion Mass, The Units Shown Imply Increments of 10 Running From 130 to 220. (b) For Emittance, The Units Imply Increments of 1 Running From 23 to 32.

parameter range is defined as one which contains a high percentage
(~80%) of the cases having COE's within 5% of the minimum for that
target. It should be emphasized that there are cases lying outside
of the parameter ranges indicated that have COE's near the minimum.
However, the indicated ranges are definitely preferred for the
particular set of options being considered. It should also be noted
that points within the indicated parameter ranges are not necessarily
independent of one another, eg., low repetition rates would require
high beam energies. Finally, we point out that there can be
significant cost variations over the different target and cavity
types. Thus, in some instances, none of the cases in the near-
optimum COE range for one target or cavity type would fall within
this range for another. This is an important point to keep in mind,
but it is not relevant to the discussion presented here.

The most significant aspect of Figure 3 is that, for each target
option, there are combinations of system parameters that can produce
COE's nearly equal to the minimum over a wide range of LIA parameter
space. For example, the single-shell target (SS) has cases with
near-equal COE at repetition rates from 3-7 Hz, beam energies from
4-7 MJ, ion energies from 7-12 GeV, and ion masses from 130-200 amu.
The only restricted parameter is emittance which tends to lie near
its upper bound of 30 urad-m. Similar results are obtained for the
Granular Wall cavity, but the Magnetically Protected and Liquid
Waterfall cavities impose more restrictive demands on the driver due
to their design limitations. For example, the Liquid Waterfall
cavity can only operate at repetition rates below 2 Hz which raises
the preferred range of beam energy for that concept to 7-10 MJ. The
driver is 250 M$ more expensive for this case, but it is still able
to deliver the required energy to the target over a range of
conditions with minimal change in cost.

The flexibility of LIA's is particularly evident with regard to
ion type (mass). As is indicated in the figure, our studies have
shown that near-optimum cases are distributed almost uniformly in ion
mass. The only exception to this observation is the symmetric target
which shows a strong preference for higher masses. This seems to be
a result of the fact that the symmetric target requires more beams
for final transport than other target options. Higher masses impact
this by allowing the ion energy to be greater for a fixed ion range.
This in turn reduces the number of beams needed for final transport,
as discussed earlier, which lowers the cost. The weak dependence of
the final COE on ion mass has suggested an alternative to multiply-
charged ions in the form of three times as many singly-charged ions
with one-third the mass.[7] This will produce the same total beam
energy and represent the same current in the LIA. However, it will
also increase the range of the ions in the target (potentially
lowering the gain) and reduce beam stability during final transport.
Work is underway to quantify these issues in order to determine if
intermediate mass ions can be considered a viable alternative to
higher charge states.

The last implication of Figure 3 relative to LIA design involves the strong preference for large emittances. The numbers following the emittance bars indicate the percentage of cases which are pinned at the 30 μrad-m limit imposed on emittance in ICCOMO. This limit is imposed due to restrictions in the emittance range (15-30 μrad-m) chosen for the LIACEP database which was the basis for the LIA model in ICCOMO. Over 80% of the near-optimum cases run into this limit. This occurs because the final focus system is able to satisfy the spot-size requirements at the target for significantly larger emittances out of the accelerator than were initially anticipated. Once this was realized, it was too late in the study to modify the accelerator model. However, further reductions in the LIA cost can presumably be generated by increasing the beam emittance and future work will consider this. Alternatively, we could argue that an additional safety margin has been included in the event that the final focusing of +3 ions is not as effective as our model predicts.

A final aspect of LIA's involves their capability to operate at much higher repetition rates than those found to be required for 1000 MWe power plants, with minimal increase in cost. This results in a cost scaling with net electric power, P, that is much more favorable than that for other advanced energy concepts. Our studies indicate that the COE for HIF power plants scales as $(P/P_0)^{-0.7}$ due to the near-constant cost of the accelerator, whereas MCF scales only as $(P/P_0)^{-0.3}$. This makes HIF power plants particularly well suited for a staged installation procedure where successive reactor cavities are added to a single LIA as required to meet increasing demand. Table 2 summarizes the principal costs for the staged installation of a 3000 MWe plant and compares them to those for single-cavity 1000 and 3000 MWe plants. The staged-installation reactor costs are completely independent in this comparison, hence reactor cost increases linearly with the number of cavities. Conversely, the driver and target manufacturing costs are assumed to be shared. They are thus sized to accommodate the full 3000 MWe requirement. The single cavity 3000 MWe target manufacturing costs are lower than the 3 cavity case due to its lower repetition rate, but its driver costs are higher because it must deliver more beam energy. Finally, driver cost increases in the staged-installation case are due to additional beam delivery systems.

Table 2. Staged Installation Cost Comparison
(Costs in M$ except COE in Mills/kWh)

COST ACCOUNT	1 CAVITY 1000 MWe	STAGE 1 1000 MWe	STAGE 2 2000 MWe	STAGE 3 3000 MWe	1 CAVITY 3000 MWe
Reactor	381	381	762	1143	835
Target Mfg	128	261	261	261	178
Driver	1008	1079	1138	1196	1212
Total Direct	2204	2685	3604	3997	3403
COE	59.2	70.4	44.0	34.9	30.1

Comparing the Stage 1 costs to the single-cavity plant, we see that the driver and target manufacturing account for nearly half of the 480 M$ premium the first cavity must pay for staged installation. This causes the COE for Stage 1 to be 19% higher than that for a single-cavity plant. By Stage 2 the COE is 26% lower than the single-cavity plant and the COE reduction increases to 41% after the addition of Stage 3. This corresponds to a savings of ~3.7 B$ in direct and indirect costs compared to the installation of three seperate plants. However, it is ~500 M$ more expensive than a single-cavity 3000 MWe plant. The future utility environment will ultimately dictate which of these options is more viable, but it is clear that the staged-installation scenario is an attractive alternative to separate plants if reasonable growth in demand is anticipated.

CONCLUSIONS

The HIFSA study has identified several LIA design innovations that result in significant reductions in capital cost and in the plant size required for competitively-priced HIF power. The primary innovation involves the use of multiply-charged ions. This leads to a 40% reduction in driver cost for a typical HIF power plant and lowers the COE to the extent that HIF can compete with other advanced energy sources both in a reactor embodiment, and from the perspective of R&D costs. One particularly encouraging aspect of the resulting accelerator designs is the broad extent of the optimum region of parameter space. This increases our confidence in the capability of HIF to realize the performance levels predicted by the systems analysis presented here. Finally, the encouraging prospects for incremental installation HIF power plants, due to the repetition rate capability of the LIA driver, are especially significant. It appears that this option would allow utilities to realize many of the cost benefits of large power plants without having to shoulder the risk associated with a large, one-time capital investment.

REFERENCES

1. B. Badger, et al., "HIBALL - A Conceptual Heavy Ion Beam Driven Fusion Reactor Study," KfK-3202/UWFDM-450, (1981).

2. B. Badger, et al., "HIBALL-II - An Improved Conceptual Heavy Ion Beam Driven Fusion Reactor Study," KfK-3840/FPA-84-4/UWFDM-625, (1984)

3. D. Dudziak, "Heavy Ion Fusion Systems Assessment," these proceedings.

4. D. Zuckerman, et al., "Performance and Cost Modeling of a Linac-Driven HIF Power Plant," these proceedings.

5. E. Lee, "Induction Accelerator Model for HIF Systems," these proceedings.

6. J. Hovingh, et al., "The Cost of Induction Linac Drivers for Inertial Fusion for Various Target Yields," these proceedings.

7. W. Herrmannsfeldt, Private Communication (1986).

REPETITION RATES IN HEAVY ION BEAM DRIVEN FUSION REACTORS

Robert R. Peterson
University of Wisconsin, Madison, WI 53706-1687

ABSTRACT

The limits on the cavity gas density required for beam propagation and condensation times for material vaporized by target explosions can determine the maximum repetition rate of Heavy Ion Beam (HIB) driven fusion reactors. If the ions are ballistically focused onto the target, the cavity gas must have a density below roughly 10^{-4} torr (3×10^{12} cm^{-3}) at the time of propagation; other propagation schemes may allow densities as high as 1 torr or more. In some reactor designs, several kilograms of material may be vaporized off of the target chamber walls by the target generated x-rays, raising the average density in the cavity to 100 torr or more. A one-dimensional combined radiation hydrodynamics and vaporization and condensation computer code has been used to simulate the behavior of the vaporized material in the target chambers of HIB fusion reactors.

INTRODUCTION

The economic feasibility of Heavy Ion Beam (HIB) driven fusion reactors as power plants depends on the ability to achieve a high rate of target shots. The required shot rate depends on the cost of the plant, the desired cost of electricity, and the target gain. A high total repetition rate for the plant can occur through a high rate for each target chamber, multiple target chambers, or a combination of the two. The allowable repetition rate for various target chamber designs is the topic of this paper.

The repetition rate for a given target chamber is determined by the required cavity gas conditions at the time of the next shot and the length of time needed to achieve these conditions. If there is no material vaporized off of the chamber walls, which is the case in designs where the target energy density on the walls is low,[1-3] very high repetition rates may be possible. However, this type of design requires very large cavities or small target yields, either of which can bring along certain penalties in the design. Another approach is to allow a thin layer from the first wall of the cavity to be vaporized and recondensed back onto the wall.[4-6] The advantage of this is that the cavities can be smaller and cheaper, or so the designers hope. Also, one could use higher gain targets that improve the economy of power production. On the other hand, one must wait until the vapor density in the cavity has fallen to the point where beam propagation is possible before firing the next shot and there is the chance that the vapor could condense on the wrong spot and damage something.

There is some uncertainty over the limits on the target chamber vapor density imposed by beam transport.[7] If the beam ions are ballistically focused onto the target by magnets that are several meters away from the target, conventional knowledge says that the density of gas in the cavity should be less than about 3×10^{12} cm^{-3} (10^{-4} torr). There are other possible ways of propagating the ions to the target that allow densities in the 1 to 10 torr range. Some of these schemes involve using electrons from the cavity gas to neutralize the beam ion space-charge, while hoping that the vapor density is high enough to damp out detrimental plasma instabilities. Others use magnetic fields created in z-pinch type plasma channels to keep the ion beams confined to small radii until they reach the target. The methods of propagating beams in higher density gases are generally much less well understood than ballistic focusing, but, as calculations presented in this paper will show, the very low densities required for ballistic focusing may lead to very low repetition rates for some of the target chamber designs.

The vaporization of first wall material and its condensation back onto the walls can be a very complicated process.[8] The target generated x-rays rapidly vaporize the wall material in an as yet poorly understood way: the x-rays raise some of the material to an energy density above that required to raise it to the boiling point but not enough to overcome the latent heat of vaporization and it is unclear what happens to this material. The vaporized material forms a hot and dense layer of plasma near the surface, which is further heated by target generated ions, that may exist long enough for some rather unusual chemistry to take place.[9] The initially very nonuniform pressure profile in the vapor causes a shock wave moving towards the center of the target chamber that eventually collides with other similar shocks, resulting in very complicated hydrodynamic motion on the target chamber gases and vapors. While this motion is occurring, the gas is radiating energy back to the first walls and is condensing. Both of these processes put significant surface heat fluxes onto the wall that can cause evaporation of wall material. Unusual molecular species formed shortly after the vaporization may have a rather low sticking coefficient or may even sputter more material than is condensed. Eventually, the vapor cools enough and enough energy has been conducted away through the walls that condensation proceeds to the point that the ion beam can be propagated through the gas and the next shot is fired.

In this paper, I will present calculations of the time-dependent average gas density in a target chamber. I will do this for three target chamber designs that allow the first wall to partially vaporize: HIBALL,[4] FIRST STEP,[5] and CASCADE.[6] I will begin with a discussion of the physics that goes into the computer code used for these calculations. I will then present and compare the results of the calculations for the three designs. I will conclude with a consideration of what can be done to improve the repetition rates for the designs.

COMPUTER MODELING

To simulate the complex physics of the vaporization and condensation of material in HIB fusion reactor target chambers, a computer code, CONRAD,[10] has been used. This code attempts to model the behavior of a radiating, moving vapor and a material that is vaporizing or on which vapor is condensing by dividing the problem into two separate regions. The vapor, one of the regions, is modeled with Lagrangian hydrodynamics and multigroup radiative heat transfer. The unvaporized material, the other region, is modeled with a standard finite difference heat transfer method. From this point on, the term "material" will refer to the unvaporized material. Each of these sections is treated with rather standard numerical techniques. There is little experience in how to model the heat and mass transfer between the two regions. For this reason, there have been some options written into the code that allow the user to choose, for example, what model to use for rapid vaporization. Once the initial rapid vaporization is finished, there is no longer any volumetric energy deposition and the additional vaporization is calculated with a standard kinetic expression for the rate that atoms leave a surface at a given temperature.

The vapor section of the problem is modeled as a one-dimensional fluid with multigroup radiation diffusion. The hydrodynamics is modeled with a Lagrangian mesh and a finite difference solution to Newton's first law. The multigroup radiation diffusion is done using a fully implicit finite difference technique, where absorption and emission terms are calculated from opacities provided by the MIXERG[11] code. The energy equation for the vapor is also solved fully implicitly, and the equation-of-state also comes from MIXERG. Heat transfer in the material is also calculated with an implicit finite difference method.

COMPUTATIONAL RESULTS

Condensation calculations have been carried out for three target chamber designs with the CONRAD computer code. The three all allow partial vaporization of the first walls by target generated x-rays. Typical parameters for the three design are listed in Table I. HIBALL uses a coating of liquid lithium-lead eutectic, $Pb_{83}Li_{17}$, on a substrate of silicon-carbide fabric to protect the rest of the structure from the target generated x-rays and ions. The chamber radius is 5 m and the hoped for repetition rate is 5 Hz. FIRST STEP uses liquid lithium that is rapidly flowing so that centrifugal force holds it up against a metal wall. The radius is only 2 m but the target yield is only 25 MJ, compared with 396 MJ for HIBALL. The designers of FIRST STEP hope to run at 10 Hz. CASCADE is 3 m in radius and the first wall is made of flowing graphite pellets that are also held against the walls by centrifugal force. Some versions of the CASCADE design use beryllium-oxide in place of graphite, but it has since been learned that BeO will dissociate and the beryllium will condense, leaving a great deal of oxygen gas in the the cavity that must be pumped out.[9] The target yield is 334 MJ. As one can see from Table I, the x-ray and ion target energy per unit area varies considerably between the three designs and the designs have different wall materials. However, there are similarities as well. For example, all of the calculations have been done using the

Table I

	HIBALL	FIRST STEP	CASCADE
First wall material	Liquid Pb-Li	Liquid Li	Graphite
Target Yield (MJ)	396	25	334
Target Design	"HIBALL"	"HIBALL"	"HIBALL"
Fraction of Yield in X-rays and Ions	0.27	0.27	0.27
Distance from Target to First Wall (m)	5	2	3
X-ray and Ion Energy per Unit Area (MJ/m^2)	0.340	0.134	0.797
Desired Rep Rate per Cavity (Hz)	5	10	5

"HIBALL" target design, scaled to the proper yield. This design is based on a Livermore design that was published several years ago.[12] A burn calculation was done for a variant of this design to provide the required x-ray and ion spectra.[13]

The average gas densities in the HIBALL, FIRST STEP, and CASCADE target chambers, as simulated by CONRAD, are shown in Figs. 1, 2, and 3 respectively. The results for HIBALL show that after 0.2 seconds the density is still 5×10^{14} cm^{-3}, more than 2 orders of magnitude higher than that required for ballistic focusing. This is because the thermal speed of the vapor atoms is very low because most of the energy has been radiated away and the high mass of the lead atoms. One should notice the vapor density is actually increasing very early in the calculation, which is due to the high radiant heat flux. The vapor density in FIRST STEP initially falls very rapidly to below 10^{14} cm^{-3} but then condensation ceases. At this point the evaporation rate is equal to the condensation rate. The evaporation rate is fairly high because the temperature of the surface of the liquid lithium is 540°C at 0.1 second. The condensation could continue if the bulk temperature of the liquid lithium were lowered below 420°C or if some other way of increasing the heat transfer could be found. The thermal conductivity has already been increased over classical values in an attempt to account for convective heat transfer. A set of three calculations have been done for CASCADE, for three values of the sticking coefficient for vapor atoms striking the surface. If all of the atoms striking the surface stick to it, a sticking coefficient of 1, the density of vapor in the cavity falls to the level required for ballistic focusing, 3×10^{12} cm^{-3}, in less than 0.1 second. It has been found,

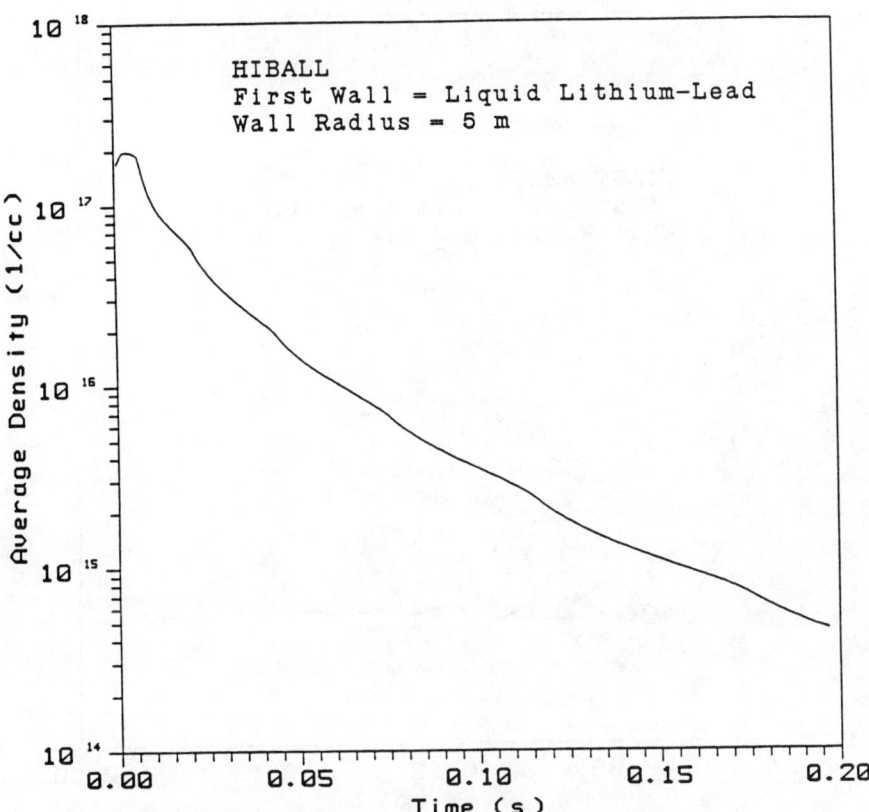

Fig. 1. Average Vapor Density in HIBALL Target Chamber versus Time.

however, that because of the chemistry of vaporized carbon the sticking coefficient may be about 0.7.[9] This leads to a density of 1×10^{13} cm^{-3} at 0.1 second and should lead to a level acceptable for ballistic focusing by 0.2 second. If the correct value is actually 0.5, condensation occurs too slowly to allow a 5 Hz repetition rate and ballistic focusing.

There has been some indication that ballistic focusing may indeed be possible at densities of more than 1×10^{14} cm^{-3}.[14] If this is true, there is no problem for FIRST STEP and CASCADE in running at 10 Hz. HIBALL may marginally be able to run at 5 Hz.

CONCLUSIONS

Computer simulations of the condensation of target explosion created vapor in three designs of HIB target chambers have been carried out. If the relatively hard vacuum of 10^{-4} torr is required

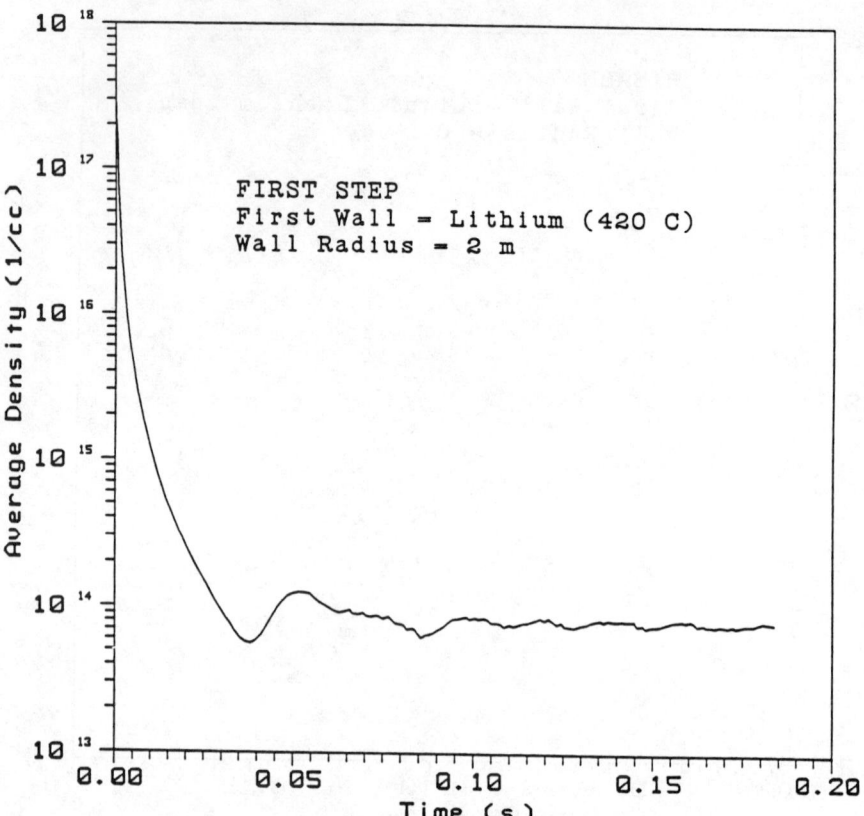

Fig. 2. Average Vapor Density in FIRST STEP Target Chamber versus Time.

for ballistic ion beam focusing, one has three different concerns for the three designs, each of which could make the repetition rate unacceptably high. In the case of HIBALL, the vapor can cool rapidly due to radiation so that the thermal speed can become very low and the cavity is so large that it takes too long for the vapor atoms to reach the surface. In FIRST STEP, the vapor pressure of the liquid lithium is high at fairly low temperatures so that the condensation can be greatly slowed if the surface temperature of the lithium is even as high as 540°C. In CASCADE, the chemistry of the vapor causes the sticking coefficient of the vapor on the surface to be significantly below 1.

Adjustments to the designs may improve the repetition rates. In HIBALL, the rate may be increased by making the cavity smaller, and in FIRST STEP, increasing the flow rate of the lithium may lower the vapor pressure by lowering the bulk temperature of the lithium

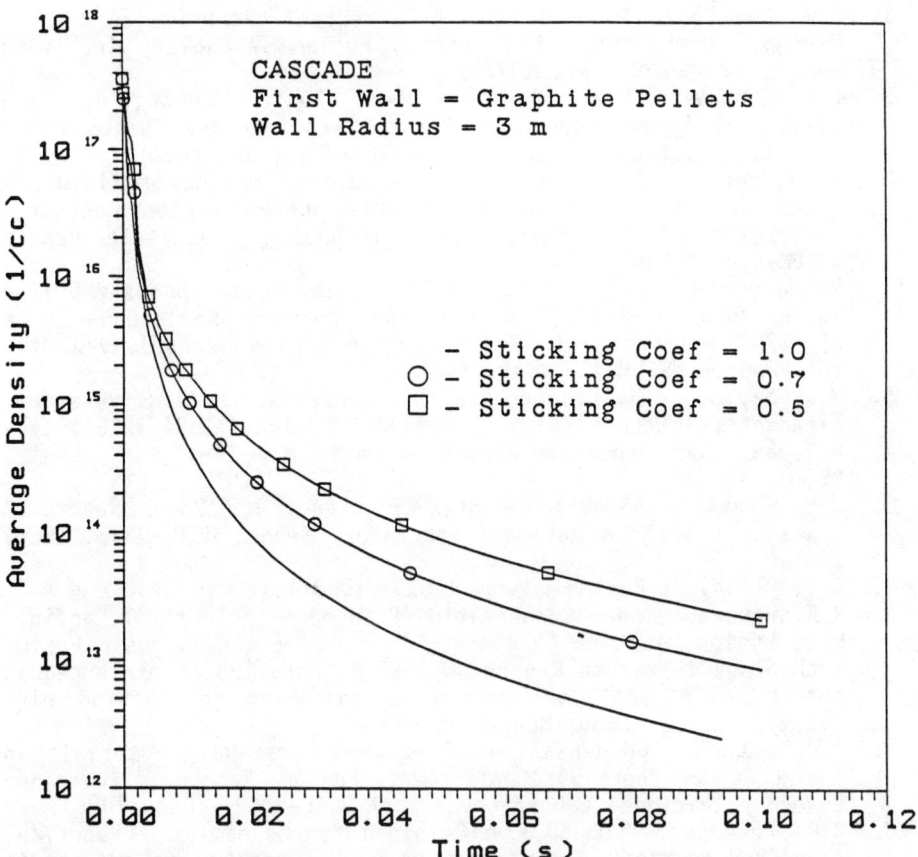

Fig. 3. Average Vapor Density in CASCADE Target Chamber versus Time.

and increasing convective heat transfer. It is harder to say what can be done to CASCADE to change the chemistry of the vapor, but the search for improvement must begin by gaining understanding of the physics of such hot and dense vapors. If it is indeed possible to focus the ion beams through denser gases, all three designs show promise of allowing reasonable repetition rates.

ACKNOWLEDGEMENTS

Parts of this work have been supported by Los Alamos National Laboratory, Lawrence Livermore National Laboratory, U.S. DOE, and Kerforschungszentrum Karlsruhe.

REFERENCES

1. R.W. Conn et. al., "SOLASE - A Conceptual Laser Fusion Reactor Design," University of Wisconsin Fusion Technology Institute Report UWFDM-220 (Dec. 1977).
2. E.W. Sucov, "Inertial Confinement Fusion Central Station Electrical Power Generating Plant," Westinghouse Fusion Power Systems Department Report WFPS-TME-81-001 (Feb. 1981).
3. B. Badger et al., "Preliminary Conceptual Design of SIRIUS, A Symmetric Illumination, Direct Drive Laser Fusion Reactor," University of Wisconsin Fusion Technology Institute Report UWFDM-568 (March 1984).
4. B. Badger et al., "HIBALL - A Conceptual Heavy Ion Beam Driven Fusion Reactor Study," Kernforschungszentrum Karlsruhe Report KfK-3202 and University of Wisconsin Fusion Technology Institute Report UWFDM-450 (Dec. 1981).
5. W.W. Saylor, J.H. Pendergrass, D.J. Dudziak, and R.R. Peterson, "Tradeoffs in the FIRST STEP Facility Design," 1984 IEEE International Conference on Plasma Science, May 14-16, St. Louis, MO, p. 35.
6. J.H. Pitts, "CASCADE: A High-Efficiency ICF Power Reactor," Lawrence Livermore National Laboratory Report UCRL-93554 (Oct. 1985).
7. C.L. Olson, J. Fusion Energy $\underline{1}$, 309 (1981).
8. R.R. Peterson, "Gas Condensation Phenomena in Inertial Confinement Fusion Reaction Chambers," University of Wisconsin Fusion Technology Institute Report UWFDM-654 (Oct. 1985) (presented at the 1985 International Symposium on Laser Interaction with Plasma, October 1985, Monterey, CA).
9. A.J.C. Ladd, "Condensation of Ablated First-Wall Materials in the CASCADE Inertial Confinement Fusion Reactor," Lawrence Livermore National Laboratory Report UCRL-53697 (Dec. 1985).
10. R.R. Peterson, "CONRAD - A Combined Hydrodynamics - Vaporization/Condensation Computer Code," University of Wisconsin Fusion Technology Institute Report UWFDM-670 (April 1986).
11. R.R. Peterson and G.A. Moses, "MIXERG - An Equation of State and Opacity Computer Code," Computer Physics Communications $\underline{28}$, 405 (1983).
12. R.O. Bangerter and D. Meeker, "Ion Beam Inertial Fusion Target Designs," Lawrence Livermore National Laboratory Report UCRL-78474 (1976).
13. G.A. Moses, R.R. Peterson, M.E. Sawan, and W.F. Vogelsang, "High Gain Target Spectra and Energy Partitioning for Ion Beam Fusion Reactor Design Studies," University of Wisconsin Fusion Technology Institute Report UWFDM-396 (Nov. 1980).
14. P. Stroud, "Streaming Modes in HIF Beam Final Transport," Los Alamos National Laboratory Report LA-UR-85-2809 (1985).

ECONOMIC STUDIES FOR HEAVY-ION-FUSION ELECTRIC POWER PLANTS*

Wayne R. Meier, William J. Hogan, Roger O. Bangerter
Lawrence Livermore National Laboratory
Livermore, CA 94550

ABSTRACT

We have conducted parametric economic studies for heavy-ion-fusion electric power plants. We examined the effects on the cost of electricity of several design parameters: cost and cost scaling for the reactor, driver, and target factory; maximum achievable chamber pulse rate; target gain; electric conversion efficiency; and net electric power. Using the most recent estimates for the heavy-ion-driver cost along with the Cascade reactor cost and efficiency, we found that a 1.5 to 3 GWe heavy-ion-fusion power plant, with a pulse rate of 5-10 Hz, can be competitive with nuclear and coal power plants.

INTRODUCTION

The Heavy-ion-fusion Systems Assessment Project has produced new estimates of the capital cost and performance of an induction linac accelerator for inertial fusion.[1] These estimates have been incorporated in an economic model of a heavy-ion-fusion (HIF) electric power plant. This model has been used to do parametric studies to identify the features of the plant that have the highest leverage for improving HIF economics. One of the factors that will be used to judge the desirability of an ICF power plant is its economic competitiveness compared to other power sources, and one of the several figures of merit for economic competitiveness is the projected cost of electricity (COE). While we caution against using any of the COEs calculated in this paper in an absolute sense, we believe the results are useful in relative comparisons. The model should primarily be used to identify the technical developments that would lead to the greatest reduction in COE. The model can also be used (if a great deal of care is taken) to grossly compare HIF to other future power sources. The same economic assumptions must be used in the comparison. This paper updates our previous analyses of HIF economics.[2,3]

COST AND PERFORMANCE MODELS

The COE in ¢/kW$_e$h is given by

$$COE = (RC_T + M + F) / (0.0876 \alpha P_n) , \qquad (1)$$

---

*Work performed under the auspices of the U.S. DOE by the Lawrence Livermore National Laboratory under Contract No. W-7405-ENG-48.

where R = annual fixed charge rate (yr^{-1}),
C_T = total capital cost of the plant (G$),
M = annual operation and maintenance cost (G$),
F = annual fuel cycle cost (G$),
α = plant availability factor, and
P_n = net electric power (GW$_e$).

The evaluation of this expression for future fission and coal plants is discussed at length in the Nuclear Energy Cost Data Base.[4] Based on the methods and financial parameters given in Ref. 4, the constant dollar fixed charge rate we use is 8.3%. We assume that the annual fuel cost is negligible, and that the O&M cost is 3% of C_T. Thus, the numerator of Eq. 1 is equal to 0.113C_T. The total capital cost is 1.83 times the direct capital cost. This factor includes home and field office construction and engineering services, owner's cost, a project contingency, and interest during construction in constant dollars. The ratio of total capital cost to direct capital cost (1.83) is midway between the value for coal plants (1.53) and the value for nuclear plants with the best experience in holding down costs (2.07). (The average ratio for fission plants is significantly higher at 2.75)

The capital cost of the power plant is broken into three items; the reactor, the driver and the target factory.

$$C_T = 1.83 \, (C_{rd} + C_{dd} + C_{tfd}) \qquad (2)$$

where C_{rd}, C_{dd}, and C_{tfd} are the direct capital costs of the reactor, driver and target factory, respectively.

Reactor Cost

The cost of the reactor, which includes the the fusion chamber and balance of plant, is based on the recent cost estimate for the Cascade reactor.[5] The Cascade reference design has an estimated direct capital cost of 0.66 G$ at a plant thermal power of 1.67 GW$_t$ and a gross electric power of 0.905 GW$_e$. The auxiliary power requirement of the Cascade plant are 1.7% of the gross electric power. The direct capital cost can be scaled to other sizes by

$$C_{rd} = C_r \, (P_t/1.67)^a \, (0.72 N_u + 0.28) \quad G\$ \qquad (3)$$

where C_r is reactor cost coefficient (G$), P_t is thermal power (GW$_t$), a is the power scaling exponent, and N_u is the number of reactor units (i.e., chamber, structures, and associated balance of plant equipment) in the plant. The reference case scaling exponent is 0.49. The last term accounts for savings in both direct and indirect costs when more than one unit is built at a single site.

Target Factory Cost

The direct capital cost of the target factory is given by

$$C_{tfd} = C_{tf} \, v^f \quad , \qquad (4)$$

where C_{tf} is the target factory cost coefficient, υ is the pulse rate (Hz), and f is the scaling exponent. The base case parameters are C_{tf} = 0.1 G$ and f = 0. That is, we assume a constant direct capital cost of 0.1 G$. The effects of scaling with pulse rate are examined in the sensitivity studies described later. There are currently no definitive studies on target factory costs; Eq. 4 is simply an estimate based on analogies to current facilities that mass produce precision products; e.g., semiconductor micro-chips.

Driver Cost

The cost of the heavy ion driver is based on an accelerator with the following characteristics: charge state = +3; ion mass = 210 amu; emittance = 30 μm-rad; ion kinetic energy = 10 GeV; number of beamlets = 8; initial tune = 85°. The direct capital cost of the driver is given by

$$C_{dd} = (0.32 + 0.088E_d)(1.25 + 0.05N_c)(1 + 0.0088(\upsilon - 5)) \text{ G\$} \qquad (5)$$

where E_d is the driver energy (MJ), N_c is the number of chambers, and υ is driver pulse rate (Hz). The first factor is the cost of the high energy part of the accelerator, which increases linearly with driver energy. The second factor includes additional costs for the low energy front end (25%) and for the beam transport and final focusing (5% per chamber). The last factor accounts for the scaling of the cost with pulse rate. The cost increases by 4.4% for each 5 Hz above the base case pulse rate of 5 Hz. Equation 5 gives driver costs that are consistent with the more detailed driver cost model developed by McDonnell Douglas.[6]

Driver Efficiency

The efficiency of the driver is a function of both its energy and its pulse rate. Hovingh's data[7] (converted to charge state +3) is fit by the following expressions at υ = 5 and 10 Hz:

$$\eta(E_d,5) = 0.208 + 0.0284 \ln(E_d - 1.7) \text{ , and} \qquad (6)$$

$$\eta(E_d,10) = 0.221 + 0.0273 \ln(E_d - 1.7) \text{ ,} \qquad (7)$$

where E_d is in MJ. At 10 Hz the efficiency is slightly higher. To scale to other pulse rates, the above equations are used in the following expression.[7]

$$\eta(E_d,\upsilon) = \frac{(1.55)(5\upsilon)\eta(E_d,5)\eta(E_d,10)}{[10\ \eta(E_d,5)(\upsilon - 5)] - [5\ \eta(E_d,10)(\upsilon - 10)]} \qquad (8)$$

The factor of 1.55 converts the charge state +1 results to the charge state +3 case. That is, the Q = 3 driver is a factor of 1.55 times more efficient than the Q = 1 driver.

Target Gain

The target gain versus driver energy relationship used in our analysis is taken from Ref. 8. The following expression is used for the gain curve for single-shelled, cryogenic targets with a $r^{3/2}R$ parameter of 0.02. [r is the focal radius (cm) of the beam and R is the range (g/cm^2) of the ions.]

$$G = 20 + 157 \ln(E_d/3.0) \quad , \tag{9}$$

where E_d is in MJ. In our parametric studies, we looked at the effects of an advanced target design with improved performance. The advanced gain is given by a multiplier times G from Eq. 9.

$$G_{adv} = 2.56 \, G \, \exp(0.36/E_d) \quad . \tag{10}$$

RESULTS OF PARAMETRIC STUDIES

Parametric studies where carried out to examine the COE as a function of the driver pulse rate. We also examined the sensitivity of the results to variations from the reference case in the cost and performance models. The reference case was a single-unit plant with a constant net electric power of 1.0 GW$_e$. (Other reference case parameters are indicated in Table II.)

Economy of Scale

The COE as a function of the driver pulse rate is shown in Fig. 1 for the reference case power of 1.0 GW$_e$ as well as for 0.5 and 1.5 GW$_e$ plant. There are several things to note. First, and

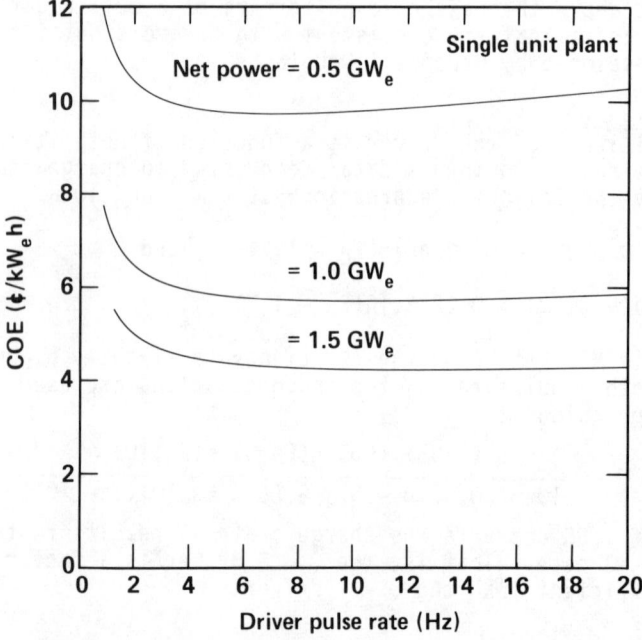

Fig. 1. Cost of electricity as a function of driver pulse rate for plants with net electric powers of 0.5, 1.0 and 1.5 GW$_e$.

most striking, is the stong economy of scale with increasing net electric power. The COE at 5 Hz is reduced by 41% (from 9.8 to 5.8 ¢/kW$_e$h) when the net power is increased from 0.5 to 1.0 GW$_e$. All of the curves show the same general shape; the COE initially decreases with increasing pulse rate and then goes through a broad, shallow minimum. The COE is insensitive to pulse rate above about 4 hz. For the 1.0 GW$_e$ case, the optimum pulse rate is about 10 Hz. The COE at 4 Hz, however, is within 5% of the minimum. At 1.5 GW$_e$, the COE is competitive with the projected COE from a coal-fired power plant (4.4 ¢/kW$_e$h).[4]

Multi-unit Power Plants

Figure 2 shows the COE as a function of the total net power for plants made up of different size units: 0.5, 1.0, and 1.5 GW$_e$ net. A multi-unit power plant consists of a single driver and target factory associated with more than one complete reactor units (i.e., chamber, structures, and balance of plant equipment). As indicated the COE decreases with increasing net power and with increasing unit size. The results shown here are all at a driver pulse rate of 5 Hz per unit.

The COE from a 1.0-GW$_e$ power plant consisting of two 0.5-GW$_e$ units is 6.4 ¢/kW$_e$h. This is 10% higher than the COE from a single-unit power plant generating 1.0 GW$_e$. Hence the

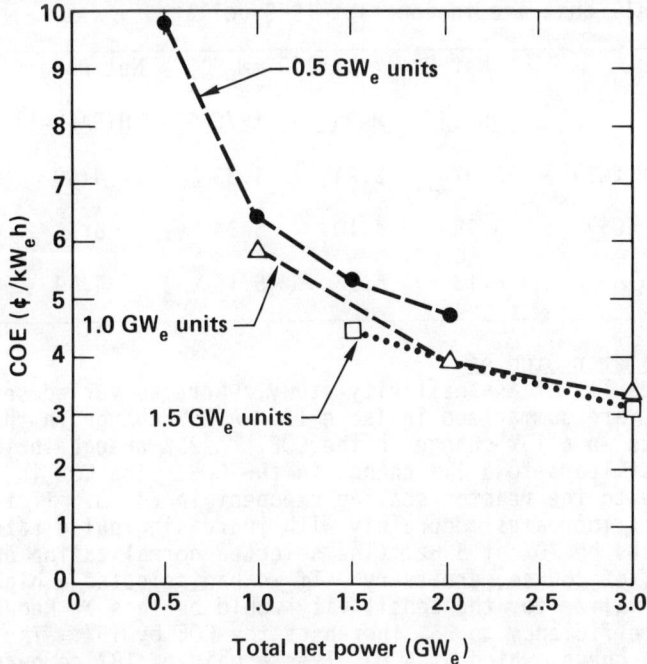

Fig. 2. Cost of electricity versus total net power for plants made up of different size units.

multi-unit plant achieves most of the economy of scale benefits of
the single unit plant. This is important since there may be other
reasons for choosing a multi-unit plant (e.g., phased construction
could more closely match load growth and spread out capital
requirements). The single-unit plant has a reactor that is less
expensive than the sum of the two smaller reactors due to the
scaling given in Eq. 3. Also the two-unit plant has additional
costs for beam transport and focusing. The COE for a four-unit,
2-GW$_e$ plant is 4.7 ¢/kW$_e$h while a two-unit, 2 GW$_e$ plant
results in a COE of 3.9 ¢/kW$_e$h. The 3 GW$_e$ plants at 3.0 and
3.3 ¢/kW$_e$h are competitive with the COE from future 1-GW$_e$
nuclear power plants (3.5 ¢/kW$_e$h).[4]

The COE from the HIF plant is compared to other fusion reactor
studies in Table I. The costs from the other studies[9-11] were
inflated to January 1985 dollars, and the economic assumptions
previously described were applied in all cases. That is, we used
the same contingency, indirect cost factors, time-related cost
factor, fixed charge rate, capacity factor, etc. As indicated, the
COE projected for the 1200 MW$_e$ HIF plant using the Cascade reactor
is 16 to 19% less than the COE for STARFIRE and MARS, respectively.
Compared to HIBALL-II at 3784 MW$_e$, the HIF/Cascade plant has a COE
that is 33% lower.

Table I. Comparison of results with other fusion reactor studies.
All cost are in constant 1985 dollars.

	Net Power = 1200 MW$_e$			Net Power = 3784 MW$_e$	
	STARFIRE	MARS	HIF/CAS	HIBALL-II	HIF/CAS
Direct cost (G$)	2.07	2.24	1.83	4.80	3.43
Total cost (G$)	3.78	4.10	3.34	8.78	6.30
COE (¢/kW$_e$h)	6.13	6.33	5.14	4.59	3.07

Other Parmameter Studies

The results of a sensitivity study, where we varied several
parameters, are summarized in Table II. A 25% change in the driver
cost results in a 13% change in the COE. A 25% change in the
reactor cost leads to a 11% change in the COE. The COE is
insensitive to the reactor scaling exponent in Eq. 3. If the target
factory cost increases moderately with increasing pulse rate, the
COE increases by 10% at 5 Hz. (The selected normalization pulse rate
of 1 Hz is, of course, arbitrary. If we had selected a higher pulse
rate to normalize to, the sensitivity would be less.) Reducing the
conversion efficiency to 35% increases the COE by 15%. The advanced
target gain curve, which at 4 MJ gives a gain of 182 compared to 65
with the reference gain curve, leads to a 8% reduction in the COE.
The economy of scale effects previously discussed are also indicated
in the table and are seen to cause the largest effects.

Table II. Changes in the COE (¢/kW$_e$h) resulting from changes in the reference case parameters. The reference case COE is 5.84 ¢/kW$_e$h. The reference case parameters are given in square brackets. Fractional changes are given in parenthesis.

Parameter	COE	Parameter	COE
Pulse rate [5 Hz]		T. F. scaling [0]	
2 Hz	6.5 (+ 11%)	0.6	6.4 (+ 10%)
10 Hz	5.7 (- 2%)	Conversion eff. [54%]	
Driver cost [C_d]		45%	6.2 (+ 6%)
0.75 C_d	5.1 (- 13%)	35%	6.7 (+ 15%)
1.25 C_d	6.6 (+ 13%)	Target gain curve [G]	
Reactor cost [C_r]		G_{adv}	5.4 (- 8%)
0.75 C_r	5.2 (- 11%)	Net electric power [1.0 GW$_e$]	
1.25 C_r	6.5 (+ 11%)	0.5	9.8 (+ 68%)
Reactor scaling [0.49]		1.5	4.4 (- 25%)
0.70	5.9 (+ 2%)	Units per plant [1]	
0.90	6.0 (+ 3%)	2	3.9 (- 33%)
		3	3.3 (- 43%)

CONCLUSIONS

We have examined the COE of HIF power plants in order to determine the economic impact of various design and system improvements. Our conclusions fall into three major areas: (1) optimum pulse rate, (2) economy of scale, and (3) various design improvements.

Optimum Pulse Rate

For the single-unit, 1-GW$_e$ power plant, the optimum pulse rate (for both driver and chamber) is about 10 Hz, resulting in a COE of 5.7¢/kW$_e$h. However, the COE is not very sensitive to pulse rate. For example, if the chamber pulse rate is limited to 4 Hz, the COE increases by less than 5%. This weak dependence of COE on pulse rate is important because major technological uncertainties are associated with predicting the maximum achievable chamber pulse rate.

Economy of Scale

As expected, the COE decreases with increasing net electric power. The reduction is significant, whether the power is increased by increasing the power of a single unit, or by using a single driver to operate several units. For a given net power, a multi-unit plant achieves most of the economy-of-scale advantage of a single large plant. These savings result from the driver being such a large fraction of the plant cost. Only if the cost of the driver is significantly reduced will the penalty for small power units be significantly affected. The HIF power plant can be competitive with coal or nuclear with a single unit 1.5 GW_e plant or with multi-unit, 2-3 GW_e plants. The COE projected in these studies indicate that the COE is 16 to 19% lower than from STARFIRE and MARS, and 33% lower than from HIBALL-II.

Various Design Improvements

Other than changing the net power of the plant, each of the variations we considered resulted in a change in the COE on the order of 10%. To make dramatic improvements in the COEs reported here will require a combination of success in reducing costs and improving performance in a variety of areas.

REFERENCES

1. J. Hovingh, et al., <u>The Cost of Induction Linac Drivers for Inertial Fusion for Various Target Yields</u>, to be published in the proceedings of this meeting.
2. W. R. Meier and W. J. Hogan, <u>ICF Reactor Economics: Identifying the High Leverage Design Features</u>, Fusion Technol. <u>8</u>, 1820 (1985).
3. W. R. Meier and W. J. Hogan, <u>Identifying Heavy-Ion-Beam Fusion Design and System Features with High Economic Leverage</u>, LLNL Report UCID-20470 (Mar. 1985).
4. J. G. Delene, et al., <u>Nuclear Energy Cost Data Base - A Reference Data Base for Nuclear and Coal-Fired Power Plant Power Generation Cost Analysis</u>, U. S. Department of Energy Report, DOE/NE-0044/2 (March 1984).
5. I. Maya, et al., <u>Final Report: Inertial Confinement Fusion Reaction Chamber and Power Conversion System Study</u>, GA Technologies Report GA-A17842 (oct. 1985).
6. D. S. Zuckerman, et al., <u>Performance and Cost Modeling of a Linac-Driven HIF Power Plant</u>, to be published in the proceedings of this meeting.
7. J. Hovingh, LLNL, private communication (Jan. 1986).
8. J. D. Lindl, J. W-K. Mark, <u>Recent Livermore Estimates of the Energy Gain of Cryogenic Single-Shell Ion Beam Targets</u>, Laser and Particle Beams, <u>3</u>, 37 (1985).
9. C. C. Baker, et al., "STARFIRE - A Commercial Tokamak Fusion Power Plant Study, Argonne National Laboratory Report ANL/FPP-80-1 (Sept. 1980).
10. B. G. Loagan, et al., <u>MARS - Mirror Advanced Reactor Study</u>, LLNL Report UCRL-53480 (June 1984).
11. B. Badger, et al., <u>HIBALL-II - An Improved Conceptual Heavy Ion Fusion Reactor Study</u>, Univiversity of Wisconsin Report UWFDM-625 (Dec. 1984).

ENERGY ANALYSIS OF HIF REACTOR "HIBLIC-I"

T. Nagai†
Department of Earth Sciences, Nagoya University, Nagoya 464, Japan

H. Obayashi
Institute of Plasma Physics, Nagoya University, Nagoya 464, Japan

T. Yamaki
Department of Physics, Nagoya University, Nagoya 464, Japan

ABSTRACT

An energy analysis of the HIF power plant "HIBLIC-I" is crudely carried out. In the present analysis, the required input energies for the driver system and reactor cavities are calculated on the basis of energy intensities for the component materials, while those for other subsystems, such as turbine plant equipment and thermal conversion equipment, which are mostly similar to or in common with the usual fission power plants, are taken from the energy analyses for LWR power plants. As the result of the present study, we obtain an energy ratio for HIBLIC-I of 16.2 for the plant lifetime of 30 years.

INTRODUCTION

Inertial confinement fusion driven by heavy ion beams (HIF) is one of the promising candidates for an ICF power plant because of the inherently high efficiency, repetition rate and operational reliability of the accelerator system. On the other hand, the cost of construction and maintenance of an HIF driver system is thought to be very expensive because of the gigantic scale of existing high energy particle accelerators. We have been asked, "Does HIF have any possibility to be a power plant comparable with conventional power plants, or even MCF power plants in an economic aspect ?" An answer to this question has come from the monetary cost estimation in the HIBALL study.[1]
Energy analysis can give another answer from the viewpoint of energy cost which is the important figure of merit of a power plant design. Energy analyses in the early stage of HIF study were carried out by Miyahara et al.[2] We estimated the energy cost of HIBLIC-I as an example of an HIF power plant, based on the Japanese data. This work was achieved as a part of the conceptual design "HIBLIC-I",[3] whose main purpose was to study the feasibility of HIF. Major parameters of HIBLIC-I are listed in Table I.

---

† Present address: Computation Center, Nagoya University, Nagoya 464, Japan.

Table I   Parameter of HIBLIC-I

DT thermal power	4000 MW	Target	Pb-Aℓ-DT
Gross thermal power	4700 MW	DT mass/target	7.37 mg
Gross electrical output	1800 MWe	Target yield	400 MJ
Net electrical output	1500 MWe	Target gain	100
Plant availability factor	75 %		
Driver type	RF-Linac	Target shot rate/cavity	1 Hz
Driver efficiency	25 %	No. of cavity	10
Ion Species	^{208}Pb$^+$	Coolant and breeder	Li
Ion energy	15 GeV	Maximum coolant temp.	470°C
		Tritium breeding ratio	1.55
Beam energy	4 MJ	Tritium inventory	5 kg
Beam power	160 TW	Tritium recovery	Y-getter
Driver repetition rate	10 Hz	Structural material	HT-9
No. of beams/cavity	6	First wall(Sacrificial wall)	
Current intensity/beam	1.78 kA	Protection scheme	Li-Curtain
		Max. dpa rate in 1st wall	36.6/FPY
Cavity pressure at RT	10^{-4} Torr	Max. dpa rate in 2nd wall	0.93/FPY
		Life time of 1st wall	2 FPY
		Life time of 2nd wall	30 FPY
		Wall loading(1st wall)	5 MW/m^2

METHODOLOGY OF ENERGY ANALYSIS AND DATABASE

Energy analysis is defined in general as the computation and measurement of energy flow in society, and in particular, as the quantification of volume of energy resources embodied, directly and indirectly, in various commodities.[4] The basic element of energy analysis is the energy intensity which represents the energy directly and indirectly required to produce a unit amount of a certain product. We have two basic methods to determine the energy intensity, i.e., energy process analysis and energy input-output analysis. In the process analysis, we have to follow the production processes of a specified product step by step, in order to determine the amount of energy required at each step, and finally to sum up all the energy inputs to calculate the energy intensity of the product. This method needs detailed investigations of the industrial processes. In the input-output analysis which was developed by Bullard and Herendeen,[5] an input-output table is utilized to calculate the energy intensity. Since interindustry transactions in an economic system are described in the input-output table, it is convenient to calculate the energy intensities for various sectors in the industry simultaneously.

To obtain energy intensities of materials used in a fusion reactor system in practice, a combination of the above two methods is employed. The energy intensities of basic materials, such as crude steel and copper, can be calculated by the input-output

Table II Energy intensities of materials

Material	Energy intensity (kcal/t)
Fe (crude steel)	5.59×10^6
Cu (electrolytic cathode copper)	1.13×10^7
SUS304 (stainless steel)	3.13×10^7
SUS316 (stainless steel)	3.21×10^7
NbTi superconducting cable	8.31×10^8
Graphite	5.0×10^7
Lithium	4.02×10^8
Cement	1.41×10^6

Table III Energy intensities of plant equipment, buildings and material transportation of an LWR[7,8]

	Energy intensity (10^9 kcal/MWe)
Plant equipment	
Materials	0.487
Fabrication	0.344
Buildings	0.987
Material transportation	0.270

analysis. Those of specific materials, which cannot be obtained from the input-output analysis, such as stainless steel and superconductor are determined by the process analysis.

It is noted that the energy intensities reflect the current industrial system and are, therefore, strongly dependent on natural and social environments. To perform the energy analysis of HIBLIC-I, the energy intensities of the materials for the cavities and the energy driver are taken from the study by Nagai and Shimazu,[6] and are listed in Table II. They obtained these values from the input-output table for 554 sectors of Japan in 1980. In the present paper, energy requirements for energy driver and reactor cavities, which are the most characteristic subsystems in the HIF plant, are evaluated on the basis of material inventory. The energy requirements for the other plant equipment, which is similar to or in common with that in existing fission power plants, are estimated from the energy intensities of materials of LWRs,[7,8] as given in Table III. In fabrication, the energy intensity for the driver system is also assumed to be equal to that of the conventional equipment in the LWR, and the energy intensity for the reactor cavities to be equal to that of a pressure vessel of the LWR. We obtain the values of 3.14×10^7 kcal/t for the driver system and 5.96×10^7 kcal/t for the cavities from Refs.7 and 8, respectively. The energy intensity required for the buildings of the driver system is assumed to be 1.41×10^6 kcal/m^2, which is an energy intensity of reinforced concrete apartments,[7] and that for tunnels for the driver system to be 2.16×10^{10} kcal/km, which is an energy intensity of tunnels for roads.[9] For buildings of other subsystems including reactor cavities and for material transportation, energy intensities are also taken from Table III. The energy intensity of site improvements of a plant is estimated to be 3.74×10^9 kcal/ha, referring to the construction of Kansai International Airport.[10]

Table IV Material and energy requirements for one cavity

	Material	Mass (t)	Energy required (10^9 kcal)
Blanket	Li	50	20
Shield/Reflector	Graphite	42	2
	Concrete	805	1
Structure material	HT-9	156	5
	SUS304	200	6
Piping	HT-9	116	4
Total		1369	38

RESULTS

Energy inputs to the materials for one reactor cavity are shown in Table IV, in which the energy intensities of HT-9 and concrete are assumed to be the same as those of SUS316 and cement, respectively. Energy required for the materials used in the driver system is shown in Table V. The superconducting cable, although there are choices of NbTi, Nb$_3$Sn and other advanced materials in the conceptual design, is typically represented by a single material of 50 % NbTi in wire form of 2 mm × 3 mm cross section and 50 kg/km of line weight density.

We calculate fabrication energy of the cavities and the energy driver based on the weights given in Tables IV and V, and we obtain

790×10^9 kcal and 2095×10^9 kcal, respectively. Energy required for the buildings of the driver system (2.3×10^5 m^2) becomes 324×10^9 kcal and that for the tunnels (15 km) 324×10^9 kcal. Energy requirements for the other plant equipment, buildings (except for those of the driver system) and material transportation are calculated with the energy intensities given in Table III. The energy for the site improvements becomes 203×10^9 kcal. In summary, the initial requirements for the HIBLIC-I power plant are shown in Table VI.

Table V Material and energy requirements for the energy driver

	Material	Mass or length	Energy required (10^9 kcal)
Linac	Fe	24963 t	140
	Cu	5811 t	66
	SUS304	1850 t	58
	Subtotal		264
Accumulator ring	NbTi	4312 km	179
	Fe	3820 t	21
	SUS316	70 t	2
	SUS304	110 t	3
	Subtotal		205
Buncher ring	NbTi	3250 km	135
	Fe	2870 t	16
	SUS316	42 t	1
	SUS304	70 t	2
	Subtotal		154
Final beam transport	NbTi	27300 km	1136
	Fe	23400 t	131
	SUS316	1200 t	39
	SUS304	780 t	24
	Subtotal		1330
	Total	66729 t	1953

DISCUSSIONS AND CONCLUDING REMARKS

The electric output, material and energy requirements per cavity of HIBLIC-I are compared with those of HIBALL,[1] as shown in Table VII. Energy requirements per unit electric output for the

Table VI Summary of the energy requirements for HIBLIC-I

	Energy required (10^9 kcal)	(% of total)
Materials		
Reactor (10 cavities)	380	3.9
Energy driver	1953	19.9
Other plant equipment.	877	8.9
Subtotal	3210	32.7
Fabrication		
Reactor (10 cavities)	790	8.0
Energy driver	2095	21.3
Other plant equipment	619	6.3
Subtotal	3504	35.6
Buildings		
Energy driver	648	6.6
Other plant facilities	1777	18.1
Subtotal	2425	24.7
Material transportation	486	4.9
Site improvements	203	2.1
Total	9828	100.0

Table VII Single cavity comparison of HIBLIC-I and HIBALL (Values given in parentheses are normalized ones with respect to HIBLIC-I.)

	HIBLIC-I	HIBALL
Net electric power (MWe)	150 (1)	942 (6)
Reactor weight*(t)	1319 (1)	9887 (7)
Energy required (10^9 kcal)	117 (1)	718 (6)

*Breeding material is not included.

reactor cavity may not strongly depend on the scale of cavity. As is shown in Table VI, the fraction of energy cost of the driver system is comparable with the rest of the plant, as expected.[2] It is noted that the energy requirements for the final beam transport occupy 68 % and the energy requirements of 1450×10^9 kcal for the superconductor become 74 % of the total investment for the driver system, as shown in Table V. This means that the optimization of beam energy, number of beams and number of reactors is especially important in respect to the energy cost of an HIF power plant. Energy required for annual maintenance and replacement of the plant facilities is unknown. However, it is assumed to be 2 % of the initial requirement in the present paper.

Table VIII Energy balances of HIBLIC-I and STARFIRE

	HIBLIC-I	STARFIRE
Net electric power (MWe)	1500	1200
Availability factor (%)	75	75
	Energy required, unit: 10^9 kcal (%)	
Initial requirements		
Reactor	1170 (8)	3037 (31)
Energy driver	4048 (26)	—
Others	4610 (29)	3169 (32)
Subtotal	9828 (63)	6206 (63)
Maintenance	5897 (37)	3724 (37)
Total	15725 (100)	9930 (100)
Energy ratio	16.2	20.5

We define an energy ratio (R) as follows:

$$R = \frac{\text{(Net electric output)} \times \text{(Operation time)}}{\text{(Total energy inputs)}}.$$

The energy ratio of HIBLIC-I becomes 16.2 for a plant lifetime of 30 years, as shown in Table VIII. In this paper, important items, such as pellet factory, disposal of radioactive wastes, land and water use, etc. are not included. Energy requirements for those items, however, are guessed to be less than 10 % of the total energy requirements.[1,2,11] The energy ratio of STARFIRE[12], a tokamak reactor, is calculated in the same manner and also shown in Table VIII. Though the ratio of HIBLIC-I is somewhat smaller than that of STARFIRE, it is still comparable to those of LWRs which are 14.6 ~ 16.8 for 75 % of plant availability and 30 years of lifetime.[8] Here, we can conclude that HIF power plant is feasible in the aspect of energy cost.

ACKNOWLEDGEMENTS

The authors would like to express their thanks to all members of the HIBLIC working group. This work was financially supported by the Grant-in-Aid for Fusion Research from the Ministry of Education, Science and Culture.

REFERENCES

1. B. Badger et al., UWFDM-450, KfK-3202 (1981) and KfK-3840, FPA-84-4, UWFDM-625 (1985).
2. A. Miyahara et al., Proc. 11 th Symp. on Fusion Tech. (Oxford 1980) Vol.2, p.1141.
   A. Miyahara and S. Kawasaki, Proc. of the Symp. on Accelerator Aspects of HIF, GSI-82-8 (1978), p.76.
3. T. Yamaki et al., IPPJ-663, (1984).
   T. Yamaki, Laser and Particle Beams 3, 29 (1985).
4. F. J. Alessio et al., EPRI EA-504 (1978).
5. C. W. Bullard and R.A. Herendeen, Proceedings of the IEEE, 63, 484 (1975).
6. T. Nagai and Y. Shimazu, Journal of Earth Sciences, Nagoya Univ. 32, 1 (1984).
7. Resources Research Committee, Life cycle energy of food, clothing and shelter. (Science and Technology Agency, 1978). (in Japanese)
8. Institute for Policy Sciences, Structure of energy utilization and energy analysis. CR-76-12 (1977). (in Japanese)
9. Y. Kaya, Energy analysis. (Denryoku-Shinposha, 1980). (in Japanese)
10. Kansai International Airport Corporation, Environmental impact assessment report on the construction of Kansai International Airport (1985). (in Japanese)
11. N. Tsoulfanidis, Nucl. Tech./Fusion 1, 234 (1981).
12. C. C. Baker et al., ANL/FPP-80-1, Vol.1-2 (1980).

HEAVY-ION FUSION REACTOR CONCEPTS, REQUIREMENTS, AND ATTRACTIVE FEATURES

J. H. Pendergrass
Los Alamos National Laboratory, Los Alamos, NM 87545

ABSTRACT

Four generic reactor-plant concepts were included in the Heavy Ion Fusion Systems Assessment to provide a wide range of pulse-repetition-rate/target-yield capability--1 to 20 Hz/150 to 3000 MJ. This permitted exploration of the impact on heavy-ion fusion cost of electricity (COE) of the repetition-rate and efficiency advantages of induction linacs relative to other ICF drivers. Minimum COE results are encouraging--from 55 to 75 mills/kWh for 1000-MWe plants to < 45 mills/kWh for 1500 MWe. HIF with induction-linac drivers can compete using more than one reactor plant concept. Minimum COE corresponds to reactor repetition rates < 10 Hz for 1000-MWe, one-reactor plants. Contrary to pre-HIFSA expectations, large pulse repetition rates are optimum only for very large plants. COE's are within 5% of minimum COE over wide ranges of repetition rate and target yield. Cost or technological problems in one part of plant parameter space need not be fatal for HIF because of the large low-COE parameter space.

THE HIFSA PROJECT

The Heavy-Ion Fusion Systems Assessment (HIFSA) examined the prospects for commercial pure heavy-ion fusion (HIF) electric power generation using induction linear accelerator (linac) drivers. Although led by Los Alamos National Laboratory, success required substantial contributions from all members of a large interdisciplinary team with members from Lawrence Livermore National Laboratory, Lawrence Berkeley Laboratory, Stanford Linear Accelerator Center, the University of Wisconsin, and McDonnll Douglas Astronautics Co.

Many of the dozens of inertial confinement fusion (ICF) reactor-plant and balance-of-plant (BOP) concepts[1-8] developed for laser fusion apparently require only minor modifications for HIF. There are also a few reactor concepts developed specifically for HIF[3-4]. A large fraction of ICF applications-studies resources has been devoted to their development. Limited HIFSA resources were therefore concentrated on (1) cost/performance models for and innovation to improve HIF targets, induction linacs, and final beam transport/focusing and (2) an HIF commercial power plant code for identification of key cost/performance issues, examination of tradeoffs and sensitivities, and search for global optima for integrated plants.

HIFSA REACTOR STUDY OBJECTIVES AND SCOPE

Only adaptation of laser fusion reactor plant concepts with which HIFSA team members had experience was considered. HIFSA reactor studies focused on (1) identification and quantification of requirements for HIF additional to those for laser fusion, areas where HIF requirements are less constraining, and required HIF reactor-driver, reactor-fuel cycle, and reactor-BOP interfaces; (2) identification and exploration of significant tradeoffs between HIF reactor and driver, fuel cycle, and BOP design requirements and desirable features; and (3) formulation of cost/performance models for HIFSA reactor concepts suitable for incorporation in the McDonnell Douglas HIF power plant systems code. All reactor plant structures and equipment were treated--not just cavity concepts--including equipment required to interface reactors with the driver, fuel cycle, and BOP. A combination of reactor concepts providing a wide range of repetition rates and capable of accommodating a wide range of target yields was desired to permit thorough exploration of the potential advantages of high-repetition-rate (> 50 Hz at little additional cost), high-efficiency induction linacs. Before HIFSA, it was thought that these characteristics might permit competitive HIF with smaller target gains, and hence, smaller driver pulse energies with attendant driver and reactor cost savings.

HIFSA REACTOR CONCEPTS

Four generic ICF reactor plant concepts were selected for inclusion in the HIFSA reactor studies: (1) a granular-wall/Brayton-cycle concept (a variant of the Livermore CASCADE[5] concept);(2) a liquid-metal-jet/steam-cycle concept (a variant of the Livermore HYLIFE[6] concept); (3) a wetted-wall/steam-cycle concept (a variant of the Los Alamos wetted-wall[7] concept); and (4) a magnetically-protected dry-wall/steam-cycle/direct conversion concept (a variant of the Los Alamos magnetically protected[8] concept). Space limitations permit only brief descriptions of these concepts. The references cited provide additional details and illustrations.

HIFSA is not a final contest between reactor plant concepts. These concepts involve different degrees of optimism with respect to high-efficiency advanced electric power generation, required safety systems, and target chamber operating characteristics. The differing degrees of optimism were largely retained to permit study of the potential benefits or penalties. Some of the HIFSA reactor plant concepts are more highly developed than others. Differences in state of development of reactor concepts were ignored.

<u>Generic Granular-Wall/Brayton-Cycle Concept</u> The granular-wall concept employs a rotating (50 rpm for a 5-m-radius cavity to maintain the central cavity and move the particles at appropriate speeds) biconical vessel and centrifugal forces to move a thick particulate blanket through the reaction chamber. The particle bed has a graphite inner layer to reduce evaporation by target x rays and enhance condensation for cavity clearing, a beryllium oxide intermediate layer to multiply neutrons (required for tritium breeding ratio > 1.0), and a lithium aluminate outer layer for tritium breeding. It provides complete protection of the reactor vessel from x rays and ions and substantial protection from neutrons for long service life. Condensation of evaporated/sputtered graphite limits reactor pulse repetition rate to about 10 Hz. For a 1000-MWe plant operating at 5 Hz, target yields of about 500 MJ must be accommodated. Driver beams can be injected only through the ends of the vessel.

Particles introduced continuously through openings at the smaller-radius ends of the reactor move to the larger-radius mid-plane of the reactor. They then move to rotating "shelves" on the outside of the reactor vessel, from which they are extracted by stationary scoops and flung upward into a hopper. From there, they are moved by gravity through heat exchangers, transfer heat to a high-temperature Brayton cycle using helium as the working fluid to give a 55% net plant efficiency, and are returned to the blanket.

The reactor, granule circulation system, and primary heat exchanger are enclosed in a vacuum chamber, avoiding seals for connecting driver beam tubes and the particle transport system to the rotating reactor, vacuum valves in the granule transport lines, and other complications. The clearance around the reactor and closely associated equipment is minimized to reduce the size of the expensive vacuum containment structure and systems. The rotating vessel is made of trapezoidal silicon-carbide (SiC) tiles held in compression by composite SiC/aluminum tendons. This design can operate at high inner bed surface temperatures (> 2000 K may be possible) with inorganic insulation and thermal radiation to the surroundings keeping the tendons outside the insulation at 700 K. SiC is a low-activation material which reduces maintenance and radwaste disposal costs. The concept developers are attempting to make a case for inherent safety that will permit elimination of costly containment structures. To this end, the concept uses non-flammable, low-activation materials wherever feasible.

<u>Generic Liquid-Metal-Jet/Steam-Cycle Concept</u> The liquid-metal-jet concept uses an array of liquid-metal jets falling through an orifice plate in the top of the reactor to protect reactor structure from neutrons, as well as from x rays and ions, for long service life. Graphite in a thin steel envelope surrounds the reaction chamber to further moderate neutrons not stopped by liquid metal and reflect them back into the reaction chamber. The jet array is configured to permit injection of driver beams between jets angled to open up clear paths. One to a few horizontal beam clusters or a single vertical beam bundle can be injected, but uniform illumination of targets is not feasible. The jet array reduces pressure buildup by venting, rather than confining, the hot vapor generated by evaporation of liquid metal by x-ray, ion, and neutron heating. The jet surface area available for condensation of liquid metal vapor is large to give a safety margin for cavity clearing following a target microexplosion. Evacuation of the reactor cavity is primarily through condensation of the liquid metal. Conventional vacuum systems are required for the driver beam tubes and target injection system, into which streaming of liquid-metal vapor is reduced by quick-opening/closing shutter valves.

The reactor structure must resist fatigue due to cyclic internal pressurization, impact of disrupted liquid-metal jet material, and pulsed heating and buckling when the cavity is at subatmospheric pressures between microexplosions. The low-alloy-steel reaction-chamber pressure vessel has a radius of 5 m and a thickness of 4 cm for operation at 1.5 Hz, 1800 MJ, and 1000 MWe net.

Two liquid metals, both good tritium breeders, have been proposed for use in liquid-metal-jet reactors: (1) pure lithium and (2) a lead-lithium eutectic mixture (83 atomic % Pb, 17 atomic % Li). Lithium has a very low density, which results in lower pumping power required for recirculation, is not activated by fusion neutrons, and holds tritium better. Lead-lithium eutectic has a lower vapor pressure and poses less of a fire hazard than lithium, but is activated by neutrons and is very dense. The laser-fusion version is operated at high temperature with lithium to couple with a high-efficiency steam cycle for a 35% net plant efficiency, also assumed for HIF. If ballistic focusing at 10^{-5} to 10^{-4} torr is required, then either the operating temperature with lithium must be lowered or a switch to a liquid metal with a lower vapor pressure, such as the lead-lithium eutectic, will be required.

The liquid-metal jets are disrupted by target microexplosions and must be reformed after each one. The large pumping power required to reform the jets and the rapid increase in pumping power and cost of the recirculating system with repetition rate limit practical rates to about 2 Hz, so that large target yields are required for 1000 MWe.

A liquid-metal side stream transfers heat to a sodium intermediate loop included for safety at substantial cost. The sodium is circulated through liquid-metal steam generators. Liquid-metal-jet reactors can't be coupled with high-efficiency, advanced power generation systems. The replacement of the intermediate loop with double-wall-tube steam generators currently being developed at commercial scale could substantially reduce reactor plant cost. An inert gas is used within the reactor containment building and other buildings to reduce the risk of liquid-metal and other fires. Steel building liners prevent contact with concrete, which reacts with liquid metals, if leaks develop. Liquid metal can be rapidly dumped into a catch tank beneath the reactor containing a large mass of iron balls for rapid cooling in emergencies. Molten-salt extraction is used to recover tritium.

<u>Generic Wetted-Wall/Steam-Cycle Concept</u> In wetted-wall reactors, liquid-metal films are injected at high speed tangentially downward through slit nozzles onto curved (cylindrical or spherical) first-wall surfaces to protect from x rays and ions and are held there by centrifugal force. Optimum film injection velocities, thicknesses, and number of injection points depend on reactor size, geometry, surface recoverage requirements, heat transport capabilities, liquid metal properties, first-wall material, driver beam propagation and focusing requirements, and the power generation cycle. Dislodgement of the films from first-wall surfaces by target microexplosions, especially in the lower half, is not expected. Thus, film injection to carry off energy trapped in the cavity with the desired temperature rise and to recover the upper half after each target are assumed. The 50 to 100 m/s velocities required for single-nozzle, top-half recoverage at 10 Hz after every target in few-m-diameter reactors are compatible with low-alloy steels at efficient steam-cycle temperatures (up to 550 °C). Pumping power is low. Repetition rates up to 10 Hz are determined by evaporated/sputtered liquid-metal condensation times.

The wetted-wall concept is versatile. First-wall-protection and blanket functions are separated. Wetted-wall cavities can be combined with blankets that operate at low, moderate, and high temperatures using solid or liquid breeding materials and gaseous, liquid, or solid heat transport materials and with a variety of high-efficiency power generation systems. Hybrid blankets suffer no substantial loss of breeding or energy multiplication due to the thin protective liquid-metal film and first-wall structure. The HIFSA wetted-wall reactor plant has conservative liquid-lithium first-wall-protection and blanket primary coolant loops coupled with a steam cycle through a sodium intermediate loop for enhanced safety giving a net plant efficiency of 36%. Target debris and unburnt tritium is transported from the reactor cavity in a modest liquid-metal stream whose characteristics can be adjusted to permit easier cleanup. The temperature and composition of the first-wall-protection liquid metal can be selected to provide cavity clearing rates and residual atmospheres to match driver-beam focusing and propagation requirements. Wetted-wall reactors are compatible with single-sided and double-sided target illumination and also with some more complex illumination geometries. Arrays of driver beams injected through vertical "orange-slice" openings between lines of longitude in a spherical reactor are particularly convenient.

Wetted-wall reactor cavities are sized and first-wall thicknesses (about 1 cm) are selected to prevent failure by fatigue. First-wall structure exposed to intense neutron irradiation is minimized and designed for rapid changeout, required at 3 to 5 year intervals, to reduce costs and allow replacement during scheduled outages for other plant maintenance. The first wall is backed by longitudinal stiffening ribs for resistance to buckling when the cavity is evacuated. The reactor vessel is protected from neutrons and serves as the vacuum barrier. An actively cooled close-in concrete shield/containment structure with removable lid and inert cover-gas systems is provided. Other safety and tritium-recovery systems are similar to those for the liquid-metal-jet concept.

<u>Generic Magnetically Protected Dry-Wall/Steam-Cycle Concept</u> Dry-wall reactors may be crucial for the success of ICF with direct-drive targets which require "uniform" illumination with "many" driver beams. Reactors which use liquid metals or mobile layers of solid particles can't be easily adapted for injection of many driver beams uniformly distributed in solid angle; dry-wall reactors can. Dry-wall reactors may also be important for success of HIF if high reactor repetition rates (> 10 Hz) are required to take advantage of accelerator high repetition-rate capability to improve HIF economics. The magnetically protected concept was selected for HIFSA as a representative dry-wall reactor for exploration of parameter space accessible only with dry-wall reactors.

If dry-wall reaction chamber size is sufficiently large, then material thickness loss rates due to x-ray and ion evaporation/ sputtering will be small enough to give long first-wall sacrificial-cladding service lifetimes. Unfortunately, dry-wall reactors must be very large--10 to 20-m radii are required for 150-MJ yields at 20 Hz. This results in large costs for shielding, containment, and other reactor plant equipment, as well as for the reactor.

Magnetic fields can deflect target-debris ions away from surfaces and out of dry-wall reactor cavities in a controlled manner through convenient openings. This feature distinguishes magnetically protected reactors from other dry-wall concepts. Only modest field strengths (0.1 to 0.2 T, so that superconducting magnets are not necessary) and simple field configurations (modified solenoidal) are required. One drawback of some dry-wall concepts--lack of assured target debris removal--is solved. The magnetic field may impede the motion of ionized evaporated species away from the surfaces from whence they came, thereby reducing effective material loss rates. Target-debris cleanup and recovery of unburnt tritium should be easier because they are more concentrated.

Thickness loss rates from graphite liners in magnetically protected reactors with no redeposition are estimated to be < 0.3 cm/y closest to 150-MJ-yield targets in a 10-m-radius cavity at 10 Hz, so that a few-cm-thick liner could last for many years. Rates of neutron damage to first-wall structure would be nearly an order of magnitude less than in wetted-wall reactors of the same fusion power. Other structure is protected by the blanket.

Liquid-metal spray condensers to absorb diverted ions avoid solid-surface ion damage, getter tritium, act as vacuum pumps, and transport tritium, target debris, and heat. Magnetohydrodynamic and electrostatic direct conversion has been proposed for converting ion energy into electrical energy at roughly double steam-cycle efficiencies and modest additional cost. Electrostatic direct conversion was assumed for HIFSA. Reactor structure, shielding, containment, other safety system, and tritium recovery system designs are similar to those for the wetted-wall concept. For HIFSA, delivery of blanket thermal energy to a conventional steam cycle was assumed giving a composite net generation efficiency of 40%. Cavity and blanket functions are separated as in the wetted-wall concept, allowing coupling with a variety of blanket and power-generation systems.

Allowance for the effects of the protective magnetic field on heavy-ion beams traversing the reactor cavity using the final focusing magnets seem feasible, provided the field returns sufficiently rapidly after each target microexplosion to the undisturbed configuration. How-ver, magnetically protected reactors must operate at low residual-gas number density, corresponding to vacua of 10^{-5} to 10^{-4} torr to prevent uncompensated deflection of ions in the driver beams by reactor magnetic fields due to differential neutralization or stripping (lower or higher charge states).

SELECTED REACTOR-RELATED HIFSA RESULTS

The dependence of total cost of electricity (COE) on pulse repetition rate for the four HIFSA reactor plant concepts is shown in Fig. 1 for representative values of other plant operating parameters. Note the large repetition-rate range (with correspondingly large target yield range) over which the granular-wall, wetted-wall, and magnetically protected concepts are near-optimum. The magnetically protected concept suffers from large reactor size and the liquid-metal-jet concept from repetition-rate limitation. Potential cost savings by higher-efficiency electric power generation and reduction of safety systems is indicated by the COE differences for the generic granular-wall and wetted-wall concepts.

Only COE's for single-reactor configurations are given because they gave lower COE's than multi-reactor plants up to 1500 MWe. COE's for only one target type are presented because the differences were surprisingly modest for substantial differences in gain relationships, except for direct-drive targets whose uniform illumination requirement drives up COE more.

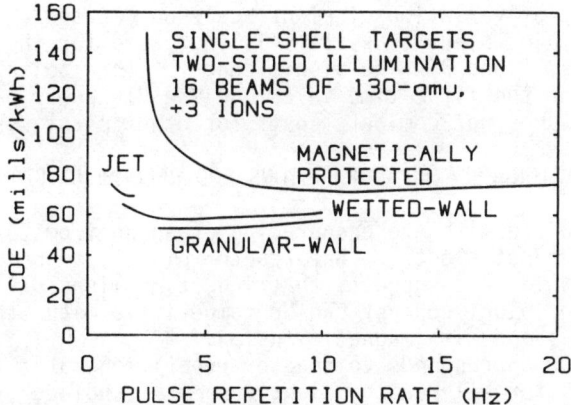

Fig. 1. COE as function of repetition rate and reactor-plant type for 1000-MWe induction-linac HIF power plant.

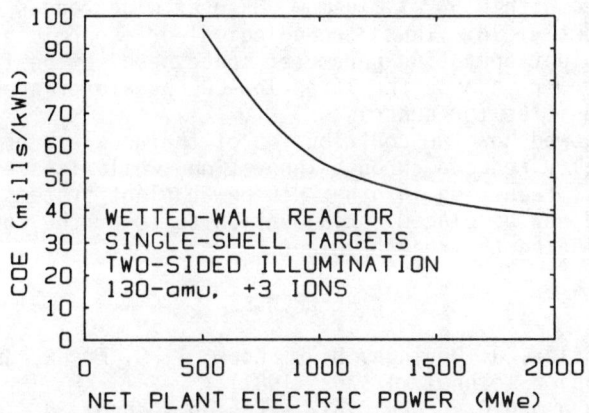

Fig. 2. COE as function of plant size for single-reactor induction-linac HIF power plants with single-shell targets illuminated from two sides by 130-amu, +3 ions.

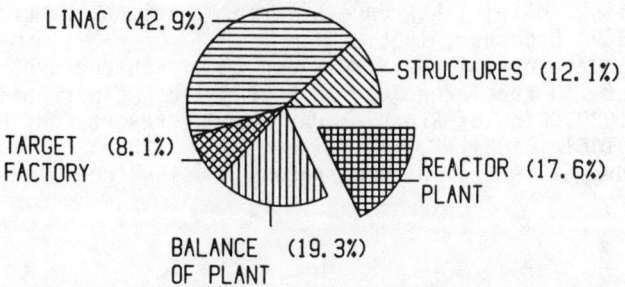

Fig. 3. Capital-cost breakdown for 1000-MWe induction-linac HIF power plant with single 5-Hz wetted-wall reactor and single-shell targets illuminated from two sides by 16 beams of 130-amu, +3 ions (containment under STRUCTURES).

The effect of variation of plant scale on COE and a capital cost breakdown for major plant subsystems is indicated for the wetted-wall concept in Figs. 2 and 3. The significance of the capital cost breakdown is that for the first time in a 1000-MWe HIF power plant, the accelerator cost does not dominate costs for other plant systems.

REACTOR-RELATED CONCLUSIONS AND RECOMMENDATIONS

Minimum COE results are encouraging, ranging from 55 to 75 mills/kWh for 1000-MWe-net-electric, one-reactor plants, to below 45 mills/kWh for 1500 MWe. This suggests that induction-linac HIF with more than one reactor plant concept can be competitive with other HIF and ICF technologies and with magnetic fusion.

Minimum COE corresponds to reactor repetition rates < 10 Hz for 1000-MWe plants for all reactor plant concepts studied. This result implies, contrary to pre-HIFSA expectations, that large accelerator pulse repetition rates are required for low COE only for very large plants.

COE's are within 5% of minimum COE over wide ranges of repetition rate and target yield. Thus, technological or economic problems in one part of plant operating parameter space need not be fatal for commercial HIF power because the large low-COE parameter space provides other opportunities for success.

HIFSA showed how the contribution of the accelerator to COE could be substantially reduced through innovation, while less emphasis was placed on cost reduction in other HIF power plant systems. More emphasis should now be placed on innovation to lower the cost of other systems, including the reactor plant.

REFERENCES

1. M. J. Monsler, J. Hovingh, D. L. Cook, T. G. Frank, and G. A. Moses, Fusion Technol. $\underline{1}$, 302 (1981).
2. W. J. Hogan and G. L. Kulcinski, Fusion Technol. $\underline{8}$, 717 (1985).
3. B. Badger, et.al., University of Wisconsin report UWFDM-450/Kfk-3202 (June, 1981).
4. Nagoya University report IPPJ-663 (January, 1984).
5. J. H. Pitts, Fusion Technol. $\underline{8}$, 1198 (1985).
6. J. A. Blink, et al., Lawrence Livermore National Laboratory report UCRL-53559 (December, 1985).
7. J. H. Pendergrass, T. G. Frank, and I. O. Bohachevsky, Proc. 4th Top. Meet. on the Technology of Controlled Nuclear Fusion (October 14-17, 1980, King of Prussia, PA), US DOE report Conf-801011, V.2, p. 1131 (July, 1981).
8. J. B. Cornwell and J. H. Pendergrass, Fusion Technol. $\underline{8}$, 1861 (1985).

INERTIAL CONFINEMENT - CONCEPT AND EARLY HISTORY

J. G. Linhart Department of Physics
University of Ferrara
I-44100 Ferrara, Italy

ABSTRACT

The concept of inertial confinement is linked to the general theme of energy compression and staging. It is shown how it arose from the ideas and experiments on dynamic pinches towards the end of the fifties and how the important key concept of a liner was further developed during the sixties. The various attempts at driving liners to speeds in excess of 1 cm/μs are reviewed in chronological order, mentioning the important impetus given to this field by the consideration of laser as a driver. It is concluded that the field of inertial confinement fusion (ICF) is becoming ever richer in possibilities, and the understanding of the physics of high-energy density has reached now a satisfactory level.

THE YEARS 1956 - 1958

When one looks forward to an exciting future, as it is the case today with ICF, it may appear to be a waste of time to talk about the past. There is, however, some merit in making a brief survey of the last 30 years or so. It enables us to see the main concepts of ICF more clearly than when we struggled to solve some specific problem and had no time to see that our particular solution to that problem was a part of a more general group of ideas.

I am not sure that I will succeed in this talk to trace such a detached and clear vision - partly because I was, and still am, a partisan of some of the approaches to ICF, partly because I am standing too near to the trees to be able to see the forest. However, I will try and I hope that others, later on, will correct and complete the picture I am presenting to you now.

In the history of science there are periods of more than usual ferment, periods when the scientific community enters a sort of "Oktoberfest" and the members of the community are allowed to dream somewhat crazy dreams without being ostracised.

One such period was the years 1956-1958. It was marked by the 1956 conference on accelerators and by the 1958 conference on peaceful uses of atomic energy, both taking place in Geneva. These conferences evolved in an extraordinary atmosphere of optimism, of new daring ideas encouraged and kindled by such people as Veksler,

Budker, Adams, Christofilos, Tuck and many, many others. I was then working at CERN and had the good fortune to encounter most of these brilliant physicists and engineers.

The majority of the papers on CTR read at the 1958 conference on peaceful uses of atomic energy were devoted to the theory and experiments on the magnetic confinement of plasmas. There were very few remarks on fusion energy from unconfined devices. Edward Teller bravely mentioned the Project Plowshare, and James Tuck (and to some extent Artsimovich) devoted a few "back-of-the-envelope" calculations to fusion from dynamic Z-pinches. Tuck calls it the "brute strength approach" and says:[1] "To produce a reactor in this way is somewhat dismaying. The thermonuclear energy release takes the form of an explosion equivalent to 1 ton TNT/cm."

I was greatly attracted by this task of making an "internal combustion engine" out of monstrous nuclear explosions. Since 1957 L. Ornstein and myself were playing with hollow Z-pinches, the purpose being to compress microwaves in a resonant cavity.[2] This experience with fast, hollow pinches made me believe that Tuck's conclusion was too pessimistic, and I repeated and refined his calculations.

The result was a new criterion for a zero energy-output reactor:

$$N \geqslant 3 \times 10^{22} \, r_{min} \quad (ions/cm)$$

at an optimum temperature of about 11 keV. In terms of plasma-energy the criterion reads: $W_p \geqslant 80 \, r_{min}$ (MJ/cm). Thus, e.g. for a pinch whose radius is 1 mm and length 10 cm the total plasma-energy required is 80 MJ, equivalent to 16 kg of TNT.

The criterion has been derived for a cylindrical geometry because at that time one had no idea how to concentrate, to drive, plasma into a spherical shape, whereas a thin cylinder as an end-product of a pinch implosion was conceivable.

The formula indicated that manageable energies can be obtained only if very thin pinches can be produced-the smaller the minimum radius r_{min} the better. It was known that in the ordinary dynamic pinch the achievement of very small radii was prevented by the formation of a shock-wave. We expected that such a wave will not appear in a hollow pinch. Starting with a thin cylindrical plasma-shell the final minimum radius of a pinch was shown to be substantially smaller than the thickness of the shell.

FROM SUPERPINCH AND LINER TO "INERTIAL CONFINEMENT"

A paper[3] on such a "superpinch" was published in 1960. Other workers must have come to similar conclusions, and experiments were reported later describing Z-pinches starting within a hollow cylinder of gas injected between electrodes.[4]

At the same time J. W. Mather and N. V. Filipov suggested another way in which a super pinch can produced – the dense plasma focus.[5,6] Somewhat later A. I. Morozov found yet another, apparently more efficient, method – the axial flow pinch.[7]

Although the plasma energies required for a fusion reactor based on a superpinch were not, as was originally suspected, in the range of tons of TNT, they were still rather formidable and it was, therefore, natural to seek some means for reducing them.

It was clear that no external magnetic fields or pressures could slow down the expansion of the minimum radius pinch – only some inertia coupled to the plasma column could. Cogitating in this way one arrives naturally at the concept of a liner or tamper. Such a concept is not new. There are many examples of its use in connection with both chemical and nuclear explosives. In mechanical engineering it appears in the form of hammers and anvils.

The idea of a liner as an agent for "inertial confinement" became connected with the research programme of our group in Frascati practically since the adoption by Euratom of the two main experiments: MIRAPI and MAFIN (minimum radius pinch and magnetic field intensification) in 1959. The theory and experiments associated with this programme were described in three papers at the IAEA Salzburg conference in 1961. To my knowledge, the term "inertial confinement" in the sense we understand it today, was used and described for the first time in the second of these papers.[8]

The introduction of a liner, imploding on a hot plasma improves the prospects of an unconfined reactor in two ways:

1) The expansion of the hot plasma is slowed down by a factor $x = [(M_L + M_p)/M_p]^{1/2}$, the inertial confinement factor. This represents the tamping role of the liner.

2) The implosion of the liner concentrates the energy of an energy source such as a condenser bank, transferring a portion of it to the plasma target. The liner is acting not only as a tamper (anvil) but also as a hammer.

Using x and defining ε as the efficiency of energy transfer from some primary source to the plasma target, the criterion for a zero-energy reactor becomes

$$W_p = 80 \ r_{min} (x\varepsilon)^{-1} \quad (MJ/cm)$$

This formula can be rewritten in spherical geometry, assuming that the plasma sphere corresponds to a section of length $2r_{min}$ of the cylindrical target. Using the plasma density n (rather than the line density N) we get

$$W_p \text{ III } 10 \, (x\varepsilon)^{-3} \left(\frac{n_s}{n}\right)^2 \quad (MJ)$$

where $n_s = 5 \times 10^{22}$ ions/cm^3.

It is clear that there are two ways to minimize the energy W_p.

The first is to use a liner much heavier than the plasma-target, in which case $x \gg 1$ and the liner is used principally as a tamper. This will result in a long inertial confinement and consequently it will be necessary to insulate the plasma thermally from the liner. Some magnetic fields will be required and the liner and the target cannot have a spherical symmetry.

The second way is to aim at $n \gg n_s$. This implies enormous plasma pressures. We shall, therefore, require that our hammer exercises an equivalent pressure, i.e.

$$p = 2nkT = \rho_L v_L^2 \, .$$

The speed of the liner v_L must, therefore be comparable with the mean thermal velocity v_t of the ions in the hot plasma, i.e. 2×10^8 cm/s. In this case the liner is being used mainly as a hammer, not so much as a tamper.

During the 1960´s the relatively modest effort on ICF moved mainly along the first line of approach. Both our Frascati group and the Russians (S. G. Alikhanov et al.) concentrated on the acceleration of liners by means of intense Z-pinch magnetic fields.[9,10] In the USA similar experiments were carried out by E. C. Cnare using θ-pinch geometry. Explosives were also used in the attempt to drive liners to compress a θ-pinch (Los Alamos and Frascati), whereas A. E. Robson (NRL) proposed projects on slow, non-destructive liners.

THE MIRAPI EXPERIMENT AT FRASCATI

In order to dare to use the second way, that of the superdense plasmas, it was necessary to discover how to accelerate liners to speeds of at least 20 cm/μsec. The search in this direction was linked with the subject of acceleration of "macroparticles" or "macrons," which today has other connotations apart from the ICF one. I remember E. R. Harrison talking about macrons and fusion back in 1960 when he came for one year to CERN, but it was only in 1963 that he published a paper on an alternative approach to thermonuclear power.[12] Following this paper there was a spate of suggestions on how to achieve such hypervelocities; however, the technical difficulties appeared insuperable. Moreover, it was shown that most efforts in this direction result in the vaporisation of the projectile before the desired speed of 20 cm/μsec is reached.

However, one knew how to accelerate plasma liners to speeds even higher than the above mentioned limit. Such liners, or at least plasma sheets, were produced in experiments on fast Z-pinches and in the form of a cylindrical shell in the MIRAPI experiment which, almost by accident, demonstrated the possibility of inertial confinement.

As I mentioned before, MIRAPI was conceived as an experiment on the implosion of a hollow, thin shell of deuterium plasma. In fact, we did not see any way of how to do it using pure deuterium and finally opted to use the "dirty" $LiAlD_4$. This was used in the form of a fine powder, injected by an electrostatic diode into the main discharge chamber.[13] The powder grains were irradiated by a flash of light which produced a partial vaporisation, mainly of D_2, creating thus a continuous gaseous bridge between the electrodes. A condenser bank (50 kJ) was then switched on initiating a Z-pinch in the powder curtain.

The first thing that happened was a prepinch involving the tenuous D_2 atmosphere between the grains. This was then followed by the massive collapse of the Li, Al plasma-liner, which, in turn, compressed the D prepinch. We observed two neutron bursts - the first, a short and weak one (0,1 μsec, 10^6 neutrons) from the prepinch alone and the second one, whose duration (1 μsec) was compatible with the dynamics of the heavy plasma shell, giving somewhat less than 10^8 neutrons. This was encouraging as a demonstration of the feasibility of ICF but it did not indicate any simple way of scaling up the experiment. Above all, it did not suggest how to achieve similar implosions with solid liners.

1986 PERSPECTIVES

Those first 10 years of research on ICF had revealed the principal problem which is and will be: the compression of energy, i.e. how can one transfer energy efficiently from a source of low energy density to a high energy density target. Any such transfer results always in energy loss. If the top of such an energy density pyramid is of scientific, technological or some other interest one may be prepared to invest a large amount of energy in order to reach it.

A pyramid similar to one encountered in ICF is exemplified by particle accelerators. Concepts which are particularly relevant are those of impact, collective and stochastic acceleration. In all these processes a portion of the energy of many slow particles is converted into the energy of a few fast ones.

In the pyramids such as that of the accelerators or that of the biological evolution, the total efficiency is not of supreme importance - the important point is to obtain the desired final product. The situation is different in ICF because the final goal is the production of energy and, therefore, one cannot aim at that regardless of the energy expenditure (see the factor ε^{-3} in W_p). It follows, therefore, that our energy-transfer pyramid has to be constructed intelligently. It can be shown quite generally, that the number of energy transfers (from the source to the target) cannot be too small (e.g. it will be inefficient to transfer directly from a condenser bank to a dense D-pellet) nor can it be too large ($\varepsilon_{tot} = \varepsilon_1 \ldots \varepsilon_n$).

It is therefore important to find efficient and appropriate energy transfers, to optimise their succession and their number. The theoretical basis for such an optimum staging has been worked out only relatively recently1.[14]

The great step towards the discovery of an efficient transfer was, undoubtedly, the concept of the rocket drive of a liner.

I have mentioned the difficulties of accelerating solid liners to speeds in excess of 10 cm/sec, i.e. that the force pushing the liner was, at the same time, responsible for its vaporisation (explosion). In this case the motto "if you can't beat them, join them" appeared to be correct. It was suggested to divide the liner into two parts: ablator (mass M_a) and pusher (mass M). The ablator is exploded, attaining a jet speed v_j driving the pusher to a speed

$$v = v_j \ln\left(\frac{M_a + M}{M}\right)$$

which is the well known Tsiolkovski formula.

It is clear that the gain expressed by the logarithmic term is limited to less than 10 for reasons of efficiency and therefore, in order to get high v, the v_j itself must be high, i.e. the ablator must behave as a superexplosive. The problem is then: where does the energy for such a superexplosion come from?

The first indications that a reply to this question was forthcoming appeared during the 1969 Varenna Summer School on "Physics of High Energy Density" with the participation of E. Teller, R. Kidder, Krokhin, members of our Frascati group and many others. R. Kidder gave two lectures with the suggestive titles of "Pressures resulting from the light-induced blow-off" and "Light-supported detonations." Other contributions dealt with the rocket drive and related subjects.

Soon afterwards a series of papers was published, starting with the well known one by J. Nuckolls, et al.[15] describing the use of lasers as energy-sources (drivers) for the explosion of an ablator. The great merit of a laser as a driver of liners then was to break down the "mental" barrier of $n \gg n_s$ and later on to break this barrier experimentally.

In the meantime other drivers have been suggested - one of them, and a very promising one, the heavy ion beam. All kinds of new ideas emerged on the outer and inner structure of targets and it is also not excluded that the concept of $x \gg 1$ (i.e. the magnetised target) will bear fruit.

TABLE I

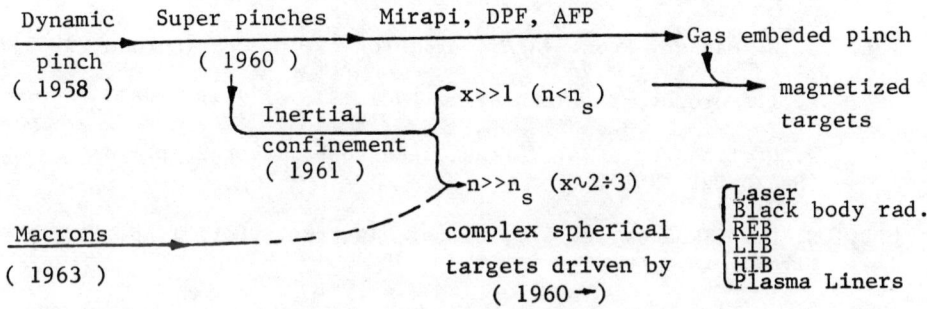

I have tried to sketch the main concepts that contributed to ICF in Table 1.

It is truly extraordinary that whilst the wealth of approaches to ICf is steadily increasing, very few have been definitively thrown on the scrap heap. Although this may appear to contribute more to confusion rather that to fusion, it has the merit, in the meantime, to make the research on ICF an adventurous and a very colourful enterprise. However, it would not be correct to finish on this frivolous note. I should rather tell you what I really think about ICF now. I have a feeling, for the first time and after having worked in this field for 30 years, that ICF is rather near the reactor-demonstration, nearer than some believe and nearer than some are allowed to admit.

The situation resembles a chess problem. All the necessary chess pieces are here - we know the number of energy transfers (stages) will be four or five.

So it's a mate in four - it's up to you to find the most elegant solution!

References

1. J. L. Tuck, Proc. 2nd Int. Conf. on Peaceful Uses of Atomic Energy, Vol. 32, p. 3 (Geneva 1958).

2. J. G. Linhart and L. Th. Ornstein: Compression of radiation fields by a magnetically driven plasma-shell, Proc. 4th Conf. Ion. Phen. in Gases, p. 774 (Uppsala, 1959).

3. J. G. Linhart, Il Nuovo Cimento, Vol. 17, p. 850 (1960).

4. O. A. Bazilevskaya, et al., Proc. 5th Int. Conf. Ion. Phen. in Gases, Vol. 2, p. 2213 (1962).

5. A. M. Andrianov, et al., Proc. 2nd Int. Conf. on Peaceful Uses of Atomic Energy, Vol. 32, p. 348 (Geneva, 1958).

6. J. W. Mather, Proc. IAEA Conf., Vol. 2, p. 389 (Culham, 1965).

7. A. I. Morozov, Zh. Tekh. Fiz., Vol. 37, p. 2147 (1967).

8. J. G. Linhart, et al., Proc, IAEA Conf., Part 2, p. 733 (Salzburg, 1961).

9. S. G. Alikhanov, et al., J. Sci. Instrum., Vol. 1, p. 543 (1968).

10. J. G. Linhart and G. Schenk, Exploding Wires, Vol. 3, p. 223 (1964).

11. E. C. Cnare, J. Appl. Phys., Vol. 37, p. 3812 (1966).

12. E. R. Harrison, Phys. Rev. Lett., Vol. 11, p. 535 (1963).

13. Ch. Maisonnier, et al., Proc. IAEA Conf., Vol. 2, p. 345 (Culham, 1965).

14. J. G. Linhart, Laser and Particle Beams, Vol. 2, p. 87 (1984).

15. J. Nuckolls, et al., Nature, Vol. 239, p. 139 (1972).

TARGETS FOR LASER AND ION BEAM DRIVERS*

Roger O. Bangerter
Lawrence Livermore National Laboratory, Livermore, CA 94550

ABSTRACT

At the two previous heavy ion fusion symposia, researchers from Livermore presented their best estimates of target energy gain. The results presented at Tokyo differed significantly from those presented at Darmstadt. The Livermore estimates were again revised for this symposium. The new estimates are given in an accompanying paper by Lindl et al. and in additional detail in this paper. The new estimates are similar to the results presented at Darmstadt. The implications of the new results are discussed.

INTRODUCTION

Recent results on targets for laser and ion beam drivers were reported by J. D. Lindl on the first day of this Symposium.[1] Lindl's talk generated some discussion during the remaining days of the Symposium because:

1. The new Livermore gain curves for lasers and ions are lower than the Livermore curves presented at the 1984 Symposium in Tokyo.[2]

2. Lindl emphasized that high target gain requires high irradiance.

The new gain curves were not available in time to be used in the Heavy Ion Fusion Systems Assessment (HIFSA),[3] which was one of the main topics reported at the Symposium. It is not suprising that this situation has already led to discussion regarding the validity of the HIFSA Study in light of the new gain curves. The emphasis on high irradiance is interesting and important because high irradiance imposes stringent requirements on the quality of the beam emerging from the driver.

The remainder of this paper gives a discussion of the new gain results and their implications and also some additional discussion on the importance of high irradiance.

---

*Work performed under the auspices of the U.S. Department of Energy by the Lawrence Livermore National Laboratory under contract number W-7405-ENG-48.

ENERGY AND POWER REQUIREMENTS

In Refs. 1 and 2, target gain is given as a function of driver energy and a variable $r^{3/2}R$ where r is focal spot radius (cm) and R is ion range (g/cm^2). This representation of gain is an approximation that is valid in a restricted domain of r given by $0.1 \leq rE_D^{-1/3} \leq 0.2$ where E_D is driver energy (MJ). To a poorer approximation, the domain of applicability may be extended to $rE_D^{-1/3} \leq 0.33$. To eliminate this approximation and restriction gain must be given as a function of the three independent variables E_D, r, and R. Figure 1 gives new gain curves as a function of these three variables. These curves are based on the same calculations as those in Lindl et al.[1]

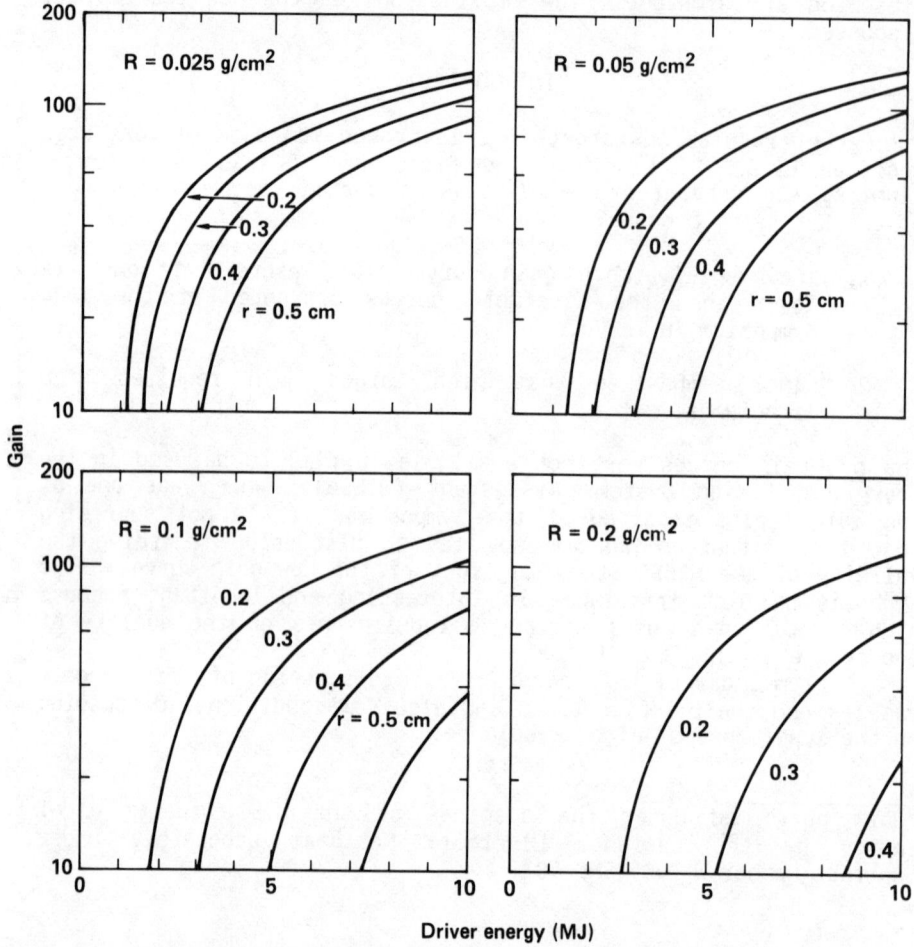

Fig. 1. Target gain as a function of driver energy for a variety of focal spot radii and ion ranges.

The peak power required by the targets is given in Fig. 2. The results given in these figures are based on single-shell targets. The targets must be illuminated from two sides, and they require an accurately shaped pulse. Double-shell targets have lower power requirements,[2] but the physics uncertainties are larger and target fabrication is more difficult. The Livermore curves for double-shell targets have not been revised.

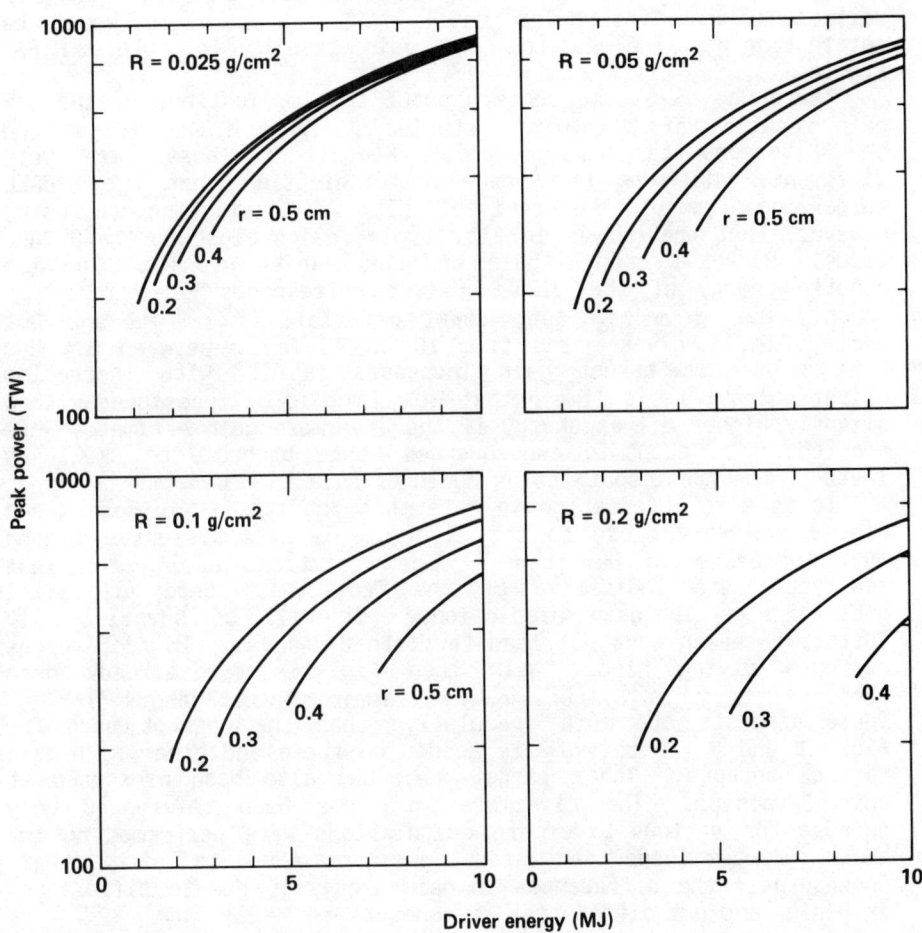

Fig. 2. Peak power requirement as a function of driver energy for a variety of focal spot radii and ion ranges.

IMPLICATIONS

In order to determine the implications of the new gain results for the economics of heavy ion fusion, the economic analysis presented at the Symposium[4] was redone by Wayne Meier using the new gain curves. The cost of electricity increased slightly less than 10%.[5] Meier's analysis fixed the product $r^{3/2}R$ at 0.02; the HIFSA study did not fix $r^{3/2}R$. Because of this additional flexibility, the HIFSA model may be even less sensitive to gain variations than the Meier model. In any case it would be interesting to include the new gain results in future HIFSA studies.

There was also some concern about the implications of the new gain results for the HIBALL studies[6,7]. The HIBALL target and the Livermore targets on which Fig. 1 is based are very different. No work has been done indicating that the HIBALL target gain should be revised. It is interesting to note, however, that the curves in Fig. 1 give gains close to the HIBALL value. HIBALL-I used a gain of about 80 at 4.8 MJ. The ion kinetic energy of the HIBALL driver corresponds to a range of roughly 0.1 g/cm^2 in hot, low-Z material. At 5 MJ and 0.1 g/cm^2, Fig. 1 gives gains from 10 to 70 for r between 0.4 and 0.2 cm. Since target gain increases rapidly with increasing driver energy, it is likely that HIBALL could be redesigned with a slightly higher driver energy if the Livermore gain estimates were adopted. The economic consequences would probably be small--at least if the focal spot radius could be held to $\leq$ 0.3 cm.

It is also interesting to note that the results in Ref. 1 and Fig. 1 are very close to the Livermore results presented at the 1982 Symposium in Darmstadt.[8] The variations during the last few years are indicative of the fact that there are still uncertainties in gain predictions. It will be surprising if further research does not lead to further changes. In fact target concepts giving higher gain than Fig. 1 have already been presented at this Symposium by Livermore investigators.[9,10] These concepts are more speculative than the concept used for Figs. 1 and 2. More work is needed on these and other high-gain target concepts. Other target work has also been presented at this Symposium. The Livermore work has been referenced only because the various Livermore calculations were performed on the same computer code and are thus normalized to each other. Consequently the differences in gain are truly due to differences in design and not differences in codes.

In summary, the curves in Fig. 1 give the current Livermore best estimates of target gain. Additional work on more speculative target concepts may lead to higher estimates in the future.

IRRADIANCE

The need for high irradiance is not new. Consider the following example. Figure 2 gives a power requirement of 400 TW for $E_D = 5$ MJ, $r = 0.3$ cm, and $R = 0.1$ g/cm^2. For two-sided illumination, the irradiance is 7.1×10^{14} W/cm^2. Reference 2 gives about 430 TW and 7.6×10^{14} W/cm^2 for the same E_D, r, and R. Thus, estimates of required irradiance have changed little in recent years. It does seem important, however, to emphasize the rapid decrease in gain (Figs. 1 and 2) with decreasing irradiance (increasing r) particularly for $R \gtrsim 0.05$ g/cm^2.

LASER TARGETS

The gain curves for lasers have also been revised (Ref. 1). Both gain and irradiance are also issues for lasers. A reduction in gain is more serious for lasers than for accelerators because the lower efficiency of lasers requires higher target gain in order to obtain an acceptably low recirculating power fraction.

Beam quality is an issue for lasers because, in some reactor designs, the final optical elements are placed 30 to 100 m from the target to minimize damage. For some high-gain laser targets, r should be < 1 mm, so that the angular divergence of rays emanating from any point on the surface of the final optical elements must be < 10^{-5} rad. This is an important constraint, particularly for lasers operating at the high average power needed for a power plant. High average power inevitably leads to nonuniform temperature, density, and stress. These nonuniformities will have to be carefully controlled to avoid excessive beam distortion.

CONCLUSIONS

In conclusion, new best-estimate gain curves have been presented at this Symposium. The new results are expected to have little effect on the economics of heavy-ion fusion principally because the high efficiency of accelerators endows heavy ion fusion with a relative insensitivity to modest variations in target performance.

Irradiance remains an important issue for inertial confinement fusion. Since achievable irradiance is closely related to beam emittance, the increased understanding of emittance growth reported at this Symposium is gratifying.

REFERENCES

1. J. D. Lindl, R. O. Bangerter, J. W-K. Mark, Y-L. Pan, elsewhere in these Proceedings.

2. J. D. Lindl and J. W-K. Mark, Proceedings of the 1984 INS International Symposium on Heavy Ion Accelerators and their Applications to Inertial Fusion, Tokyo, Jan 13-27, 1984, p. 629.

3. Donald J. Dudziak and W. B. Herrmannsfeldt, and other papers referenced therein, elsewhere in these Proceedings.

4. Wayne R. Meier, William J. Hogan, and Roger O. Bangerter, elsewhere in these Proceedings.

5. Wayne R. Meier, private communication.

6. HIBALL-I "A Conceptual Heavy Ion Beam Driven Fusion Reactor Study," University of Wisconsin Fusion Technology Institute Report UWFDM-450, Kernforschungszentrum Karlsruhe Report KfK-3702 (1981).

7. HIBALL-II "An Improved Heavy Ion Beam Driven Fusion Reactor Study," University of Wisconsin Fusion Technology Institute Report UWFDM-625, Kernforschungszentrum Karlsruhe Report KfK-3840, Fusion Power Associates Report FPA-84-4 (1984).

8. J. W-K. Mark, Proceedings of the Symposium on Accelerator Aspects of Heavy Ion Fusion, Gesellschaft für Schwerionenforschung Report GSI-82-8 (1982), p. 454.

9. James W-K. Mark and Yu-Li Pan, elsewhere in these Proceedings.

10. James W-K. Mark and John D. Lindl, elsewhere in these Proceedings.

PRESENT STATUS AND FUTURE PROSPECTS
OF GAS LASERS AS DRIVERS FOR ICF

Yoshiaki Kato
Institute of Laser Engineering, Osaka University
Suita, Osaka 565, Japan

ABSTRACT

The KrF laser would be a very suitable driver for ICF if a high power, high efficiency system could be developed. Various advances that have been made recently are reviewed. Significant progress is expected in near future.

INTRODUCTION

Theoretical and experimental investigations have shown that laser radiation of shorter wavelength is absorbed in plasmas at higher density, resulting in better absorption efficiency, less hot electron production, high hydrodynamic efficiency, and better conversion to thermal x-rays in high Z targets. These are all favorable conditions for ICF using direct-drive targets as well as indirect-drive targets. Therefore in this review among various gas lasers, we discuss on the KrF laser which is a high efficiency, short wavelength laser (λ=248 nm) with the potential capability of scaling to a high power system.

According to the recent study by Gardner and Bodner[1] on the optimization of direct-drive targets, hydrodynamic efficiency as high as 15% can be attained by irradiation with a 1/4 micron laser. When a coupling efficiency (product of absorption efficiency and hydrodynamic efficiency) of 10% is attained, we can expect a pellet gain of G=200 in the ideal case with a laser energy of $\sim$2.5 MJ.[2,3] If we limit the recirculating power in an ICF reactor to $\sim$10% of the total power, the driver efficiency has to be $\eta_D \gtrsim 5\%$.

The pulse width required to compress a direct-drive, single-shell target of $\sim$4mm diameter with a laser energy of a few MJ is estimated to be $\sim$5 ns.[4] A longer laser pulse, thus lower laser power, could be used with direct irradiation of a double-shell target.[3]

In the case of direct-drive targets, illumination uniformity has to be strictly controlled. Illumination nonuniformities arising from wavefront aberration and intensity nonuniformity of the incident laser beam can be controlled by randomizing the wavefront[5] (Fig. 1) and furthermore eliminating the spatial coherence[6] of the laser beam. Experimental verification of these proposals could be made by using the short wavelength, symmetric irradiation facilities such GEKKO XII at ILE, Osaka University and OMEGA at LLE, University of Rochester. In the case of indirect-drive targets, the requirement for laser beam control is significantly reduced. However more laser energy is required to attain the same pellet gain in comparison to the direct-drive target.

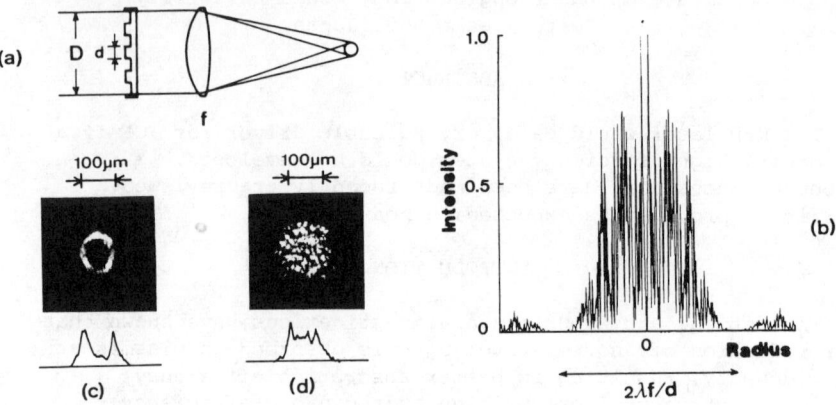

Fig. 1. Random phasing of laser beam for uniform target illumination.

HIGH POWER KrF LASER SYSTEM DESIGN

As an ICF driver, the KrF laser has various advantages such as short wavelength (good target coupling), gas laser medium (high repetition rate and low nonlinear refractive index), broad gain bandwidth (suitable for generating incoherent radiation for direct-drive target) and high intrinsic efficiency (10%).

The major disadvantage is that the KrF laser is not an energy storage device due to the short life time of the excited state (<10 ns). Therefore laser energy has to be continuously extracted during the entire pumping duration. The pumping duration is determined by the optimization of the pumping density in the laser medium and the electrical characteristics of the electron beam generator. These lead to a pumping duration of longer than 100 ns. Therefore novel methods are required to produce a short pulse of $\lesssim 10$ ns without loosing the high efficiency of the KrF laser amplifier.

Extensive study was undertaken at Lawrence Livermore National Laboratory (LLNL) to evaluate the KrF laser as a fusion driver, and a conceptual design of a 1.5 MJ, 2 Hz KrF fusion laser system was reported in 1980.[7] For generating a short laser pulse, a combination of pulse stacking[8] and backward Raman pulse compression[9] was invented. The basic design architecture is shown in Fig. 2.

According to this system design, the overall system efficiency is 3.0% and the cost is $250/J of laser output in a 20 ns pulse. A few technical developments such as long shot-life switches ($\gtrsim 10^7$ shots) and high damage threshold optical coatings ($4\sim 6$ J/cm^2) are needed before the system can be built as specified.

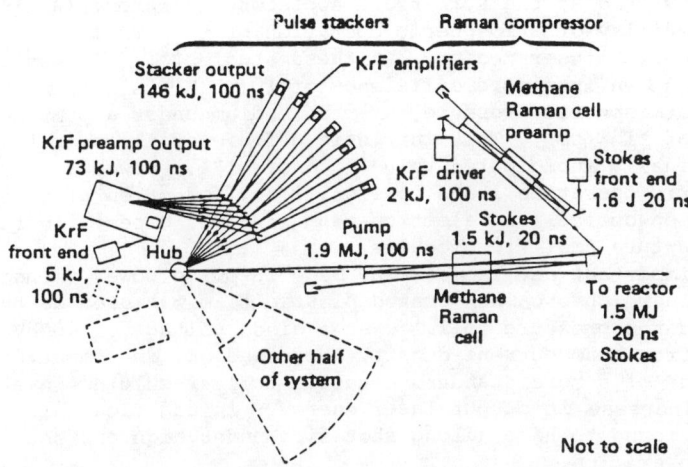

Fig. 2. Architecture of one-half of the hybrid KrF pulse-stacker/Raman compressor laser design[7].

RECENT PROGRESS IN HIGH POWER KrF LASERS

We can use the LLNL design as the reference system. However it was based on various uncertainties which were unknown at that time. Research has been conducted up to the present in order to clarify the critical aspects of the high power KrF laser system. Furthermore a multi-kilojoule, 5 ns KrF laser system AURORA is being constructed at Los Alamos National Laboratory (LANL).[10] In this section we review some of the recent progress made at various laboratories.

I. KrF LASER AMPLIFIER

There are two important aspects in amplifier design. The first is the molecular kinetics to define the operation condition and the second is the pulse power technology to maximize the efficiency.

The KrF laser amplifier is characterized by high gain, low saturation intensity, and limitation in the maximum intensity due to saturated loss in the laser medium. Output energy of 850 J in 150 ns was obtained with 100 mJ input (stage gain 8000) from an amplifier gain volume of 30x40x120 cm^3 at LLNL.[11] A mixture of 95% argon, 5% krypoton and 0.3% fluorine with a total pressure of 2 atm was used, resulting in an intrinsic efficiency of ~6%. This value should be compared with the kinetic code predictions which average ~7.5% for this case.

Recently detailed and careful investigations were undertaken at the Institute for Laser Science (ILS), Chofu University, and at the Rutherford Appleton Laboratory (RAL) on the feasibility of atmospheric operation of the KrF laser with a krypton-rich laser medium. In these studies[12,13] it was proven that an intrinsic efficiency of 10(~12%) is obtained with an atmospheric-pressure, Kr-rich medium under a pumping density of $\gtrsim 1MW/cm^3$. This intrinsic efficiency is lower than the originally anticipated value[14,15] of ~17%. However atmospheric operation leads to significant reduction of stresses on optical and electron-beam windows, especially for large aperture amplifier modules.

An important development was made in pulse power technology at ILS, in which a carbon-coated plastic film was used as the anode and the pressure foil.[16] At a diode voltage of 400kV, the electron beam current density increased 38% in comparison to the use of a more standard metallic foil, resulting in a 3-fold increase in output laser energy. In addition this foil was found to have a long shot life under high current denstiy operation.

II. PULSE COMPRESSION

Generation of a short pulse by angular-multiplexed pulse stacking requires a complex optical system and is susceptible to inducing a large amount of spontaneous emission before the main pulse. Pulse compression with nonlinear optical techniques such as Raman amplification and stimulated Brillouin scattering (SBS) reduces the complexities in the optics and generates short pulses of high beam quality. However various factors such as energy conversion efficiency and pulse compression ratio have to be examined to implement these concepts into the system design.

Backward Raman Amplification was studied in detail at LLNL.[17] It was demonstrated that 75%-85% of the pump photons which may have low optical quality are converted to the Stokes photons of high beam quality. The intensity gain of the Stokes pulse is 2-3, which is limited by the growth of the second Stokes pulse.

Forward Raman amplification, which has been studied at RAL, is viewed as an optical beam combiner where the many pump beams are combined into a single Stokes beam of controlled beam quality.[18] Up to 56 pump beams confined in a light guide were combined into a single beam with a power conversion efficiency of 70%. However in this case intensity gain, thus pulse compression can not be achieved.

Another pulse compression technique, studied at the University of Alberta,[20] is backward Brillouin scattering. Since the SBS process is not limited by the growth of the second Stokes amplification, a large compression ratio is attained compared with backward Raman amplification. With the backward Brillouin amplification in SF_6 gas, a 390 ps pulse was generated from a 24 ns pump pulse with an energy extraction efficiency of 40%.[21]

Recently a very innovative device was developed at ILS[22], which might lead to a new design concept of the KrF laser system. At short wavelength, the Verdet constant which is inversely proportional to λ^2 becomes ~10 times larger than the value at visible wavelength. In fact it was found that H_2O is a very good material for Faraday rotation at 1/4 micron. As shown in Fig. 3, the Faraday rotator was placed in the cavity to produce a laser pulse of alternating polarization in the cavity by injection locking. By developing a fast-respose Faraday rotator, it could be used to switch-out a single pulse of short duration which has efficiently extracted the laser energy during long pumping duration.

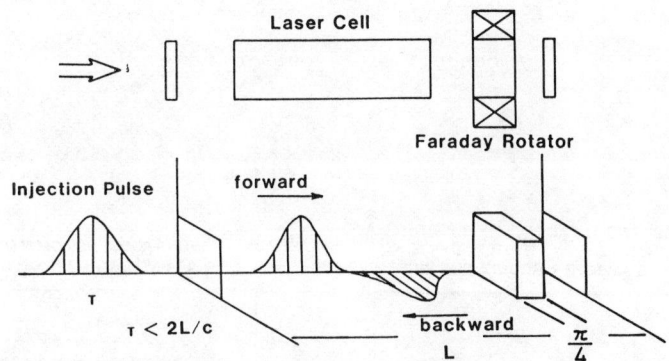

Fig. 3 Polarization-rotated energy extraction with an intra-cavity Faraday rotator

FUTURE PROGRESS

One fo the key factors which determine the future prospects of the KrF laser may be the AURORA system being developed at LANL.[10] The main design feature of AURORA is to test angular multiplexing for generating multi-kilojoule laser pulses of 5 ns duration as shown in Fig. 4. Since this system employs various new aspects in pulse power technology, optics design, beam propagation, and laser amplification, it will provide very important and useful data for future design of high power KrF laser systems.

Recently a new design proposal for a single pulse test facility using a 2-4 MJ KrF laser was presented by LANL.[23]

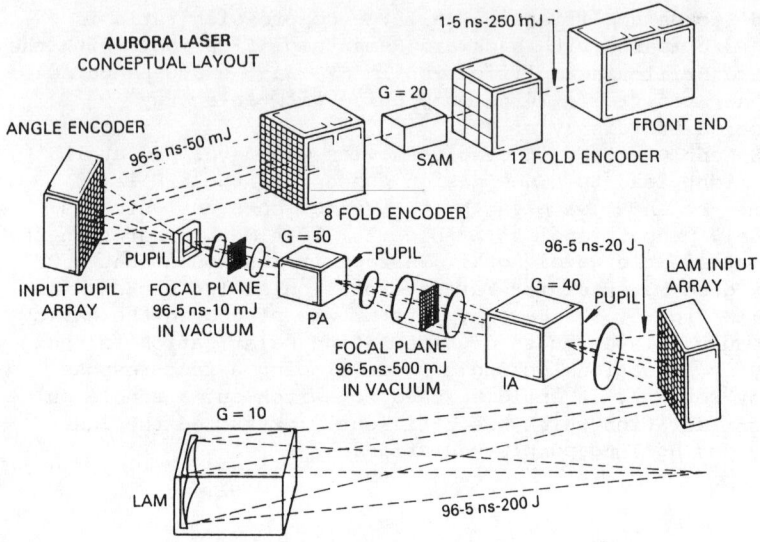

Fig. 4 A conceptual layout of Aurora laser

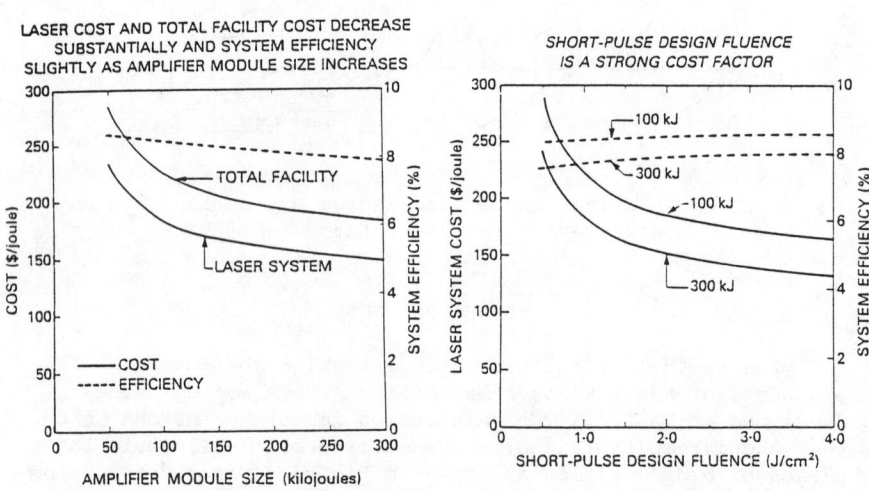

Fig. 5
Laser system cost and total facility cost vs. size of the amplifier module.

Fig. 6
Laser system cost vs. optical fluence design.

The design goal includes a total system efficiency of 7-10% and a total facility cost of ≲$200/J. In this design a high intrinsic efficiency of 18% is assumed by using a Kr-rich laser medium. This efficiency might be optimistic based on the recent results quoted in the previous section[12,13]. Fig. 5 and 6 show the dependence of the system cost on the amplifier module size and the short pulse design fluence. Developments in the large-aperture optics and high damage-threshold coating[24,25] are clearly needed. In addition, control of parastic oscillation and amplified spontaneous emission in the large aperture amplifiers will become a very important issue.

Although almost ten years have elapsed since the KrF laser was called a new laser, still KrF is a new laser in the sense that there are various unknown factors and new ideas are evolving to improve the performance. Considering the importance of developing a reliable and high performance short wavelength laser for ICF, it would be desirable to advance steady and innovative research for developing a high power KrF laser system.

ACKNOWLEDGEMENT

The author is indebted to Prof. H. Takuma. Prof. A.A. Offenberger, Dr. J.R.Murray, Dr. D.C.Cartwright and Dr. M.J.Shaw for sending invaluable research reports for preparation of this review.

REFERENCES

1. J.H.Gardner and S.E.Bodner, Phys.Fluids, to be pulbished.
2. J.Mayer-ter-Vehn, Nucl.Fusion, 22, 561(1982).
3. J.D.Lindl and J.W-K.Marks, Laser and Particle Beams, 3, 37(1985).
4. H.Takabe, private communication.
5. Y.Kato, K.Mima, N.Miyanaga, S.Arinaga, Y.Kitagawa, M.Nakatsuka and C.Yamanaka, Phys.Rev.Lett. 53, 1057(1984).
6. R.H.Lehmberg and S.P.Obenschain, Opt.Commun. 46, 27(1983).
7. J.Caird, W.O.Allen, H.G.Hipkin, J.Benford, Y.G.Chen, D.Dakin, G.Frazier, T.Naff, T.S.T.Young, M.R.Flannery, G.M.Perron, S.N.Suchard, D.E.Vandenburg, E.V.George, R.A.Hass, W.F.Krupke, M.J.Monsler, L.Pleasance and L.Seppala, Lawrence Livermore National Laboratory, UCRL-53077, 1980(unpublished).
8. J.J.Ewing, R.A.Hass, J.C.Swingle, E.V,George and W.F.Krupke, IEEE J.Quantum Electon., QE-15, 368(1979).
9. J.R.Murray, J.Goldhar, D.Eimerl and A.Szöke, IEEE J.Quantum Electron., QE-15, 342(1979).
10. L.A.Rosocha, P.S.Bowling, M.D.Burrows, M.Kang, J.Hanlon, J.McLeod and G.W,York, Jr., Laser and Particle Beams, 4, 55(1986).
11. J.Goldhar, K.S.Jancaitis and J.R.Murray, CLEO'84, June 21,

1984.
12. F.Kannari, M.J.Shaw and F.O'Neill, J.Appl.Phys., submitted.
13. K.Ueda, H.Nishioka and H.Takuma, Fusion Technology, submitted.
14. F.Kannari, A.Suda, M.Obara and T.Fujioka, Appl.Phys.Lett. $\underline{45}$, 305(1984).
15. E.T.Selesky and W.D.Kimura, IQEC'84, June 18-21, 1984.
16. K.Ueda, H.Nishioka and H.Takuma, Proc.Internat.Conf. Lasers'84, P.598(1984).
17. J.Goldhar, M.W,Taylor and J.R.Murray, IEEE J.Quantum Electron., $\underline{QE-20}$, 772(1984).
18. J.P.Partanen and M.J.Shaw, J.Opt.Soc.Am.B, submitted.
19. M.J.Shaw, J.P.Partanen, Y.Owadano, I.N.Ross, E.Hodgson, C.B.Edwards and F.O'Neill, J.Opt.Soc.Am.B, submitted.
20. A.A.Offenberger, D.C.D.McKen and R.Fedosejevs, Final Report to Energy Resources Fund, June 1984.
21. R.Fedosejevs and A.A.Offenberger, IEEE, J.Quantum Electron., $\underline{QE-21}$, 1558(1985).
22. K.Ueda, H.Nishioka, H.Hisano, T.Kaminaga and H.Takuma, Rev.Laser Engineering, $\underline{13}$, 805(1985) (in Japanese).
23. R.J.Jensen, et al., Laser and Particle Beams, $\underline{4}$, 3(1986).
24. D.Milam, W.H.Lowdermilk, J.G.Wilder and I.M.Thomas, CLEO'84, Technical Digest THB3, 136(1984).
25. K.Yoshida, H.Yoshida, Y.Kato and C.Yamanaka, Appl.Phys.Lett. $\underline{47}$, 911(1985).

PRESENT STATUS AND FUTURE PROSPECTS FOR DIRECT DRIVE LASER FUSION

Stephen E. Bodner
U.S. Naval Research Laboratory, Washington, D.C. 20375

ABSTRACT

If one assumes that the best short wavelength laser will have an efficiency of 5-7%, and if one assumes that reasonable cost electricity requires that the product of laser efficiency and pellet gain be greater than 10-15, then pellet gains for laser fusion must be at least 150- 300. The only laser fusion concept with any potential for energy applications then seems to be directly driven targets with moderately thin shells and 1/4 micron KrF laser light. This direct drive concept has potential pellet energy gains of 200-300.

Although heavy ion reactor chambers can use two narrow incoming beams, all existing laser fusion concepts require a multiplicity of incoming beams. This excludes most existing reactor chamber designs. But an integration of the best features of the Solase and Sirius chamber concepts[1] seems very attractive. With dry walls made of aluminum plus a graphite liner, there would be negligible long-term radioactivity. The main problem with laser fusion reactor chambers may be the coating on the last mirror; it is easily damaged by neutrons.

INTRODUCTION

The direct drive concept is inherently attractive for two reasons. Since the laser energy is deposited directly on the pellet, one energy conversion step has been eliminated, and therefore it could in principle be more efficient. Second, because the direct drive pellet consists of just the pellet, there is less physical complexity and perhaps fewer modes of failure. But historically the direct drive pellet concept has been a high risk approach because of the difficulty in achieving sufficient symmetry of illumination. Without good laser illumination symmetry on the pellet, the energy gains are low.

We have now solved this problem in principle with an optical technique called "ISI", or induced spatial incoherence, that can produce the very extreme laser smoothness that is needed for direct drive laser fusion.[2] We have recently shown that this ISI technique also controls various plasma instabilities.[3] We have also found with our computer codes and some supporting experiments that the growth rate of the Rayleigh-Taylor instability is much less than the classical value when one uses short 1/4 micron laser light, because of the high mass ablation rate.[4] The control of this fluid instability means that we can design pellets with thinner shells. And finally we have found that the overall

coupling efficiency of laser energy to imploding fuel can be
fairly high, of order 10%, when one uses a short laser wavelength
and a moderately thin pellet shell.[5] As a result the net pellet
gain can be quite high, of order 200-300 using Meyer-Ter-Vehn's
analytic model for pellet gain.

With these recent breakthroughs, direct drive laser fusion
may finally reach its original potential for civilian energy
applications. It then becomes worthwhile to reevaluate the
engineering feasibility of the various reactor chamber concepts.
The following sections will outline these physics advances of
symmetry, preheat, stability, and efficiency, followed by some
comments on reactor design.

SYMMETRY

The ISI concept is based on the realization that laser light
can be focused to the size of a few wavelengths, while high gain
pellets will be several millimeters in radius. The ISI concept
takes advantage of this fact by trading off the ability to focus
light for better beam quality.[2] Fig. 1 illustrates the concept

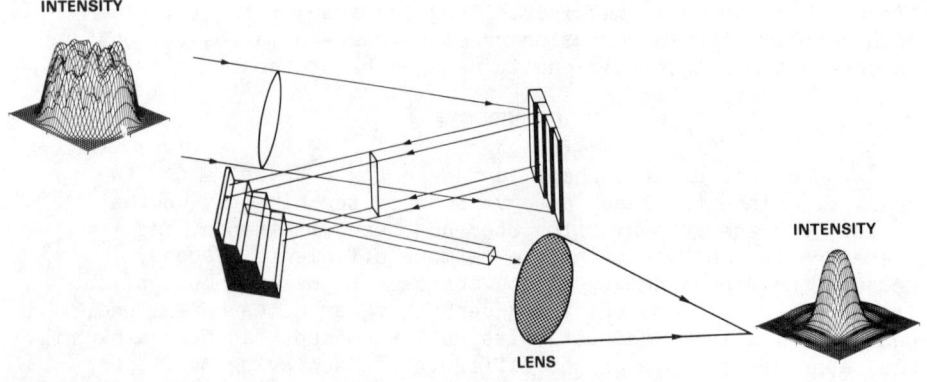

Fig 1. ISI mirror echelons

Starting with a moderate quality laser beam, we divide it
with mirrors into a large number of much smaller beamlets
(typically a few hundred using a 20x20 array of mirrors). If the
original beam is divided into enough small beamlets, then each one
will be nearly perfect in beam uniformity. Each of these beamlets
can then be made statistically independent of the others if the
time delay produced by the mirrors is longer than the laser's
coherence time. (Normally one thinks of a laser as being

perfectly coherent, but it is easy to increase the bandwidth of a KrF laser and thereby reduce the coherence time to about one psec.)

When the beamlets are focused with a lens, they all overlap. This produces an interference pattern, but if one averages over a few hundred coherence times, then the beamlets average out to produce an extremely smooth laser profile. Since a reactor-sized pellet responds hydrodynamically on time scales of about a nsec, the pellet is accelerated inward by this time-averaged smooth laser profile. A recent example of the use of ISI is shown in Fig. 2.

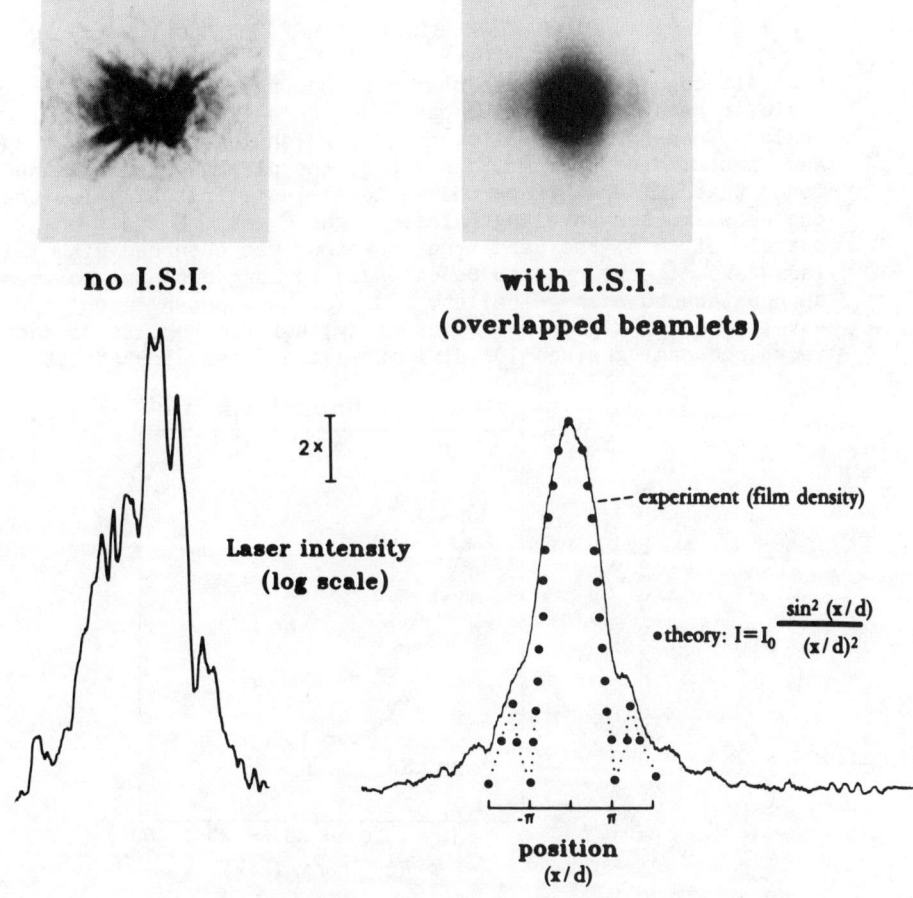

Fig. 2 Focal distributions with and without ISI, using the frequency-doubled PHAROS III laser

It has been shown analytically that a finite number of laser beams can produce a perfectly uniform implosion, if each of the beams produces a $\cos^2(\theta)$ pressure profile on the pellet shell. For example, one can produce perfect uniformity with 6, 8, 10, 12, etc. beams. However real laser beams cannot produce such a perfectly controlled profile, and misalignments are inevitable. Therefore it is better to use more laser beams. We have found that 32 beams, each with ISI optics, suffices for a high gain pellet.[6] Calculations on reactor-sized spherical targets show that the shell can be driven inward with net pressure nonuniformities of one to two percent, and with pressure nonuniformities in any given legendre polynomial of a few tenths of a percent.

PLASMA INSTABILITIES AND PREHEAT

Although hydrodynamic phenomena occur on a nanosecond time scale, plasma instabilities can occur on a picosecond time scale. Therefore the ISI technique, which leads to rapid temporal and spatial fluctuations, can modify the plasma coupling. We have found that ISI diminishes the various plasma instabilities when one uses shorter wavelength laser light.[3] Fig. 3 shows one example of this; the hard x-ray spectrum was quenched with ISI. These experiments need to be extended to larger plasma volumes appropriate to a large pellet, and also to shorter laser wavelength and higher laser intensity, but the results to date are very encouraging since ISI diminishes the plasma instabilities.

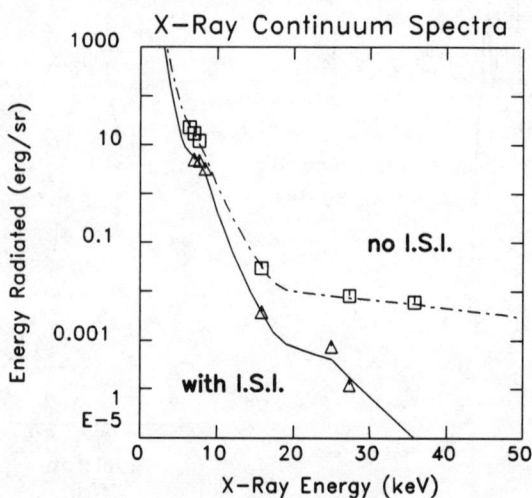

Fig. 3 With a 527 nm laser, ISI drastically reduced the hard x-rays over that obtained with the plain laser focused to the same diameter (0.3 mm).

HYDRODYNAMIC INSTABILITY

If the Rayleigh-Taylor instability grows at the classical growth rate, then the various inertial fusion schemes will not succeed because the pellet shell will not implode uniformly. However we have found with our 2D hydrodynamic simulation code that the growth rate is much less (about 30% of classical) because the ablation effect can provide convectively stabilization.[4] We find that this stabilization is a function of the laser's wavelength. The laser wavelength must be 1/4 micron or less in order to drive moderately thin pellet shells (with an initial aspect ratio R/ Δ R about 10:1). There have been some initial experiments at our lab with one micron laser light;[7] although consistent with the code's growth rate they did not explicitly measure this growth rate. Experiments are under way with both one and one- half micron laser light to explicitly measure the growth rate; further experiments will be necessary at 1/4 micron using ISI optics.

COUPLING EFFICIENCY AND PELLET GAIN

The rocket efficiency in direct drive laser fusion is dependent upon many variables, such as the laser wavelength and pulse shape, the pellet shell thickness, and even the zone resolution of the computer code. Generally, rocket efficiencies have been in the range of 5-10% for large high gain pellets. With two exceptions. In 1974 Afanase'ev et al found in their simulations that the rocket efficiency can be as high as 15%, using one micron light on a very thin pellet shell.[8] More recently we have found a similar rocket efficiency with quarter micron light on a moderately thin pellet shell.[5] With the stabilization of the Rayleigh-Taylor mode predicted by our computer simulations, we expect this latter case to implode with sufficient stability. With the 70% absorption predicted by our codes, the net coupling efficiency is then 15% x 70% = 10%.

With a 10% overall coupling efficiency, an ignition temperature of 3.5 to 5 keV, an ignitor density of 50 gm/cc, and a cold fuel isentrope of about twice the fermi-degenerate value, we can calculate a pellet gain using the isobaric model of Meyer-ter-Vehn,[9] and ignoring any two-dimensional degradation. The ignition energy is then about 0.5 to 1.0 megajoules, and the pellet gain at a few megajoules is in the range of 200-300. This should suffice for energy applications, as stated in the introduction. There are two provisos. First, this analytic model of pellet gain ignores any two-dimensional degradation due to hot/cold fuel mix, etc. Second, there is not yet any detailed point design of a direct drive target that can reach the conditions listed above. Further target design studies are needed.

LASER AND REACTOR DESIGNS

The use of ISI requires a broad laser bandwidth. Control of fluid instabilities and plasma instabilities, and high coupling efficiency, require a UV laser wavelength. All of these constraints can be satisfied with a KrF laser. It has a short quarter-micron wavelength. It also has a naturally broad bandwidth, if it is pulse-compressed with angular multiplexing rather than Raman compression. KrF is also scalable to the megajoule size, and has a high efficiency, probably in the range of 5-7%, and is estimated by some to cost about $200/joule in the megajoule range. Thus KrF seems to meet all of the requirements of direct drive laser fusion, and perhaps is the only laser that can meet these requirements. But KrF is also still in the development stage, and therefore it is not yet clear that it can meet the promises of its proponents.

In the ISI technique described above, the echelons must be placed at the output of the laser. But when extrapolated to a megajoule laser for a fusion reactor, this configuration would require a large number of optically coated echelon steps operating at high optical fluence levels. For example, a recent conceptual study concluded that one would require 240 steps in each transverse direction for each of the 32 drive beams. Also, 15% of the laser energy is diffracted into side lobes that do not interact with the pellet. There is now an alternate ISI scheme in which the echelons are completely eliminated.[10] Basically, the image is created at the output of the oscillator, using partially coherent laser light and this image is then optically relayed onto the pellet. Initial experiments with a small KrF discharge oscillator-preamplifier system have recently demonstrated that such smooth laser profiles can be obtained.

Inertial fusion is attractive as a reactor concept because of its inherent simplicity and because the high technology driver is located far away from the reaction chamber. This simplicity must be applied to the reactor chamber also. This chamber, in my view, should use simple dry or wetted walls, and not liquid lithium waterfalls or other complex designs. The solid angle subtended by the laser beams should be very small -- a few percent at most -- in order to effectively use the fusion neutrons. The chamber should be modular, so that one can repair a part of it without unraveling all 32 beam lines. There should be only very low long-term radioactivity, not just a few orders of magnitude less that fission reactors, so that maintanence costs will be low.

I think that all of the above requirements can be met by a combination of two reactor concepts developed at the University of Wisconsin,[2] called Sirius and Solase. The Sirius concept uses a dry wall chamber that is built in sections much like a soccer ball. The laser beams enter the chamber at the corners of these sections. With ISI, the solid angle subtended by the laser beams is very small -- about a percent for all 32 beams. In front of the chamber wall is another layer, made of graphite, that protects the main wall from radiation damage, and which will not itself become radioactive. For the main chamber wall and heat exchange system, I like the Solase design. It is made of an aluminum alloy and graphite which have negligible long term radioactivity, and the heat exchange medium consists of solid grains of lithium oxide. The chamber interior is filled with a neon gas, to provide additional protection against target debris. As I indicated in the Introduction, the primary uncertainty in laser fusion seems to be the final turning mirror. It probably must be coated, in order to have high reflectance, and optical coatings may be very sensitive to neutron damage by color center formation.

SUMMARY

This review has been fairly optimistic in tone. There have been a number of true breakthroughs that make the direct drive concept now interesting for energy applications. Although I have not had space to review all of the data, there is a substantial data base to support the concept.

In order to complete the evaluation of this direct drive concept we need additional effort in several areas. First, we need to develop better techniques to measure Rayleigh-Taylor growth rates, and to verify that the growth rate is as low as predicted for quarter micron KrF laser light. We need better rules for the maximum allowable source term for Rayleigh-Taylor, arising from pellet fabrication and laser nonuniformity. We also need to verify that the plasma instabilities are indeed below threshold with ISI and quarter micron light. We need an integrated test of ISI in spherical geometry using a frequency-doubled Nd:glass laser and a pellet that is large enough so that the laser-plasma coupling mechanism is similar to that expected in a larger system. We need a better understanding of the burn degradation to be expected from hot/cold fuel mix. And of course we need to know if KrF lasers are indeed a practical efficient driver for ICF.

Perhaps the most important summary comment I can make concerns the development of new laser drivers. The direct drive laser fusion concept only makes sense if one has a laser with both short wavelength and broad bandwidth, the latter being a requirement for ISI. The only laser that meets these requirements is KrF. There is a large investment in Nd: glass lasers throughout the world, but we do not know how to frequency triple or quadruple its infrared light with broad bandwidth. We should obviously utilize the large Nd: glass laser facilities, but the future of laser fusion also clearly lies with KrF.

REFERENCES

1. "Solase", Univ. of Wisc. Report VWFDM-220, (1977); "SIRIUS-M," Univ. of Wisc. Report FDM-651, (1985).
2. R.H. Lehmberg and S.P. Obenschain, Opt. Comm. $\underline{46}$, 27 (1983).
3. S.P. Obenschain, et. al., Phys. Rev. Lett., to be published.
4. M. Emery, et. al., Phys. Rev. Lett., to be published.
5. J. Gardner and S. Bodner, Phys. Fluids, to be published.
6. A. Schmitt and J. Gardner, J. Appl. Phys, to be published, 1 July 1985.
7. J. Grun, et. al., Phys. Rev. Lett., $\underline{53}$, 1352 (1984).
8. Y.V. Afanas'ev, et. al., Sov. JETP Phys., $\underline{21}$, 68 (1975).
9. J. Meyer-ter-Vehn, Nucl. Fusion, $\underline{22}$, 561 (1982).
10. R.H. Lehmberg and J. Goldhar, submitted to Fusion Tech.

PRESENT STATUS AND FUTURE PROSPECTS OF LIGHT IONS
AS DRIVERS FOR INERTIAL FUSION*

J. Pace VanDevender
Sandia National Laboratories, Albuquerque, NM 87185

ABSTRACT

Light ion beams offer a low cost (50-80 $/J for repetitive driver), 20% efficient driver for Inertial Confinement Fusion (ICF). Since the ions deposit their energy in a dense (10^{22} electrons/cm^3) plasma and are shielded from each other, collective effects and the associated preheating electrons are avoided. Power concentration has been the dominant problem. In 1984, a proof-of-principle experiment demonstrated an intense ion beam with adequate beam divergence (15 mrad) at the source current and charge density required for fusion. In 1985, the result was scaled to the total current and source dimensions required for ignition, and the Particle Beam Fusion Accelerator II (PBFA II) was completed. PBFA II is designed to generate and focus a lithium ion beam onto a spherical target at 100 TW/cm^2, investigate ignition, and explore the technology required for high gain targets.

While lasers were being developed for near-term experiments in Inertial Confinement Fusion, the ICF program developed alternate technology for a low-cost, high-energy, efficient source of power with charged particle beams. The initial approach was using electron beams, but that was abandoned in 1979 after the efficient production of light ions was demonstrated.[1,2,3]

The principal research topics for the light ion approach have been the generation of a sufficiently powerful electrical pulse, the transformation of that electrical energy into particle beam energy, the focusing of the beam to the few-millimeter diameter, and the transport of the beam to a target at some distance from the source. Substantial progress has been made on all of these areas and the perceived risk of the light ion approach has been substantially reduced.

KEY PHYSICS ISSUES

The principal uncertainty has been whether or not the electromagnetic energy could be converted efficiently to an ion beam and whether that beam could be focused onto the target. Beam production and focusing occur in an ion diode as shown in Figure 1. Electromagnetic energy from the pulsed power generator is fed through the vacuum-insulated transmission lines and create a potential difference of many millions of volts between the anode and the cathode. The anode is a source of ions and the cathode is a source of electrons. The electrons would normally flow quickly to the anode and consume most of the energy of the device if they were

---
*This work was supported by the U.S. Department of Energy under contract DE-AC04-76DP00789.

not inhibited by an applied magnetic field from the field coil shown in Figure 1. The magnetic field is sufficiently intense to inhibit the electrons from reaching the anode. The more massive ions, however, can easily cross this magnetic field and continue to the target.

Figure 1.

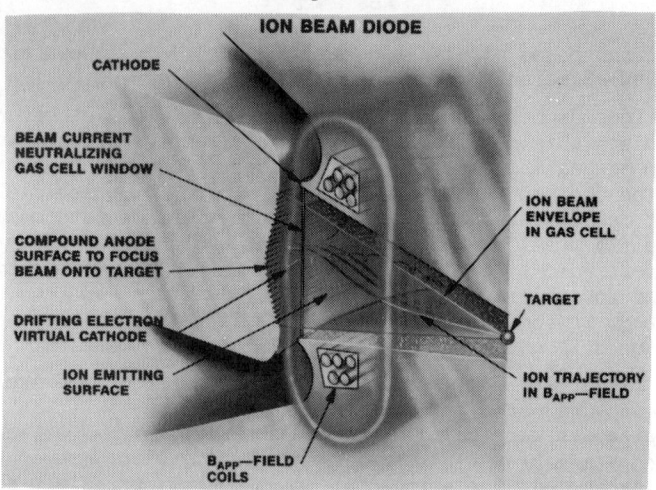

Because these intense beams of ions have millions of amperes of current, the number density of ions is very large. This large number density introduces new collective instabilities which might defocus the ion beam. These potentially damaging instabilities can grow very quickly for these diodes and most of the instabilities can reach saturation during the beam pulse. Because of these potentially damaging instabilities, the light ion approach to ICF was considered to be very high risk. However, no one knew the strength of these instabilities in saturation and, therefore, the effects of the instabilities on the ion focusing problem.

In 1983, a crucial experiment[4] was performed on the Proto I accelerator at Sandia National Laboratories (SNL) that showed a factor of two improvement in a focal spot size and demonstrated the beam divergence adequate for driving PBFA II targets in the future. Since the beam current density, which is directly related to the charge density, is a principal source of the electron and ion driven instabilities in the diode, and since the Proto I experiment was at the same current density required for ignition experiments on PBFA II, it is considered[5] a proof-of-principle experiment of ion beam focusing. New diagnostics were applied to map the beam focusing from various parts of the anode. D. J. Johnson, et al.,[4] found that the beamlets from different heights on the anode were focusing at different times. The apparent cause of this astigmatic behavior was imperfect distribution of electron charge within the A-K gap. By empirically shaping a non-spherical anode surface, all of the ions from all portions of the anode were focused onto the same central spot at the same time. The full-width-at-half-maximum of the beam focus was 1.3 mm from a 4.5 cm radius diode. This divergence of 14.6 mrad includes the steering error, the scattering from the cathode foil, and the intrinsic divergence from the ion production and acceleration process. This divergence is adequate for initial experiments on PBFA II and represents a successful proof-of-principle experiment: large electron and ion current densities in the Applied-B diode do not spoil the focus of intense beams.

The proof-of-principle experiment on Proto I was extended to higher total current and the larger diode size required for PBFA II on the PBFA I accelerator at SNL. PBFA I had intially been built for electron beam fusion experiments and was first operated in June 1980. However, by the time the accelerator was built, we had learned to produce intense ion beams, and the favorable energy deposition with intense ion beams had led to the switch to ions. PBFA I was, therefore, converted (at some cost in efficiency) to the world's largest ion beam accelerator to generate 2 MV and produce the PBFA II current of 5 MA. The diagnostics utilized on Proto I were extended to PBFA I and the technology was further developed: a new method of neutralizing the ion charge and current was devised. Thin plastic filaments were placed near the edge of the cathode and inside the transport cell where the applied magnetic field lines returned. Electrons from the filaments controlled the steering of the ion beam and provided charge neutralization and partial current neutralization. Since this diode did not have a gas cell, the ion beam was not scattered on its way to the target. In March 1985, the PBFA I scaling experiments were successful at the current, diode radius, and the source current density required for PBFA II. The measured power density on target was 1.5 TW/cm^2 with protons comprising approximately 22% of the diode charge.[6]

The importance of this achievement with respect to PBFA II can be best understood by the equation for the power density on target as a function of the various parameters of the diode.

$$P_t = \left(\frac{JV}{\theta^2}\right)\left(\frac{\Omega}{4\pi}\right) f_1 f_2$$

The equation is simply the ratio of the beam power that hits the target of a radius r to the target area $4\pi r^2$. The power on target is the current in the diode times the voltage of the diode, which is the diode power, times the fraction f_1 of the diode current that is in focusable ion species times the fraction f_2 of the focusable ion species that actually hit the target. A conservative estimate of f_2 is approximately 0.5. The fraction f_1 is measured by lithium activation to infer the total charge of protons compared to the electrical measurement of the charge in the diode above a certain threshold voltage for the reaction. The divergence θ of the ion beam is defined as the radius of the target divided by the radius of the diode for a target size in which f_2 is approximately 0.5. The solid angle Ω is the solid angle subtended by the anode as seen from the target and is equal to the area of the anode divided by the square of the diode radius. The ion current density J at the anode source is equal to the total diode current I divided by the area of the anode. Consequently, the divergence can be inferred from measurements of (1) the power density on target, (2) the fraction of diode charge in focusable ions measured from nuclear activation, (3) the diode voltage, (4) the diode current, and (5) the geometry of the experiment. The PBFA I experiment yields a divergence of 13.5 mrad. When this result is applied to PBFA II at 30 MV, 90% beam purity, and slightly larger solid angle of the diode, the projected power density on target is in excess of 100 TW/cm^2.

These results were all done with protons while PBFA II will use lithium ions. Other experiments[7-10] have indicated that the beam divergence is reduced as the ion mass is increased. Consequently, the Li$^+$ ion sources being developed for PBFA II should result in even lower beam divergence.

In addition, the general scaling of beam brightness, JV/θ^2, which is the maximum power density that could be achieved on target, with the voltage from all the different diodes indicates that the beam divergence may be improving as approximately the reciprocal of the square root of the voltage.[5] If we were to assume that the divergence improves because of either higher voltage or ion mass, the projected power density for PBFA II would be much greater than required. The next major step in demonstration of a high power density on target clearly requires the larger voltage, the improved ion source purity, and the heavier ion of PBFA II.

ION SOURCE DEVELOPMENT

The development of a high purity, preformed, uniform source of Li$^+$ ions has been a major technical challenge for the light ion fusion program. The singly-charged lithium ion is desirable for many reasons. Its charge-to-mass ratio is small, so magnetic bending in time-varying, self-magnetic fields of the diode is minimized. In addition, the low ionization potential of lithium neutrals to make the Li$^+$ ion means that a high density Li$^+$ plasma can be formed without significant contribution of the ion thermal velocity to the beam divergence. The large ionization energy of Li$^+$ to make Li^{++} means that the population of Li^{++} can be very small, and preheat from the more energetic Li^{++} particles should be negligible. Since approximately 10^{15} lithium ions/cm^2 must be supplied to the beam by the lithium source, and since depletion of the lithium ion source should be a small perturbation on the plasma, a density of 10^{16} ions/cm^2 in a 1 mm thick plasma has been defined as the goal for PBFA II. The uniformity must be excellent over an area of approximately 1000 cm^2 and the source of ions should be established before the main accelerator pulse arrives. These constraints place severe requirements on the Li$^+$ ion source. None of the ususal ion sources met these requirements when the ion source was identified as a major initiative in 1983.

Since that time, research has been carried out for many ion sources. At least three will probably be utilized on PBFA II, and the best performer will be chosen for the eventual target experiments.

The Li^+ source[11] is being addressed by five parallel approaches: (1) LiF, glow-discharge-cleaned and UV-enhanced flashover source, (2) an electrohydrodynamically generated liquid-lithium ion source,[12] (3) lithium vapor source that is injected from the cathode to the anode and ionized upon stagnation, (4) a flash-heated LiAg source[13] that produces a dense lithium vapor on the anode for subsequent ionization by the LIBORS process (visible laser light excites neutral lithium, which collides inelastically to heat electrons, which ionize the lithium neutrals), and (5) direct vaporization and ionization of the lithium with photons from a ArF laser. Options 1, 2, and 4 are being pursued by Sandia. Option 3 is being pursued by GT-Devices, Alexandria, VA, USA, and Option 5 is being pursued by Western Research Corporation, San Diego, CA, USA. All options have generated promising ion sources on some scale. Tests on PBFA II are planned in 1986-1987 to determine which ion source will produce the best beam.

PBFA II

PBFA II is shown in an artist's conception in Figure 2. It is the first super-power accelerator specifically designed and built for light ion fusion experiments. The requirements of high voltage

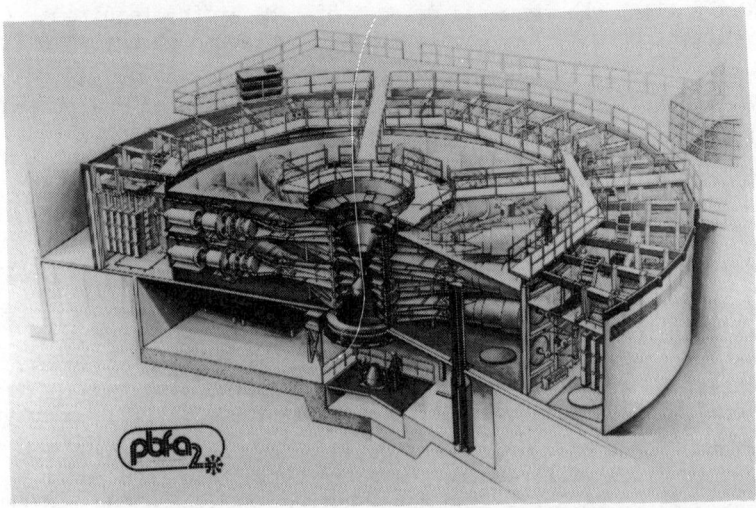

Figure 2. Artist's Configuration of PBFA II.

and high power have led to a four-tiered, 36 module, pulsed power configuration. The reliability of these high-voltage, high-current components exceeds the previous state-of-the-art by large factors. The technology of PBFA II has required many major advances over the past fifteen years. The device is a major experiment in accelerator physics.

The energy store consists of 1 Marx generator for each module. Each of these devices is composed of 60 high-energy density, 100 kV capacitors charged in parallel over two minutes, and then switched into a series configuration when the accelerator is fired. The output pulse duration of the Marx generators is 1 μs. The power pulse is conditioned in the 36 modules in parallel by successively charging and discharging the coaxial transmission lines in the water section to form a 50 ns power pulse. All the modules are synchronized by a newly developed, laser-triggered, multi-stage gas switch. The power pulses from the 36 modules are added in parallel and series combinations to produce a 130 TW power pulse at 15 MV. The power flows into the central vacuum chamber for a final stage of pulse compression and voltage amplification.

A new technology was developed for this final stage of pulse compression—the plasma erosion opening switch[14,15]—just outside the ion diode in PBFA II. Plasma is injected across the magnetically-insulated transmission lines to electrically short these lines during the first 55 ns of the power pulse. During this time, the energy is stored magnetically in the vacuum region between the conductors of the magnetically-insulated line. As the energy stored nears its maximum, the plasma opening switch develops an electron sheath near the cathode and the plasma ions are eroded away to produce a pathway for the electromagnetic wave. The major pulsed power uncertainty in PBFA II is the successful operation of the 30 MV plasma opening switch. Voltage gains of 2 to 3 have been observed at several laboratories[16-18] with a maximum voltage of up to 6 MV.[17] Also, the required current of 5 MA at the required switch current density has been successfully tested on the PBFA I accelerator.[18] The extension of this technology to 30 MV and 150 TW on PBFA II experiments will be the first major piece of new pulsed power physics to be obtained. The Naval Research Laboratory (NRL), one of the two major collaborators in the development of light ion beam fusion (along with Cornell University), is primarily responsible for developing the plasma opening switch technology.

The ion diode in PBFA II is the Applied-B ion diode, which was introduced at Cornell University[3] and further developed on Proto I[4] and PBFA I.[6] However, the magnetic field geometry has been altered substantially to reduce the size of the capacitor banks for establishing the magnetic field for electron insulation at 30 MV. The multi-coil design has been thoroughly explored with two-dimensional, fully electromagnetic, particle-in-cell (PIC) computer simulations,[19] but has never been tested on any accelerator. The test of this diode will require the current, voltage, and power available only on PBFA II. The total current and the diode radius are the same as we have tested on PBFA I, but the addition of the

high purity ion source, the new magnetic field configuration, and
the very high voltage will extend the regime of ion diode physics
significantly. These diode experiments will be the second major set
of experiments to be conducted on PBFA II.

The first shot for PBFA II occurred on December 11, 1985.
Exploratory experiments are being conducted with the plasma opening
switch and ion diode while the energy-storage and pulse-forming
sections are being optimized. It now operates reliably at 100% of
its designed energy storage. "Shakedown" shots indicated that the
design is sound and no major retrofits will be required. The
optimization of the accelerator for its maximum energy and power
output will require at least a year. The initial ion diode
experiments will be done during this early optimization phase, but
the major efforts on beam focusing will continue for at least
another year. With the power and energy delivered to the target, we
anticipate requiring several years to achieve ignition of
thermonuclear fuel in the laboratory. A variety of target
experiments will be conducted by the personnel from Sandia National
Laboratories (SNL) in collaboration with the scientists from the
other major institutions participating in the national ICF program,
i.e., Lawrence Livermore National Laboratory (LLNL), Los Alamos
National Laboratory (LANL), KMS Fusion, the University of
Rochester's Laboratory for Laser Energetics, and the Naval Research
Laboratory (NRL). The goal of these experiments will be to show
that a very small fuel mass required for both military and energy
applications can be imploded satisfactorily, compressed to adequate
density, and raised to sufficient temperature for igniting the
thermonuclear fuel.

THE POSSIBILITY OF GOING PAST IGNITION

PBFA II is designed for ignition and possibly breakeven
experiments. The power and energy of the single pulse from PBFA II
should be adequate for achieving these goals if the implosion
hydrodynamics is as modeled by the most sophisticated computer codes
presently available. To produce much more energy than the
accelerator consumes necessitates precise pulse shaping of the ion
beam to produce carefully tailored density and temperature profiles
within the target.[20] These high-gain targets can only be explored
with pulse shaping. However, PBFA II was designed with the goal of
reaching ignition, not the goal of pulse-shaping and high-target
gain.

A recent development in the understanding of ion diodes could
potentially permit PBFA II to be retrofitted for pulse-shaping and
an initial venture into high-gain physics. The PBFA II diode has
the target centrally located. This type of diode is called a barrel
diode and the ions are generated from all of the anode surfaces
around the target. The explosive yield from a high-gain target in
the center of this diode would result in the destruction of a large
part of the central accelerator. Clearly, another approach is
required for high-gain. The transport of intense ion beams through
current carrying channels[21-23] has been studied experimentally and
theoretically and shown to be a feasible concept. This approach
will be necessary for a future reactor application to produce

standoff. The ion beam must be extracted from the diode, injected into the channel, and transported to the target.

The problem of providing a focusable extractor ion diode had been a major unsolved problem for light ion fusion. Experiments[24] on this type of diode resulted in electron losses near the outer edge of the diode, non-uniform ion emission, and very poor focusing. The electrons were not adequately controlled in this type of extraction diode. Three individuals (P. L. Dreike while still at Cornell University, J. P. Quintenz and S. A. Slutz both of SNL) independently realized the problem was establishing the right magnetic field configuration within the diode. Slutz and Seidel have modeled this analytically for a focusing extraction ion diode and tested the model with a series of two-dimensional, fully self-consistent, relativistic, PIC, electromagnetic simulations.[25] The simulations showed the electron loss pattern previously observed in experiments unless the new coil geometry was incorporated. With the new coil geometry, the electrons were well behaved and the beam appears to be quite focusable. Experiments with this new coil configuration are being planned and the extractor diode will be developed for the pulse-shaping option on PBFA II.

In principle, PBFA II could be modified to test beam extraction, beam transport, and pulse-shaping required for a reactor. All levels of PBFA II would be added in series to power a single extractor ion diode. The ion beam would be injected into a current carrying channel for transport to the target. The speed of the ions at each instant would be selected through carefully controlling the voltage so that the beam could form the appropriate pulse shape by bunching and debunching on the way to the target. In order to handle the full power of PBFA II, the channel would probably have to be a wall confined discharge[21] and perhaps have a final focusing cell[26] at the output of the channel to produce the high power density onto the target. Key technical issues include the voltage control on the ion diode and the modification of the PBFA II pulse-forming network to provide the correct pulse shape with appropriate means for fine-tuning that pulse shape.

The Target Development Facility (TDF)[27] is proposed as a logical step past PBFA II. It is presently conceived as a 6 to 10 MJ facility that can test high-gain, thermonuclear targets at a shot rate of up to 10 per day. TDF would be used to cover the range of power and energy necessary for full development of reliable, well-characterized, high-gain targets at acceptably low cost. In addition, TDF could be used for the military applications of ICF including simulation of nuclear weapon effects using the x-rays and gamma-rays from a small thermonuclear source, and study of nuclear-driven directed energy weapon physics. The facility would be a full nuclear facility and, therefore, would be able to meet this wide spectrum of ICF applications.

REFERENCES

1. J. W. Poukey, Appl. Phys. Lett. <u>26</u>, 145 (1975).
2. S. A. Goldstein and R. Lee, Phys. Rev. Lett. <u>35</u>, 1160 (1975).
3. P. L. Dreike, C. Eichenberger, S. Humphries, and R. Sudan, J. of Appl. Phys. <u>47</u>, 85 (1976).
4. D. J. Johnson, et al., J. of Appl. Phys. <u>58</u>, 12 (1985).
5. J. P. VanDevender, et al., Lasers and Particle Beams <u>3</u>, 93 (1985).
6. J. E. Maenchen, et al., to be submitted (1986).
7. C. W. Mendel, Jr., et al., Proc. of the 5th Intl. Conf. on High-Power Particle Beams, San Francisco, CA, Sept. 12-14, 1983, pg. 218.
8. J. M. Neri, "Production and Characterization of Ion Beams from Magnetically Insulated Ion Diodes," PhD Thesis, Cornell University, Laboratory of Plasma Studies, Ithaca, NY.
9. K. Imasaki, S. Miyamoto, T. Ozaki, H. Fujita, N. Yugami, S. Nakai, and C. Yamanaka, "Present Status of Research for LIB-ICF at ILE Osaka University," ILE 8412P, Institute of Laser Engineering, Osaka University, Japan (1985).
10. Y. Maron, M. D. Coleman, D. A. Hammer, and H. S. Peng, "Spectroscopic Measurements of the Electric Fields and Ion Transverse Velocities in an Intense Ion Beam Diode," Cornell University, to be published.
11. R. A. Gerber, et al., IEEE Trans. in Nucl. Science NS-32, <u>5</u>, 1718 (1985).
12. A. L. Pregenzer, J. Appl. Phys. <u>58</u>, 4509 (1985).
13. P. L. Dreike and G. C. Tisone, J. Appl. Phys. <u>59</u>, 371 (1986).
14. C. W. Mendel, Jr., and S. A. Goldstein, J. of Appl. Phys. <u>48</u>, 1004 (1977).
15. B. V. Weber, R. J. Commisso, R. A. Meger, J. M. Neri, W. F. Oliphant, and P. F. Ottinger, Appl. Phys. <u>45</u>, 1043 (1984); and P. F. Ottinger, S. A. Goldstein, and R. A. Meger, J. of Appl. Phys. <u>56</u>, 774 (1984).
16. H. Karow, et al., Proc. of the 5th IEEE Pulsed Power Conf., Arlington, VA, June 10-12, 1985.
17. K. Imasaki, et al., Proc. of the IAEA 10th Intl. Conf. on Plasma Physics and Controlled Nuclear Fusion Research, London, England, September 12-19, 1984, Vol. 3, p. 71, IAEA, Vienna (1985).
18. R. W. Stinnett, et al., to be submitted for publication (1986).
19. S. A. Slutz, D. B. Seidel, and R. S. Coats, accepted for publication in J. of Appl. Phys. (1986).
20. J. D. Lindl and J. W-K Mark, Lasers and Particle Beams <u>3</u>, 37 (1985).
21. F. L. Sandel, Naval Research Laboratory, unpublished.
22. J. N. Olsen and R. J. Leeper, J. Appl. Phys. <u>53</u> (1982).
23. J. J. Watrous and R. E. Olson, to be presented at the 7th Topical Meeting on the Tech. of Fusion Energy, Reno, NV, June 15-19, 1986, and to be published in Fusion Technology (1986).
24. D. J. Johnson, J. P. Quintenz, and M. A. Sweeney, J. Appl. Phys. <u>57</u>, 794 (1985).
25. S. A. Slutz and D. B. Seidel, J. Appl. Phys. <u>59</u>, 2685 (1986).
26. P. F. Ottinger, S. A. Goldstein, and D. Mosher, Naval Research Laboratory Report NRL-4948 (1982).

27. R. E. Olson, D. L. Cook, R. R. Peterson, R. L. Engelstad, E. G. Lovell, D. L. Henderson, and G. A. Moses, Sandia National Laboratories Report SAND85-1216; and D. L. Cook, Proc. of IEEE 9th Symp. on Engineering Problems of Fusion Research, October 26-29, 1981, Vol. 1, p. 664.

PRESENT STATUS AND FUTURE PROSPECTS OF HEAVY ION BEAMS AS DRIVERS FOR ICF

Terry F. Godlove
U.S. Department of Energy
Office of Basic Energy Sciences
Washington D.C. 20545

ABSTRACT

A candidate driver for a practical inertial fusion reactor system must, among other characteristics, be cost effective and reliable for the parameters required by the fusion target and the remainder of the system. Although the history of large particle accelerators provides abundant evidence of their reliability at high repetition rates, their capital cost for the fusion application has been open to question. Attempts to design cost effective systems began with accelerators based on currently available technology such as RF linacs and storage rings. The West German HIBALL and the Japanese HIBLIC are examples of this initial effort. These designs are sufficiently credible that a strong argument can be made for the heavy ion method in general, but to reduce the cost per unit power it was found necessary to design for large scale, hence high capital cost. Emphasis in the U.S. shifted to newer technologies which offer hope of significant improvement in cost. In this paper the status of various heavy ion driver designs are compared with currently perceived requirements in order to illustrate their potential and assess their development needs.

INTRODUCTION

It is now one decade since serious study began[1] of heavy ion fusion (HIF), and a good time to take stock. National programs exist in the United States, West Germany and Japan. Smaller specialized programs are conducted in Great Britain, France and Spain. Although funding has generally been constrained compared to other major inertial confinement fusion (ICF) drivers, nevertheless a great deal of work has been done,[2] and indeed much of the effort has not been directly funded by the national programs.

This review is confined to the heavy ion driver. The accelerator, as well as the research effort itself, can be divided into several major categories: ion sources, the injector as a whole, means for acceleration, means for current multiplication including final bunching, and final focusing, which may necessarily include gas in the reactor chamber and perhaps special means for ameliorating the effects of space charge. All of these basic elements are put together in a conceptual design, which must then be interfaced with a target design and reactor system. Conceptual design studies developed during the past decade, partly from prior knowledge, partly from ongoing research, provide the primary basis for this status review.

Design studies must be understood from a balanced perspective. On the one hand they are vital to identify weak spots and difficult interfaces, and to show the approximate size and cost of future systems, but on the other hand since reactor size facilities are many years distant, one must recognize that the design is subject to change with ongoing research and innovation. The situation is different for high energy and nuclear physics facilities, where the extrapolation from one facility to the next is always less than one order of magnitude and is relatively near term. For heavy ion fusion the required extrapolation in intensity is many orders of magnitude, and little full scale development work has been done comparable to other large facilities. Thus designs are certain to be unique, and reviewers must take care to separate those features which must be demonstrated to be convincing from other features which simply require hard work to stretch the limits of known methods. Parenthetically, it is to the credit of the heavy ion fusion community that they already have the ability to publish detailed designs, including costs, in spite of the relatively low level of development funding to date.

For RF accelerator drivers, the West German HIBALL study[3] and the Japanese HIBLIC[4] illustrate the current status of design. For simplicity we use the HIBALL study as the basis for review. The induction linac driver was first described in terms of a conceptual design by the Lawrence Berkeley Laboratory (LBL) group[5] and considerably extended by work described in this Symposium. Our purpose is to review these designs for technical credibility and cost as well as to point the way to research and development tasks which will be necessary to demonstrate the design approaches.

RF LINAC WITH STORAGE RINGS

Figure 1 shows a schematic diagram for HIBALL II. A 5-km RF linac provides the means for accelerating the beam at constant current. The other major components - the linac tree with frequency doubling at each combination, the transfer rings, storage rings and buncher rings, all provide means for current multiplication. In this design the multiplication factor is 1.3 kA/0.16 A = 8,000 excluding the linac tree. The maximum storage time, reduced to 4 msec from the 40 msec time of HIBALL I, is kept to a minimum to reduce the longitudinal microwave instability in the storage rings. The final momentum spread, 1%, is a compromise between the difficulty of accelerator design and the difficulty of final focus. The choice of 8 ion sources represents a cost-benefit tradeoff to obtain the desired 165 mA in the main linac. Emittance growth in all of the various beam manipulations was carefully studied to keep within the allowed emittance budget, in this case a factor of 30. To reduce the cost of electricity (COE) the beams are sequenced through four reactor chambers, each operating at 5Hz and 900 MW electric, giving a total electric output of 3800 MW. By making the output this large, they were able to reduce the estimated COE to 48 mills/kWhr. However the total direct cost of the driver is large, $2.4 billion (1984 dollars).

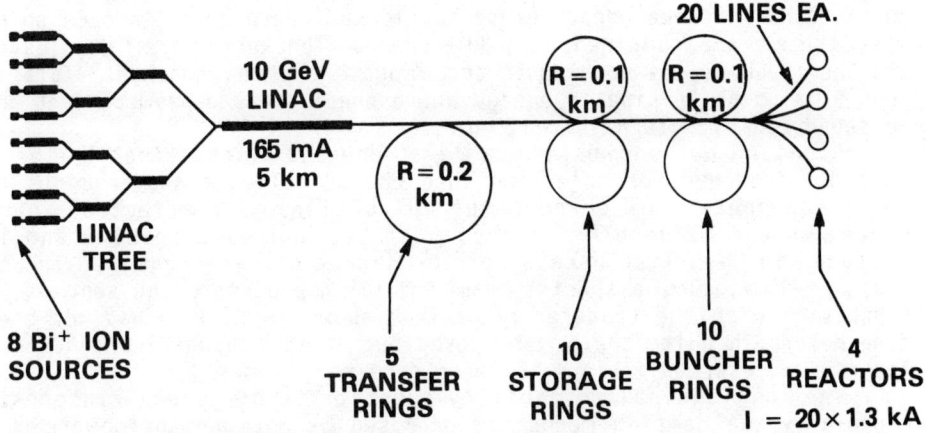

Fig. 1. Schematic of HIBALL-II (RF-based) reactor system design.

Table I gives a list of some of the more important issues for the RF driver method. Most of these are taken directly from the HIBALL design report.[3]

Table I. Selected RF Driver Issues

o Funneling with frequency doubling
o Debunching of linac beam - effect of space charge
   on momentum spread
o Combining beams with septum
o Longitudinal microwave instability in rings
   - higher charge/mass allowed?
   - verify nonlinear saturation mechanism
o Prepulse formation - preliminary ideas
o Verify beam transport over long distance
o Final bunch compression
o Overall emittance growth
o Cost of RF power generation

Commenting on each issue in turn, funneling at these beam currents has not been done before but looks tractable.[6] At each point where beams are combined, the RF bunches ("buckets") are interleaved and the frequency is doubled to accommodate the new bunches. This is done at a point in kinetic energy where the increased beam current can be supported by external focusing.

Although not shown in Fig. 1, at the end of the linac the beam must be debunched for injection into the so-called transfer rings to begin the process of current multiplication. The effect of space charge on the debunching process must be handled carefully and is included in this list because of the degree of care required in the design. Similarly, a special beam-combining process using septums in connection with the transfer rings (not shown in Fig. 1) has not been done before in quite the way proposed but looks reasonable, since the inverse process (splitting) has been done many times.

The longitudinal instability question is of general interest, especially the damping mechanism proposed by Hofmann and coworkers.[7] Calculations of the lifetime of the beam in the storage rings prompted the HIBALL team to reduce the ion charge state from 2 to 1. However, since a significant increase in the cost of the linac is thereby incurred it would be desirable to try to return to the higher charge state. Experiments are especially needed here and would benefit the induction linac method as well as the RF. In a paper at this Symposium Rees reports on progress on such measurements at the Rutherford Appleton Laboratory (RAL).[8] Funds are provided in the U.S. program for R.L. Martin of Argonne National Laboratory to collaborate with Rees and co-workers at RAL.

The HIBALL team mentions adding a few RF cavities to obtain a low intensity prepulse, which would be very desirable for optimum target performance. Little work has been done so far on the subject of pulse shaping. Conversely, a great deal of excellent work has been done on the next issue, that of beam transport. What remains is to verify the space charge limits over long distances, and to complete the characterization of emittance growth for beam bending and in the presence of various nonlinearities.

A major issue common to both methods is the final current multiplication by bunch compression. In low current accelerators a relatively straightforward compromise is made between high momentum spread, to allow the accelerator to function with reasonable longitudinal phase space (hence lower cost), and low momentum spread, to ease the optics of the final lenses. For high currents the problem is compounded by the presence of space charge and by neutralization, if used or required. A complete analytic theory to aid the designer in this highly nonlinear situation is probably impossible. However a very convenient measure of the effect of space charge alone is given by the generalized beam perveance, K, given by

$$K = 2(I/I_0)(m/M)(Z^2/\beta^3 \gamma^3),$$

where the symbols in order are beam particle current, Alfven current, (17kA), electron and ion mass, charge state, ion velocity/c, and $\gamma = (1-\beta^2)^{-1/2}$.

In an idealized geometry where emittance and aberrations can be neglected, the perveance determines the degree of difficulty of focusing. Put another way, if a focusing condition can be found either by a realistic calculation or by experiment that provides an acceptable spot size for a given combination of energy, current and ion mass, then the result can be scaled to any other combination having the same perveance. As a gross rule of thumb it appears that values from roughly 2×10^{-5} to 2×10^{-4} span the range from moderate to difficult.[9] In this connection simulation calculations for complete focusing systems will continue to be very helpful but of course are design specific.

Perhaps the single most important issue is the overall emittance growth. The total number of beam manipulations is large, and all of the emittance growth factors must be multiplied together to calculate the final emittance. Finally, I have added an engineering issue, namely the cost of RF power. Since it is a large fraction of the cost of the driver, lower cost methods would be very desirable.

INDUCTION LINAC

Moving on to the induction linac, Figure 2 shows a schematic of a reactor driver proposed by the LBL group.[10] The schematic is simpler in concept because current multiplying rings are not employed. Current multiplication, in this case a factor of about 2000, is combined with acceleration in the linac itself by means of ramping the voltage pulses. This prevents the head of the bunch from running away from the tail. Indeed, in most designs the spatial length of the bunch is held roughly constant. The low energy end, below say 50 MeV, remains a thorny problem as it has from the beginning. Although it is a relatively small fraction of a full-scale driver, it represents a very large fraction of the cost of getting started in an experimental heavy ion program.

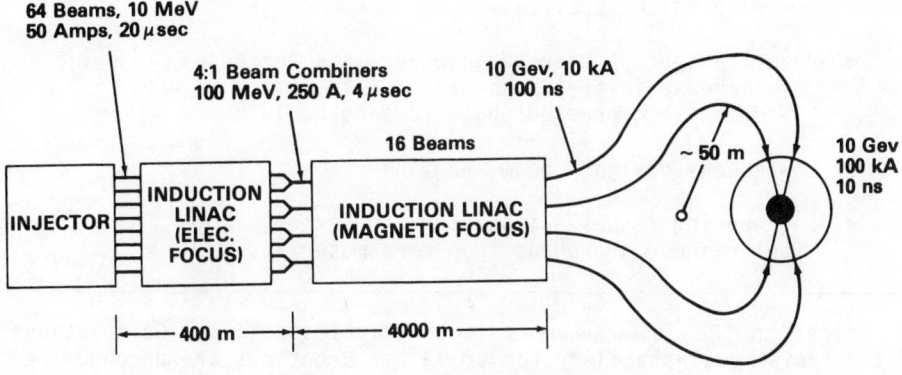

Fig. 2. Schematic of induction linac driver design. Beam currents shown are total electrical current, for charge state 3.

The driver parameters shown are from recent thinking of the Berkeley group. They propose a 64-beam injector which is then merged by 4 to 1 to give 16 beams through the main body of the accelerator. In this connection the 16-beam, 2-MV injector under development at the Los Alamos National Laboratory (LANL) is an important contribution.[11] It provides the starting point for any conceivable next phase facility in the U.S. program.

Multiple beams, separately focused but threaded through common accelerating gaps, are at present an established strategy to reduce the cost of an induction linac for HIF. Fessenden et al. summarize recent LBL work on multiple beam experiments.[12] Electrostatic focusing quadrupoles, now rather routine at LBL after several years of research, are used at low velocity, while magnetic quadrupoles are appropriate for the higher velocity found in the main accelerator.

The induction linac was adopted for the U.S. program for a combination of reasons, including the fact that Europe and Japan were pursuing the RF method, but also based on an important thrust of the program, that of cost reduction. Although a large linear induction accelerator for ions has never been demonstrated, considerable experience exists with kiloampere electron beams, and no one doubts that it will work at some ratio of cost to performance. Table II lists some of the questions which will affect that ratio, and by implication, the kind of research which is necessary.

Table II. Selected Induction Linac Driver Issues

- Longitudinal stability
    - longitudinal resistive instability
    - stability against high beam loading

- Final focus in vacuum and in reactor gas
    - momentum spread limit including space charge?
    - maximum allowed gas pressure?
    - external neutralization?

- Beam transport (phase advance per lens period)
    - increase initial phase advance to $85°$?
    - decrease depressed phase advance to $10°$?

- Low energy design - beam merging?

- Engineering issues - insulators, switches, tolerances, amorphous iron core material.

First on the issue list is longitudinal stability. Calculations of the resistive instability for small perturbations are encouraging[13] but need to be checked experimentally. Also of some concern are the details of the interaction of the beam with the high-power modulators under conditions of high beam loading. Is a tradeoff required between efficient coupling to the cavities on the one hand and stability on the other?

The final focus remains an exacting problem and is similar to that described for the RF method. Two interrelated questions have been studied for some years: the final focusing elements need to be designed for maximum momentum spread, including space charge, to reduce the cost of the accelerator; and current thinking is to employ neutralization in the chamber as a vital step to overall cost reduction. Calculations tend to be cumbersome not only due to space charge but also because of the large number of design options, including the possibility of externally supplied neutralization such as cold electrons. Ionization of the reactor gas and stripping of the beam ions must be included in realistic designs. More research is needed in these areas, especially on defining the limits on neutralization. Whether, for example, 90% neutralization is a valid and realizable goal remains an open question until good experiments are available. Olson gives a review of the topic at this Symposium.[14]

The next question is the limits of beam transport. At issue is the ratio of the betatron phase shift per period (sometimes called "tune") for very small beam current to the full current case. It is a convenient measure of the degree to which the space charge limits can be stretched in high-current beam transport and has a strong impact on the design and cost of an induction driver. It has thus been an important issue for the induction method from the beginning of the program. Until recently LBL employed design values of 60° initial, and 24° depressed tune. It has become clear, however, that a higher ratio, for example, 85° to 10° would be very desirable provided the beam is stable. The recent trend of both experiments and calculations is encouraging in this connection.

The injector and low energy part of the accelerator, up to perhaps 50 MeV, become a larger fraction of the driver as the rest of the system is lowered in cost. LBL proposes merging beams to get higher total current at the source while keeping the number of beams in the main accelerator at an optimum value. Research will be needed to determine whether merging is really necessary and if so whether the penalty in emittance growth is acceptable. Both simulation and experiments are necessary. The paper by Celata at this Symposium is relevant to this question.[15]

Finally, although some work has been done on various engineering issues, most need considerably more development. Among these are cost effective insulators, high-power switches, details of the core material and numerous fabrication techniques.

Improvements in the above systems, particularly final focus and tune depression, allow higher beam currents. This allows a higher ion charge/mass ratio, which leads directly to a lower voltage accelerator, which means lower cost. In this manner the natural ability of induction cavities to accelerate high currents is exploited. Ultimately, however, the advantage of higher charge/mass in a complete system must be weighed against the disadvantage of slightly lower target gain. System studies, discussed below, indicate that this will not be a serious penalty in most cases.

INDUCTION LINAC SYSTEM ASSESSMENT

One of the most interesting new reports at this Symposium is the Heavy Ion Fusion Systems Assessment, or HIFSA, in the U.S. program.[16] HIFSA was motivated by the fact that a comprehensive system study for the induction method was lacking in the U.S. program, by the need to assess and reduce driver costs, and, not least, to guide future U.S. HIF research. The assessment was conducted by a team led by LANL and including Lawrence Livermore National Laboratory (LLNL) and LBL. System integration was performed by McDonnell Douglas Astronautics with assistance from Titan Systems. The method employed was to develop analytic or curve-fitted models for each subsystem of the reactor plant, including targets, then integrate them in an overall system code to determine the optimum operating parameters, costs and efficiency. Funding for HIFSA was shared by three offices in the Department of Energy and by the Electric Power Research Institute. Accelerator design and cost estimates were provided by LBL based on the LIACEP code originally developed in 1980.[5]

A major purpose of the HIFSA was parameter variation by computing a large number of cases and studying their systematics. Using the COE as the "bottom line" a number of interesting conclusions were derived. For example, for a 1000 MWe plant surprisingly broad optima were found for the repetition rate, the kinetic energy, and the number of beams in the accelerator. These were 5 Hz, about 8 GeV, and 16 beams, respectively, but a factor of two variation in each of these parameters made little difference in the COE provided other parameters were allowed to vary. Table III shows three cases out of a large number computed which illustrate some of the results.

Table III. Three cases from the induction-based study HIFSA.
Major parameters: 1 GWe, 5 Hz, 7-8 GeV, 16 beams.

	Case A	Case B	Case C
Charge State	+1	+3	+3
First Wall	Wet	Wet	Granular
Target	1 Shell	1 Shell	Advanced
Illumination	2 Sided	2 Sided	1 Side
Pulse Energy (MJ)	5.9	5.8	3.4
Target Gain	88	86	106
Micro-Perveance	40	150	150
Driver Cost (G$)	1.6	0.9	0.8
Other Direct (G$)	1.3	1.3	1.0
COE (mills/KwHr)	75	59	49

The encouraging tolerance to major parameter variation is apparently primarily due to the high electrical efficiency of the accelerator, typically over 30%. As suspected, however, no cases could be found which would remove the sensitivity of the COE to plant size. The COE for a 500 MWe plant was always substantially higher, although this result is somewhat ameliorated by the fact that a utility could if desired build a larger plant in stages based on upgrading the same driver and/or adding more reaction chambers.

Cases A and B illustrate the advantage in changing to +3 ion charge state. The direct driver cost is dropped almost in half and the COE drops from 75 to 59 mills/kWhr. However, the final focus is made more difficult as confirmed by the perveance increasing by a factor of four, to 1.5×10^{-4}. (Microperveance is $10^6 \times$ perveance.) Neutralization in the final focus will probably be required for this perveance.

Case C represents the "optimized optimum". It not only uses charge state 3, but also an advanced granular wall reactor similar to LLNL's CASCADE design,[19] and an advanced target with higher gain and one-sided illumination. With these assumptions the main improvement is in the other direct cost, which drops from 1.3 to 1.0 billion dollars, lowering the COE to 49 mills/kWhr. This is a very encouraging result for a 1 GWe plant. While it must be viewed in the context of the assumptions and years of required R&D, it nevertheless places the heavy ion method squarely in an acceptable range of cost.

OTHER STUDIES

It is interesting to compare the results of the HIFSA project to the 1979 driver/reactor assessment sponsored by the Electric Power Research Institute (EPRI), summarized in Table IV for a 1 GWe plant.

Table IV. Summary of "Assessment of Drivers and Reactors for Inertial Confinement Fusion."*

	KrF Laser	Heavy Ion	Light Ion
Pulse Energy (MJ)	7.0	8.5	8.5
Pellet Gain	185	100	100
Thermal Power (GW)	1.7	1.1	1.1
Driver Cost (M$)	950	940	415
Bal. of Plant (M$)	1440	1015	1015
COE (mills/KwHr)	75	61	45

* EPRI AP—1371, K. Brueckner et al (1980)

The EPRI study was performed by an independent panel led by K. Brueckner.[18] Although the level of detail and the depth of the system modeling for the EPRI study do not compare with the HIFSA project, nevertheless the overall results are similar for the heavy ion driver. The EPRI study has the added feature that an attempt was made to compare the three major ICF drivers on a consistent basis. It is the only such documented comparison available, to my knowledge.

The general conclusion indicated by the EPRI study for the krypton fluoride laser is that the lower efficiency of this driver has a major impact on the system in two respects: an advanced high gain target must be assumed in order to obtain the required gain-efficiency product, in most studies >10; and the balance of plant cost, and COE, are substantially increased due to the larger recirculating power and larger microexplosion in the reactor chamber.

In the case of light ions, there is little doubt that the high voltage pulse generators developed for that method are cost effective, especially for the current DOE program which emphasizes single-pulse weapons applications. However, a number of extremely difficult obstacles must be overcome before the light ion method can be considered a viable energy driver. Among these are appropriate ion sources, means for focusing the required megamperes of ions, means for protecting the apparatus from the microexplosion, and repetitive firing of the system. The status of both laser and light ion drivers are thoroughly reviewed elsewhere in this Proceedings. The reason for commenting on them here is that comparisons, although necessarily difficult at this stage of research, are a vital part of national program planning. As noted above, the EPRI study remains the only documented study of its kind.

A qualitative evaluation of ICF drivers was published in 1982 by J.H. Nuckolls.[19] It is shown in Table V.

Table V. Inertial fusion driver evaluation by J.H. Nuckolls.

	KrF Laser	Heavy Ion	Light Ion
Efficiency (10–20%)	?	++	+
Focusing*	+	+	?
Target Coupling (10%)	++	++	+
Rep. Rate (10–20 Hz)	+	++	+
Cost ($200/J at 2MJ)	+	+	++

* 100–1000 TW/sq.cm. at 5 meters

For simplicity only the three major current drivers are reproduced from Nuckolls' article. He rates them according to five key attributes, with plus signs indicating a positive attribute.

Two opposing viewpoints on heavy ions can be found in recently published extensive reviews. In a discussion comparing laser, light ion and heavy ion drivers, T.H. Johnson writes:[20]

"Heavy ion drivers would probably be a factor of
five to ten more expensive."

In a similar discussion, C. Deutsch writes:[21]

"Clearly, heavy ion drivers are likely to win the race."

The National Academy of Sciences recently conducted a panel review of the DOE ICF program. The charter of the panel was limited to programs supported by DOE's Defense Programs, hence did not include Heavy Ion Fusion Accelerator Research (HIFAR), which is managed within the Office of Energy Research. However, briefings were given to the panel by HIFAR representatives and the final report includes the comments:[22]

" - - the heavy ion beam driver appears to offer considerable advantages in efficiency, reliability, focusing, and classical energy deposition. - - - Heavy ion beams may well be the best eventual driver for energy applications - - ."

The author is indebted to many members of the heavy ion fusion community, in particular to the HIFSA team for permission to use portions of their results prior to publication.

REFERENCES

1. R.O. Bangerter, W.B. Herrmannsfeldt, D.L. Judd and L. Smith, Eds., ERDA Summer Study of Heavy Ions for Inertial Fusion Report LBL-5543 (1976).
2. Recent program summaries and technical contributions may be found elsewhere in these Proceedings.
3. HIBALL-II, An Improved Conceptual Heavy Ion Beam Driven Fusion Reactor Study, Kernforshungszentrum Karlsruhe Report KfK 3840 (also FPA-84-4 and UWFDM-625), July 1985.
4. HIBLIC-I, Conceptual Design of Heavy Ion Fusion Reactor, Research Information Center, Institute of Plasma Physics, Nagoya University, Report IPPJ-663, January 1985.
5. A. Faltens, E. Hoyer and D. Keefe, Proc.4th Int'l Topical Conf. on High-Power Electron and Ion-Beam Research and Technology, Eds., H.J. Doucet and J.M. Buzzi, Ecole Polytechnique, Palaiseau, p. 751 (1981); see also J. Hovingh et al, Ref. 16.
6. K. Bongardt, Heavy Ion Accelerators and Their Applications to Inertial Fusion, Inst. for Nuclear Study, Univ. of Tokyo, Proc. INS Int'l Symposium, p. 470 (1984); also F.W. Guy, R.H. Stokes, and T.P. Wangler, elsewhere in these Proceedings.
7. I. Hofmann, I. Boszik, and A. Jahnke, IEEE Trans. Nucl. Sci. NS-30, 2546 (1983).

8. G. Rees, elsewhere in these Proceedings.
9. J.D. Lawson, The Technology of Heavy Ion Fusion, invited paper, 11th Symposium on Fusion Technology, Oxford, England, Sept., 1980.
10. D. Keefe, elsewhere in these Proceedings.
11. D.C. Wilson, K.B. Riepe, E.O. Ballard, E.A. Meyer, R.P. Shurter, F.W. Van Haaften, and S. Humphries, Jr., elsewhere in these Proceedings.
12. T.J. Fessenden, D.L. Judd, D. Keefe, C. Kim, L.J. Laslett, L. Smith, and A.I. Warwick, elsewhere in these Proceedings.
13. J. Biscognano, I. Haber and L. Smith, IEEE Trans. Nucl. Sci. NS-30, 2501 (1983).
14. C. Olson, elsewhere in these Proceedings.
15. C. Celata, elsewhere in these Proceedings.
16. Papers in these Proceedings related to the HIFSA project are given by D. Dudziak and W.B. Herrmannsfeldt; E.P. Lee; V.O. Brady, A. Faltens, J. Hovingh, D. Keefe, and E.P. Lee; three by D.E. Driemeyer, L.M. Waganer, and D.S. Zuckerman; R.R. Peterson; W.R. Meier, W.J. Hogan and R.O. Bangerter; and J.H. Pendergrass; see also J. Hovingh et al., "Cost/Performance Analysis of an Induction Linac Drive System for Inertial Fusion," Proc. 11th Symp. on Fusion Engineering, Austin, Texas (November, 1985) (to be published).
17. J.H. Pitts, Fusion Tech. $\underline{8}$, 1198 (1985).
18. Assessment of Drivers and Reactors for Inertial Confinement Fusion, La Jolla Inst. Rept. AP-1371 (1980).
19. J.H. Nuckolls, Physics Today, Sept. 1982, p. 29.
20. T.H. Johnson, Proc. IEEE $\underline{72}$, 590 (1984).
21. C. Deutsch, Ann. Phys. Fr. $\underline{11}$, 18 (1986).
22. "Review of the Department of Energy's Inertial Fusion Program," Nat. Academy Press, Wash., D.C. 20418 (1986), pg. 24.

REACTOR DESIGN ASPECTS FOR INERTIAL CONFINEMENT FUSION

G. Keßler, U. von Möllendorff
Nuclear Research Center, P.O. Box 3640, D-7500 Karlsruhe, FRG

G.A. Moses, R. Peterson
University of Wisconsin, 1500 Johnson Drive, Madison, Wisconsin, USA

ABSTRACT

Some aspects of reactor chamber engineering for power production by heavy ion fusion (and inertial confinement fusion in general) are discussed, mainly from the viewpoint of pellet-chamber and driver-chamber interface problems.

INTRODUCTION

Power reactor plants are presently built worldwide in block sizes of 800 to 1500 MWe for reasons of economy. In many cases several (4 to 6) of these power plants are collocated which leads to energy centers of 3000-6000 MWe at one site (Europe, Japan, USSR, Canada, USA). This development is mentioned because it is especially important in view of HIF power plants where several reactor chambers can be driven by one accelerator.

TASKS OF ICF REACTOR CHAMBER

The main task of an inertial confinement fusion (ICF) power reactor chamber is to convert the pulsed fusion energy into steady thermal power. While the pellets will explode at frequencies of one to several Hertz the power supplied to the steam turbine and generator must be steady. The reactor chamber must, therefore, have sufficient heat capacities to equalize temperature peaks. To achieve high thermal efficiency of energy conversion the reactor chamber must operate at sufficiently high coolant temperatures of about 500 °C.

The inner surfaces and structures of the reactor chamber which are exposed to radiation from the pellet micro-explosion must withstand this pulsed radiation load (neutrons, X-radiation, ion debris from pellet) over sufficiently long operating times. The inner surface of the reactor chamber may suffer stresses due to ablation and evaporation. The pulsed radiation load causes thermal cycling and pulsed radiation damage to the structural materials.

After each pellet micro-explosion adequate conditions in the chamber atmosphere must be reestablished for both pellet injection and beam propagation. This condition limits the repetition frequency at which the chamber can operate.

Last not least the reactor chamber must contain breeding materials (Li, LiO_2, $LiSO_4$ etc.) in its blanket to breed at least one triton per triton burned within the pellet. A neutron multiplying medium (Pb, Be etc.) may be required to achieve sufficient breeding ratios. The tritium inventory of the reactor chamber should be small for safety and environmental reasons.

PARTICLE BEAM ENERGY AND NUMBER OF BEAM CHANNELS

Conceptual studies for future ICF power reactors in most cases aim at a block size of about 1000 MWe per reactor chamber. Laser or light ion beam drivers would have to feed one reactor chamber each. Only in the case of heavy ions it is possible that one driver with its high (20-30 Hz) repetition rate feeds 4-6 reactor chambers which would form an energy center.

One characteristic design feature of ICF power plants is the decoupling of the power producing reactor chamber from its driver. This, however, necessitates the transport of the beam energy to the reactor chamber and the focusing of the beam onto the pellet in the center of the reactor chamber.

According to the present status of pellet physics a beam energy of at least 5 MJ has to be transported through a number of channels from the driver to the chamber to ignite the pellet. The number of beam channels depends on:
- the transport characteristics and performance limits of the individual beam channel and beam guide devices and
- symmetry requirements for pellet implosion.

From a design point of view, reasonable numbers of beam channels will vary between 4 to 6 as a lower limit and about 20 to 24 as an upper limit. The lower number assumes very high performance characteristics for beam power transport and beam focussing but also low symmetry conditions for pellet physics, whereas the higher number of beam channels is based on reverse conditions. In the HIBALL study[1] 20 beam channels were assumed for reasons of beam energy transfer, beam focussing and symmetry in pellet implosion. However, the HIBLIC study[2] assumes only 6 beam channels per reactor chamber.

RADIATION LOAD TO AN EVACUATED REACTOR CHAMBER

When ignited the pellet burns in less than one nanosecond. Only about 30% of the DT contained in the pellet is burnt. The fusion energy (Fig. 1) is released in the form of 14 MeV neutrons (carrying e.g. 72% of the released energy), soft X-rays (e.g., 22%) and ions (e.g., 5%). These percentages depend somewhat on the pellet design. The total energy released per microexplosion depends on the pellet gain. In present conceptual studies a pellet gain of 80-150 on the basis of theoretical pellet studies is assumed. These gains still have to be proven experimentally in the future.

The soft X-rays from the pellet explosion will arrive first at the inner surface of the reactor chamber (Fig. 1). Their energy is deposited in a thin surface layer leading to evaporation or sputtering processes. The energy of the 14 MeV neutrons is absorbed in the relatively large volume of the thick blanket. The debris of ions from the microexplosion arrives last and can be absorbed in the vapor generated by the soft X-rays if the chamber has a liquid-metal first surface; in a dry-wall chamber, the ions are stopped in a thin surface layer like the X-rays.

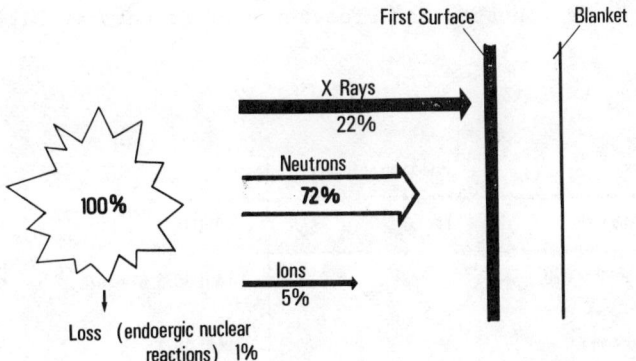

Fig. 1. Partition of Energy Yield (HIBALL Reference Values)

OVERVIEW OF REACTOR WALL CONCEPTS

Table 1 shows the basic concepts which have evolved for protection of the structures of the reactor chamber over the past fifteen years[3].

Replaceable or sacrificial layers, wetted walls, or a gas filling of the reactor chamber offer protection against soft X-rays and ions only. A filling gas in the reactor chamber would absorb the X-rays and ion debris and radiate the energy to the chamber surface over relatively longer time spans.

When using the protection concept of sacrificial layers the sputtered or vaporized materials of the first surface may have to be pumped out of the reactor chamber in addition to the vaporized material and gas of the pellet. This is a serious disadvantage. In addition the sacrificial parts will have to be replaced at least every 2-3 years. In all of these concepts, the volume damage by high energy neutrons to the steel of the blanket structure becomes very high. As a consequence the blanket steel structures also will have to be replaced every 2-3 years. The conclusion is: Although in case of dry-wall, sacrificial layer protection concepts[4] the repetition rate can be higher (about 10 Hz) in principle, this advantage is offset by serious disadvantages.

These problems are avoided by the liquid wall concept where the first surface is regenerated through vaporization and recondensation. In addition, a thick liquid wall (e.g., 2 m thick at 33% packing fraction) moderates the high energy neutrons and shields the steel structures from the high radiation damage. In this case the structural wall may not need replacement over the whole life time of the plant.

Table 1. Basic Concepts for Protection of Structural Wall

Concept	Protection against	
Magnetic Field	Ions	LANL 1974
Sacrifical Dry Wall or Layer	Ions, X-Rays	LANL 1973, KFA 1973
Wetted Wall		LANL 1973
HIBLIC (Sacr. Wall + Liquid Curtain)		U Nagoya 1984
Gas Filling		UW 1973
Liquid Wall Falling rain Waterfall HYLIFE (jets) HIBALL (INPORTS)	Ions, X-Rays, Neutrons	ORNL 1971 ANL 1974 LLNL 1977 LLNL 1978 UW 1981

REPETITION RATE

The repetition rate of a liquid-wall or wetted-wall vacuum reactor chamber depends on:
- the time interval in which the vaporized material of the thin surface layer can recondense on the surface,
- the time interval in which adequate vacuum for beam propagation can be reestablished and
- the time interval in which the pellet can be injected to the center of the chamber.

Detailed analysis[1] shows that vaporization and recondensation phenomena are decisive and limit the repetition rate to values between 1 and 5 Hz for all wetted or liquid wall concepts.

The most advanced design concepts today, accounting for all protection requirements appear to be the HYLIFE[5], the HIBALL[1] and the HIBLIC[2] concepts. HYLIFE and HIBLIC have 1 Hz repetition rate whereas HIBALL has 2 Hz.

Table 2 compares the main design data of reactor chambers for the Westinghouse (sacrificial layer, dry wall) concept[4] and the wetted and liquid wall concepts of HIBALL and HIBLIC.

Table 2. Examples of HIF Plant Designs (RF-LINAC)

Design	Westinghouse	HIBALL	HIBLIC
Chamber Concept	bare wall/ sacr. liner	liquid wall/ wetted wall	liquid wall + sacrificial liner
Driver pulse energy (MJ)	2	5	4
Minimum radius (to first surface), m	10	5	1
Hot Spot Load MW/m²	2.8	6.3	32
Driver Rep. Rate, Hz	10	20	10
Chamber Rep. Rate, Hz	10	5	1
Pellet gain	175	80	100
Fusion Power per Chamber, MW	3500	2500	400
Breeding Ratio	1.2	1.2	1.65
Min. lifetime of structural parts (full power years)	5	2	2

THE HIBALL REACTOR CHAMBER CONCEPT

The HIBALL[1] chamber (Fig. 2) uses a thick liquid wall of $Li_{17}Pb_{83}$ metal serving as neutron shield, coolant, breeder material and neutron multiplier. The metal flows along the slightly conical roof and down the cylindrical wall inside porous structures braided from silicon carbide (SiC) fiber. Throttle orifices can be provided at the inlet and outlet of each channel or tube to control the flow velocity. The SiC is protected from the short-range fusion energy by a coolant film (1 mm or thicker) seeping through the braid and covering the outside. This principle has been denoted INPORT, for inhibited flow in porous tubes. The bottom of the chamber is covered by a pool of the $Li_{17}Pb_{83}$.

The HIBALL chamber concept is actually a combination of the liquid-wall and wetted-wall principles. The advantage of the higher repetition rate of 5 Hz achieved by this guided and velocity controlled coolant flow over free-falling liquid jets as in HYLIFE is bought at the price of exposing some structural material, the innermost SiC INPORT tubes, to the neutron irradiation. SiC can withstand high neutron fluences. The innermost INPORT units were estimated to

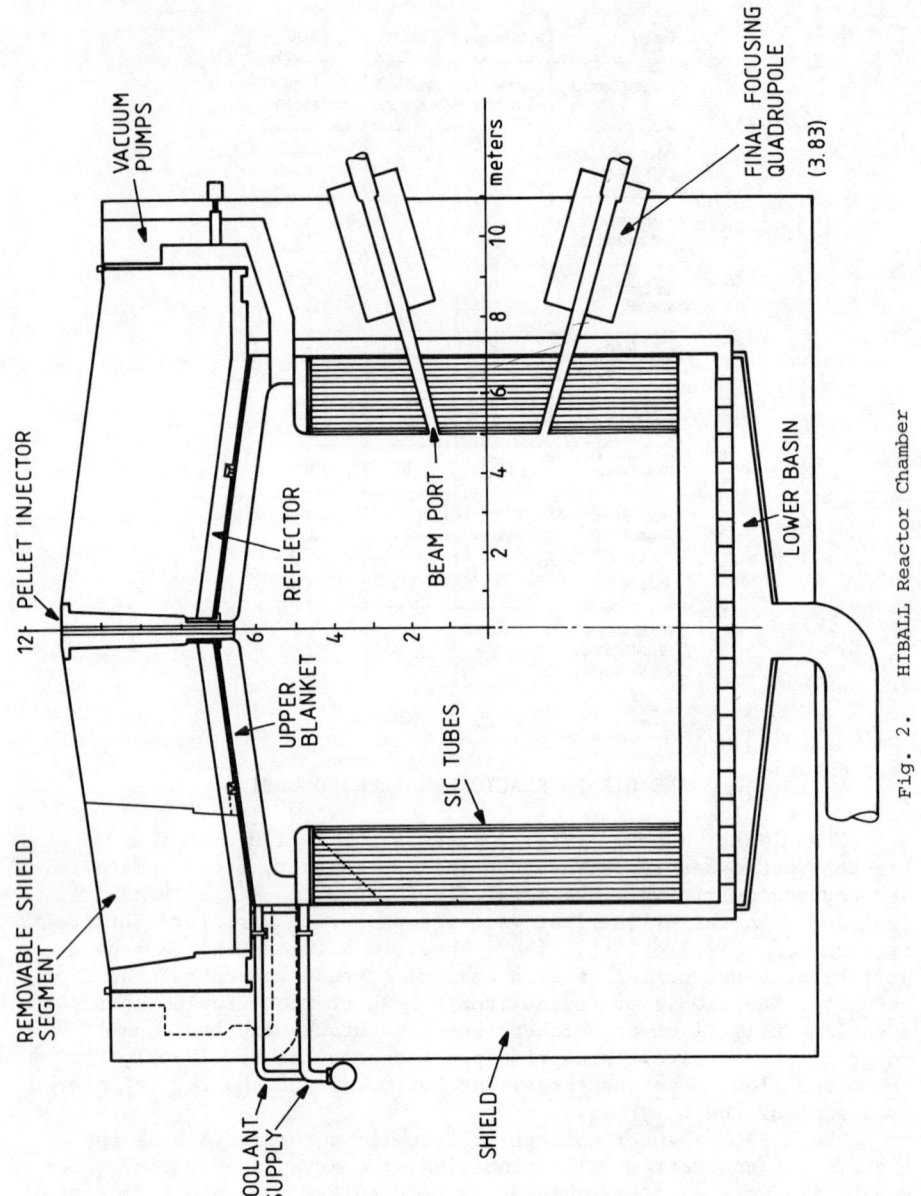

Fig. 2. HIBALL Reactor Chamber

withstand operation for 2 full-power years. Obviously, with a suitable design they can be exchanged much easier, cheaper and faster than a structural and pressure-holding chamber wall as it was foreseen in the Westinghouse dry wall concept using steel tubes clad with tantalum. The INPORT concept with guided flow also allows for the geometrical arrangement needed for the beams to enter the reactor chamber.

The recondensation process to reestablish beam propagation conditions and the temperature or vapor pressure conditions of the $Li_{17}Pb_{83}$ coolant and breeding material are strongly interdependent and fit together in the HIBALL concept only in a narrow design window. The soft X-rays from the exploding pellet generate LiPb vapor which at a later time absorbs energy from the ion debris. The vapor is sufficiently recondensed within 0.2 sec. The slight pressure increase from the material of the burning pellet has to be tackled by cryopumps with effective vacuum pumping speeds of about 10^6 ltr/sec for D-T and He each. Penetrations of about 20 m^2 in the upper part of the reactor chamber have to be provided for the pumping.

The tritium bred by the fusion neutrons in lithium is mostly diffusing out of the coolant into the chamber atmosphere, then pumped out of the reactor chamber and processed in a purification system. About 100 g tritium remain dissolved in the $Li_{17}Pb_{83}$ coolant. Because of the high tritium permeability through steel all primary coolant pipes, coolant pumps etc. have to be double walled. An intermediate liquid-metal circuit will have to be provided or, alternatively, the steam generators must have double walled pipes. A leak containment system around the reactor chamber must assure a tritium release of only about 10 Ci/(GWe·d). ICF reactors will also need decay heat and emergency cooling systems as are provided for in fission reactors.

MAIN DESIGN CHARACTERISTICS OF A CONCEPTUAL HIF POWER REACTOR CHAMBER

Fig. 3 summarizes the main design parameters of a 1000 MWe HIF reactor chamber and explains the ensuing values of chamber size and pellet gain.
- A liquid wall protects the steel structures from X rays, ion debris and neutrons in order to attain 25 years operation lifetime of the reactor chamber. This, however, limits the repetition rate to 5 Hz or less.
- A driver energy of at least 5 MJ, a chamber gain (blanket energy multiplication) of 1.25 and a postulated output power of 1000 MWe then require a pellet gain of about 80 at 5 Hz or of about 270 at 1.5 Hz.
- An allowable design heat load of 5 MW_{th}/m^2 (containing 1.7 MW_{th}/m^2 for X-rays and Ion debris) leads to a chamber size of
  - 6 m radius first surface,
  - 8 m radius steel shell (2 m space for liquid wall),
  - 10-11 m radius of the reactor chamber at its outer boundary.

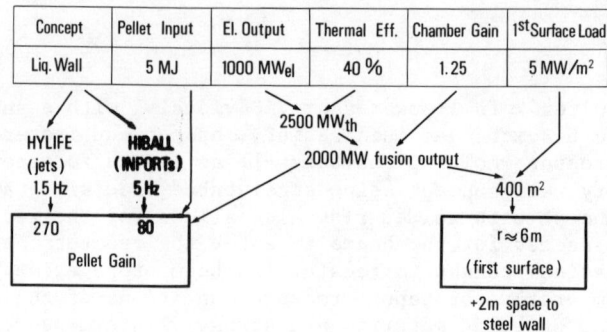

Fig. 3. Design Characteristics of a Reactor Chamber

This size of the reactor chamber is comparable to the size of the reactor tank of a liquid metal fast breeder reactor (LMFBR) of about 1000 MWe. Like the LMFBR, the liquid metal cooled ICF reactor chamber also needs secondary coolant circuits with coolant pumps, intermediate heat exchangers, steam generators etc. The overall size of the HIF power plant (excluding the driver) therefore will be similar to an LMFBR plant, e.g. the French Superphenix.

REFERENCES

1. B. BADGER et al.: KfK 3202 (2 vols.), Kernforschungszentrum Karlsruhe (1981) or UWFDM-450, University of Wisconsin (1981); B. BADGER et al.: KfK 3840 or FPA-84-4 or UWFDM-625 (1985)
2. The Working Group on HIBLIC-I: IPPJ-663, Institute of Plasma Physics (Nagoya University, Japan) (1984)
3. M.J. MONSLER, J. HOVINGH, D.L. COOK, T.G. FRANK and G.A. MOSES: Nuclear Technology/Fusion 1, 302 (1981)
4. E.W. SUCOV (ed.): WFPS-TME-81-001 or DOE/DP/40086-1 (2 vols.), Westinghouse Electric Corporation (1981)
5. See e.g. J.H. PITTS and J. HOVINGH in: Mechanical Engineering Department Technical Review (W.B. SIMECKA, R.M. DENNEY and C. TALABER, eds.), UCRL-50016-80-2, p. 35, Lawrence Livermore National Laboratory (1980)

TEN YEARS OF HIF RESEARCH

J D Lawson
Rutherford Appleton Laboratory, Chilton, Oxon, UK

ABSTRACT

In this paper, written after the meeting in place of the normal closing address, the development of ideas for HIF during its first decade is sketched. This provides a framework for discussion of new work presented at the meeting.

INTRODUCTION

It is just ten years since the first meeting was called to consider the possibility of using energetic heavy ions as drivers for inertial confinement fusion. A paper presented at the US National Accelerator conference a year earlier by Maschke and an internal report from the Argonne National Laboratory by Martin and Arnold had aroused considerable interest in the accelerator community. Targets for laser driven fusion had already been under study for several years, and more recently the possibility of using electrons and light ions was being explored at laboratories operating pulsed high current accelerators in the MeV range. Consideration of heavy ion targets was a natural extension to this work.

The meeting, entitled 'ERDA Summer Study of Heavy Ions for Inertial Fusion' took place in July 1976, and represented the first large scale discussion between specialists from the main accelerator laboratories in the USA. Also present were target experts, plasma physicists interested in high current beam propagation, and engineers who had been studying reactor requirements for inertial fusion.

The situation as seen at the end of the meeting was summarized in the digest. "The central result was an affirmation that ion beam fusion power merits serious attention. The accelerator experts found no fatal flaws in the systems they studied ..... Target experts developed pellet requirements in which they have high confidence, and also less demanding targets that may be acceptable. Reactor designers began to consider a wide range of concepts.... Considerable enthusiasm was generated". This quotation continues with the cautionary statement: "However, it is clear that present information is inadequate to establish the technical feasibility of ion beam fusion with reasonable confidence, or to select the optimum type of accelerating system..." Six "areas requiring research and development" are then identified.

Interest and activity soon spread to Europe, especially England, Germany and France, and to Japan and U.S.S.R. The challenging problem of concentrating energy in space and time, several megajoules over about a square centimetre in tens of nanoseconds, and the design of efficient targets and reactors to contain the explosion generated a whole range of new and interesting problems.

Two kinds of activity developed, the first was concerned with evaluating the physics in unfamiliar areas. In the accelerator field greater understanding of beam propagation near the space charge limit was required, and substantial extrapolations of existing technique in many directions were indicated. This gave rise to a very fruitful theoretical and experimental study of beam propagation near the space charge limit. New insights into the behaviour of such beams have been obtained, and higher currents can be propagated than originally anticipated. Target physics gave rise to interesting problems of energy deposition from heavy ions, where the mechanisms are very different from those being studied for laser fusion. In addition to the workshops mentioned earlier, there have been two international workshops, in France and England, devoted to the topic of "Atomic Physics for Ion Fusion", dealing with numerous aspects of basic physics relevant to targets and reactors. A further topic of importance, relevant to systems using storage rings, is the loss of particles due to charge exchanging collisions. An experimental programme has confirmed general expectations, and provided some useful basic data. Reactor design and vacuum considerations again differed from those appropriate for lasers. Several studies for reactor systems had already been done for laser fusion; adaptation to heavy ions is not trivial, since vacuum conditions are more stringent. Important questions of vacuum physics throughout the system, and in particular at interfaces between beam lines and reactor have been identified, but more study is needed. Questions still open are the number of beamlets focused on the target, and their geometrical configuration, which may be different from that required for lasers. Simultaneous consideration of target, accelerator, and reactor design is needed.

The second class of activity may be termed 'scenario building'. Early papers show many ingenious arrangements, built from linacs of various types, storage rings and synchrotrons. Numerous combinations were proposed, and these were subject to critical scrutiny at the workshops that followed at one or two year intervals, of which this is the seventh. Synchrotrons, favoured in earlier proposals, were later found to be unsuitable at the rather lower energies and higher currents now favoured for efficient target design. At present two basic configurations are being considered; one uses linacs for acceleration and storage rings for accumulating the charge and producing short bunches; in the other, very high currents, shared between several parallel beams are accelerated, and compressed during acceleration in an induction linear accelerator.

The first scenario represents an extrapolation of present technique, and it is easier to see where the problems arise. These are associated with space-charge, which seems now to limit the permissible charge state that can be accelerated to singly charged ions, and with the problem of maintaining the optical beam quality through a large number of transfer operations. The most complete and up-to-date study is the German-Wisconsin joint 'HIBALL II'; which illustrates something of the complexity required to achieve the conditions required.

The second approach relies on the induction linac for the main accelerator; there is no previous practical experience of such machines for heavy ion acceleration, and it is here that most innovation is required. The basic configuration is much simpler, and it is in the accelerator itself rather than in transfer operations that beam quality must be carefully controlled. Progress is now being made, and it is possible to construct scenarios adequate for comparative costing. These points are detailed later.

All schemes face the problem of 'final focusing', an operation requiring considerable precision. Beam quality must be good, aiming at the target accurate, and vacuum conditions at the interface between beam-line and reactor carefully controlled. There is awareness of these problems, but reliable detailed prediction is not yet possible in such an unfamiliar area.

Before reporting in more detail specific progress and changes in perspective since the Tokyo meeting two years ago, it is of interest to note some features of the evolution of the enterprise until that time. This has not always been smooth, difficulties of establishing a satisfactory organizational framework, and limited funding, plus complications connected with the classification of much of the target physics initially caused some frustrations and disappointment. The situation has now stabilised, however; morale is good, but funding still inadequate to keep a balanced coverage of all the problems involved. Nevertheless, there is a general feeling that 'value for money' with available resources has been good.

Some problems, notably acceleration at low energies, and the provision of ion sources with adequate quality now look less formidable, thanks to developments during the last few years, particularly of the radiofrequency quadrupole, (RFQ). Others, associated with space-charge effects and a beam energy lower than that originally envisaged, look more severe. Apart from the initial move to lower energy beams, estimates of target performance have not greatly changed; later work in the USA, Europe and Japan has increased confidence in earlier estimates.

One of the special difficulties with HIF is that of devising a convincing staged programme with steps that are not too large and expensive. With lasers, but not with ions, small scale experiments, yielding high power density with modest input power and very small targets are feasible. Nevertheless, after much discussion, programmes leading to a 'high temperature experiment', designed to concentrate a few kilojoules on a target, have been defined in the USA and Germany.

Many reactor schemes have been studied, most of which have lithium inside the reactor vessel in the form of liquid or granules of lithium based ceramic. Since these all require large extrapolations from existing experience, it will necessarily be a long time before a complete realistic design can be made. Nevertheless, it is important that different concepts should be critically examined, and used as a basis for preliminary costing.

NEW RESULTS AND NEW DIRECTIONS

The material presented at the conference may be divided into several classes; among these are reports of steady progress in directions already being pursued, reports of plans for future work, and reports of new results or thoughts that suggest a change of emphasis or direction.

The latter type of contribution usually attracts most attention and certainly there was considerable interest in the proposal to use triply rather than singly ionized ions in the induction linac. The advantage of doing this emerged clearly in the 'system optimization' studies carried out by the McDonnell Douglas Astronautics Co in conjunction with the Berkeley, Livermore, and Los Alamos groups. The quantitative output from these studies was produced by a comprehensive computer programme incorporating cost estimates based on the technology of induction accelerators, targets, and reactors as understood at the present time. Predictions were made of capital cost and cost of electricity for various reactor types, allowing the known significant parameters, for example repetition rate, input energy to target, beam emittance, type of target, beam configuration etc to vary over a wide range. The results showed a considerable advantage for charge state 3 compared to 1, but rather weak dependence on most other parameters; in particular, repetition rate in the range of 3 - 9 pulses per second. Costs associated with a staged installation plan, starting with a system of only 500 MW were found to be comparable with those anticipated with fission, or from the more optimistic magnetic fusion studies. This was felt to be a satisfactory and encouraging conclusion.

Operation with charge state 3 had not earlier been thought practical. Certainly in the conventional linacs and storage ring scenario there are problems with values greater than 1, a fact reflected in the change from Bi^{++} to Bi^{+} in the HIBALL 2 scenario. The constraints in the induction linac are different, however, and progress was reported by Brown on metal vapour vacuum arc (MEVVA) ion sources which show promise of intense high quality beams of higher ion states. Very successful initial experiments at Berkeley on the acceleration and compression, to give 'current amplification', of multiple beams (4) of Cs + ions in an induction linac, encouraged the view that a full scale system for 3+ ions would be achievable.

A higher charge state means that the ions are more readily focused in the 'final focusing' lens but that the defocusing effect of space-charge between the lens and pellet is more severe. An appreciation of these problems was given by Lee, and the desirability of beam neutralization was stressed by Olson, in a paper in which he discussed the vacuum requirements for high charge state ions.

Steady progress was made in the physics of targets, in particular the details of the stopping of heavy ions in hot matter; experiments are planned in France and Germany, and theory was presented by several groups, including a contribution from Spain. Updated curves of target performance were presented by the Livermore group. The overall picture remains the same, but newly released information showing comparative conversion efficiencies of ions and lasers for indirectly driven targets illustrates the advantage of heavy ions at high irradiance.

Many papers on particular topics were presented. One area where understanding has increased since the last conference concerns the propagation of beams in periodic systems. Earlier ideas, based on the conservation of transverse energy of Lapostolle have been clarified and extended by Reiser, Wangler and others and some interesting new insights provided by Anderson. Correlation is found between beam profiles and r.m.s. emittance. These studies, together with experiments at Berkeley and Maryland, give confidence in our understanding of the limits and possibilities of transporting space-charge dominated beams. Work at Los Alamos on the electrostatic injector for the induction linac, and theoretical papers on the combining of beams by funnelling were presented.

Papers relevant to the linac and storage ring schemes included reports of progress on RFQ design, measurements in existing machines and future plans for 'high temperature' experiments in Germany and Japan. There were detailed papers on several aspects of beam, accelerator, reactor and target physics that will contribute to basic understanding and future design, and raise further questions to be studied.

In the final 'round table' discussions an overview of the position regarding targets was given by Bangerter, and a number of important questions to be answered before a reactor design can be considered realistic for an operational power station were emphasized by Kessler. Alternative approaches to ICF using lasers were covered by Kato and Bodner. Both agreed that short wavelengths are required, and that efficiency seems to be limited. Bodner advocated a multi-beam direct illumination scheme, using a KrF laser and the 'random phasing' technique developed at the Naval Research Laboratory to produce a well defined spot, and use of a 'SOLASE' type reactor scheme. Progress towards a light ion experiment on 'PBFA-2' to demonstrate particle driven fusion with light ions with power of order 100 TW/cm^2 was described by Vandevender from Sandia. Several difficult problems still remain to be solved, however, if this power level is to be obtained within the next few years. Although light ions may be the first to produce high power on a target, development to a power producing reactor is difficult, and not yet envisaged. Finally, Godlove outlined the status and prospects of heavy ion drivers, quoting several recent reviews and listing some outstanding problems.

In the discussion which followed a number of points concerning reactors were raised. Kessler particularly emphasized neutron shielding, the need for long life of essential components, and safety features; he did not believe that a dry wall reactor like the modular SOLASE type scheme proposed by Bodner was practical.

Another topic treated also in the earlier paper by Lindl and emphasized by Linhart in the discussion, was the problem of aiming the beam at the targets with sufficient accuracy to produce an adequately spherical implosion. Differences of opinion regarding the relative merits of direct and indirect illumination were expressed. It was evident that tolerances are tight, and adequate control of beam position and target injection present a challenging engineering problem.

CONCLUDING REMARKS

Looking back over the first ten years it is apparent that much progress has been made towards understanding the requirements for a fusion reactor based on HIF. The comment made ten years ago, quoted earlier, still applies, though considerable progress has been made in "selecting the optimum type of accelerating system". The structure of the framework of constraints is becoming clearer. Comparing with the first ten years of magnetic fusion development, this status of HIF is quite encouraging. Nevertheless, the effort on such an ambitious enterprise is thinly spread; many features are still unclear and important questions are not being studied. Several groups now have well-defined plans, and the way ahead to 'high-temperature' experiments for both induction linacs and conventional linac plus storage ring schemes has been charted. Since world effort is limited, it is especially important that complete scenarios, looking at all aspects, and particularly problems at the interfaces between accelerator, reactor and target, should be kept up to date. An important component of a complex for power production is a factory to make some 10^8 cheap pellets a year. This was not discussed at the meeting.

Heavy ion fusion is now of interest in many countries. It is encouraging to hear a major presentation on target physics from the Spanish group, and to learn of interest in the USSR. Although there were no delegates present, a paper by Kapchinskij, inventor of the RFQ, and colleagues was contributed to the proceedings. This deals with ion sources, and the successive acceleration of Bi^{++} ions in RFQ's and linacs to 20 GeV.

Clearly there is scope for many individual judgements concerning the ultimate feasibility of heavy ion fusion, and the part it might play in meeting our energy needs. It is also clear, ten years later, that the "considerable enthusiasm" referred to in the 1976 comment is still very much in evidence.

Author Index

A

Anderson, O. A., 253
Aragonés, J. M., 381
Arai, S., 271
Arnold, R., 371
Artemov, V. S., 49

B

Ballard, E. O., 323
Bangerter, R. O., 89, 243, 515, 547
Barchewitz, R., 360
Batalin, V. A., 49
Beau, M., 360
Beuve, M. A., 360
Beynon, T. D., 100
Billman, K. W., 482, 489
Bimbot, R., 408
Bock, R., 23
Bodner, S. E., 561
Bolle, J., 156
Brandon, S. T., 227, 243
Brady, V. O., 474
Brown, I. G., 207
Burkhart, C., 314

C

Celata, C. M., 278
Chang, C-L., 227
Choe, J. Y., 295
Crandall, K. R., 166
Cukier, M., 360

D

Damiltzev, E. N., 49
Decker, J. F., 1
Dei-Cas, R., 360
Della-Negra, S., 408
Deutsch, C., 408, 416
Djadin, A. Ju., 49

Driemeyer, D. E., 482, 489, 498
Drobot, A. T., 227
Dudziak, D. J., 111
Dumax, B., 408

F

Faltens, A., 227, 474
Ferch, M., 346
Fessenden, T. J., 145
Fleurier, C., 408
Forslund, D. W., 330
Fukushima, T., 271

G

Gago, J. A., 381
Gajewski, R., 20
Gámez, L., 381
Gardés, D., 408
Gerhard, A., 346
Glicenstein, J. F., 360
Godlove, T. F., 579
González, M. C., 381
Grebogi, C., 309
Guihaumé, J. M., 360

H

Haas, C. R., 391
Haber, I., 236
Herrmannsfeldt, W. B., 111
Hirao, Y., 39, 271
Ho, D. D.-M., 227, 243
Hoffmann, D. H. H., 391, 408
Hofmann, I., 74
Hogan, W. J., 515
Honrubia, J. J., 381
Hovingh, J., 474, 489
Humphries, S., Jr., 314, 323

I

Iosseliani, D. D., 49

J

Jacoby, J., 391
Jones, M. E., 287
Judd, D. L., 145, 264
Junior, P., 338

K

Kapchinskiy, I. M., 49
Katayama, T., 126
Kato, Y., 553
Keefe, D., 63, 145, 474
Kehne, D., 186
Keller, R., 156
Kessler, G., 591
Kim, C. H., 145, 264
Klein, H., 338, 346
Kopf, U., 156
Kozodaev, A. M., 49
Krafft, G. A., 227
Kurs, A. R., 49
Kurz, M., 346
Kuschin, V. V., 49

L

Laget, J. P., 360
Langbein, K., 338
Laslett, L. J., 145, 264
Lawson, J. D., 186, 599
Lazarev, N. V., 49
Lee, E. P., 227, 243, 461, 474, 489
Lemons, D. S., 287
Len, L. K., 314
Lindl, J. D., 89, 441
Linhart, J. G., 539
Lipkin, I. M., 49
Low, K., 186

M

Magelssen, G. R., 330, 428
Mark, J. W-K., 89, 227, 243, 435, 441
Martínez-Val, J. M., 381

Maynard, G., 408
McAdoo, J., 186
Meier, W. R., 515
Melchert, F., 400
Meyer, E. A., 323
Meyer-ter-Vehn, J., 371, 448
Mills, R. S., 166
Mínguez, E., 381
Mizobuchi, A., 271
Moreau, C., 360
Moses, G. A., 591
Mosnier, J. P., 360
Müller, R. W., 156

N

Nagai, T., 523
Nakanishi, T., 271
Namkung, W., 295
Niu, K., 454
Noll, R., 391

O

Obayashi, H., 523
Ocaña, J. L., 381
Olivier, M., 271
Olson, C. L., 215
Otero, R., 381

P

Pan, Y.-L., 89, 435
Parshin, I. O., 49
Pendergrass, J. H., 531
Perlado, J. M., 381
Peter, Th., 371
Peterson, R. R., 507, 591
Plotnikov, S. V., 49
Prior, C. R., 186, 302

R

Ramis, R., 448
Rees, G. H., 138
Reiser, M., 186
Renaud, M., 360
Riehl, G., 338
Riepe, K. P., 323
Rink, K., 400
Rinn, K., 400
Rivet, M. F., 408

S

Salzborn, E., 400
Santolaya, J. M., 381
Schempp, A., 338, 346
Schevchenko, V. G., 49
Schmalz, R., 448
Schriever, R. L., 13
Serrano, J. F., 381
Shurter, R. P., 323
Skachkov, V. S., 49
Smith, L., 145, 264
Spädtke, P., 156

T

Tatsumi, S., 271

U

Ueda, N., 271
Ugarov, S. B., 49
Uhm, H. S., 295, 309

V

Van Haaften, F. W., 323
Vanderhaegen, D., 416
VanDevender, J. P., 569
Velarde, G., 381
Velarde, P. M., 381
von Möllendorff, U., 591

W

Waganer, L. M., 482, 489, 498
Wahl, H., 391
Wangler, T. P., 166
Warwick, A. I., 145, 264
Weikl, B., 391
Weyrich, K., 391, 408
Wilson, D. C., 323

Y

Yamada, S., 271
Yamaki, T., 523

Z

Zarubin, A. B., 49
Zoubek, N., 338
Zuckerman, D. S., 482, 489, 498

List of Participants

Dr. Oscar A. Anderson
Lawrence Berkeley Laboratory
1 Cyclotron Rd., Bldg. 4
Berkeley, CA 94720

Dr. N. Angert
GSI
D–6100 Darmstadt
W. GERMANY

Mr. Evan O. Ballard
E533
Los Alamos National Laboratory
Los Alamos, NM 87545

Dr. Roger Bangerter
Lawrence Livermore National Laboratory
P.O. Box 808, L-477
Livermore, CA 94550

Prof. D. Beynon
Department of Physics
University of Birmingham
Birmingham B152TT
UNITED KINGDOM

Dr. Joseph Bisognano
CEBAF
12070 Jefferson Ave.
Newport News, VA 23606

Mr. David Bixler
Department of Energy
DP–23, GTN
Washington, DC 20545

Prof. Rudolf M. Bock
GSI
P.O. Box 110541
D–6100 Darmstadt
W. GERMANY

Dr. Steve Bodner
Naval Research Laboratory-Code 4730
4555 Overlook Ave., S.W.
Washington, DC 20375

Ian Brown
Lawrence Berkeley Laboratory
1 Cyclotron Rd., Bldg. 53
Berkeley, CA 94720

Dr. C. Celata
Lawrence Berkeley Laboratory
1 Cyclotron Rd.
Berkeley, CA 94720

Dr. Benjamin Cooper
Committee on Energy and Natural Resources
SD–360
US Senate
Washington, DC 20510

Dr. David H. Crandall
Department of Energy
ER–542, GTN
Washington, DC 20545

Prof. Ronald C. Davidson
MIT Plasma Fusion Center
167 Albany St. (NW16-202)
Cambridge, MA 02139

Dr. Stephen Dean
Fusion Power Associates
2 Professional Dr. #249
Gaithersburg, MD 20879

Dr. James Decker
Department of Energy
ER–2, FORS
Washington, DC 20585

Dr. Renato Dei-Cas
Centre d'Etudes de
  Bruyeres-le-Chatel BP No. 12
91680-Bruyeres-le-Chatel
FRANCE

Prof. C. Deutsch
Physique des Plasma
Universite Paris-Sud
91405 Orsay
FRANCE

D. E. Driemeyer
McDonnell Douglas Astro Co.
P.O. Box 516, Bldg. 278/1
St. Louis, MO 63166

Dr. T. Fessenden
Lawrence Berkeley Laboratory
1 Cyclotron Rd.
Berkeley, CA 94720

Dr. Claude Fleurier
GREMI – UER/SFA
Univ d'Orleans
45046 Orleans Cedex
FRANCE

Dr. Ryszard Gajewski
Department of Energy
ER–16, GTN
Washington, DC 20545

Prof. H. B. Gilbody
Department of Pure & Applied Physics
Queen's University of Belfast
Belfast BT7 1NN
NORTHERN IRELAND

Dr. Terry Godlove
Department of Energy
ER–16 GTN
Washington, DC 20545

Dr. Celso Grebogi
Naval Surface Weapons Center
White Oak
Silver Spring, MD 20903-5000

Dr. Hans R. Griem
Laboratory for Plasma & Fusion Energy
University of Maryland
College Park, MD 20742

Dr. Irving Haber
Naval Research Laboratory–Code 4790
4555 Overlook Ave., S.W.
Washington, DC 20375

Dr. William B. Herrmannsfeldt
Stanford Linear Accelerator Center
Stanford, CA 94305

Dr. Dennis Hewett
Lawrence Livermore National Laboratory
P.O. Box 5508, L-472
Livermore, CA 94550

Prof. Yasuo Hirao
Institute for Nuclear Study
University of Tokyo
3–2–1 Midori-cho
Tanashi, Tokyo 188
JAPAN

Dr. Darwin Ho
L–477
Lawrence Livermore National Laboratory
P.O. Box 808
Livermore, CA 94550

Dr. N. J. Hoffman
ETEC
POB 1449
Canoga Park, CA 91304

Dr. Dieter H. H. Hoffmann
GSI–Darmstadt
Postfach 110541
D–6100 Darmstadt
W. GERMANY

Dr. Ingo Hofmann
GSI
D–6100 Darmstadt
W. GERMANY

Dr. Jack Hovingh
Lawrence Berkeley Laboratory
47/112
1 Cyclotron Rd.
Berkeley, CA 94720

Dr. Richard F. Hubbard
Naval Research Laboratory
Code 4790
4555 Overlook Ave., S.W.
Washington, DC 20375

Prof. P. Junior
Inst. f. Angewandte Physik
University of Frankfurt
D–6000 Frankfurt/Main
W. GERMANY

S. L. Kahalas
Office of Inertial Fusion
DP 23.1
US Department of Energy
Washington, DC 20545

Dr. Takeshi Katayama
Institute for Nuclear Study
University of Tokyo
Tanashi, Tokyo 188
JAPAN

Prof. Yoshiaki Kato
Institute of Laser Engineering
Osaka University, 2–6 Yamada-oka
Suita, Osaka 565
JAPAN

Dr. Denis Keefe
Lawrence Berkeley Laboratory
1 Cyclotron Rd.
Berkeley, CA 94720

Mr. David Kehne
Laboratory for Plasma & Fusion Energy
University of Maryland
College Park, MD 20742

Prof. G. Kessler
INR Kernforschungszentrum
P.0. Box 3640
D–7500 Karlsruhe
W. GERMANY

Dr. Charles H. Kim
Lawrence Berkeley Laboratory
1 Cyclotron Rd.
Berkeley, CA 94720

Dr. J. Klabunde
GSI
D–6100 Darmstadt
W. GERMANY

Prof. Horst Klein
Inst. f. Angewandte Physik
University of Frankfurt
D–6000 Frankfurt/Main
W. GERMANY

Geoffrey A. Krafft
CEBAF
12070 Jefferson Ave.
Newport News, VA 23606

Dr. William L. Kruer
Lawrence Livermore National Laboratory
P.O. Box 5508, L-472
Livermore, CA 94550

Dr. A. Bruce Langdon
Lawrence Livermore National Laboratory
P.O. Box 5508, L-472
Livermore, CA 94550

Prof. Pierre M. Lapostolle
3 rue Victor Daix
92200 Neuilly sur Seine
FRANCE

Dr. John D. Lawson
Rutherford Appleton Laboratory
Chilton, Didcot
Oxfordshire OXll OQX
UNITED KINGDOM

Dr. Ed Lee
Lawrence Berkeley Laboratory
47/112
1 Cyclotron Rd.
Berkeley, CA 94720

Dr. Jim Leiss
13013 Chestnut Oak Dr.
Gaithersburg, MD 20878

Dr. Don Lemons
B259
Los Alamos National Laboratory
Los Alamos, NM 87545

Dr. John D. Lindl
Lawrence Livermore National Laboratory
P.O. Box 808, L–477
Livermore, CA 94550

Prof. J. G. Linhart
Dipartimento Di Fisica
Dell' Universitá Di Ferrara
I–44100 Ferrara
ITALY

Prof. Howard Long
Box 1388
8398 S. W. Mohawk St.
Tualatin, OR 97062

Dr. Peter Loschialpo
Naval Research Laboratory
Code 4710
4555 Overlook Ave., S.W.
Washington, DC 20375

Mr. Keng Low
Laboratory for Plasma & Fusion Energy
University of Maryland
College Park, MD 20742

Dr. Glenn R. Magelssen
E531
Los Alamos National Laboratory
Los Alamos, NM 87545

Dr. Jim Mark
Lawrence Livermore National Laboratory
P.O. Box 808
Livermore, CA 94550

Dr. Ronald L. Martin
Argonne National Laboratory
Argonne, IL 60439

Dr. Alfred W. Maschke
TRW MS 01/1261
Redondo Beach, CA 90278

Dr. John McAdoo
Laboratory for Plasma & Fusion Energy
University of Maryland
College Park, MD 20742

Mr. Earl Meyer
H821
Los Alamos National Laboratory
Los Alamos, NM 87545

Dr. Juergen Meyer-ter-Vehn
Max Planck Inst. Quantenoptik
D–8046, Garching
W. GERMANY

Prof. E. Minguez
ETS. I. Industriales (UPM)
Inst. de Fusion Nuclear (Denim)
Castellana, 80 28046 Madrid
SPAIN

Michael Monsler
SATOR 1 Technology, Inc.
315 E. Eisenhower Pkwy.
Ann Arbor, MI 48104

Dr. G. Moses
Department of Nuclear Engineering
University of Wisconsin
Madison, WI 53706

Dr. Rolf Müller
GSI
D–6100 Darmstadt
W. GERMANY

Dr. Won Namkung
Research & Technology Center R43
Naval Surface Weapons Center
Silver Spring, MD 20910

Prof. Keishiro Niu
Tokyo Institute of Technology
Nagatsuta, Midori-ku,
Yokohama 227
JAPAN

Prof. Haruo Obayashi
Institute of Plasma Physics
Nagoya University
Chikusa-Ku, Nagoya 464
JAPAN

Dr. Craig Olson
Sandia National Laboratory
Division 1241
Albuquerque, NM 87185

Prof. A. Pascolini
Departimento di Fisica
Universita di Padova
I35131 Padova
ITALY

Dr. John H. Pendergrass
Los Alamos National Laboratory
P.O. Box 1663, Mail Stop–F611
Los Alamos, NM 87544

Mr. Gerald J. Peters
Department of Energy
ER-224 GTN
Washington, DC 20545

Dr. Robert Peterson
Nuclear Engineering Department
University of Wisconsin
1500 Johnson Dr.
Madison, WI 53706

Walter M. Polansky
Department of Energy—ER–16
Washington, DC 20545

Dr. C. Prior
Rutherford Appleton Laboratory
Chilton, Didcot
Oxfordshire OX11 OQX
UNITED KINGDOM

Dr. H. Rabin
Associate Dean of Engineering
University of Maryland
College Park, MD 20742

Mr. Robert Rader, Dir. (ER–33)
Research & Technology Ass. Div.
US Department of Energy
MS G–226 GTN
Washington, DC 20545

Dr. Grahame Rees
Rutherford Appleton Laboratory
Chilton, Didcot
Oxfordshire OX11 OQX
UNITED KINGDOM

Prof. Martin P. Reiser
Laboratory for Plasma & Fusion Energy
University of Maryland
College Park, MD 20742

Helen Dantsker Rudd
Code 4790
Naval Research Laboratory
Washington, DC 20375

Prof. E. Salzborn
Inst. f. Kernphysik-U Giessen
Leihgesterner Weg 217
6300 Giessen
W. GERMANY

Mark Edward Savage
2809 Cagua N.E.
Albuquerque, NM 87110

Dr. A. Schempp
Inst. f. Angewandte Physik
University of Frankfurt
6000 Frankfurt
W. GERMANY

Prof. W. Schmidt
Scientific Liaison Office
1 Farragut Square S.
Washington, DC 20006

Dr. Richard Schriever
Department of Energy
DP–23 GTN
Washington, DC 20545

Dr. Robert Scott
Electric Power Research Institute
P.O. Box 10412
Palo Alto, CA 94303

Dr. J. Struckmeier
GSI
P.O. Box 110541
D–6100 Darmstadt
W. GERMANY

Dr. Michael Tiefenback
Lawrence Berkeley Laboratory
1 Cyclotron Rd.
Berkeley, CA 94720

Dr. Han S. Uhm
Naval Surface Weapons Center
Silver Spring, MD 20903

Dr. Pace VanDevender
ONG 1200
P.O. Box 5800
Sandia National Laboratory
Albuquerque, NM 87185

Mr. Les Waganer
McDonnell Douglas Astro Co.
P.O. Box 516, Bldg. 278
St. Louis, MO 63166

Dr. Tai Sen Wang
AT–6 MS H829
Los Alamos National Laboratory
Los Alamos, NM 87545

Dr. Thomas Wangler
H817
Los Alamos National Laboratory
Los Alamos, NM 87545

Dr. Anthony Warwick
AFRD 47–112
Lawrence Berkeley Laboratory
Berkeley, CA 94720

Oren A. Wasson
RADP BIL9
National Bureau of Standards
Gaithersburg, MD 20899

Dr. Doug Wilson
Los Alamos National Laboratory
P.O. Box 1663, MS E527
Los Alamos, NM 87545

Dr. Mark Wilson
Center for Radiation Research
National Bureau of Standards
Gaithersburg, MD 20899

Prof. Hermann Wollnik
2 Physik Institut
H. Buffring 16
DG3 Giessen
W. GERMANY

Dr. Tetsuji Yamaki
Department of Physics
Nagoya University
Chigusa Nagoya 464
JAPAN

Dr. Dave Zuckerman
McDonnell Douglas Astro Co.
P.O. Box 516, Bldg. 278
St. Louis, MO 63166

AIP Conference Proceedings

		L.C. Number	ISBN
No. 1	Feedback and Dynamic Control of Plasmas – 1970	70-141596	0-88318-100-2
No. 2	Particles and Fields – 1971 (Rochester)	71-184662	0-88318-101-0
No. 3	Thermal Expansion – 1971 (Corning)	72-76970	0-88318-102-9
No. 4	Superconductivity in d- and f-Band Metals (Rochester, 1971)	74-18879	0-88318-103-7
No. 5	Magnetism and Magnetic Materials – 1971 (2 parts) (Chicago)	59-2468	0-88318-104-5
No. 6	Particle Physics (Irvine, 1971)	72-81239	0-88318-105-3
No. 7	Exploring the History of Nuclear Physics – 1972	72-81883	0-88318-106-1
No. 8	Experimental Meson Spectroscopy –1972	72-88226	0-88318-107-X
No. 9	Cyclotrons – 1972 (Vancouver)	72-92798	0-88318-108-8
No. 10	Magnetism and Magnetic Materials – 1972	72-623469	0-88318-109-6
No. 11	Transport Phenomena – 1973 (Brown University Conference)	73-80682	0-88318-110-X
No. 12	Experiments on High Energy Particle Collisions – 1973 (Vanderbilt Conference)	73-81705	0-88318-111-8
No. 13	π-π Scattering – 1973 (Tallahassee Conference)	73-81704	0-88318-112-6
No. 14	Particles and Fields – 1973 (APS/DPF Berkeley)	73-91923	0-88318-113-4
No. 15	High Energy Collisions – 1973 (Stony Brook)	73-92324	0-88318-114-2
No. 16	Causality and Physical Theories (Wayne State University, 1973)	73-93420	0-88318-115-0
No. 17	Thermal Expansion – 1973 (Lake of the Ozarks)	73-94415	0-88318-116-9
No. 18	Magnetism and Magnetic Materials – 1973 (2 parts) (Boston)	59-2468	0-88318-117-7
No. 19	Physics and the Energy Problem – 1974 (APS Chicago)	73-94416	0-88318-118-5
No. 20	Tetrahedrally Bonded Amorphous Semiconductors (Yorktown Heights, 1974)	74-80145	0-88318-119-3
No. 21	Experimental Meson Spectroscopy – 1974 (Boston)	74-82628	0-88318-120-7
No. 22	Neutrinos – 1974 (Philadelphia)	74-82413	0-88318-121-5
No. 23	Particles and Fields – 1974 (APS/DPF Williamsburg)	74-27575	0-88318-122-3
No. 24	Magnetism and Magnetic Materials – 1974 (20th Annual Conference, San Francisco)	75-2647	0-88318-123-1

No. 25	Efficient Use of Energy (The APS Studies on the Technical Aspects of the More Efficient Use of Energy)	75-18227	0-88318-124-X
No. 26	High-Energy Physics and Nuclear Structure – 1975 (Santa Fe and Los Alamos)	75-26411	0-88318-125-8
No. 27	Topics in Statistical Mechanics and Biophysics: A Memorial to Julius L. Jackson (Wayne State University, 1975)	75-36309	0-88318-126-6
No. 28	Physics and Our World: A Symposium in Honor of Victor F. Weisskopf (M.I.T., 1974)	76-7207	0-88318-127-4
No. 29	Magnetism and Magnetic Materials – 1975 (21st Annual Conference, Philadelphia)	76-10931	0-88318-128-2
No. 30	Particle Searches and Discoveries – 1976 (Vanderbilt Conference)	76-19949	0-88318-129-0
No. 31	Structure and Excitations of Amorphous Solids (Williamsburg, VA, 1976)	76-22279	0-88318-130-4
No. 32	Materials Technology – 1976 (APS New York Meeting)	76-27967	0-88318-131-2
No. 33	Meson-Nuclear Physics – 1976 (Carnegie-Mellon Conference)	76-26811	0-88318-132-0
No. 34	Magnetism and Magnetic Materials – 1976 (Joint MMM-Intermag Conference, Pittsburgh)	76-47106	0-88318-133-9
No. 35	High Energy Physics with Polarized Beams and Targets (Argonne, 1976)	76-50181	0-88318-134-7
No. 36	Momentum Wave Functions – 1976 (Indiana University)	77-82145	0-88318-135-5
No. 37	Weak Interaction Physics – 1977 (Indiana University)	77-83344	0-88318-136-3
No. 38	Workshop on New Directions in Mossbauer Spectroscopy (Argonne, 1977)	77-90635	0-88318-137-1
No. 39	Physics Careers, Employment and Education (Penn State, 1977)	77-94053	0-88318-138-X
No. 40	Electrical Transport and Optical Properties of Inhomogeneous Media (Ohio State University, 1977)	78-54319	0-88318-139-8
No. 41	Nucleon-Nucleon Interactions – 1977 (Vancouver)	78-54249	0-88318-140-1
No. 42	Higher Energy Polarized Proton Beams (Ann Arbor, 1977)	78-55682	0-88318-141-X
No. 43	Particles and Fields – 1977 (APS/DPF, Argonne)	78-55683	0-88318-142-8
No. 44	Future Trends in Superconductive Electronics (Charlottesville, 1978)	77-9240	0-88318-143-6
No. 45	New Results in High Energy Physics – 1978 (Vanderbilt Conference)	78-67196	0-88318-144-4

No. 46	Topics in Nonlinear Dynamics (La Jolla Institute)	78-57870	0-88318-145-2
No. 47	Clustering Aspects of Nuclear Structure and Nuclear Reactions (Winnepeg, 1978)	78-64942	0-88318-146-0
No. 48	Current Trends in the Theory of Fields (Tallahassee, 1978)	78-72948	0-88318-147-9
No. 49	Cosmic Rays and Particle Physics – 1978 (Bartol Conference)	79-50489	0-88318-148-7
No. 50	Laser-Solid Interactions and Laser Processing – 1978 (Boston)	79-51564	0-88318-149-5
No. 51	High Energy Physics with Polarized Beams and Polarized Targets (Argonne, 1978)	79-64565	0-88318-150-9
No. 52	Long-Distance Neutrino Detection – 1978 (C.L. Cowan Memorial Symposium)	79-52078	0-88318-151-7
No. 53	Modulated Structures – 1979 (Kailua Kona, Hawaii)	79-53846	0-88318-152-5
No. 54	Meson-Nuclear Physics – 1979 (Houston)	79-53978	0-88318-153-3
No. 55	Quantum Chromodynamics (La Jolla, 1978)	79-54969	0-88318-154-1
No. 56	Particle Acceleration Mechanisms in Astrophysics (La Jolla, 1979)	79-55844	0-88318-155-X
No. 57	Nonlinear Dynamics and the Beam-Beam Interaction (Brookhaven, 1979)	79-57341	0-88318-156-8
No. 58	Inhomogeneous Superconductors – 1979 (Berkeley Springs, W.V.)	79-57620	0-88318-157-6
No. 59	Particles and Fields – 1979 (APS/DPF Montreal)	80-66631	0-88318-158-4
No. 60	History of the ZGS (Argonne, 1979)	80-67694	0-88318-159-2
No. 61	Aspects of the Kinetics and Dynamics of Surface Reactions (La Jolla Institute, 1979)	80-68004	0-88318-160-6
No. 62	High Energy e^+e^- Interactions (Vanderbilt, 1980)	80-53377	0-88318-161-4
No. 63	Supernovae Spectra (La Jolla, 1980)	80-70019	0-88318-162-2
No. 64	Laboratory EXAFS Facilities – 1980 (Univ. of Washington)	80-70579	0-88318-163-0
No. 65	Optics in Four Dimensions – 1980 (ICO, Ensenada)	80-70771	0-88318-164-9
No. 66	Physics in the Automotive Industry – 1980 (APS/AAPT Topical Conference)	80-70987	0-88318-165-7
No. 67	Experimental Meson Spectroscopy – 1980 (Sixth International Conference, Brookhaven)	80-71123	0-88318-166-5
No. 68	High Energy Physics – 1980 (XX International Conference, Madison)	81-65032	0-88318-167-3
No. 69	Polarization Phenomena in Nuclear Physics – 1980 (Fifth International Symposium, Santa Fe)	81-65107	0-88318-168-1

No. 70	Chemistry and Physics of Coal Utilization – 1980 (APS, Morgantown)	81-65106	0-88318-169-X
No. 71	Group Theory and its Applications in Physics – 1980 (Latin American School of Physics, Mexico City)	81-66132	0-88318-170-3
No. 72	Weak Interactions as a Probe of Unification (Virginia Polytechnic Institute – 1980)	81-67184	0-88318-171-1
No. 73	Tetrahedrally Bonded Amorphous Semiconductors (Carefree, Arizona, 1981)	81-67419	0-88318-172-X
No. 74	Perturbative Quantum Chromodynamics (Tallahassee, 1981)	81-70372	0-88318-173-8
No. 75	Low Energy X-Ray Diagnostics – 1981 (Monterey)	81-69841	0-88318-174-6
No. 76	Nonlinear Properties of Internal Waves (La Jolla Institute, 1981)	81-71062	0-88318-175-4
No. 77	Gamma Ray Transients and Related Astrophysical Phenomena (La Jolla Institute, 1981)	81-71543	0-88318-176-2
No. 78	Shock Waves in Condensed Mater – 1981 (Menlo Park)	82-70014	0-88318-177-0
No. 79	Pion Production and Absorption in Nuclei – 1981 (Indiana University Cyclotron Facility)	82-70678	0-88318-178-9
No. 80	Polarized Proton Ion Sources (Ann Arbor, 1981)	82-71025	0-88318-179-7
No. 81	Particles and Fields –1981: Testing the Standard Model (APS/DPF, Santa Cruz)	82-71156	0-88318-180-0
No. 82	Interpretation of Climate and Photochemical Models, Ozone and Temperature Measurements (La Jolla Institute, 1981)	82-71345	0-88318-181-9
No. 83	The Galactic Center (Cal. Inst. of Tech., 1982)	82-71635	0-88318-182-7
No. 84	Physics in the Steel Industry (APS/AISI, Lehigh University, 1981)	82-72033	0-88318-183-5
No. 85	Proton-Antiproton Collider Physics –1981 (Madison, Wisconsin)	82-72141	0-88318-184-3
No. 86	Momentum Wave Functions – 1982 (Adelaide, Australia)	82-72375	0-88318-185-1
No. 87	Physics of High Energy Particle Accelerators (Fermilab Summer School, 1981)	82-72421	0-88318-186-X
No. 88	Mathematical Methods in Hydrodynamics and Integrability in Dynamical Systems (La Jolla Institute, 1981)	82-72462	0-88318-187-8
No. 89	Neutron Scattering – 1981 (Argonne National Laboratory)	82-73094	0-88318-188-6
No. 90	Laser Techniques for Extreme Ultraviolt Spectroscopy (Boulder, 1982)	82-73205	0-88318-189-4

No. 91	Laser Acceleration of Particles (Los Alamos, 1982)	82-73361	0-88318-190-8
No. 92	The State of Particle Accelerators and High Energy Physics (Fermilab, 1981)	82-73861	0-88318-191-6
No. 93	Novel Results in Particle Physics (Vanderbilt, 1982)	82-73954	0-88318-192-4
No. 94	X-Ray and Atomic Inner-Shell Physics – 1982 (International Conference, U. of Oregon)	82-74075	0-88318-193-2
No. 95	High Energy Spin Physics – 1982 (Brookhaven National Laboratory)	83-70154	0-88318-194-0
No. 96	Science Underground (Los Alamos, 1982)	83-70377	0-88318-195-9
No. 97	The Interaction Between Medium Energy Nucleons in Nuclei – 1982 (Indiana University)	83-70649	0-88318-196-7
No. 98	Particles and Fields – 1982 (APS/DPF University of Maryland)	83-70807	0-88318-197-5
No. 99	Neutrino Mass and Gauge Structure of Weak Interactions (Telemark, 1982)	83-71072	0-88318-198-3
No. 100	Excimer Lasers – 1983 (OSA, Lake Tahoe, Nevada)	83-71437	0-88318-199-1
No. 101	Positron-Electron Pairs in Astrophysics (Goddard Space Flight Center, 1983)	83-71926	0-88318-200-9
No. 102	Intense Medium Energy Sources of Strangeness (UC-Sant Cruz, 1983)	83-72261	0-88318-201-7
No. 103	Quantum Fluids and Solids – 1983 (Sanibel Island, Florida)	83-72440	0-88318-202-5
No. 104	Physics, Technology and the Nuclear Arms Race (APS Baltimore –1983)	83-72533	0-88318-203-3
No. 105	Physics of High Energy Particle Accelerators (SLAC Summer School, 1982)	83-72986	0-88318-304-8
No. 106	Predictability of Fluid Motions (La Jolla Institute, 1983)	83-73641	0-88318-305-6
No. 107	Physics and Chemistry of Porous Media (Schlumberger-Doll Research, 1983)	83-73640	0-88318-306-4
No. 108	The Time Projection Chamber (TRIUMF, Vancouver, 1983)	83-83445	0-88318-307-2
No. 109	Random Walks and Their Applications in the Physical and Biological Sciences (NBS/La Jolla Institute, 1982)	84-70208	0-88318-308-0
No. 110	Hadron Substructure in Nuclear Physics (Indiana University, 1983)	84-70165	0-88318-309-9
No. 111	Production and Neutralization of Negative Ions and Beams (3rd Int'l Symposium, Brookhaven, 1983)	84-70379	0-88318-310-2

No. 112	Particles and Fields – 1983 (APS/DPF, Blacksburg, VA)	84-70378	0-88318-311-0
No. 113	Experimental Meson Spectroscopy – 1983 (Seventh International Conference, Brookhaven)	84-70910	0-88318-312-9
No. 114	Low Energy Tests of Conservation Laws in Particle Physics (Blacksburg, VA, 1983)	84-71157	0-88318-313-7
No. 115	High Energy Transients in Astrophysics (Santa Cruz, CA, 1983)	84-71205	0-88318-314-5
No. 116	Problems in Unification and Supergravity (La Jolla Institute, 1983)	84-71246	0-88318-315-3
No. 117	Polarized Proton Ion Sources (TRIUMF, Vancouver, 1983)	84-71235	0-88318-316-1
No. 118	Free Electron Generation of Extreme Ultraviolet Coherent Radiation (Brookhaven/OSA, 1983)	84-71539	0-88318-317-X
No. 119	Laser Techniques in the Extreme Ultraviolet (OSA, Boulder, Colorado, 1984)	84-72128	0-88318-318-8
No. 120	Optical Effects in Amorphous Semiconductors (Snowbird, Utah, 1984)	84-72419	0-88318-319-6
No. 121	High Energy e^+e^- Interactions (Vanderbilt, 1984)	84-72632	0-88318-320-X
No. 122	The Physics of VLSI (Xerox, Palo Alto, 1984)	84-72729	0-88318-321-8
No. 123	Intersections Between Particle and Nuclear Physics (Steamboat Springs, 1984)	84-72790	0-88318-322-6
No. 124	Neutron-Nucleus Collisions – A Probe of Nuclear Structure (Burr Oak State Park - 1984)	84-73216	0-88318-323-4
No. 125	Capture Gamma-Ray Spectroscopy and Related Topics – 1984 (Internat. Symposium, Knoxville)	84-73303	0-88318-324-2
No. 126	Solar Neutrinos and Neutrino Astronomy (Homestake, 1984)	84-63143	0-88318-325-0
No. 127	Physics of High Energy Particle Accelerators (BNL/SUNY Summer School, 1983)	85-70057	0-88318-326-9
No. 128	Nuclear Physics with Stored, Cooled Beams (McCormick's Creek State Park, Indiana, 1984)	85-71167	0-88318-327-7
No. 129	Radiofrequency Plasma Heating (Sixth Topical Conference, Callaway Gardens, GA, 1985)	85-48027	0-88318-328-5
No. 130	Laser Acceleration of Particles (Malibu, California, 1985)	85-48028	0-88318-329-3
No. 131	Workshop on Polarized ^{3}He Beams and Targets (Princeton, New Jersey, 1984)	85-48026	0-88318-330-7
No. 132	Hadron Spectroscopy–1985 (International Conference, Univ. of Maryland)	85-72537	0-88318-331-5

No.	Title		
No. 133	Hadronic Probes and Nuclear Interactions (Arizona State University, 1985)	85-72638	0-88318-332-3
No. 134	The State of High Energy Physics (BNL/SUNY Summer School, 1983)	85-73170	0-88318-333-1
No. 135	Energy Sources: Conservation and Renewables (APS, Washington, DC, 1985)	85-73019	0-88318-334-X
No. 136	Atomic Theory Workshop on Relativistic and QED Effects in Heavy Atoms	85-73790	0-88318-335-8
No. 137	Polymer-Flow Interaction (La Jolla Institute, 1985)	85-73915	0-88318-336-6
No. 138	Frontiers in Electronic Materials and Processing (Houston, TX, 1985)	86-70108	0-88318-337-4
No. 139	High-Current, High-Brightness, and High-Duty Factor Ion Injectors (La Jolla Institute, 1985)	86-70245	0-88318-338-2
No. 140	Boron-Rich Solids (Albuquerque, NM, 1985)	86-70246	0-88318-339-0
No. 141	Gamma-Ray Bursts (Stanford, CA, 1984)	86-70761	0-88318-340-4
No. 142	Nuclear Structure at High Spin, Excitation, and Momentum Transfer (Indiana University, 1985)	86-70837	0-88318-341-2
No. 143	Mexican School of Particles and Fields (Oaxtepec, México, 1984)	86-81187	0-88318-342-0
No. 144	Magnetospheric Phenomena in Astrophysics (Los Alamos, 1984)	86-71149	0-88318-343-9
No. 145	Polarized Beams at SSC & Polarized Antiprotons (Ann Arbor, MI & Bodega Bay, CA, 1985)	86-71343	0-88318-344-7
No. 146	Advances in Laser Science–I (Dallas, TX, 1985)	86-71536	0-88318-345-5
No. 147	Short Wavelength Coherent Radiation: Generation and Applications (Monterey, CA, 1986)	86-71674	0-88318-346-3
No. 148	Space Colonization: Technology and The Liberal Arts (Geneva, NY, 1985)	86-71675	0-88318-347-1
No. 149	Physics and Chemistry of Protective Coatings (Universal City, CA, 1985)	86-72019	0-88318-348-X
No. 150	Intersections Between Particle and Nuclear Physics (Lake Louise, Canada, 1986)	86-72018	0-88318-349-8
No. 151	Neural Networks for Computing (Snowbird, UT, 1986)	86-72481	0-88318-351-X

RAYMOND H. FOGLER LIBRARY
DATE DUE

BOOKS ARE SUBJECT TO
RECALL AFTER TWO WEEKS